Second Edition

Statistical Reasoning

for Everyday Life

Second Edition

Statistical Reasoning
for Everyday Life

Jeffrey O. Bennett
University of Colorado at Boulder

William L. Briggs
University of Colorado at Denver

Mario F. Triola
Dutchess Community College

Addison
Wesley

Boston   San Francisco   New York
London   Toronto   Sydney   Tokyo   Singapore   Madrid
Mexico City   Munich   Paris   Cape Town   Hong Kong   Montreal

Publisher:	Greg Tobin
Sponsoring Editor:	Deirdre Lynch
Editorial Assistant:	Anna Stillner
Executive Marketing Manager:	Brenda Bravener
Senior Marketing Manager:	Michael Boezi
Managing Editor:	Karen Guardino
Senior Production Supervisor:	Peggy McMahon
Senior Designer:	Barbara T. Atkinson
Cover Designer:	Leslie Haimes
Cover Photo:	Walter Bibikow/© PictureQuest
Text Designer:	Leslie Haimes
Production Services:	Lifland et al., Bookmakers
Photo Research:	Beth Anderson
Senior Prepress Supervisor:	Caroline Fell
Manufacturing Buyer:	Evelyn Beaton
Technical Illustrator:	Techsetters, Inc.
Composition:	Progressive Information Technologies

For permission to use copyrighted material, grateful acknowledgment is made to the copyright holders on pages 461 and 462 in the back of the book, which are hereby made part of this copyright page.

Library of Congress Cataloging-in-Publication Data

Bennett, Jeffrey O.

    Statistical reasoning for everyday life/Jeffrey O. Bennett, William L. Briggs, Mario F.

    Triola.–2nd ed.

      p. cm.

    Includes index.

    ISBN 0-201-77128-4

    1. Statistics. I. Briggs, William L. II. Triola, Mario F. III. Title.

QA276.12.B45 2002

519.5–dc21   2002024938

This book is dedicated to everyone
who will try to make the world a
better place.
We hope that your study of statistics
will be useful to your efforts.

And it is dedicated to those who
make our own lives brighter,
especially Lisa, Grant, Brooke, Julie,
Katie, Ginny, Marc, and Scott.

Contents

Preface

Why Study Statistics?

Science fiction writer H. G. Wells once wrote, "Statistical thinking will one day be as necessary for efficient citizenship as the ability to read and write." The future that Wells imagined is here. Statistics is now an important part of everyday life, unavoidable whether you are starting a new business, deciding how to plan for your financial future, or simply watching the news on television. Statistics comes up in everything from opinion polls to economic reports to the latest research on cancer prevention. Understanding the core ideas behind statistics is therefore crucial to your success in the modern world.

What Kind of Statistics Will You Learn in This Book?

Statistics is a rich field of study—so rich that it is possible to study it for a lifetime and still feel as if there's much left to learn. Nevertheless, you can understand the core ideas of statistics with just a quarter or semester of academic study. This book is designed to help you learn these core ideas. The ideas you'll study in this book represent the statistics that you'll *need* in your everyday life—and that you can reasonably learn in one course of study. In particular, we've designed this book with three specific purposes:

- To provide you with the understanding of statistics you'll need for **college** courses, particularly in social sciences such as economics, psychology, sociology, and political science.

- To help you develop the ability to reason using statistical information—an ability that is crucial to almost any **career** in the modern world.

- To provide you with the power to evaluate the many news reports of statistical studies that you encounter in your daily **life**, thereby helping you to form opinions about their conclusions and to decide whether the conclusions should influence the way you live.

Who Should Read This Book?

We hope this book will be useful to everyone, but it was designed primarily for students who are *not* planning additional course work in statistics. In particular, this book should provide a suitable introduction to statistics for students majoring in any field of study except mathematics or the physical sciences. The level of this text should be appropriate for anyone who has completed two years of high school mathematics.

Approach

This book takes an approach designed to help you understand important statistical ideas qualitatively, using quantitative techniques only when they clarify those ideas. Here are a few of the key pedagogical strategies that guided the creation of this book.

Start with the Big Picture.   Most people entering a statistics course have little prior knowledge of the subject, so it is important to keep sight of the overall purpose of statistics while learning individual ideas or methods. We therefore begin this book with a broad overview of statistics in Chapter 1, in which we explain the relationship between samples and populations, discuss sampling methods and the various types of statistical study, and show numerous examples designed to help you decide whether to believe a statistical study. This "big picture" overview of statistics provides a solid foundation for the more in-depth study of statistical ideas in the rest of the book.

Build Ideas Step by Step.   The goal of any course in statistics is to help students understand real statistical issues. However, it is often easier to begin by investigating simple examples in order to build step-by-step understanding that can then be applied to more complex studies. We apply this strategy within every section and every chapter, gradually building toward real examples and case studies.

Use Computations to Enhance Understanding.   The primary goal of this book is to help you understand statistical concepts and ideas, but we firmly believe that this goal is best achieved by doing at least some computation. We therefore include computational techniques wherever they will enhance understanding of the underlying ideas.

Connect Probability to Statistics.   Many statistics courses include coverage of probability, but to students the concept of probability often seems disconnected from the rest of the subject matter. This is a shame, since probability plays such an integral role in the science of statistics. We discuss this point beginning in Chapter 1, with the basic structure of statistical studies, and then revisit it throughout the book—particularly in Chapter 6, where we present many ideas of probability. For those courses in which coverage of probability is not emphasized, Chapter 6 is designed to be optional.

Stay on Goal: Applying Statistical Reasoning to Everyday Life.   Because statistics is such a rich subject, it can be difficult to decide how far to go with any particular statistical topic. In making such decisions for this book, we always turned back to the goal reflected in the title: This book is supposed to help you with the statistical reasoning needed in everyday life. If we felt that a topic was not often encountered in everyday life, we left it out. In the same spirit, we included a few topics—such as a discussion of percentages in Chapter 2 and an in-depth study of graphics in Chapter 3—that are not often covered in statistics courses but are a major part of the statistics encountered in daily life.

Pedagogical Features

Time Out to Think.   Appearing throughout every chapter, the *Time Out to Think* feature poses short conceptual questions designed to help you reflect on important new ideas. They also serve as excellent starting points for classroom discussions.

Examples and Case Studies.   Numbered examples within each section are designed to build understanding and to offer practice with the types of questions that appear in the exercises. Case studies, which always focus on real issues, go into more depth than the numbered examples.

Focus on . . . .   Appearing at the end of each chapter, the *Focus on . . .* sections go into even more depth than the case studies. The topics of these sections were chosen to demonstrate the great variety of fields in which statistics plays a role, including history, environmental studies, agriculture, and economics.

Review Questions.   Every section ends with *Review Questions* that ask students to summarize the important ideas covered in the section. These questions can be answered simply by reviewing the text.

Exercises.   Every section includes *Exercises* that are designed to offer practice with the concepts covered in the section. These exercises are intended for self-study (answers to most odd-numbered exercises appear in the back of the book) or homework assignments.

Projects for the Web and Beyond.   Most sections include one or more projects that require outside research or more significant activity. Many of the projects involve use of the Web.

In the News.   Most sections include a set of *In the News* questions that ask students to find examples of the statistical ideas presented in the section. These questions may be assigned as homework or used for class discussions.

Chapter Review Exercises.   Appearing at the end of each chapter, the *Chapter Review Exercises* tie many of the chapter ideas together. They are designed primarily for self-study, with answers to all of them appearing in the back of the book.

About the Second Edition

We've developed this Second Edition of *Statistical Reasoning for Everyday Life* with the help of many users and reviewers. Here is a brief list of key changes made for this edition:

- We've carefully edited and revised the entire book, implementing many outstanding suggestions for improving descriptions of key concepts and for simplifying terminology and jargon.

- Chapter 10 has been completely rewritten. The new Section 10.1 covers the ideas of risk and life expectancy; it is a heavily revised version of Section 6.4 from the First Edition. Section 10.2 is a new section on statistical paradoxes; it includes examples of Simpson's paradox and the famous problem of how claims of specific accuracy (such as "90% accurate") for drug or polygraph tests can be very misleading. The new Section 10.3 covers hypothesis testing with two-way tables, introducing the idea of the chi-square test.

- All of the Focus sections have been edited and updated. In addition, we've added an entirely new Focus on Sociology ("Does Daycare Breed Bullies?") in Chapter 1.

- The exercise sets at the end of each section have been completely reworked and reorganized. The new exercise sets are divided into two subsets, *Basic Skills* and *Further Applications*, designed to make it easier for instructors to differentiate problems by their level of sophistication.

- We've added a new set of questions, called *Sensible Statements?*, to the beginning of each exercise set (under the *Basic Skills* subheading). These questions are ideal for checking student understanding, because they are quite easy (requiring only a sentence or two to answer) for students who understand the key concepts but very difficult for those who don't grasp the key concepts.

- We've revised and expanded the selection of projects at the end of each section, offering more ideas and choices for instructors.

- The entire book has been redesigned, and all of the artwork re-created, in order to take full advantage of the pedagogical possibilities offered by the use of full color in this edition.

Supplements

The following supplements are available from Addison-Wesley Publishing Company. Instructors: Contact your local sales consultant or e-mail the company directly at *exam@aw.com* for examination copies.

WEB SITE: MyMathLab.com

MyMathLab.com is a complete online course for Addison-Wesley mathematics textbooks that integrates interactive, multimedia instruction correlated to the textbook content. MyMathLab is easily customizable to suit the needs of students and instructors and provides a comprehensive and efficient online course-management system that allows for diagnosis, assessment, and tracking of students' progress. MyMathLab has numerous useful features:

- Fully interactive textbooks are built in CourseCompass, a version of Blackboard™ designed specifically for Addison-Wesley.

- Chapter and section folders from the textbook contain a wide range of instructional content: software tools and electronic supplements, to name a few.

- Hyperlinks take you directly to online testing, diagnosis, tutorials, and gradebooks in MathXL—Addison-Wesley's tutorial and testing system for mathematics and statistics.

- Instructors can create, copy, edit, assign, and track all tests for their course, as well as track student tutorial and testing performance.

- With push-button ease, instructors can remove, hide, or annotate Addison-Wesley preloaded content, add their own course documents, or change the order in which material is presented.

- Using the communication tools found in MyMathLab, instructors can hold online office hours, host a discussion board, create communication groups within their class, send e-mail, and maintain a course calendar.

- *Technology Manual and Workbook* is a comprehensive, text-specific manual available on MyMathLab.com. It includes instructions on using technology to solve examples in the text. Instructions and output are provided for STATDISK, Excel, Minitab, and the TI-83 Plus calculator. Also included are projects that require use of the technology. The manual/workbook comes with STATDISK, a statistical software package for Windows and Macintosh computers, and DDXL, a software supplement that enhances the statistics capabilities of Excel. Several large data sets are also included in formats for STATDISK, Excel, Minitab, and the TI-83 Plus calculator.

For more information, visit our Web site at *www.mymathlab.com* or contact your Addison-Wesley sales representative for a live demonstration.

SUPPLEMENTS FOR INSTRUCTORS

Instructor's Guide and Solutions Manual. This comprehensive manual, by David Lund, contains solutions to all text exercises and answers to the Time Out to Think features. ISBN: 0-201-83844-3.

Computerized Test Generator. TestGen-EQ lets you view and edit test bank questions, transfer them to tests, and print in a variety of formats. The program, for Mac and Windows, also offers many options for organizing and displaying test banks and tests. A built-in random number and test generator makes TestGen-EQ ideal for creating multiple versions of tests and provides more possible test items than does the Printed Test Bank. Users can export tests and HTML files so that they can be viewed with a Web browser. ISBN: 0-201-79515-9.

Printed Test Bank. The Test Bank contains three tests for every chapter of the text. All tests probe the conceptual understanding of students, as well as computational competencies where appropriate. ISBN: 0-201-83845-1.

SUPPLEMENTS FOR STUDENTS

Student Solutions Manual. This manual, by David Lund, provides detailed, worked-out solutions to all odd-numbered text exercises. ISBN: 0-201-83846-X.

AWL Math Tutor Center. For qualified adopters, free tutoring is available to students who purchase a new copy of this textbook. The Math Tutor Center (MTC) is staffed with qualified statistics and mathematics instructors who provide students with tutoring on text examples, concepts, and odd-numbered text exercises. Tutoring assistance is provided via a toll-free telephone number, fax, and e-mail and is available five days a week, seven hours a day. Each new book can be bundled with a free six-month subscription to the service. Request ISBN 0-321-11937-1 for the text bundled with the MTC registration. Students who already have their books may purchase a subscription to the MTC by having their bookstore order ISBN 0-201-71049-8.

ACKNOWLEDGMENTS

Writing a textbook requires the efforts of many people besides the authors. This book would not have been possible without the help of many people. We'd particularly like to thank our editors at Addison-Wesley, Greg Tobin and Deirdre Lynch, whose faith allowed us to create this book. We owe a special debt to our development editor Elka Block, who edited and reviewed the First Edition in every stage of its development, and without whom the project could not have succeeded. We'd also like to thank the rest of the team at Addison-Wesley who helped produce this book, including Anna Stillner, Peggy McMahon, Brenda Bravener, Julia Coen, Marlene Thom, Beth Anderson, Barbara Atkinson, and our production manager, Sally Lifland.

For reviewing this text and the previous edition and providing invaluable advice, we thank the following individuals:

Dale Bowman, *University of Mississippi*

Patricia Buchanan, *Penn State University*

Robert Buck, *Western Michigan University*

Olga Cordero-Brana, *Arizona State University*

Terry Dalton, *University of Denver*

Jim Daly, *California Polytechnic State University*

Mickle Duggan, *East Central University*

Juan Estrada, *Metropolitan State University, Minneapolis–St. Paul*

Jack R. Fraenkel, *San Francisco State University*

Frank Grosshans, *West Chester University*

Silas Halperin, *Syracuse University*

Golde Holtzman, *Virginia Polytechnic Institute and State University*

Colleen Kelly, *San Diego State University*

Jim Koehler

Stephen Lee, *University of Idaho*

Kung-Jong Lui, *San Diego State University*

Judy Marwick, *Prairie State College*

Richard McGrath, *Penn State University*

Abdelelah Mostafa, *University of South Florida*

Todd Ogden, *University of South Carolina*

Nancy Pfenning, *University of Pittsburgh*

Steve Rein, *California Polytechnic State University*

Lawrence D. Ries, *University of Missouri–Columbia*

Larry Ringer, *Texas A&M University*

John Spurrier, *University of South Carolina*

Gwen Terwilliger, *University of Toledo*

David Wallace, *Ohio University*

Larry Wasserman, *Carnegie Mellon University*

Sheila Weaver, *University of Vermont*

Robert Wolf, *University of San Francisco*

Fancher Wolfe, *Metropolitan State University, Minneapolis–St. Paul*

Ke Wu, *University of Mississippi*

For helping to ensure the accuracy of this text, we thank the following individuals:

Nkechi Agwu, *Borough of Manhattan Community College*
David Lund, *University of Wisconsin–Eau Claire*
Kim Smith, *University of Tennessee*
Eric A. Suess, *California State University–Hayward*
Sheila Weaver, *University of Vermont*

To the Student

How to Succeed in Your Statistics Course

If you are reading this book, you probably are enrolled in a statistics course of some type. The keys to success in your course include approaching the material with an open and optimistic frame of mind, paying close attention to how useful and enjoyable statistics can be in your life, and studying effectively and efficiently. The following sections offer a few specific hints that may be of use as you study.

USING THIS BOOK

Before we get into more general strategies for studying, here are a few guidelines that will help you use *this* book most effectively.

- Before doing any assigned exercises, read assigned material *twice*.
  —On the first pass, read quickly to gain a "feel" for the material and concepts presented.
  —On the second pass, read the material in more depth and work through the examples carefully.

- During the second reading, take notes that will help you when you go back to study later. In particular:
  —*Use the margins!* The wide margins in this textbook are designed to give you plenty of room for making notes as you study.
  —Don't highlight—<u>underline!</u> Using a pen or pencil to <u>underline</u> material requires greater care than highlighting and therefore helps keep you alert as you study.

- After you complete the reading, and again when studying for exams, make sure you can answer the review questions at the end of each unit.

- You'll learn best by *doing*, so do plenty of the end-of-section exercises and the end-of-chapter review exercises. In particular, try some of the exercises that have answers in the back of the book, in addition to any exercises assigned by your instructor.

BUDGETING YOUR TIME

A general rule of thumb for college classes is that you should expect to study about 2 to 3 hours per week *outside* class for each unit of credit. Based on this rule of thumb, a student taking 15 credit hours should expect to spend 30 to 45 hours each week studying outside of class. Combined with time in class, this works out to a total of 45 to 60 hours spent on academic work—not much more than the time required of a typical job, and you get to choose your own hours. Of course, if you are working while you attend school, you will need to budget your time carefully. Here are some rough guidelines for how you might divide your studying time.

If your course is	Time for reading the assigned text (per week)	Time for homework assignments (per week)	Time for review and test preparation (average per week)	Total study time (per week)
3 credits	1 to 2 hours	3 to 5 hours	2 hours	6 to 9 hours
4 credits	2 to 3 hours	3 to 6 hours	3 hours	8 to 12 hours
5 credits	2 to 4 hours	4 to 7 hours	4 hours	10 to 15 hours

If you find that you are spending fewer hours than these guidelines suggest, you could probably improve your grade by studying more. If you are spending more hours than these guidelines suggest, you may be studying inefficiently; in that case, you should talk to your instructor about how to study more effectively.

GENERAL STRATEGIES FOR STUDYING

- Don't miss class. Listening to lectures and participating in discussions is much more beneficial than reading someone else's notes. Active participation will help you retain what you are learning.

- Budget your time carefully. Putting in an hour or two each day is more effective, and far less painful, than studying all night before homework is due or before exams.

- If a concept gives you trouble, do additional reading or problem solving beyond what has been assigned. And if you still have trouble, ask for help; you surely can find friends, colleagues, or teachers who will be glad to help you learn.

- Working together with friends can be valuable in helping you to solve difficult problems. However, be sure that you learn *with* your friends and do not become dependent on them.

PREPARING FOR EXAMS

- Rework exercises and other assignments; try additional exercises to be sure you understand the concepts. Study your performance on assignments, quizzes, or exams from earlier in the semester.

- Study your notes from lectures and discussions. Pay attention to what your instructor expects you to know for an exam.

- Reread the relevant sections in the textbook, paying special attention to notes you have made in the margins.

- Study individually before joining a study group with friends. Study groups are effective only if *every* individual comes prepared to contribute.

- Don't stay up too late before an exam. Don't eat a big meal within an hour of the exam (thinking is more difficult when blood is being diverted to the digestive system).

- Try to relax before and during the exam. If you have studied effectively, you are capable of doing well. Staying relaxed will help you think clearly.

Applications Index

In the center column, CS = Case Study, E = Example, F = Focus, IE = In-text Example, P = Problem, and PR = Project. In the right column, the section number is followed by the page number.

Second Edition

Statistical Reasoning
for Everyday Life

As a general rule, the most successful [person] in life is the [person] who has the best information.

—Benjamin Disraeli

Speaking of Statistics

Statistical reasoning is one of the most practical and relevant abilities that you can develop. In this first chapter, we'll see why statistics is so important to everyone living in a modern society. We'll lay the foundations for the study of statistics in the rest of this book. We'll discuss a few of the basic ideas behind statistics and study many examples of how these ideas are applied. In the process, we'll even see how the use of statistics can intentionally or unintentionally mislead us.

LEARNING GOALS

1. Understand the two meanings of the term *statistics* and the basic ideas involved in any statistical study, including the relationships among the study's population, sample, sample statistics, and population parameters.

2. Understand the importance of choosing a representative sample and become familiar with several common methods of sampling.

3. Understand the different ways in which a statistical study can be conducted: observational study, experiment, or meta-analysis.

4. Use the concepts of this chapter in evaluating the results of a statistical study described in the news.

1.1 What Is/Are Statistics?

If you are like most students using this textbook, you are new to the study of statistics. You may not be sure why you are studying statistics, and you may be feeling anxious about what lies ahead. But as you begin reading, put your concerns aside and prepare to be pleasantly surprised.

Let's begin by discussing what the subject of statistics is all about. First and foremost, it's about almost everything in modern society. It's about how we learn whether a new drug is effective in treating cancer. It's about how agricultural inspectors ensure the safety of the food supply. It's about opinion polls, pre-election polls, and exit polls. It's about market research and the effectiveness of advertising. It's about sports, where we rank players and teams primarily through their statistics. Indeed, you'll be hard-pressed to think of any topic that is not linked with statistics in some important way.

The primary goal of this book is to help you learn the core ideas behind statistical methods. These basic ideas are not difficult to understand, although mastery of the details and theory behind them can require years of study. One of the great things about statistics is that even the small amount of theory covered in this book will give you the power to understand the statistics you encounter in the news, in your classes or workplace, and in your everyday life.

A good place to start is with the term *statistics* itself, which can be either singular or plural and has different meanings in the two cases. When it is singular, *statistics* is the *science* that helps us understand how to collect, organize, and interpret numbers or other information about some topic; we refer to the numbers or other pieces of information as *data*. When it is plural, *statistics* are numerical measurements that describe some characteristic. For example, if there are 30 students in your class and they range in age from 17 to 64, the numbers "30 students," "17 years," and "64 years" are all statistics that describe your class in some way.

> To understand God's thoughts, we must study statistics, for these are the measure of His purpose.
>
> – Florence Nightingale

TECHNICAL NOTE

You'll sometimes hear the word *data* used as a singular synonym for *information*, but technically *data* is plural: One piece of information is called a *datum*, and two or more pieces are called *data*.

Two Definitions of Statistics

- Statistics is the *science* of collecting, organizing, and interpreting data.
- Statistics are the *data* that describe or summarize something.

How Statistics Works

Did you watch the last Super Bowl? Advertisers want to know, because the cost of commercial time during the big game is now more than $2 million for a 30-second spot. This advertising can be well worth the price if enough people are watching. For example, news reports stated that 86.8 million Americans watched the New England Patriots win Super Bowl XXXVI. You may be wondering: Who counted all these people?

The answer is *no one*. The claim that 86.8 million people watched the Super Bowl came from statistical studies conducted by a company called Nielsen Media Research. This company publishes the results of its studies as the famous *Nielsen ratings*. Remarkably, Nielsen compiles these ratings by monitoring the television viewing habits of people in only 5,000 homes.

If you are new to the study of statistics, Nielsen's conclusion may seem like quite a stretch. How can anyone draw a conclusion about millions of people by studying just a few thousand? However, statistical science shows that this conclusion can be quite accurate, as long as the statistical study is conducted properly. Let's take the Nielsen ratings of the Super Bowl as an example, and ask a few key questions that will illustrate how statistics works in general.

By the Way...

Statistics originated with the collection of census and tax data, which are affairs of state. That is why the word *state* is at the root of the word *statistics*.

What Is the Goal of the Research?

First, consider Nielsen's goal, which in this case is to determine the total number of Americans who watched the Super Bowl. In the language of statistics, we say that Nielsen is interested in the **population** of all Americans. The number that Nielsen hopes to determine—the number of people who watched the Super Bowl—is a particular characteristic of the population. In statistics, the characteristics of the population that we hope to estimate are called **population parameters**.

Although we usually think of a population as a group of people, a statistical population can be any kind of group—people, animals, or things. For example, in a study of automobile safety, the population might be *all cars on the road*. Similarly, the term *population parameter* can refer to any characteristic of a population. In the case of automobile safety, the population parameters might include the total number of cars on the road, the accident rate among cars on the road, or the range of weights of cars on the road.

By the Way...

Arthur C. Nielsen founded his company and invented market research in 1923. He introduced the Nielsen Radio Index to rate radio programs in 1942 and extended his methods to television programming in the 1960s.

> **Definitions**
>
> The **population** in a statistical study is the *complete* set of people or things being studied.
>
> **Population parameters** are specific characteristics of the population.

EXAMPLE 1   Populations and Population Parameters

For each of the following situations, describe the population being studied and identify some of the population parameters that would be of interest.

a. You work for Farmers Insurance and you've been asked to determine the cost of claims from automobile accidents in which airbags inflate.

b. You've been hired by McDonald's to determine the weights of the potatoes delivered each week for French fries.

c. You are a business reporter covering Genentech Corporation and you are investigating whether their new treatment is effective against childhood leukemia.

SOLUTION

a. The population consists of people who have made claims from accidents in which airbags inflated. The relevant population parameter is the mean (average) cost of claims for accidents with air bags.

b. The population consists of all the potatoes delivered each week for French fries. Relevant population parameters include the mean weight of the potatoes and the variation of the weights (for example, are most of them close to or far from the mean?).

c. The population consists of all children with leukemia. Important population parameters are the percentage of children who recover *without* the new treatment and the percentage of children who recover with the new treatment.

TECHNICAL NOTE

The *mean* is what most of us think of as the *average*. We will discuss the mean in Chapter 4, where we will also see other ways of describing an average.

What Actually Gets Studied?

If researchers at Nielsen were all-powerful they might determine the number of people watching the Super Bowl by asking every individual American. But no one can do that, so instead they try to estimate the number of Americans watching by studying only a relatively small group of people. In other words, Nielsen attempts to learn about the population of all Americans by carefully monitoring the viewing habits of a much smaller **sample** of Americans.

More specifically, Nielsen has devices (called "people meters") attached to televisions in 5,000 homes, so the roughly 13,000 people who live in these homes are the sample of Americans that Nielsen studies.

The individual measurements that Nielsen collects from the people in the 5,000 homes constitute the **raw data**. Nielsen collects much raw data—for example, when and how long each TV in the household is on, what show it is tuned to, and who in the household is watching it. Nielsen then consolidates these raw data into a set of numbers that characterize the sample, such as the percentage of viewers in the sample who watched each individual television show or the total number of people in the sample who watched the Super Bowl. These numbers are called **sample statistics**.

> ### Definitions
>
> A **sample** is a subset of the population from which data are actually obtained.
>
> The actual measurements or observations collected from the sample constitute the **raw data**.
>
> **Sample statistics** are characteristics of the sample found by consolidating or summarizing the raw data.

EXAMPLE 2   Unemployment Survey

The U.S. Labor Department defines the *civilian labor force* as all those people who are either employed or actively seeking employment. Each month, the Labor Department reports the unemployment rate, which is the percentage of people actively seeking employment within the entire civilian labor force. To determine the unemployment rate, the Labor Department surveys 60,000 households. For the unemployment reports, describe the

a. population   b. sample   c. raw data   d. sample statistics   e. population parameters

SOLUTION

a. The *population* is the group that the Labor Department wants to learn about, which is the civilian labor force.

b. The *sample* consists of all the people among the 60,000 households surveyed.

c. The *raw data* consist of all the information collected in the survey.

d. The *sample statistics* summarize the raw data for the sample. In this case, the relevant sample statistic is the percentage of people in the sample who are actively seeking employment. (The Labor Department also calculates similar sample statistics for subgroups in the population, such as the percentages of teenagers, men and women, and veterans who are unemployed.)

e. The *population parameters* are the characteristics of the entire population that correspond to the sample statistics. In this case, the relevant population parameter is the actual unemployment rate. Note that the Labor Department does *not* actually measure this population parameter, since data are collected only for the sample. ∎

How Do Sample Statistics Relate to Population Parameters?

Suppose Nielsen finds that 14% of the people in the 5,000 homes in its sample watched the Super Bowl. This "14%" is a sample statistic, because it characterizes the sample. But what Nielsen really wants to know is the corresponding population parameter, which is the percentage of all Americans who watched the Super Bowl.

By the Way...

According to the Labor Department, someone who is not working is not necessarily unemployed. For example, stay-at-home moms and dads are not counted among the unemployed unless they are actively trying to find a job, and people who tried to find work but gave up in frustration are not counted as unemployed.

Unfortunately, there is no way for Nielsen researchers to know the population parameter, because they've studied only a sample. However, Nielsen researchers hope that they've done their work so that the sample statistic is a good estimate of the population parameter. In other words, they would like to conclude that because 14% of the sample watched the Super Bowl, approximately 14% of the population also watched the Super Bowl. One of the primary purposes of statistics is to help researchers assess the validity of this conclusion.

Statistical science provides methods that enable researchers to determine how well a sample statistic estimates a population parameter. We will discuss some of these methods in Chapter 8, but you may already be familiar with one technique. Results from surveys or opinion polls often include something called the **margin of error**. By adding and subtracting the margin of error from the sample statistic, we find a range, or interval, of values *likely* to contain the population parameter. More technically, we can have 95% confidence that this range contains the population parameters. We'll discuss the precise meaning of "likely" and "95% confidence" in Chapter 8, but for now you can look at the newspaper description in Figure 1.1. In the case of the Nielsen ratings, the margin of error is about 1 percentage point. Thus, if 14% of the sample was watching the Super Bowl, then the range 13% to 15% is likely to contain the actual percentage of the population watching the Super Bowl.

How the Poll Was Conducted

The latest New York Times/CBS News Poll of New York State is based on telephone interviews conducted Oct. 23 to Oct. 28 with 1,315 adults throughout the state. Of those, 1,026 said they were registered to vote. Interviews were conducted in either English or Spanish.

In theory, in 19 cases out of 20 the results based on such samples will differ by no more than three percentage points in either direction from what would have been obtained by seeking out all adult residents of New York State. For smaller subgroups, the potential sampling error is larger.

Figure 1.1   The margin of error in a survey or opinion poll usually describes a range that is likely (with 95% confidence) to contain the population parameter. This excerpt from the *New York Times* explains a margin of error of 3 percentage points.

It is important to realize that while the margin of error is a very useful number, it is meaningful only if the study has been conducted with care. For example, the calculations that lead to a margin of error are based on the assumption that *the sample represents the population fairly*. We'll discuss methods for drawing samples in the next section, but in most cases this means selecting sample members randomly from the population. Another important point is that *the margin of error is generally smaller for a larger sample*. As we'll discuss in Chapter 8, precise margin of error calculations are a bit subtle. Typically, however, the margin of error for a well-conducted poll will be about

- 10 percentage points for a sample size of 100
- 5 percentage points for a sample size of 400
- 3 percentage points for a sample size of 1,000
- 1 percentage point for a sample of 10,000

Time out to think

Is it really possible to learn anything meaningful about the entire population of the United States by conducting a survey or poll of just 400 people? Explain.

Definition

The **margin of error** in a statistical study is used to describe the range of values likely to contain the population parameter. We find this range of values by adding and subtracting the margin of error from the sample statistic obtained in the study. That is, the range of values likely to contain the population parameter is

from (sample statistic − margin of error)

to (sample statistic + margin of error)

EXAMPLE 3 People on Mars?

The Pew Research Center for the People and the Press conducted a survey that involved interviewing 1,546 adult Americans about their attitudes toward the future. In response to a question that asked whether humans would land on Mars within the next 50 years, 76% of these 1,546 people said either *definitely yes* or *probably yes*. The margin of error for the poll was 3 percentage points. Describe the population and the sample for this survey, and explain the meaning of the sample statistic of 76%. What can we conclude about the percentage of the population that thinks humans will land on Mars within the next 50 years?

SOLUTION   The population about which the researchers hoped to learn was all adult Americans, and the sample they studied consisted of the 1,546 people who were interviewed. The sample statistic of 76% is the *actual* percentage of people in the sample who answered that humans would definitely or probably land on Mars in the next 50 years. We do not know the actual population parameter (the percentage of all adult Americans who think people will definitely or probably land on Mars) because the researchers did not ask everyone. However, using the margin of error of 3 percentage points, we find that the range of values

$$\text{from } 76\% - 3\% = 73\%$$
$$\text{to } 76\% + 3\% = 79\%$$

is likely to contain the true percentage of adult Americans who think humans will definitely or probably land on Mars within the next 50 years. ∎

EXAMPLE 4 Cloning Humans?

The same Pew survey also asked people whether they believed that humans would be cloned within the next 50 years. On this question, 51% answered either *definitely yes* or *probably yes*. Again, the margin of error was 3 percentage points. Is it accurate to say that a majority of adult Americans think that humans will be cloned in the next 50 years?

SOLUTION   No. To find the range of values likely to contain the percentage of *all* adult Americans who think human cloning will definitely or probably occur, we add and subtract the margin of error of 3 percentage points from the sample statistic of 51%. This gives a range of values from 48% to 54%. Because this range includes values on both sides of 50%, we cannot conclude that the majority (that is, greater than 50%) of adult Americans think that humans will be cloned in the next 50 years. ∎

By the Way...

A clone is an exact genetic copy of its parent. For example, a clone of *you* would be genetically identical to you, but would be born as a baby and have different life experiences than you. The first successful cloning of an adult mammal came in 1997, when Ian Wilmut and his colleagues in Scotland cloned a sheep. The clone, named Dolly, has no father, since *all* her genes came from the mother.

Time out to think

Look for a report on an opinion poll in this week's news. Does the report give a margin of error? What does it mean in this case?

Putting It All Together: The Process of a Statistical Study

The process used by Nielsen Media Research is similar to that used in many statistical studies. The following summary gives the five basic steps in statistical studies, and Figure 1.2 summarizes the general relationships among a population, a sample, the sample statistics, and the population parameters.

Basic Steps in a Statistical Study

1. State the goal of your study precisely; that is, determine the population you want to study and exactly what you'd like to learn about it.

2. Choose a sample from the population. (Be sure to use an appropriate sampling technique, as discussed in the next section.)

3. Collect raw data from the sample and summarize these data by finding sample statistics of interest.

4. Use the sample statistics to make inferences about the population.

5. Draw conclusions; determine what you learned and whether you achieved your goal.

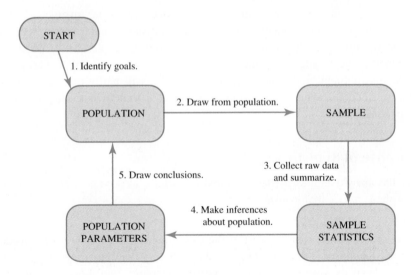

Figure 1.2   Process of a statistical study.

Of course, these steps are somewhat idealized, and the actual steps may be different in a real study. Moreover, as you might imagine, the details hidden in the basic steps are critically important. For example, a poorly chosen sample in Step 2 can render the entire study meaningless. Another important detail involves how the study is set up to collect the raw data; we'll discuss the general types of statistical studies in Section 1.3. Later in the book, we'll discuss techniques for making inferences about a population and drawing reliable conclusions.

EXAMPLE 5   Identifying the Steps

Consider the Pew Research Center survey described in Examples 3 and 4. Identify how researchers applied the five basic steps in a statistical study.

By the Way...

Statisticians often divide their subject into two major branches. *Descriptive statistics* is the branch that deals with *describing* raw data in the form of graphics and sample statistics. *Inferential statistics* is the branch that deals with *inferring* (or estimating) population parameters from sample data.

Statistical thinking will one day be as necessary for efficient citizenship as the ability to read and write.

— H. G. Wells

SOLUTION   The steps apply as follows.

1. The researchers had a goal of learning about specific attitudes of Americans toward the future. They chose adult Americans as the population, deliberately leaving out children.

2. They chose 1,546 adult Americans for their sample. Although we are not told how the sample was drawn, we will assume that it was drawn so that the 1,546 adult Americans are typical of the entire adult American population.

3. They collected the raw data by asking carefully chosen questions of the people in the sample. The raw data are the individual responses. They summarized these data with sample statistics, such as the overall percentages of people in the sample who answered yes or no to each question.

4. Techniques of statistical science allowed the researchers to infer population characteristics. In this case, the inference consisted of estimating the population parameters and calculating the margins of error.

5. By making sure that the study was conducted properly and interpreting the estimates of the population parameters, the researchers drew overall conclusions about Americans' attitudes toward the future. ∎

Statistics: Decisions for an Uncertain World

The Nielsen ratings and most of the examples we've discussed so far involve surveys or polls of some type. However, the subject of statistics encompasses much more. Statistics is invoked almost any time we need to make a decision about a complex issue. We use statistics to decide whether new cancer treatments are effective, whether global warming poses a threat to the environment, and whether going to college will really help you earn more. In fact, it's fair to say that the primary purpose of statistics is to help us make good decisions in an uncertain world.

By the Way...

Prior to the Salk vaccine, other vaccines had been found to *cause* polio instead of preventing it. Researchers therefore developed a slow, cautious approach to testing new vaccines. This approach makes sense–but also means that some people contracted polio during the time that the Salk vaccine was being tested and approved.

A global effort to vaccinate children against polio, begun at the urging of the World Health Organization in May 1998, seeks to eradicate polio in the early 21st century.

> **The Purpose of Statistics**
>
> Statistics has many uses, but perhaps its most important purpose is to help us make good decisions about issues that involve uncertainty.

This purpose will be clear in most of the case studies and examples we consider in this book, but occasionally we'll have to discuss a bit of theory that may seem somewhat abstract at first. If you keep the overall purpose of statistics in mind, you'll be rewarded in the end when you see how the theory helps us understand our world. The following case study will give you a taste of what lies ahead. It involves several important theoretical ideas that led to one of the 20th century's greatest accomplishments in public health.

CASE STUDY   The Salk Polio Vaccine

If you had been a parent in the 1940s or 1950s, one of your greatest fears would have been the disease known as polio. Each year during this long polio epidemic, thousands of young children were paralyzed by the disease. In 1954, a large experiment was conducted to test the effectiveness of a new vaccine created by Dr. Jonas Salk (1914–1995). The experiment involved a sample of 400,000 children chosen from the population of all children in the United States. Half of these 400,000 children received an injection of the Salk vaccine. The other half received an injection that contained only salt water. (The salt water injection was a *placebo*, as we will discuss later.) Among the children receiving the Salk vaccine, only 33 contracted polio. In contrast, there were 115 cases of polio among the children who did not get the Salk vaccine. Using techniques of statistical science that we'll study later, the researchers concluded that the vaccine was effective at preventing polio. They therefore decided to launch a major effort to improve the Salk vaccine and distribute it to the population of *all* children. As a result, children in the United States and other developed countries began to get the vaccine routinely, and the horror of polio is now largely a thing of the past. ∎

Review Questions

1. Why do we say that the term *statistics* has two meanings? Describe both meanings.

2. Define the terms *population* and *population parameter* as they apply to statistical studies. How are these ideas related to the goals of a study?

3. Define the terms *sample*, *raw data*, and *sample statistics*.

4. Briefly explain why Nielsen ratings give accurate results for the sample statistics but only estimates for the population parameters.

5. What is the margin of error in a statistical study and why is it important?

6. Describe the five basic steps in a statistical study. Explain how these steps apply to the Nielsen ratings.

7. What is the primary purpose of statistics? Briefly explain how a decision made with the help of statistics led to a vaccine that prevents polio.

Exercises

BASIC SKILLS AND CONCEPTS

SENSIBLE STATEMENTS? For Exercises 1–6, determine whether the given statement is sensible and explain why it is or is not.

1. When the IRS decided to determine how many people were cheating on their taxes, they did a study in which the sample consisted of every adult in the United States.

2. My professor conducted a study in which he was unable to measure any sample statistics, but he succeeded in determining the population parameters with a very small margin of error.

3. A poll conducted two weeks before the election found that Smith would get 70% of the vote, with a margin of error of 3%, but he ended up losing the election anyway.

4. The goal of a new startup company is to compete with Nielsen Media Research in compiling television ratings. They intend to succeed by providing data with a larger margin of error than Nielsen's while charging television networks the same price for their service.

5. The goal of the research is to learn about depression among people who have suffered through a family tragedy, so the population of interest is everyone who has been sick in the past month.

6. We know for certain that a majority of Americans support the President's position on this issue because an opinion poll found support from 65% of Americans, with a margin of error of 5%.

7. PRE-ELECTION POLL. In order to gauge public opinion about the candidates in a recent election for Governor of California, CBS News conducted telephone interviews with 1,026 people who were registered to vote in California. Describe the population, sample, raw data, sample statistics, and population parameters for this poll.

8. GALACTIC DISTANCES. Astronomers typically determine the distance to a galaxy (a galaxy is a huge collection of billions of stars) by measuring the distances of just a few stars within it and taking the mean of these distance measurements. Describe the population, sample, raw data, sample statistics, and population parameters involved in measuring a galaxy's distance.

9. COMPUTER SPEED. *Byte* magazine rates computers in part according to their speed in "benchmark" tests. For example, to test the speed of a new computer model from Dell, the magazine editors will purchase one of the new models and see how rapidly it completes a series of tasks. Describe the population, sample, raw data, sample

statistics, and population parameters involved when *Byte* rates a new computer model from Dell.

10. **RANGE OF VALUES.** Based on the following sample statistics and margins of error, identify the range of values likely to contain the actual population parameter.

   a. 62 inches with a margin of error of 0.5 inch
   b. 45% with a margin of error of 1 percentage point
   c. 158 pounds with a margin of error of 4.5 pounds

11. **RANGE OF VALUES.** Based on the following sample statistics and margins of error, identify the range of values likely to contain the actual population parameter.

   a. 4.5 months with a margin of error of 0.4 month
   b. 95% with a margin of error of 0.1 percentage point
   c. 457 miles with a margin of error of 1.2 miles

12. **INTERPRETING THE MARGIN OF ERROR.** A marketing survey finds that 65% of the movie-goers surveyed prefer "stadium style" seating in theaters to traditional seating, with a margin of error of 5 percentage points. Interpret the survey results.

13. **ABOUT TO WIN?** A poll is conducted the day before a state election for senator. There are only two candidates running. The poll shows that 53% of the voters surveyed favor the Republican candidate, with a margin of error of 2.5 percentage points. Based on this poll, should the Republican plan a victory party? Why or why not?

14. **ABOUT TO LOSE?** A poll is conducted the day before an election for U.S. Representative. With only two candidates running, the poll shows that 48.5% of the voters surveyed favor the Democratic candidate, with a margin of error of 2.0 percentage points. Based on this poll, should the Democratic candidate expect to lose the election? Why or why not?

FURTHER APPLICATIONS

15. **DO PEOPLE LIE ABOUT VOTING?** In a survey of 1,002 people, 701 (which is 70%) said that they voted in a recent presidential election (based on data from ICR Research Group). The margin of error for the survey was 3 percentage points. However, actual voting records show that only 61% of all eligible voters actually did vote. Does this necessarily imply that people lied when they answered the survey? Explain.

INTERPRETING REAL STUDIES. For each of Exercises 16–21, do the following:

a. Based on the given information, state what you think was the goal of the study. Identify a possible population and the population parameter of interest.

b. Briefly describe the sample, raw data, and sample statistic for the study.

c. Based on the sample statistic and the margin of error, identify the range of values likely to contain the population parameter of interest.

16. In a *Time*/CNN poll, 748 adults were asked whether they believed their children would have a higher standard of living than they have; 63% of those polled said yes. The margin of error was 3.6 percentage points.

17. A Yankelovich poll determined that 70% of 4,000 people surveyed in the United States agreed with the statement "People have to realize that they can only count on their own skills and abilities if they're going to win in this world." The margin of error in the poll was 1.6 percentage points.

18. Based on its survey of 60,000 households (see Example 2), the U.S. Labor Department reported an unemployment rate of 5.6% in January 2002. The margin of error for the report was 0.2 percentage point.

19. Among a sample of 772 male recruits between the ages of 18 and 24, the U.S. Marine Corps found that the mean height was 69.7 inches. The margin of error for the study was 0.20 inch (USDHEW publication 79–1659).

20. Researchers studying a sample of 1,525 women aged 18 to 24 found a mean cholesterol level of 191.7. The margin of error for the study was 2.06 (USDHEW publication 78–1652).

21. A Roper Organization survey of 2,000 adults in the United States revealed that 64% of those surveyed kept money in a regular savings account. The margin of error for the survey was 2.0 percentage points.

STEPS IN A STUDY. Describe how you would apply the five basic steps in a statistical study to the issues in Exercises 22–27.

22. You want to determine the average amount of pizza consumed each year by students at your school.

23. You want to predict who will win the next election for class president.

24. You want to know the mean height of adult American women.

25. You want to know the typical percentage of the bill that is left as a tip in restaurants.

26. You want to know how long alkaline camera batteries last.

27. You want to know the percentage of high school students who are vegetarians.

PROJECTS FOR THE WEB AND BEYOND

For useful links, select "Links for Web Projects" for Chapter 1 at www.aw.com/bbt.

28. **CURRENT NIELSEN RATINGS.** Find the Nielsen ratings for the past week. What were the three most popular television shows? Explain both the "rating" and the "share" for each show.

29. **ATTITUDE UPDATE.** The Pew Research Center for the People and the Press studies public attitudes toward the press, politics, and public policy issues. Go to its Web site and find the latest survey about attitudes. Write a summary of what was surveyed, how the survey was conducted, and what was found.

30. **LABOR STATISTICS.** Use the Bureau of Labor Statistics Web page to find monthly unemployment rates over the past year. If you assume that the monthly survey has a margin of error of about 0.2 percentage point, has there been a noticeable change in the unemployment rate over the past year? Explain.

31. **COMPARING AIRLINES WITH STATISTICS.** The U.S. Department of Transportation routinely publishes on-time performance, lost baggage rates, and other statistics for different airline companies. Find a recent example of such statistics. Based on what you find, is it fair to say that any particular airline stands out as better or worse than others? Explain.

IN THE NEWS

1. **STATISTICS IN THE NEWS.** Identify at least three stories from the past week that involve statistics in some way. In each case, write one or two paragraphs describing the role of statistics in the story.

2. **STATISTICS IN YOUR MAJOR.** Write a description of some ways in which you think the science of statistics is important in your major field of study. (If you have not chosen a major, answer this question for a major that you are considering.)

3. **STATISTICS IN SPORTS.** Choose a sport and describe at least three different statistics commonly tracked by participants in or spectators of the sport. In each case, briefly describe the importance of the statistic to the sport.

4. **ECONOMIC STATISTICS.** The government regularly publishes many different economic statistics, such as the unemployment rate, the inflation rate, and the surplus or deficit in the federal budget. Study recent newspapers and identify five important economic statistics. Briefly explain the purpose of each of these five statistics.

1.2 Sampling

The only way to know the true value of a population parameter is to observe *every* member of the population. For example, to learn the exact mean height of all students at your school, you'd need to measure the height of every student. A collection of data from every member of a population is called a **census**. Unfortunately, conducting a census is often impractical. In some cases, the population is so large that it would be too expensive or time-consuming to collect data from every member. In other cases, a census may be ruled out because it would interfere with a study's overall goals. For example, a study designed to test the quality of candy bars before shipping could not involve a census because that would mean testing a piece of every candy bar, leaving none intact to sell.

Not everything that can be counted counts, and not everything that counts can be counted.

—Albert Einstein

Definition

A **census** is the collection of data from *every* member of a population.

Fortunately, most statistical studies can be done without going to the trouble of conducting a census. Instead of collecting data from every member of the population, we collect data from a sample and use the sample statistics to make inferences about the population. Of course, the inferences will be reasonable only if the members of the sample represent the population fairly, at least in terms of the characteristics under study. That is, we seek a **representative sample** of the population.

> ### Definition
>
> A **representative sample** is a sample in which the relevant characteristics of the sample members are generally the same as the characteristics of the population.

EXAMPLE 1   A Representative Sample for Heights

Suppose you want to determine the mean height of all students at your school. Which is more likely to be a representative sample for this study, the men's basketball team or the students in your statistics class?

SOLUTION   The men's basketball team is not a representative sample for a study of height, both because it consists only of men and because basketball players tend to be taller than average. The mean height of the students in your statistics class is much more likely to be close to the mean height of all students, so the members of your class make a more representative sample than the members of the men's basketball team. ∎

Bias

Imagine that, for the 5,000 homes in its sample, Nielsen chose only homes in which the primary wage earners worked a late-night shift. Because late-night workers aren't home to watch late-night television, Nielsen would find that none of the late-night television shows were popular among the homes in this sample. Clearly, this sample would *not* be representative of all American homes, and it would be wrong to conclude that late-night shows were unpopular among all Americans. We say that such a sample is *biased* because the homes in the sample differed in a specific way from "typical" American homes. (In reality, Nielsen takes great care to avoid such obvious bias in the sample selection.) Generally, the term **bias** refers to any problem in the design or conduct of a statistical study that tends to favor certain results. Thus, we cannot trust the conclusions of a biased study.

> ### Definition
>
> A statistical study suffers from **bias** if its design or conduct tends to favor certain results.

Bias can arise in many ways. For example:

- A sample is biased if the members of the sample differ in some specific way from the members of the general population. In that case, the study will tend to get results that reflect the unusual characteristics of the sample rather than the actual characteristics of the population.

- A researcher is biased if he or she has a personal stake in a particular outcome. In that case, the researcher might intentionally or unintentionally distort the true meaning of the data.

- The data set itself is biased if its values were collected intentionally or unintentionally in a way that makes the data unrepresentative of the population.

- A graph representing the data is biased if it tells only part of the story or depicts the data in a misleading way (see Section 3.4).

Preventing bias from creeping into a study is one of the greatest challenges in statistical research. Looking for bias is therefore one of the most important steps in evaluating a statistical study, and we will encounter the topic of bias throughout this book.

EXAMPLE 2   Why Use Nielsen?

Nielsen Media Research earns money by charging television stations and networks for its services. For example, NBC pays Nielsen to provide ratings for its television shows. Why doesn't NBC simply do its own ratings, instead of paying a company like Nielsen to do them?

SOLUTION   The cost of advertising on a television show depends on the show's ratings. The higher the ratings, the more the network can charge for advertising. Thus, NBC has a financial stake in the ratings, giving it a clear bias if it conducts its own ratings. Advertisers would therefore not trust ratings that NBC produced on its own. By hiring an independent source, such as Nielsen, NBC can provide information that advertisers will believe.   ∎

Time out to think
The fact that NBC pays Nielsen for its services might seem to give Nielsen a financial incentive to make NBC look good in the ratings. How does the fact that Nielsen also provides ratings for other networks help prevent it from being biased toward NBC?

Sampling Methods

A good statistical study *must* have a representative sample. Otherwise the sample is biased and conclusions from the study are not trustworthy. Let's examine a few common sampling methods that, at least in principle, can provide a representative sample.

SIMPLE RANDOM SAMPLES

In most cases, the best way to obtain a representative sample is by choosing *randomly* from the population. With **simple random sampling**, every possible sample of a particular size has an equal chance of being selected. For example, to choose a simple random sample of 100 students from all the students in your school, you could assign a number to each student in your school and choose the sample by drawing 100 of these numbers from a hat. As long as each student's number is in the hat only once, every sample of 100 students has an equal chance of being selected. As a faster alternative to using a hat, you might choose the student numbers with the aid of a computer or calculator that has a built-in *random number generator*.

Time out to think
Look for the random number key on a calculator. (Nearly all scientific calculators have one.) What happens when you push it? How could you use the random number key to select a sample of 100 students?

Because simple random sampling gives every sample of a particular size the same chance of being chosen, it is very likely to provide a representative sample. In addition to obtaining a simple random sample, another primary concern is making sure that the sample is large enough. The larger the simple random sample, the more likely it is to be representative of the population.

EXAMPLE 3   Telephone Book Sampling

You want to conduct an opinion poll in which the population is all the residents in a town. Could you choose a simple random sample by randomly selecting names from the local telephone book?

SOLUTION   A sample drawn from a telephone book does not give everyone in the town the same chance of being selected because many people are not listed in the phone book. Many phone numbers are shared by two or more people but have only one listing, some people have unlisted phone numbers, and some people (such as the homeless) don't have telephones. Thus, a sample drawn from the telephone book is not a simple random sample.   ■

SYSTEMATIC SAMPLING

Simple random sampling is effective, but in many cases we can get equally good results with a simpler technique. Suppose you are testing the quality of microchips produced by Intel. As the chips roll off the assembly line, you might decide to test every 50th chip. This ought to give a representative sample because there's no reason to believe that every 50th chip has any special characteristics compared to other chips. This type of sampling, in which we use a system such as choosing every 50th member of a population, is called **systematic sampling**.

EXAMPLE 4   Museum Assessment

When the National Air and Space Museum wanted to test possible ideas for a new solar system exhibit, a staff member interviewed a sample of visitors selected by systematic sampling. She interviewed a visitor exactly every 15 minutes, choosing whoever happened to enter the current solar system exhibit at that time. Why do you think she chose systematic sampling rather than simple random sampling? Was systematic sampling likely to produce a representative sample in this case?

SOLUTION   Simple random sampling might occasionally have selected two visitors so soon after each other that the staff member would not have had time to interview each of them. The systematic process of choosing a visitor every 15 minutes prevented this problem from arising. Because there's no reason to think that the people entering at a particular moment are any different from those who enter a few minutes earlier or later, this process is likely to give a representative sample of the population of visitors during the time of the sampling.   ■

EXAMPLE 5   When Systematic Sampling Fails

You are conducting a survey of students in a co-ed dormitory in which males are assigned to odd-numbered rooms and females are assigned to even-numbered rooms. Can you obtain a representative sample when you choose every 10th room?

SOLUTION   No. If you start with an odd-numbered room, every 10th room will also be odd-numbered (such as room numbers 3, 13, 23, . . .). Similarly, if you start with an even-numbered room, every 10th room will also be even-numbered. Thus, you will obtain a sample consisting of either all males or all females, neither of which is representative of the co-ed population. ∎

Time out to think

Suppose you chose every 5th room, rather than every 10th room, in Example 5. Could the sample then be representative?

CONVENIENCE SAMPLES

Systematic sampling is easier than simple random sampling, but may also be impractical in many cases. For example, suppose you want to know what proportion of the students at your school are left-handed. It would take much effort to select either a simple random sample or a systematic sample, because both require drawing from all the students in the school. In contrast, it would be very easy to use the students in your statistics class as your sample—you could just ask the left-handed students to raise their hands. This type of sample is called a **convenience sample** because it is chosen for convenience rather than by a more sophisticated procedure. In trying to find the proportion of left-handed people, the convenience sample of your statistics class is probably fine; there is no reason to think that there would be a different proportion of left-handed students in a statistics class than anywhere else. But if you were trying to determine the proportions of students with different majors, this sample would be biased because some majors require a statistics course and others do not. In general, convenience sampling tends to be more prone to bias than most other forms of sampling.

EXAMPLE 6   Salsa Taste Test

A supermarket wants to find out whether it should carry a new brand of salsa, so it offers free tastes at a stand in the store and asks people what they think of the salsa. Explain why this is a convenience sample. What population is being studied? Is this sample likely to be representative of that population?

SOLUTION   The sample of shoppers stopping for a taste of the salsa is a convenience sample because these people are easy to interview about the product. (This type of convenience sample, in which people choose whether or not to be part of the sample, is called a *self-selected sample*. We will study self-selected samples further in Section 1.4.) This sample is unlikely to be representative of *all* shoppers because it probably includes only people who like salsa. However, because the store wants to determine whether to carry the new brand of salsa, the population of interest consists only of salsa eaters. Thus, this convenience sample may well be representative of the population of interest. ∎

CLUSTER SAMPLES

Cluster sampling involves the selection of *all* members in randomly selected groups, or *clusters*. Imagine that you work for the Department of Agriculture and wish to determine the percentage of farmers who use organic farming techniques. It would be difficult and costly to collect a simple random sample or a systematic sample because either sample would require visiting many individual farms that are located far from one another. A convenience sample of farmers in a single county would be biased because farming practices vary from region to region. Thus, you might decide to select a few dozen counties at random from across the United States and then survey *every* farmer in each of those counties. Each county contains a cluster of farmers, and the sample consists of *every* farmer within the randomly selected clusters.

EXAMPLE 7   Gasoline Prices

You want to know the mean price of gasoline at gas stations located within a mile of rental car return locations at airports. Explain how you might use cluster sampling in this case.

SOLUTION   You could randomly select a few airports around the country. For these airports, you would check the gasoline price at *every* gas station within a mile of the rental car return location.

STRATIFIED SAMPLES

Suppose you are conducting a poll to predict the outcome of the next U.S. presidential election. The population under study is all likely voters, so you might choose a simple random sample from this population. However, because presidential elections are decided by electoral votes cast on a state-by-state basis, you'll get a better prediction if you determine voter preferences within each of the 50 states. Thus, your overall sample ought to consist of separate random samples from the 50 states. In statistical terminology, the populations of the 50 states represent subgroups, or **strata**, of the total population. Because your overall sample consists of randomly selected members from each stratum, we say that you've used **stratified sampling**.

EXAMPLE 8   Unemployment Data

The Labor Department surveys 60,000 households each month to compile its unemployment report (see Example 2 in Section 1.1). To select these households, the Department first groups cities and counties into about 2,000 geographic areas. It then randomly selects households to survey within these geographic areas. How is this an example of stratified sampling? What are the strata? Why is stratified sampling important in this case?

SOLUTION   The unemployment survey is an example of stratified sampling because it first breaks the population into subgroups. The subgroups, or strata, are the people in the 2,000 geographic regions. Stratified sampling is important in this case because unemployment rates are likely to differ in different geographic regions. For example, unemployment rates in rural Kansas may be very different from those in the Silicon Valley. By using stratified sampling, the Labor Department ensures that its sample fairly represents all geographic regions.

Summary of Sampling Methods

Each of the different sampling methods may be more or less appropriate for different studies. Some studies combine two or more types of sampling; for example, a convenience sample might be chosen first and then simple random sampling might be done to reduce the size of the sample further. Here are three key ideas to keep in mind:

- Regardless of how a sample is chosen, the study can be successful only if the sample is representative of the population.
- A biased sample is very unlikely to be a representative sample.
- A well-chosen sample has a good chance of being representative, but still may turn out to be biased just because of bad luck in the actual drawing of the sample.

Figure 1.3 illustrates the sampling methods we have discussed.

By the Way…

As mandated by the U.S. Constitution, voting for the President is actually done by a small group of people called *electors*. Each state may select as many electors as it has members of Congress (counting both Senators and Representatives). When you cast a ballot for President, you actually cast a vote for your state's electors, each of whom has promised to vote for a particular presidential candidate. The electors cast their votes a few weeks after the general election.

Simple Random Sampling:
Every sample of the same size has an equal chance of being selected. Computers are often used to generate random telephone numbers.

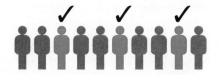

Systematic Sampling:
Select every kth member.

Convenience Sampling:
Use results that are readily available.

Hey! Do you support the death penalty?

Election precincts in Carson County

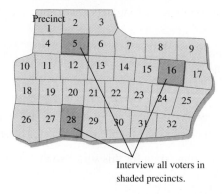

Interview all voters in shaded precincts.

Cluster Sampling:
Divide the population area into sections, randomly select a few of those sections, and then choose all members in them.

Stratified Sampling:
Partition the population into at least two strata, then draw a sample from each.

Figure 1.3

> ### Common Sampling Methods
>
> **Simple random sampling:** We choose a sample of items in such a way that every sample of the same size has an equal chance of being selected.
>
> **Systematic sampling:** We use a simple system to choose the sample, such as selecting every 10th or every 50th member of the population.
>
> **Convenience sampling:** We use a sample that happens to be convenient to select.
>
> **Cluster sampling:** We first divide the population into groups, or clusters, and select some of these clusters at random. We then obtain the sample by choosing *all* the members within each of the selected clusters.
>
> **Stratified sampling:** We use this method when we are concerned about differences among subgroups, or *strata*, within a population. We first identify the strata and then draw a random sample within each stratum. The total sample consists of all the samples from the individual strata.

EXAMPLE 9   Sampling Methods

Identify the type of sampling used in each of the following cases:

a. The apple harvest from an orchard is collected in 1,200 baskets. An agricultural inspector randomly selects 25 baskets and then checks every apple in each of these baskets for worms.

b. An educational researcher wants to know whether, at a particular college, men or women tend to ask more questions in class. Of the 10,000 students at the college, she interviews 50 randomly selected men and 50 randomly selected women.

c. In trying to learn whether planetary systems are common, astronomers conduct a survey by looking for planets among 100 of the nearest stars.

d. To determine who will win autographed footballs, a computer program randomly selects the ticket numbers of 11 people in a stadium filled with people.

SOLUTION

a. The apple inspection is an example of cluster sampling because the inspector begins with a randomly selected set of clusters (baskets) and then checks every apple in the selected clusters.

b. The groups of men and women represent two different strata for this study, so this is an example of stratified sampling.

c. The astronomers focus on nearby stars because they are easier to study, so this is an example of a convenience sample.

d. Because the computer selects the 11 ticket numbers at random, every ticket number has an equal chance of being chosen. This is an example of simple random sampling. ∎

Review Questions

1. What is a *census*? Why is conducting a census often impractical?

2. Why is it so important that a statistical study use a *representative sample*?

3. What is *bias*? How can it affect a statistical study? Give examples of several different types of bias.

4. What is *simple random sampling*? What makes it a good sampling method? Why is it sometimes difficult or costly?

5. What is *systematic sampling*? Give an example in which systematic sampling will tend to give a representative sample, and give another in which it will not produce a representative sample.

6. Why is *convenience sampling* particularly prone to bias? Can it ever produce a representative sample? Explain.

7. Describe an example of *cluster sampling*, and explain why it is sometimes a useful sampling technique.

8. What is *stratified sampling*, and when is it useful? What do we mean by the *strata*?

9. Suppose that a sample in a study is selected in the best possible way. Are we then guaranteed that the sample is representative, or is it possible that the sample might still be biased? Explain.

Exercises

BASIC SKILLS AND CONCEPTS

SENSIBLE STATEMENTS? For Exercises 1–6, determine whether the given statement is sensible and explain why it is or is not.

1. For a high school project, I conducted a census to determine the average rate that teenagers charge for babysitting.

2. Even though the study used a convenience sample, its results may be meaningful.

3. The study must have been biased, because it concluded that 75% of Americans are more than 6 feet tall.

4. We obtained a simple random sample of milk-producing cows in Jefferson County by drawing the names of 50 dairy farms from a hat and asking the owners of those farms to select three cows for us to study.

5. A good strategy for stratified sampling involves first using simple random sampling to choose 500 people, then randomly dividing them into ten groups of 50 to represent the ten strata.

6. Although the study was conducted with a representative sample and careful analysis, the conclusions still reflect the researcher's anti–death penalty bias.

CENSUS. Is a census practical in the situations described in Exercises 7–10? Explain your reasoning.

7. You want to determine the mean GPA of the 50 students in a particular high school mathematics class.

8. You want to determine the mean GPA of all high school seniors in the United States.

9. You want to determine the mean annual energy costs of all homes in Missouri.

10. You want to determine the mean temperature of coffee served at Starbucks.

11. REPRESENTATIVE SAMPLE? You want to determine the mean number of calories eaten daily by girls on high school track teams. Which of the following is most likely to be a representative sample? Why? Also explain why each of the other choices is *not* likely to make a representative sample for this study.

   • The seniors on your high school's girls' track team

   • The shot putters and javelin throwers on your high school's girl's track team

   • The sprinters on your high school's girls' track team

   • The entire girls' track team at your high school

12. REPRESENTATIVE SAMPLE? You want to determine the typical dietary habits of students at a college. Which of the following would make the best sample? Why? Also explain why each of the other choices would *not* make a good sample for this study.

   • Students in a single dormitory

   • Students majoring in public health

   • Students who participate in intercollegiate sports

   • Students enrolled in a required mathematics class

BIAS. Are there sources of bias in the situations described in Exercises 13–18? Explain your reasoning.

13. A film critic on ABC gives her opinion of the latest movie from Disney, which also happens to own ABC.

14. President George W. Bush shows a graph illustrating how much money people will save under his plan for tax reform.

15. The *New York Times Book Review* prints a review of a new book for which it has also accepted a paid advertisement.

16. Monsanto conducts a study to determine whether its new, genetically engineered soybean poses any threat to the environment.

17. Monsanto hires independent university scientists to determine whether its new, genetically engineered soybean poses any threat to the environment.

18. Scientists hired by Greenpeace (which opposes genetically engineered crops) conduct a study to determine whether Monsanto's new, genetically engineered soybean poses any threat to the environment.

FURTHER APPLICATIONS

BIAS IN SAMPLING. Exercises 19–27 describe sampling procedures. In each case, state whether you think the procedure is likely to yield a representative sample or a biased sample and why.

19. In a quality control study of computer chips, tests are run on every 100th chip that comes off the assembly line.

20. In a survey of customer satisfaction with the local telephone company, researchers call the first person listed on each page of the local telephone directory.

21. An exit poll designed to predict the winner of a local election uses interviews with everyone who votes between 7:00 and 7:30 A.M.

22. An exit poll designed to predict the winner of a national election uses interviews with randomly selected voters in New York.

23. College administrators seek to know students' choices for their all-time best and worst professors. They mail a survey to every student at the college and ask students to mail their responses back. They then count the responses from those surveys that were returned by mail.

24. In a survey to find out whether students at a particular high school favor banning off-campus lunches, teachers interview 300 students randomly selected from the school cafeteria at lunchtime.

25. Scientists who have developed a new nonfat margarine substitute test the product for taste on a random sample of 200 employees of their company.

26. To determine the opinions on capital punishment of people in the 18- to 24-year age group, researchers survey a random sample of 500 Marines in this age group.

27. To determine what fraction of customers uses coupons, market researchers conduct a survey at a supermarket on a weekday between 10:00 A.M. and noon.

IDENTIFYING THE SAMPLING METHOD. Exercises 28–35 describe sampling methods. In each case, identify the sampling method as either simple random sampling, systematic sampling, convenience sampling, cluster sampling, or stratified sampling. Briefly explain why you think this sampling method was chosen.

28. An IRS (Internal Revenue Service) auditor randomly selects for audits 30 taxpayers whose gross income is less than $30,000 and 30 taxpayers whose gross income is more than $30,000.

29. *People* magazine chooses its "best-dressed celebrities" by looking at responses from readers who mail in a survey printed in the magazine.

30. A medical researcher at Johns Hopkins University interviews all leukemia patients in each of 20 randomly selected counties.

31. A study of sleep habits uses 50 participants whose ages are between 20 and 29, 50 participants whose ages are between 30 and 39, and 50 participants whose ages are between 40 and 49.

32. Every 100th roll of rope that is produced is given a strength test.

33. In a pre-election poll for a mayoral race, all homes on First, Third, and Fifth Streets are visited.

34. Four hundred men and four hundred women are interviewed to determine their opinions on the candidates for President.

35. Two hundred people are randomly selected from each of the 50 states for a study of how incomes vary among the states.

CHOOSING A SAMPLING METHOD. For each of Exercises 36–41, suggest a sampling method that is likely to produce a representative sample. Explain why you chose this method over other methods.

36. You want to predict the winner of an upcoming election for student body president.

37. You want to determine the percentage of people in this country in each of the four major blood groups (A, B, AB, and O).

38. You want to determine the number of smokers in this country who die from lung cancer each year.

39. You want to determine the number of lung cancer victims (each year) who smoked.

40. You want to determine the average mercury content of the tuna fish consumed by U.S. residents.

41. You want to determine whether drinking three cups of herbal tea per day reduces the chances of getting a cold.

PROJECTS FOR THE WEB AND BEYOND

For useful links, select "Links for Web Projects" for Chapter 1 at www.aw.com/bbt.

42. **SAMPLE FOR THE NIELSEN RATINGS.** Use information available on the Nielsen Media Research Web site to answer the following questions.

    a. How does Nielsen select the sample of homes to be included in a viewer survey?
    b. Describe a few ways in which Nielsen attempts to check that the results from its people meter surveys are accurate.
    c. Based on what you have learned, do you think the Nielsen ratings are reliable? If so, why? If not, why not?

43. **UNEMPLOYMENT SAMPLE.** Use the Bureau of Labor Statistics Web page to find details on how the Bureau chooses the 60,000 households in its monthly survey. Write a short summary of the procedure and why it is likely to yield a representative sample.

44. **SELECTIVE VOTING.** The Academy Awards, the Heisman Trophy, and the *New York Times* "Bestseller List"

are just three examples of selections that are determined by the votes of specially selected individuals. Pick one of these selection processes and describe who votes and how those people are chosen. Discuss sources of bias in the process.

IN THE NEWS

1. **SAMPLING IN THE NEWS.** Find a recent news report about a statistical study that you find interesting. Write a short summary of how the sample for the study was chosen, and briefly discuss whether you think the sample was representative of the population under study.

2. **OPINION POLL SAMPLE.** Find a recent news report about an opinion poll carried out by a news organization (such as *USA Today*, *New York Times*, or CNN). Briefly describe the sample and how it was chosen. Was the sample chosen in a way that was likely to introduce any bias? Explain.

3. **POLITICAL POLL SAMPLE.** Find results from a recent poll conducted by a political organization (such as the Republican or Democratic party or an organization that seeks to influence Congress on some particular issue). Briefly describe the sample and how it was chosen. Was the sample chosen in a way that was likely to introduce any bias? Should you be more concerned about bias in such a poll than you would be in a poll conducted by a news organization? Explain.

1.3   Types of Statistical Studies

Statistical studies are conducted in many different ways. In all cases, the people or objects chosen for the sample are called the **subjects** of the study. If the subjects are people, it is common to refer to them as **participants** in the study.

> ### Definition
>
> The **subjects** of a study are the people or objects chosen for the sample; if the subjects are people, they may also be called the **participants** in the study.

Nielsen Media Research *observes* the television-viewing behavior of the subjects in its 5,000 sample homes. That is, Nielsen tries to determine what the subjects are watching on TV, but does not try to influence what they watch. We therefore call this type of research an **observational study**.

You can observe a lot by just watching.
—Yogi Berra

Note that an observational study may involve activities that go beyond the usual definition of *observing*. Measuring people's weights requires interacting with them, as in asking them to stand on a scale. But in statistics, we consider these measurements to be observations because the interactions do not change people's weights. Similarly, an opinion poll in which researchers conduct in-depth interviews is considered observational as long as the researchers attempt only to learn people's opinions, not to change them.

In contrast, consider a medical study designed to test whether large daily doses of vitamin C help prevent colds. To conduct this study, the researchers must ask some people in the sample to take large doses of vitamin C every day. This type of statistical study is called an **experiment**. The purpose of an experiment is to study the effects of some **treatment**—in this case, large daily doses of vitamin C.

Traditionally, statisticians considered observational studies and experiments to be the two basic types of statistical studies. In recent years, however, a third type of study, called **meta-analysis**, has become increasingly important. A meta-analysis is essentially a study combining many previous studies. Consider the case of vitamin C and colds, which has been the topic of hundreds of studies over the past few decades. Rather than conduct a new experiment, researchers might elect to analyze the previous studies as a group. In some cases, a meta-analysis can reveal trends more clearly than individual studies.

> If I have ever made any valuable discoveries, it has been owing more to patient attention, than to any other talent.
>
> —Isaac Newton

Three Basic Types of Statistical Studies

1. In an **observational study,** researchers observe or measure characteristics of the subjects, but do not attempt to influence or modify these characteristics.

2. In an **experiment,** researchers apply some **treatment** and observe its effects on the subjects of the experiment.

3. In a **meta-analysis,** researchers study a topic that has been the subject of many previous studies. The meta-analysis considers the previous studies as a combined group, with the aim of finding trends that were not evident in the individual studies.

EXAMPLE 1   Type of Study

> I feel the greatest reward for doing is the opportunity to do more.
>
> —Jonas Salk

a. The Salk polio vaccine study, in which some participants were given a new vaccine and others were given no treatment (see the Case Study on page 8), was an *experiment* because researchers applied a treatment—in this case, the vaccine—and then determined the number of participants who contracted polio.

b. An opinion poll in which people are asked for whom they plan to vote in the next election is an *observational study*; it attempts to determine voting preference, but does not try to sway votes.

c. A study in which social scientists evaluate the effects of new welfare laws by poring through the results of more than 100 previous studies of welfare recipients is a *meta-analysis*; its goal is to summarize many previous studies. ∎

Issues in the Design of Experiments

For most observational studies, the most important issue is choosing a representative sample. But sample selection is only one of many important issues in designing an experiment. Let's consider a few other key issues.

THE NEED FOR CONTROLS

Consider again the study of vitamin C and colds. Suppose the people taking vitamin C daily get an average of 1.5 colds in a three-month period. How can the researchers know whether the subjects would have gotten more colds without the vitamin C? To answer this type of question, the researchers must conduct their experiment with two groups of subjects: One group takes large doses of vitamin C daily, and another group does not. The group of people who take vitamin C is called the **treatment group** because its members receive the treatment (vitamin C) being tested. The group of people who do *not* take vitamin C is called the **control group**. The researchers can be confident that vitamin C is an effective treatment only if the people in the treatment group get significantly fewer colds than the people in the control group. The control group gets its name from the fact that it helps control the way we interpret experimental results.

Treatment and Control Groups

The **treatment group** in an experiment is the group of subjects who receive the treatment being tested.

The **control group** in an experiment is the group of subjects who do *not* receive the treatment being tested.

Note: Some experiments involve only a treatment group, many involve both a treatment group and a control group, and others involve more than two groups (for example, three groups that each receive a different treatment).

EXAMPLE 2   Treatment and Control

Look again at the Salk polio vaccine Case Study (page 8). Which group of children constituted the treatment group? Which constituted the control group?

SOLUTION   The treatment was the Salk vaccine. The treatment group consisted of the children who received the Salk vaccine. The control group consisted of the children who did not get the Salk vaccine and instead got an injection of salt water. ∎

EXAMPLE 3   Mozart Treatment

A study divided college students into two groups. One group listened to Mozart or other classical music before being assigned a specific task, and the other group simply was assigned the task. Researchers found that those listening to Mozart performed the task slightly better, but only if they did it within a few minutes of listening to the music. Identify the treatment and control groups.

SOLUTION   The treatment was the classical music. The treatment group consisted of the students who listened to the music. The control group consisted of the students who did not listen to the music. ∎

BEWARE OF CONFOUNDING

Using control groups helps to ensure that we account for known factors that could affect a study's results. However, researchers may be unaware of or be unable to account for other important factors. Consider an experiment in which a statistics teacher seeks to determine whether students who study collaboratively (in study groups with other students) earn higher grades than students who study independently. The teacher chooses five students

By the Way…

The so-called Mozart effect holds that listening to Mozart can make babies smarter. The supposed effect spawned an entire industry of Mozart products for children. The state of Georgia even began passing out Mozart CDs to new mothers. However, a recent meta-analysis of studies of the Mozart effect suggests that there is no significant effect after all.

who will study collaboratively (the treatment group) and five others who will study independently (the control group). To ensure that the students all have similar abilities and will study diligently, the teacher chooses only students with high grade-point averages. At the end of the semester, the teacher finds that the students who studied collaboratively earned higher grades.

But suppose that, unbeknownst to the teacher, the collaborative students all lived in a dormitory where a curfew ensured that they got plenty of sleep. This factor—or perhaps others that were not considered in the study—could partially explain the results. In other words, the experiment's conclusion may *seem* to support the benefits of collaborative study, but this conclusion is not justified because the teacher did not account for the curfew or other factors that may have affected the results.

In statistical terminology, this study suffers from **confounding**. The higher grades could be attributed to a mixture of effects from collaborative work and curfews, and the teacher did not account for these **confounding factors**.

> ## Definition
>
> A study suffers from **confounding** if the effects of different factors are mixed so that we cannot determine the effects of the specific factors we are studying. The factors that lead to the confusion are called **confounding factors.**

CASE STUDY Confounding Drug Results

In the early 1990s, Pfizer Corporation developed a new drug (called fluconazole) to prevent fungal infections in hospital patients. Several studies found the new drug to be more effective than an older drug (called amphotericin B). However, a subsequent analysis by other researchers found that the older drug had been administered orally when it was supposed to be given by injection. This introduced a source of confounding into the studies: The original researchers thought the results showed the new drug to be more effective than the old drug, but they had not taken into account the confounding effects of how the old drug was administered. Once the new researchers took this effect into account, they found that the new drug was no more effective than the old drug. ∎

ASSIGNING TREATMENT AND CONTROL GROUPS

As the collaborative study experiment illustrates, results are almost sure to suffer from confounding if the treatment and control groups differ in any important way. Researchers generally employ two strategies to prevent such differences and thereby ensure that the treatment and control groups can be compared fairly. First, they assign participants to the treatment and control groups *at random*, using some technique that ensures that each participant has an equal chance of becoming part of either group. When the participants are randomly assigned, it is less likely that the people in the treatment and control groups will differ in some way that will affect the study results. There are other strategies for assigning subjects to groups, but random selection is often a favored procedure.

Second, researchers try to ensure that the treatment and control groups are sufficiently large. For example, in the collaborative study experiment, including 50 students in each group rather than 5 would have made it much less likely that all the students in one group would live in a special dormitory. Thus, having a larger number of subjects in the experiment helps to ensure that we can see the true nature of any effects from the treatment.

> **Strategies for Selecting Treatment and Control Groups**
>
> 1. **Use a randomized experiment (randomization).** Make sure that the subjects of the experiment are assigned to the treatment or control group at random so that each subject has an equal chance of being assigned to either group.
>
> 2. **Use a sufficiently large number of subjects or participants.** Make sure that the treatment and control groups are both sufficiently large that they are unlikely to differ in a significant way (aside from the fact that one group gets the treatment and the other does not).

EXAMPLE 4   Salk Study Groups

Briefly explain how the two strategies for selecting treatment and control groups were used in the Salk polio vaccine study.

SOLUTION   A total of about 400,000 children participated in the study, with half receiving an injection of the Salk vaccine (the treatment group) and the other half receiving an injection of salt water (the control group). The first strategy was implemented by choosing children for the two groups randomly from among all the children. The second strategy was implemented by using a large number of participants (200,000 in each group) so that the two groups were unlikely to differ by accident. ∎

THE PLACEBO EFFECT

When an experiment involves people, effects can occur simply because people know they are part of the experiment. For example, suppose you are testing the effectiveness of a new anti-depression drug. You find 500 people who are suffering from depression and randomly divide them into a treatment group that receives the new drug and a control group that does not. A few weeks later, interviews with the patients show that people in the treatment group tend to be feeling much better than people in the control group. Can you conclude that the new drug works?

Unfortunately, it's quite possible that the mood of people receiving the drug improved simply because they were glad to be getting some kind of treatment. In that case, you can't be sure whether the drug really helped. This type of effect, in which people improve because they believe that they are receiving a useful treatment, is called the **placebo effect**. (The word *placebo* comes from the Latin "to please.")

To distinguish between results caused by a placebo effect and results that are truly due to the treatment, researchers try to make sure that the participants do not know whether they are part of the treatment or control group. To accomplish this, the researchers give the people in the control group a **placebo**: something that looks just like the drug being tested, but lacks its active ingredients. For example, in a test of a drug that comes in pill form, the placebo might be a pill of the same shape and size that contains sugar instead of the real drug ingredients. In a test of an injected vaccine, the placebo might be an injection that contains only a saline solution (salt water) instead of the real vaccine.

As long as the participants do not know whether they received the real treatment or a placebo, the placebo effect ought to affect the treatment and control groups equally. (For ethical reasons, participants *should* be told that some of them will be given a placebo, but not told individually whether they belong to the treatment or control group.) If the results are significantly different for the two groups, the differences can be attributed to the drug. For example, in the study of the anti-depression drug, we would conclude that the drug was effective only if

> A strong imagination brings on the [placebo effect] . . . Everyone feels its impact, but some are knocked over by it . . . . [Doctors] know that there are men on whom the mere sight of medicine is operative.
>
> —French philosopher
> Michel de Montaigne (1533–1592)

the control group received a placebo and members of the treatment group improved much more than members of the control group. (Some experiments use three groups: a treatment group, a placebo group, and a control group. The placebo group is given a placebo while the control group is given nothing.)

By the Way…

The placebo effect can be remarkably powerful. In some studies, up to 75% of the participants receiving the placebo actually improve. For some patients, the effect is so powerful that they plead to continue their treatment even after being told they were given a placebo rather than the real treatment. Nevertheless, different researchers disagree about the strength of the placebo effect, and some even question the reality of the effect.

Definitions

A **placebo** lacks the active ingredients of a treatment being tested in a study, but is identical in appearance to the treatment. Thus, study participants cannot distinguish the placebo from the real treatment.

The **placebo effect** refers to the situation in which patients improve simply because they believe they are receiving a useful treatment.

Time out to think

In decades past, researchers often told all the participants in a study that they were receiving the real treatment, but actually gave a placebo to half the participants. Do you think such a study is ethical? Why or why not?

EXAMPLE 5   Vaccine Placebo

What was the placebo in the Salk polio vaccine study? Why did researchers use a placebo in this experiment?

SOLUTION   The placebo was the salt water injection given to the children in the control group. To understand why the researchers used a placebo for the control group, suppose that a placebo had *not* been used. When improvements were observed in the treatment group, it would have been impossible to know if the improvements were due to the vaccine or to the placebo effect. In order to remove this confounding, all participants had to believe that they were being treated in the same way. This ensured that any placebo effect would occur in both groups equally, so that researchers could attribute any remaining differences to the vaccine.

By the Way…

A related effect, known as the *Hawthorne effect*, occurs when treated subjects somehow respond differently simply because they are part of an experiment—regardless of the particular way in which they are treated. The Hawthorne effect gets it name because it was first observed in a study of factory workers at Western Electric's Hawthorne plant.

EXPERIMENTER EFFECTS

Even if the study subjects don't know whether they received the real treatment or a placebo, the *experimenters* may still have an effect. In testing an anti-depression drug, for example, experimenters will probably interview patients to find out whether they are feeling better. But if the experimenters know who received the real drug and who received the placebo, they may inadvertently smile more at the people in the treatment group. Their smiles might improve those participants' moods, making it seem as if the treatment worked when in fact the improvement was caused by the experimenter. This type of confounding, in which the experimenter somehow influences the results, is called an **experimenter effect** (or a Rosenthall effect). The only way to avoid experimenter effects is to make sure that the experimenters don't know which subjects are in which group.

Definition

An **experimenter effect** occurs when a researcher or experimenter somehow influences subjects through such factors as facial expression, tone of voice, or attitude.

EXAMPLE 6   Child Abuse?

In a famous case, two couples from Bakersfield, California, were convicted of molesting dozens of preschool-age children at their day care center. The evidence for the abuse came primarily from interviews with the children. However, the conviction was overturned—after the couple had served 14 years in prison—when a judge re-examined the interviews and concluded that the children had given answers that they thought the interviewers wanted to hear. If we think of the interviewers as experimenters, this is an example of an experimenter effect because the interviewers influenced the children's answers through the tone and style of their questioning.   ∎

BLINDING

In statistical terminology, the practice of keeping people in the dark about who is in the treatment group and who is in the control group is called **blinding**. A **single-blind** experiment is one in which the participants don't know which group they belong to, but the experimenters do know. If neither the participants nor the experimenters know who belongs to each group, the study is said to be **double-blind**. Of course, *someone* has to keep track of the two groups in order to evaluate the results at the end. In a double-blind experiment, the researchers conducting the study typically hire experimenters to make any necessary contact with the participants. Thus, the researchers themselves avoid any contact with the participants, ensuring that they cannot influence them in any way. The Salk polio vaccine study was double-blind because neither the participants (the children) nor the experimenters (the doctors and nurses giving the injections and diagnosing polio) knew who got the real vaccine and who got the placebo.

> ## By the Way...
>
> Many similar cases of supposedly widespread child abuse at day care centers and preschools are being re-examined to see if experimenter effects (by those who interviewed the children) may have led to wrongful convictions. Similar claims of experimenter effects have been made in cases involving repressed memory, in which counseling supposedly helped people retrieve lost memories of traumatic events.

> ### Blinding in Experiments
>
> An experiment is **single-blind** if the participants do not know whether they are members of the treatment group or members of the control group, but the experimenters do know.
>
> An experiment is **double-blind** if neither the participants nor any experimenters know who belongs to the treatment group and who belongs to the control group.

EXAMPLE 7   What's Wrong with This Experiment?

For each of the experiments described below, identify any problems and explain how the problems could have been avoided.

a. A new drug for attention deficit disorder (ADD) is supposed to make affected children more polite. Randomly selected children suffering from ADD are divided into treatment and control groups. The experiment is single-blind. Experimenters evaluate how polite the children are during one-on-one interviews.

b. Educational researchers wonder if listening to classical music when studying improves learning. They give two groups of students an identical 2-hour lesson, and then allow time to study for a short exam. One group, made up of 50 students who told the researchers that they like classical music, listens to classical music while they study. The other group, made up of 50 other students who told the researchers that they don't like classical music, studies in silence. The results show that the students who listened to classical music did better on the test.

c. Researchers wonder if the effects of a rare degenerative disease can be slowed by exercise. They identify six people suffering from the disease, and randomly assign three to a treatment group that exercises every day and three to a control group that avoids exercise. After six months, they compare the amounts of degeneration in each group.

d. A chiropractor performs adjustments on 25 patients with back pain. Afterward, 18 of the patients say they feel better. He concludes that the adjustments are an effective treatment.

SOLUTION

We have forty million reasons for failure, but not a single excuse.
— Rudyard Kipling

a. The experimenters assess politeness in interviews, but because they know which children received the real drug, they may inadvertently speak differently to these children during the interviews. Or, they might interpret the children's behavior differently since they know which subjects received the real drug. These are experimenter effects that can confound the study results. The experiment should have been double-blind.

b. The problem with this study is that students were not assigned to the two groups at random. By placing students who like classical music in one group and students who don't like it in the other, the researchers created a situation in which the two groups do not share the same general characteristics.

c. The results of this study will be difficult to interpret because the sample sizes are not sufficiently large. Of course, with a rare disease, finding people to participate in an experiment may be difficult.

d. The 25 patients who receive adjustments represent a treatment group, but this study lacks a control group. The patients may be feeling better because of a placebo effect rather than any real effect of the adjustments. The chiropractor might have improved his study by hiring an actor to do a fake adjustment (one that feels similar, but doesn't actually conform to chiropractic guidelines) on a control group. Then he could have compared the results in the two groups to see whether a placebo effect was involved. ∎

Case-Control Studies

Sometimes it may be impractical or unethical to create a controlled experiment. For example, suppose we want to study how alcohol consumed during pregnancy affects newborn babies. Because it is already known that consuming alcohol during pregnancy can be harmful, it would be unethical to divide a sample of pregnant mothers randomly into two groups and then require the members of one group to consume alcohol. However, we may be able to conduct a **case-control study**, in which the participants naturally form groups by choice. In this example, the **cases** consist of mothers who consume alcohol during pregnancy *by choice*, and the **controls** consist of mothers who choose *not* to consume alcohol. (Case-control studies are often called *retrospective studies* because they are usually conducted by looking back in time.)

A case-control study is *observational* because the researchers do not change the behavior of the participants. It also bears a resemblance to an experiment because the cases essentially represent a treatment group and the controls represent a control group. Note, however, that we cannot randomly assign the cases and controls in case-control studies. This can lead to confounding because the groups may differ in some way that invalidates the study results. Thus, we must be very careful in the interpretation of case-control studies because they are particularly prone to confounding.

Definitions

A **case-control study** is an observational study that resembles an experiment because the sample naturally divides into two (or more) groups. The participants who engage in the behavior under study form the **cases**, like a treatment group in an experiment. The participants who do not engage in the behavior are the **controls**, like a control group in an experiment.

By the Way…

Tobacco companies long claimed that confounding in case-control studies made it impossible to conclude that smoking causes lung cancer. Although case-control studies showed that smokers had higher rates of lung cancer than nonsmokers, the companies suggested that a genetic difference caused people both to smoke and to get lung cancer. Thus, they claimed that smoking was not the cause of lung cancer.

EXAMPLE 8   Which Type of Study?

For each of the following questions, what type of statistical study (observational study, experiment, or meta-analysis) is most likely to lead to an answer? Why? If an observational study, state whether it should be a case-control study. If an experiment, state whether it should be single- or double-blind.

a. What is the mean income of stock brokers?

b. Do seat belts save lives?

c. Can lifting weights improve runners' times in a 10-kilometer (10K) race?

d. Does skin contact with a particular glue cause a rash?

e. Can a new herbal remedy reduce the severity of colds?

f. Dozens of individual studies have given contradictory results: Can exercise increase life span?

SOLUTION

a. An *observational study* can tell us the mean income of stock brokers. We need only survey the brokers, and the survey itself will not change their incomes.

b. It would be unethical to do an experiment in which some people were told to wear seat belts and others were told *not* to wear them. Thus, a study to determine whether seat belts save lives must be *observational*. However, this observational study takes the form of a *case-control study*. Some people choose to wear seat belts (the cases) and others choose not to (the controls). By comparing the death rates in accidents between cases and controls, we can learn whether seat belts save lives. (They do.) Of course, we must watch for confounding in this study. For example, we must make sure that people who do not wear seat belts are not also more reckless drivers, in which case it would be difficult to interpret results.

c. We need an *experiment* to determine whether lifting weights can improve runners' 10K times. One group of runners will be put on a weight-lifting program, and a control group will be asked to stay away from weights. We must try to ensure that all other aspects of their training are similar. Then we can see whether the runners in the lifting group improve their times more than those in the control group. Note that we cannot use blinding in this experiment because there is no way to prevent participants from knowing whether they are lifting weights.

d. An *experiment* can help us determine whether skin contact with a particular glue causes a rash. In this case, it's best to use a *single-blind experiment* in which we apply the real glue to participants in one group and apply a placebo that looks the same, but lacks the active ingredient, to members of the control group. There is no need for a double-blind experiment because it seems unlikely that the experimenters could influence whether a person gets a rash. (However, if the question of whether the subject *has* a rash is subject to interpretation, the experimenter's knowledge of who got the real treatment could affect this interpretation.)

With proper treatment, a cold can be cured in a week. Left to itself, it may linger for seven days.
— Medical Folk Saying

e. We should use a *double-blind experiment* to determine whether a new herbal remedy can reduce the severity of colds. Some participants get the actual remedy, while others get a placebo. We need the double-blind conditions because the severity of a cold may be affected by mood or other factors that researchers might inadvertently influence. In the double-blind experiment, the researchers do not know which participants belong to which group and thus cannot treat the two groups differently.

f. Because the individual studies have given contradictory results, we should try a *meta-analysis* to see whether any trends emerge when we look at many studies together.   ∎

CASE STUDY   Drugs to Fight Depression: A Meta-Analysis

Government researchers estimate that 1 in 5 Americans suffer from depression at some time in their lives and that the annual cost to the economy in lost productivity is at least $40 billion. Until the 1990s, the only drugs available to combat depression belonged to a class known as tricyclic antidepressants. But in the past decade, several new drugs have come into common use. The most famous is Prozac, which has been prescribed so widely that some people claim we now live in a "Prozac nation." But do Prozac and the other new drugs work? To answer this question, researchers from the Agency for Health Care Policy and Research conducted a meta-analysis.

The researchers began by searching medical literature for studies about treatment of depression. They found more than 8,000 studies reported over a nine-year period. By looking for studies that met certain special criteria, such as observing patients for at least six weeks and comparing new and old types of anti-depression drugs, they narrowed this list to about 300 studies. They also considered about 600 studies dealing with side effects of the drugs. Their meta-analysis consisted of analyzing the results of all these studies as a group so as to look for trends that might not have been evident in individual studies.

They found that, overall, the new drugs are more effective than placebos for severely depressed people. About 50% of such people responded to the new drugs, compared to about 32% who responded to a placebo. However, the old tricyclic antidepressants were equally effective—and often at lower prices. The old and new drugs had different side effects, but the degrees of severity of these side effects were roughly equal. While these results can be interpreted as both good and bad news for the makers and users of the new drugs, the meta-analysis probably was more valuable for what it could *not* say. The researchers found that the data were inadequate to determine whether the new drugs were effective for mild depression or for children, despite the fact that they are commonly prescribed for both. They also found the data inadequate to evaluate herbal treatments, such as kava and St. John's wort. Thus, the meta-analysis pointed in the directions in which new research is needed.   ∎

Time out to think

Look again at the percentage responses to the new drugs (50%) and to the placebo (32%) in the meta-analysis of anti-depression drugs. Does the relatively large response to the placebo suggest that patients should be given a placebo before being given a real drug with potential side effects? Why or why not? If you were a psychologist, what would you suggest to someone who was part of the 50% who did not respond to the real drug?

Review Questions

1. Describe and contrast the three basic types of statistical study: observational study, experiment, and meta-analysis.

2. What do we mean by the *treatment group* and *control group* in an experiment?

3. What is *confounding*? Give an example of how it can occur.

4. Describe two strategies for selecting treatment and control groups in experiments, and explain why they are so important.

5. What is a *placebo*? Describe the placebo effect and how it can lead to confounding in experiments.

6. What is an *experimenter effect*, and how can it lead to confounding?

7. What do we mean by *blinding* in an experiment? Describe the basic design of single-blind and double-blind experiments.

8. What is a *case-control study*? Why is such a study considered to be observational? How is it similar to an experiment?

Exercises

BASIC SKILLS AND CONCEPTS

SENSIBLE STATEMENTS? For Exercises 1–6, determine whether the given statement is sensible and explain why it is or is not.

1. The Department of Education conducted an observational study to determine the average salary of high school teachers in each of the 50 states.

2. A paint company conducted a double-blind experiment to determine which of two types of exterior house paint was more resistant to rain.

3. The lab conducted an experiment to determine whether the throat culture was positive for strep.

4. In a study of medications designed to slow the rate of balding in men, a placebo group had better results than the control group.

5. A meta-analysis was conducted to determine the population of New Zealand.

6. A case-control experiment was used to determine the average family size in Utah.

TYPE OF STUDY. For Exercises 7–18, state whether the study is an observational study, an experiment, or a meta-analysis. If it is an experiment, describe the treatment and control groups. If it is observational, state whether it is a case-control study and, if it is, distinguish between the cases and the controls.

7. In a study of hundreds of Swedish twins, it was determined that the level of mental skills was more similar in identical twins (twins coming from a single egg) than in fraternal twins (twins coming from two separate eggs) (*Science*).

8. A Michigan State University study of starting salaries for college graduates found that computer science majors had the highest salaries ($42,500), while communications majors had the lowest ($25,600).

9. A National Cancer Institute study of 716 melanoma patients and 1,014 cancer-free patients matched by age, sex, and race found that those having a single large mole had twice the risk of melanoma; having 10 or more moles was associated with a 12 times greater risk of melanoma (*Journal of the American Medical Association*).

10. A European study of 1,500 men and women with exceptionally high levels of the amino acid homocysteine found that these individuals had double the risk of heart disease. However, the risk was substantially lower for those in the study who took vitamin B supplements (*Journal of the American Medical Association*).

11. A Gallup survey done for CNN/*USA Today* found that 80% of Americans think we're less civil than we were 10 years ago and 67% think that Americans are more likely to use vulgar language than they were 10 years ago.

12. Researchers at New York University found that the genetically modified corn known as Bt corn releases an insecticide through its roots into the soil (*Nature*).

13. A breast cancer study began by asking 25,624 women questions about how they spent their leisure time. The health of these women was tracked over the next 15 years. Those women who said they exercise regularly were found to have a lower incidence of breast cancer (*New England Journal of Medicine*).

14. A Gallup poll sponsored by the National Sleep Foundation discovered that one-third of adult Americans are "excessively sleepy during the daytime."

15. An analysis of 11 individual studies attempted to determine if there is a conclusive link between vasectomies and the incidence of prostate cancer (*Chance*).

16. A survey of 275,811 first-year college students revealed that 32.4% of these students had an A average in high school (Higher Education Research Institute).

17. A study by the National Cancer Institute analyzed the results of 72 previous studies and concluded that "high consumers of tomatoes and tomato products are at substantially decreased risk of numerous cancers" (*Journal of the National Cancer Institute*).

18. Hundreds of scientific and statistical studies have been done to determine whether high-voltage overhead power lines increase the incidence of cancer among those living nearby. A summary study based on many previous studies concluded that there is no significant link between power lines and cancer (*Journal of the American Medical Association*).

FURTHER APPLICATIONS

WHAT'S WRONG WITH THIS EXPERIMENT? For each of the (hypothetical) studies described in Exercises 19–23, identify any problems that are likely to cause confounding and explain how the problems could be avoided. Discuss whether any kind of blinding would be advisable.

19. An experiment is designed to evaluate two different kinds of plant food for geraniums. Forty geranium plants are divided into two groups of equal size. The plants in the first group are watered with a liquid plant food, while a powdered food is sprinkled around the plants in the second group.

20. One hundred people are selected for a study of the effect of regular walking on heart rate. Each person is allowed to choose whether to be in the walker or non-walker group. The resting heart rate of each individual is recorded. After six weeks, the resting heart rate of each individual is compared with his/her initial heart rate.

21. In a comparison of car waxes, 20 drivers in Florida are asked to use a solid wax, while 20 drivers in Washington are asked to use a liquid wax. After a year of use, all drivers are asked to compare before and after photographs of their cars.

22. In an experiment designed to test whether a new medication can alleviate depression, the new drug is given to two people and a placebo to two other people.

23. In a study of sleep and its effect on tempers, 50 subjects are asked to use alarm clocks to sleep 10% less than usual. Another 50 subjects are allowed to sleep as much as they like. After a week, researchers interview all of the subjects to assess whether they were more or less likely to have temper tantrums than they would have been under their normal sleep habits.

ANALYZING EXPERIMENTS. Exercises 24–29 present questions that might be addressed using an experiment. If you were to design the experiment, how would you choose the treatment and control groups? Should the experiment be single-blind, double-blind, or neither? Explain your reasoning.

24. Which of two SAT-preparation programs produces better results?

25. Can listening to jazz music while studying improve students' grades?

26. Does a brand of dog food labeled "Lean for Seniors" really prevent weight gain?

27. Does playing soccer help swimmers improve their times?

28. Does taking an aspirin a day reduce the incidence of heart attacks?

29. Does a self-proclaimed mind reader really have supernatural powers?

 ## PROJECTS FOR THE WEB AND BEYOND

For useful links, select "Links for Web Projects" for Chapter 1 at www.aw.com/bbt.

30. **EXPERIMENTER EFFECTS IN REPRESSED MEMORY CASES.** Search the Web for articles and information about the controversy regarding recovering repressed memories. Briefly summarize one or two of the most interesting cases and, based on what you read, express your own opinion as to whether the allegedly recovered memories are being influenced by experimenter effects.

31. **EXPERIMENTAL ETHICS.** Ethical standards change from era to era. One notoriously unethical case was a study of syphilis conducted in Tuskegee, Alabama, from 1932 to 1972. In this study, African American males were told that

they were receiving treatment for syphilis, but in fact they were not. The researchers' hidden goal was to study the long-term effects of the disease. Use the Web to learn about the history of the Tuskegee syphilis study. Hold a class discussion about the ethical issues involved in this case, or write a short essay summarizing the case and its ethical lessons.

32. **DEBATE: SHOULD WE USE DATA FROM UNETHICAL EXPERIMENTS?** Past research often did not conform to today's ethical standards. In extreme cases, such as research conducted by doctors in Nazi Germany, the researchers sometimes killed the subjects of their experiments. While this past unethical research clearly violated the human rights of the experimental subjects, in some cases it led to insights that could help people today. Is it ethical to use the results of unethical research?

33. **MEDICAL STUDIES.** Find a recent issue of the *Journal of the American Medical Association* or the *New England Journal of Medicine*. Select an article that has a topic of interest. Although the article may be technical, determine what kind of study was used and identify the groups that were used in the study. Comment on the use of placebos and blinding; comment on the control of confounding variables.

IN THE NEWS

1. **OBSERVATIONAL STUDIES.** Look through newspapers from the past few weeks and find an example of a statistical study that was observational. Briefly describe the study and summarize its conclusions.

2. **EXPERIMENTAL STUDIES.** Look through newspapers for the past few weeks and find an example of a statistical study that involved an experiment. Briefly describe the study and summarize its conclusions.

3. **CASE-CONTROL STUDIES.** Look through newspapers for the past few weeks and find an example of a case-control study. Briefly describe the study and summarize its conclusions.

4. **META-ANALYSIS.** Look through newspapers for the past few weeks and find an example of a meta-analysis. Briefly describe the study and summarize its conclusions.

1.4 Should You Believe a Statistical Study?

We've studied many concepts and definitions up to this point. It may take you a while to absorb them all fully. In fact, much of the rest of this book is devoted to helping you understand how they all tie together. But already you know enough to achieve one of the major goals of this book: being able to answer the question "Should you believe a statistical study?"

Most researchers conduct their statistical studies with honesty and integrity, and most statistical research is carried out with diligence and care. Nevertheless, statistical research is sufficiently complex that bias can arise in many different ways, thereby invalidating the results of a study. Thus, we must always examine reports of statistical research carefully, looking for anything that might make us question the results. There is no definitive way to answer the question "Should I believe a statistical study?" But the following eight guidelines can be especially helpful. We'll devote the rest of the chapter to exploring how they can be applied. Along the way, we'll also introduce a few more definitions and concepts that will prepare you for discussions to come later.

By the Way...

News reports do not always provide enough information for you to evaluate a statistical study. You can usually find additional information, such as original reports or press releases, on the Web. Look for clues such as "reported by NASA" or "published in the *New England Journal of Medicine.*"

Eight Guidelines for Critically Evaluating a Statistical Study

1. Identify the goal of the study, the population considered, and the type of study.

2. Consider the source, particularly with regard to whether the researchers may be biased.

3. Examine the sampling method to decide whether it is likely to produce a representative sample.

4. Look for problems in defining or measuring the variables of interest, which can make it difficult to interpret any reported results.

5. Watch out for confounding variables that can invalidate the conclusions of a study.

6. Consider the setting and wording of any survey, looking for anything that might tend to produce inaccurate or dishonest responses.

7. Check that results are fairly represented in graphics and concluding statements, because both researchers and media often create misleading graphics or jump to conclusions that the results do not support.

8. Stand back and consider the conclusions. Did the study achieve its goals? Do the conclusions make sense? Do the results have any practical significance?

Guideline 1: Identify the Goal, Population, and Type of Study

The first step in evaluating a statistical study is to understand the goal and approach of the study. Based on what you hear or read about a study, try to answer basic questions such as these:

- What was the study designed to determine?

- What was the population under study? Was the population clearly and appropriately defined?

- Was the study an observational study, experiment, or meta-analysis? If it was an observational study, was it a case-control study? If it was an experiment, was it single- or double-blind, and were the treatment and control groups properly randomized? Given the goal, was the type of study appropriate?

By the Way...

Surveys show that nearly half of Americans believe their horoscopes. However, in controlled experiments, the predictions of horoscopes come true no more often than would be expected by chance.

EXAMPLE 1   Appropriate Type of Study?

A newspaper reports: "Researchers gave each of the 100 participants their astrological horoscopes, and asked them whether the horoscopes appeared to be accurate. 85% of the participants reported that the horoscopes were accurate. The researchers concluded that horoscopes are valid most of the time." Analyze this study according to Guideline 1.

SOLUTION   Clearly, the goal of the study was to determine the validity of horoscopes. Based on the news report, it appears that the study was *observational*: The researchers simply asked the participants about the accuracy of the horoscopes. However, because the accuracy of a horoscope is somewhat subjective, this study should have been a controlled experiment in which some people were given their actual horoscope and others were given a fake horoscope.

Then the researchers could have looked for differences between the two groups. Moreover, because researchers could easily influence the results by how they questioned the participants, the experiment should have been double-blind. In summary, the type of study was inappropriate to the goal and its results are meaningless.   ∎

Time out to think

Try your own test of horoscopes. Cut out yesterday's 12 horoscopes so that each one is on a separate piece of paper, without anything identifying the sign. Shuffle the pieces of paper randomly, and ask a few people to guess which one was supposed to be their personal horoscope. How many people choose the right one? Discuss your results.

EXAMPLE 2   Does Aspirin Prevent Heart Attacks?

A study reported in the *New England Journal of Medicine* (Vol. 318, No. 4) sought to determine whether aspirin is effective in preventing heart attacks. It involved 22,000 male physicians considered to be at risk for heart attacks. The men were divided into a treatment group that took aspirin and a control group that did not. The results were so convincing in favor of the benefits of aspirin that for ethical reasons the experiment was stopped before it was completed, and the subjects were informed of the results. Many news reports led with the headline that taking aspirin can help prevent heart attacks. Analyze this headline according to Guideline 1.

SOLUTION   The study was an experiment, which is appropriate, and its results appear convincing. However, the fact that the sample consisted only of men means that the results should be considered to apply only to the population of men. Because results of medical tests on men do not necessarily apply to women, the headlines misstated the results when they did not qualify the population. A lingering question remains: Why didn't the researchers include women in the study?   ∎

Guideline 2: Consider the Source

Statistical studies are supposed to be objective, but the people who carry them out and fund them may be biased. Thus, it is important to consider the source of a study and evaluate the potential for biases that might invalidate its conclusions.

Bias may be obvious in cases where a statistical study is carried out for marketing, promotional, or other commercial purposes. For example, a toothpaste advertisement that claims "4 out of 5 dentists prefer our brand" appears to be statistically based, but we are given no details about how the survey was conducted. Because the advertisers obviously want to say good things about their brand, it's difficult to take the statistical claim seriously without much more information about how the result was obtained.

Other cases of bias may be more subtle. For example, suppose that a carefully conducted study concludes that a new drug helps cure cancer. On the surface, the study might seem quite believable. But what if the study was funded by a drug company that stands to gain billions of dollars in sales if the drug is proven effective? The researchers may well have carried out their work with great integrity despite the source of funding, but it might be worth a bit of extra investigation to be sure.

Major statistical studies are usually evaluated by unbiased experts. For example, the process by which scientists examine each other's research is called **peer review** (because the scientists who do the evaluation are *peers* of those who conducted the research). Reputable scientific journals require all research reports to be peer reviewed before the research is

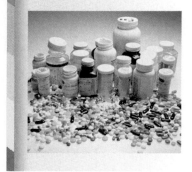

accepted for publication. Peer review does not guarantee that a study is valid, but it lends credibility because it implies that other experts agree that the study was carried out properly.

> **Definition**
>
> **Peer review** is a process in which several experts in a field evaluate a research report before the report is published.

EXAMPLE 3   Is Smoking Healthy?

By 1963, enough research on the health dangers of smoking had accumulated that the Surgeon General of the United States publicly announced that smoking is bad for health. Research done since that time built further support for this claim. However, while the vast majority of studies showed that smoking is unhealthy, a few studies found no dangers from smoking, and perhaps even health *benefits*. These studies generally were carried out by the Tobacco Research Institute, funded by the tobacco companies. Analyze these studies according to Guideline 2.

SOLUTION   Even in a case like this, it can be difficult to decide whom to believe. However, the studies showing smoking to be unhealthy came primarily from peer-reviewed research. In contrast, the studies carried out at the Tobacco Research Institute had a clear potential for bias. The *potential* for bias does not mean their research was biased, but the fact that it contradicts virtually all other research on the subject should be cause for concern.   ∎

EXAMPLE 4   Press Conference Science

The nightly TV news shows scientists at a press conference announcing that they've discovered evidence that a newly developed chemical can stop the aging process. The work has not yet gone through the peer review process. Analyze this study according to Guideline 2.

SOLUTION   Scientists often announce the results of their research at a press conference so that the public may hear about their work as soon as possible. However, a great deal of expertise may be required to evaluate their study for possible biases or other errors—which is the goal of the peer review process. Until the work is peer reviewed and published in a reputable journal, any findings should be considered preliminary.   ∎

Guideline 3: Examine the Sampling Method

A statistical study cannot be valid unless the sample is representative of the population under study. Poor sampling methods almost guarantee a biased sample that makes the study results useless.

Biased samples can arise in many ways, but two closely related problems are particularly common. The first problem, often called **selection bias** (or a **selection effect**), occurs whenever the researchers select their sample in a way that tends to make it unrepresentative of the population. For example, a pre-election poll that surveys only registered Republicans has selection bias because it is unlikely to reflect the opinions of voters of all parties (and independents).

The second problem, often called **participation bias**, occurs when people *choose* to be part of a study—that is, all the participants are volunteers. Participation bias occurs in **self-selected surveys** (or **voluntary response surveys**)—surveys in which people decide for themselves whether to be included in the survey. In such cases, people who feel strongly about an issue or desire a change are more likely to participate. Hence the opinions of the respondents are unlikely to represent the opinions of the larger population, which is less emotionally attached to the issue. A self-selected survey is therefore always subject to participation bias.

Definitions

Selection bias (or a **selection effect**) occurs whenever researchers *select* their sample in a biased way.

Participation bias occurs any time participation in a study is voluntary.

A **self-selected survey** (or **voluntary response survey**) is one in which people decide for themselves whether to be included in the survey.

CASE STUDY   The 1936 Literary Digest Poll

The *Literary Digest*, a popular magazine of the 1930s, successfully predicted the outcomes of several elections using large polls. In 1936, editors of the *Literary Digest* conducted a particularly large poll in advance of the presidential election. They randomly chose a sample of 10 million people from various lists, including names in telephone books and rosters of country clubs. They mailed a postcard "ballot" to each of these 10 million people. About 2.4 million people returned the postcard ballots. Based on the returned ballots, the editors of the *Literary Digest* predicted that Alf Landon would win the presidency by a margin of 57% to 43% over Franklin Roosevelt. Instead, Roosevelt won with 62% of the popular vote. How did such a large survey go so wrong?

The sample suffered from both selection bias and participation bias. The selection bias arose because the *Literary Digest* chose its 10 million names in ways that favored affluent people. For example, selecting names from telephone books meant choosing only from those who could afford telephones back in 1936. Similarly, country club members are usually quite wealthy. The selection bias favored the Republican Landon because affluent voters of the 1930s tended to vote for Republican candidates.

The participation bias arose because return of the postcard ballots was voluntary, so people who felt strongly about the election were more likely to be among those who returned their ballots. This bias also tended to favor Landon because he was the challenger—people who did not like President Roosevelt could express their desire for change by returning the postcards. Together, the two forms of bias made the sample results useless, despite the large number of people surveyed.

With proper sampling, a large sample is more likely to be representative than a smaller sample. But the *Literary Digest* case shows that poor sampling gives poor results, even when the sample is unusually large. ∎

EXAMPLE 5   Self-Selected Poll

The television show *Nightline* conducted a poll in which viewers were asked whether the United Nations headquarters should be kept in the United States. Viewers could respond to the poll by paying 50 cents to call a "900" phone number with their opinions. The poll drew 186,000 responses, of which 67% favored moving the United Nations out of the United States. Around the same time, a poll using simple random sampling of 500 people found that 72% wanted the United Nations to *stay* in the United States. Which poll is more likely to be representative of the general opinions of Americans?

SOLUTION   The *Nightline* sample was severely biased. It had selection bias because its sample was drawn only from the show's viewers, rather than from all Americans. The poll itself was a self-selected survey in which viewers not only chose whether to respond, but also had to *pay* 50 cents to participate. This cost made it even more likely that respondents would

By the Way...

A young pollster named George Gallup conducted his own survey prior to the 1936 election. Sending postcards to only 3,000 randomly selected people, he correctly predicted not only the outcome of the election, but also the outcome of the *Literary Digest* poll, to within 1%. Gallup went on to establish a very successful polling organization.

By the Way...

More than a third of all Americans routinely shut the door or hang up the phone when contacted for a survey, thereby making self-selection a problem for legitimate pollsters. One reason people hang up may be the proliferation of selling under the guise of market research (often called "sugging"), in which a telemarketer pretends you are part of a survey in order to try to get you to buy something.

be those who felt a need for change. Thus, despite its large number of respondents, the *Nightline* survey was unlikely to give meaningful results. In contrast, a simple random sample of 500 people is more likely to be representative, so the finding of this small survey has a better chance of representing the true opinions of all Americans. ∎

EXAMPLE 6   Planets Around Other Stars

Until the mid-1990s, astronomers had never found conclusive evidence for planets outside our own solar system. But improving technology made it possible to begin finding such planets, and nearly 80 had been discovered by early 2002. The technology makes it easier to find large planets than small ones, and easier to find planets that orbit close to their stars than planets that orbit far from their stars. According to the leading theory of solar system formation, large planets that orbit close to their stars should be very rare. But many of the new discoveries involve large planets with close orbits. Does this mean there is something wrong with the leading theory of solar system formation?

SOLUTION   Although the theory suggests that the types of planets in the early discoveries should be rare, the technology makes these rarities the easiest ones to find. Thus, the finding may in part represent a *selection effect* that biases the sample (of discovered planets) toward a rare type, in which case there may be nothing wrong with the leading theory. Until the technology allows astronomers to discover a less biased sample, we cannot use the results to draw conclusions about the theory.

By the Way...

By 2012, NASA hopes to build an orbiting telescope (called the Terrestrial Planet Finder) that will be able to detect Earth-size planets around other stars.

The sizes of the planets in our solar system and the Sun, to scale. (Distances are not to scale.) ∎

Guideline 4: Look for Problems in Defining or Measuring the Variables of Interest

Statistical studies usually attempt to measure what we call **variables**. The term *variable* simply refers to an item or quantity that can vary or take on different values. For example, variables in the Nielsen studies of television viewing habits include *show being watched* and *number of viewers*. The variable *show being watched* can take on different values such as

"Super Bowl," "*60 Minutes*," or "*The X Files*." The variable *number of viewers* depends on the popularity of a particular show. In essence, the raw data in any statistical study are the different values of the variables under study.

Definitions

A **variable** is any item or quantity that can vary or take on different values.

The **variables of interest** in a statistical study are the items or quantities that the study seeks to measure.

Results of a statistical study may be especially difficult to interpret if the variables under study are difficult to define or measure. For example, imagine trying to conduct a study of how exercise affects resting heart rates. The variables of interest would be *amount of exercise* and *resting heart rate*. Both variables are difficult to define and measure. In the case of *amount of exercise*, it's not clear what the definition covers—does it include walking to class? Even if we specify the definition, how can we measure *amount of exercise* given that some forms of exercise are more vigorous than others?

Time out to think

How would you measure your resting heart rate? Describe some difficulties in defining and measuring resting heart rate.

The following two examples describe real cases in which defining or measuring variables caused problems in statistical studies.

EXAMPLE 7   Can Money Buy Love?

A Roper poll reported in *USA Today* involved a survey of the wealthiest 1% of Americans. The survey found that these people would pay an average of $487,000 for "true love," $407,000 for "great intellect," $285,000 for "talent," and $259,000 for "eternal youth." Analyze this result according to Guideline 4.

SOLUTION   The variables in this study are very difficult to define. How, for example, do you define "true love"? And does it mean true love for a day, a lifetime, or something else? Similarly, does the ability to balance a spoon on your nose constitute "talent"? Because the variables are so poorly defined, it's likely that different people interpreted them differently, making the results very difficult to interpret. ∎

EXAMPLE 8   Illegal Drug Supply

A commonly quoted statistic is that law enforcement authorities succeed in stopping only about 10% to 20% of the illegal drugs entering the United States. Should you believe this statistic?

SOLUTION   There are essentially two variables in a study of illegal drug interception: *quantity of illegal drugs intercepted* and *quantity of illegal drugs NOT intercepted*. It should be relatively easy to measure the quantity of illegal drugs that law enforcement officials intercept. However, because the drugs are illegal, it's unlikely that anyone is reporting the quantity of drugs that are *not* intercepted. How, then, can anyone know that the intercepted drugs are 10% to 20% of the total? In a *New York Times* analysis, a police officer was quoted as saying that his colleagues refer to this type of statistic as "PFA" for "pulled from the air." ∎

Guideline 5: Watch Out for Confounding Variables

Often, variables that are *not intended* to be part of the study can make it difficult to interpret results properly. In Section 1.3, we referred to such variables as *confounding factors*; they are more formally called *confounding variables*. As we saw earlier, it's not always easy to discover confounding variables. Sometimes they are discovered only years after a study is completed, and other times they are not discovered at all, in which case a study's conclusion may be accepted even though it's not correct. Fortunately, confounding variables are sometimes more obvious and can be discovered simply by thinking hard about factors that may have influenced a study's results.

EXAMPLE 9   Radon and Lung Cancer

Radon is a radioactive gas produced by natural processes (the decay of uranium) in the ground. The gas can leach into buildings through the foundation and can accumulate to relatively high concentrations if doors and windows are closed. Imagine a study that seeks to determine whether radon gas causes lung cancer by comparing the lung cancer rate in Colorado, where radon gas is fairly common, with the lung cancer rate in Hong Kong, where radon gas is less common. Suppose that the study finds that the lung cancer rates are nearly the same. Is it fair to conclude that radon is *not* a significant cause of lung cancer?

SOLUTION   The variables under study are *amount of radon* and *lung cancer rate*. However, because smoking can also cause lung cancer, *smoking rate* may be a confounding variable in this study. In particular, the smoking rate in Hong Kong is much higher than the smoking rate in Colorado, so any conclusions about radon and lung cancer must take the smoking rate into account. Careful studies have shown that radon gas *can* cause lung cancer, and the U.S. Environmental Protection Agency (EPA) recommends taking steps to prevent radon from building up indoors. ∎

Guideline 6: Consider the Setting and Wording of Any Survey

Even when a survey is conducted with proper sampling and with clearly defined terms and questions, it's important to watch out for problems in the setting and wording that might produce inaccurate or dishonest responses. Dishonest responses are particularly likely when the survey concerns sensitive subjects, such as personal habits or income. For example, the question "Do you cheat on your income taxes?" is unlikely to elicit honest answers from those who cheat, unless they are assured of complete confidentiality (and perhaps not even then).

In other cases, even honest answers may not really be accurate if the wording of questions invites bias. Sometimes just the order of a question can affect the outcome. A poll conducted in Germany asked the following two questions:

- *Would you say that traffic contributes more or less to air pollution than industry?*
- *Would you say that industry contributes more or less to air pollution than traffic?*

With the first question, 45% answered traffic and 32% answered industry. With the second question, only 24% answered traffic while 57% answered industry. Thus, simply changing the order of the words *traffic* and *industry* dramatically changed the survey results.

EXAMPLE 10   Do You Want a Tax Cut?

When Republicans in Congress proposed a 10% across-the-board cut in income taxes, the Republican National Committee commissioned a poll to find out whether Americans supported the proposal. Asked whether they favored the tax cut, 67% of respondents answered *yes*. Should we conclude that Americans would like to see the budget surplus used for a tax cut?

SOLUTION   A question like "Do you favor a tax cut?" is biased because it does not give other options. In fact, an independent poll conducted at the same time gave respondents a list of options for using surplus revenues. This poll found that 31% wanted the money devoted to Social Security, 26% wanted it used to reduce the national debt, and only 18% favored using it for a tax cut. (The remaining 25% of respondents chose a variety of other options.)   ∎

> ### By the Way...
>
> People are more likely to choose the item that comes first in a survey because of what psychologists call the *availability error*—the tendency to make judgments based on what is *available* in the mind.

EXAMPLE 11   Sensitive Survey

Two surveys asked Catholics in the Boston area whether contraceptives should be made available to unmarried women. The first survey involved in-person interviews, and 44% of the respondents answered *yes*. The second survey was conducted by mail and telephone, and 75% of the respondents answered *yes*. Which survey was more likely to be accurate?

SOLUTION   Contraceptives are a sensitive topic, particularly among Catholics (because the Catholic Church officially opposes contraceptives). The first survey, with in-person interviews, may have encouraged dishonest responses. The second survey made responses more confidential, and therefore more likely to reflect the respondents' true opinions. Thus, the second survey probably was more accurate.   ∎

Guideline 7: Check That Results Are Fairly Represented in Graphics or Concluding Statements

Even when a statistical study is done well, it may be misrepresented in graphics or concluding statements. Researchers occasionally misinterpret the results of their own studies or jump to conclusions that are not supported by the results, particularly when they have personal biases. News reporters may misinterpret a survey or jump to unwarranted conclusions that make a story seem more spectacular. Misleading graphics are especially common—so common that we will devote a large part of Chapter 3 to this topic. You should always look for inconsistencies between the interpretation of a study (in pictures and in words) and any actual data given along with it.

EXAMPLE 12   Does the School Board Need a Statistics Lesson?

The school board in Boulder, Colorado, created a hubbub when it announced that 28% of Boulder school children were reading "below grade level," and hence concluded that methods of teaching reading needed to be changed. The announcement was based on reading tests on which 28% of Boulder school children scored below the national average for their grade. Do these data support the board's conclusion?

SOLUTION   The fact that 28% of Boulder children scored below the national average for their grade implies that 72% scored at or above the national average. Thus, the school board's ominous statement about students reading "below grade level" makes sense only if "grade level" means the national average score for a particular grade. This interpretation of "grade level" is curious because it means that half the students in the nation are always below grade level—no matter how high the scores. The conclusion that teaching methods needed to be changed was not justified by these data.   ∎

Guideline 8: Stand Back and Consider the Conclusions

Finally, even if a study seems reasonable according to all the previous guidelines, you should stand back and consider the conclusions. Ask yourself questions such at these:

- Did the study achieve its goals?
- Do the conclusions make sense?
- Can you rule out alternative explanations for the results?
- If the conclusions make sense, do they have any practical significance?

Extraordinary claims require extraordinary evidence.

—Carl Sagan

EXAMPLE 13   Extraordinary Claims

A study concludes that wearing a gold chain increases your chances of surviving a car accident by 10%. The claim is based on a statistical analysis of data about survival rates and what people were wearing. Careful analysis of the research shows that it was conducted properly and carefully. Should you start wearing a gold chain whenever you drive a car?

SOLUTION   Despite the care that went into the study, the claim that a gold chain can save your life in a collision is difficult to believe. After all, how could a thin chain help in a high-speed collision? It's certainly *possible* that some unknown effect of gold chains makes the conclusion correct, but it seems far more likely that the results were either a fluke or due to an unidentified confounding variable (for example, perhaps those who wear gold chains are wealthier and drive newer cars with more advanced safety features, lowering their fatality rate). ∎

EXAMPLE 14   Practical Significance

An experiment is conducted in which the weight losses of people who try a new "Fast Diet Supplement" are compared to the weight losses of a control group of people who try to lose weight in other ways. After eight weeks, the results show that the treatment group lost an average of $\frac{1}{2}$ pound more than the control group. Assuming that it has no dangerous side effects, does this study suggest that the Fast Diet Supplement is a good treatment for people wanting to lose weight?

SOLUTION   Compared to the average person's body weight, a weight loss of $\frac{1}{2}$ pound hardly matters at all. Thus, while the statistics in this case may be interesting, they don't seem to have much practical significance. ∎

Review Questions

1. Briefly describe each of the eight guidelines for evaluating statistical studies. Give an example to which each guideline applies.

2. What is peer review? How is it useful?

3. Describe and contrast selection bias and participation bias in sampling. Give an example of each.

4. Why are self-selected surveys always prone to participation bias?

5. What do we mean by the variables of interest in a study?

6. What are confounding variables, and what problems can they cause?

Exercises

BASIC SKILLS AND CONCEPTS

APPLYING GUIDELINES. Determine which of the eight guidelines should be applied in Exercises 1–6. Explain your reasoning.

1. The public relations department of Wholesome Foods conducted a taste test of 10 different brands of yogurt.

2. A sample selected at a union meeting of steelworkers was surveyed about preferences in eye shadow.

3. Five hundred randomly selected adults were asked whether they felt good when they woke up in the morning.

4. Researchers conclude that an irrigation system used to grow tomatoes in California is more effective than a competing system used in Arizona.

5. A questionnaire distributed by an independent market research firm asked respondents if Skyline Beer is their favorite beer.

6. Under the headline "Baxter predicted to win in a landslide," it was reported that Baxter received 55% of the votes in a pre-election poll, compared with 45% for her opponent.

7. **IT'S ALL IN THE WORDING.** Princeton Survey Research Associates did a study for *Newsweek* magazine illustrating the effects of wording in a survey. Two questions were asked.

   • Do you personally believe that abortion is wrong?

   • Whatever your own personal view of abortion, do you favor or oppose a woman in this country having the choice to have an abortion with the advice of her doctor?

   To the first question, 57% of the respondents replied yes, while 36% responded no. In response to the second question, 69% of the respondents favored the choice, while 24% opposed the choice. Discuss why the two questions produced seemingly contradictory results. How could the results of the questions be used selectively by various groups?

8. **TAX OR SPEND?** A Gallup poll asked the following two questions:

   • Do you favor a tax cut or "increased spending on other government programs"? *Result*: 75% for the tax cut.

   • Do you favor a tax cut or "spending to fund new retirement savings accounts, as well as increased spending on education, defense, Medicare and other programs"? *Result*: 60% for the spending.

Discuss why the two questions produced seemingly contradictory results. How could the results of the questions be used selectively by various groups?

ACCURATE HEADLINES? Exercises 9–10 give a headline and a brief description of the statistical news story that accompanied the headline. In each case, discuss whether the headline accurately represents the story.

9. Headline: "Drugs shown in 98 percent of movies"
   Story summary: A "government study" claims that drug use, drinking, or smoking was depicted in 98% of the top movie rentals (Associated Press).

10. Headline: "Sex more important than jobs"
    Story summary: A survey found that 82% of 500 people interviewed by phone ranked a satisfying sex life as important or very important, while 79% ranked job satisfaction as important or very important (Associated Press).

WHAT DO YOU WANT TO KNOW? Exercises 11–13 pose two related questions that might form the basis of a statistical study. Briefly discuss how the two questions differ and how these differences would affect the goal of a study and the design of the study.

11. First question: What percentage of blind dates lead to a marriage?
    Second question: What percentage of marriages begin with a blind date?

12. First question: What percentage of introductory classes on campus are taught by full-time faculty members?
    Second question: What percentage of full-time faculty members teach introductory classes?

13. First question: How often do teenagers run red lights?
    Second question: How often are red lights run by teenagers?

STAT-BYTES. Politicians must make their political statements (often called sound-bytes) very short because the attention span of listeners is so short. A similar effect occurs in reporting statistical news. Major statistical studies are often reduced to one or two sentences. The summaries of statistical reports in Exercises 14–16 are taken from various news sources. Discuss what crucial information is missing and what more you would want to know before you acted on the report.

14. *USA Today* reports on a Harris poll claiming that the percentage of adults with a "great deal of confidence" in military leaders stands at 54% (up from 37% in 1997).

15. CNN reports on a Zagat Survey of America's Top Restaurants which found that "only nine restaurants achieved a

rare 29 out of a possible 30 rating and none of those restaurants are in the Big Apple."

16. A University of Michigan study concluded that drunken driving by high school seniors had declined. The report was based on two studies. In 1984, 31.2% of high school seniors reported driving after drinking sometime during the previous two weeks. By 1997, only 18.3% of high school seniors reported driving after drinking sometime during the previous two weeks.

PROJECTS FOR THE WEB AND BEYOND

For useful links, select "Links for Web Projects" for Chapter 1 at www.aw.com/bbt.

17. ANALYZING A STATISTICAL STUDY. Go to the Web address given above and explore the link for one of the studies listed there. Write a short report applying each of the eight guidelines given in this section. (Some of the guidelines may not apply to the particular study you are analyzing; in that case, explain why the guideline is not applicable.)

18. HARPER'S INDEX. Go to the Web site for Harper's Index and study a few of the recently quoted statistics. Be sure to select the option on the page that allows you to see the sources for the statistics. Choose three statistics that you find particularly interesting and discuss whether you believe them, based on the guidelines given in this section.

19. TWIN STUDIES. Researchers doing statistical studies in biology, psychology, and sociology are grateful for the existence of twins. Twins can be used to study whether certain traits are inherited from parents (nature) or acquired from one's surroundings during upbringing (nurture). Identical twins are formed from the same egg in the mother and have the same genetic material. Fraternal twins are formed from two separate eggs and share roughly half of the same genetic material. Find a published report of a twin study. Discuss how identical and fraternal twins are used to form case and control groups. Apply Guidelines 1–8 to the study and comment on whether you find the conclusions of the report convincing.

20. AMERICAN DEMOGRAPHICS. Consult an issue of *American Demographics,* a nontechnical magazine that specializes in summarizing statistical studies involving Americans and their lifestyles. Select one specific article and use the ideas of this section to summarize and evaluate the study. Be sure to cite information that you believe is missing and should be provided for you to make a complete analysis.

21. MAMMOGRAM DEBATE. For the past 10 years, major health organizations have given conflicting and changing opinions on the value of mammograms in preventing cancer deaths and major surgery. Find a recent study on the subject (*The Lancet*, October 20, 2001, and the *New York Times*, December 9, 2001) and describe the findings of the new meta-analysis. Explain specifically why this recent study found previous studies to be flawed.

IN THE NEWS

1. APPLYING THE GUIDELINES. Find a recent newspaper article or television report about a statistical study on a topic that you find interesting. Write a short report applying each of the eight guidelines given in this section. (Some of the guidelines may not apply to the particular study you are analyzing; in that case, explain why the guideline is not applicable.)

2. BELIEVABLE RESULTS. Find a recent news report about a statistical study whose results you believe are meaningful and important. In one page or less, summarize the study and explain why you find it believable.

3. UNBELIEVABLE RESULTS. Find a recent news report about a statistical study whose results you *don't* believe are meaningful and important. In one page or less, summarize the study and explain why you don't believe its claims.

4. SELF-SELECTED SURVEY. Find an example of a recent survey in which the sample was self-selected. Describe the makeup of the sample and how you think the self-selection affected the results of the survey.

5. ILL-DEFINED VARIABLES. Find an example of a recent statistical study in which the variables of interest were difficult to define or measure. Summarize the study, explain the problems, and state how you think the problems affected the conclusions.

Chapter Review Exercises

1. In 1920, only 35% of U.S. households had telephones, but the rate is now much higher. A recent survey of 1,123 households showed that 1,055 of them had telephones. Based on those results, the *Portland Sentinel* reports that "93.9% of households have telephones (margin of error is 1.3 percentage points)."

   a. Interpret the margin of error by identifying the range of values likely to contain the proportion of all households that have telephones.
   b. Identify the population.
   c. Is this study an experiment or an observational study? Explain.
   d. Is the reported value of 93.9% a population parameter or a sample statistic? Why?
   e. The report included no information about the method of sampling that was used. If you learned that survey subjects responded to an article asking readers to call the newspaper, would you consider the survey results to be valid? Why or why not?
   f. Describe a procedure for selecting similar survey subjects using stratified sampling.
   g. Describe a procedure for selecting similar survey subjects using cluster sampling.
   h. Describe a procedure for selecting similar survey subjects using systematic sampling.

2. An important element of this chapter is the concept of a simple random sample.

   a. What is a simple random sample?
   b. A housing development has 40 homes on each of 20 streets. If you randomly select a street and interview the head of the household in each of the 40 homes on that street, do you have a simple random sample?
   c. Describe a sampling plan that results in a simple random sample of five students from your class.

3. You have developed a new drug that is supposed to eliminate congestion caused by colds. You have 50 subjects who have agreed to be treated as soon as they get a cold, and all of the subjects do get a cold.

   a. Can you determine the effectiveness of the drug by treating all of the 50 subjects? Why or why not?
   b. In testing the effectiveness of the drug, what is the placebo effect?
   c. How might an experiment be designed so that it would be possible to distinguish between the placebo effect and the effectiveness of the drug?
   d. What is blinding and why is it important in testing the effectiveness of the drug?
   e. What is an experimenter effect, and how might this effect be minimized?

4. Because it closely parallels the makeup of the nation, Marion, Indiana, is a favorite choice of researchers conducting market testing. When researchers wanted to gauge reaction to potato chips made with the fat substitute Olestra, they went to Marion in search of opinions that would be likely to reflect the sense of the nation.

   a. Which type of sampling is used when residents of Marion are selected—systematic, convenience, cluster, or stratified?
   b. Write a survey question (not related to potato chips) that would require a sample group with more diversity than can be found in Marion.

FOCUS ON SOCIOLOGY

Does Daycare Breed Bullies?

Decades ago, it was almost a given that mothers would stay at home to take care of their young children. But as more women have entered the workforce, more children are spending more of their waking hours in some type of daycare. In 1975, fewer than 40% of women with children under age 6 worked full- or part-time outside the home. Today, more than 60% of such mothers do. This dramatic increase in the number of children in daycare led psychologists and sociologists to study the effects of daycare. In 1991, the National Institute of Child Health and Human Development (NICHD) started the Early Child Care study, which has tracked more than 1,300 children enrolled in daycare at 10 different research sites around the United States.

The study has been conducted carefully by outstanding researchers, but its results have proven controversial nonetheless. As an example, consider this result reported by the Early Child Care study in April, 2001: Young children who spend more time in daycare are significantly more likely to be aggressive bullies as they get older. Working mothers everywhere felt pangs of guilt as they worried about the effects of daycare on their children, and the news media reported widely that daycare breeds bullies. But should we believe this claim? There are many ways to evaluate the claim, but for practice, let's use the eight guidelines given in Section 1.4.

Guideline 1: The goal of the study seems clear (to learn about effects of daycare), as does the population under study (all children in daycare). However, the type of study raises a potential problem. The study is observational, rather than a randomized experiment, because it would be unethical to tell parents whether or not to use daycare. More specifically, it is a case-control study in which the cases are children in daycare and the controls are children cared for by stay-at-home moms. Unfortunately, while an observational study can help establish a link (or *correlation*) between variables, it cannot by itself establish cause and effect (see Chapter 7). In this case, the study provides evidence of a link between daycare and bullying, but does not tell us whether one is the cause of the other. The problem is that, because parents chose whether their children were in daycare or at home, we cannot be sure that the case and control groups are truly comparable. For example, it may be that mothers who must juggle children and work tend to be more stressed than nonworking mothers, which in turn might make daycare children more stressed than their stay-at-home counterparts. Or, it could be that mothers with rambunctious children are more likely to put their children in daycare. Thus, regardless of the validity of the link between daycare and bullying, the observational nature of the study makes it impossible to support a claim that daycare *causes* the bullying.

Guideline 2: The study has been conducted with care, and we have no reason to suspect any particular bias on the part of the researchers. However, daycare is an emotionally charged issue and researchers may have strong opinions about it, raising the potential for bias. Indeed, different researchers have interpreted the study results differently, which suggests that bias may play a role in the interpretations, if not in the study itself.

Guideline 3: Researchers used a type of stratified sampling to recruit participants for the study, because they wanted to match national demographics for factors such as socioeconomic background, race, and family structure. This type of sampling seems appropriate to the study. However, the families that were recruited had a choice about whether to participate in the study. This self-selection automatically raises the potential for participation bias, which may affect the study's results.

Guideline 4: The claim about daycare and bullying essentially involves two variables: the amount of time a child spends in daycare and the child's level of aggressiveness. Both are very difficult to define. For example, the Early Child Care study defines "daycare" as care provided by anyone other than the mother, which means that care by fathers or grandparents counts as daycare. Many people object to this definition. Similarly, defining aggression is very difficult; for example, it may be difficult to distinguish between active play and true aggression.

Guideline 5: There are many potentially confounding variables in the Early Child Care study. As noted earlier, the study does not provide data about whether mothers of children in daycare are subject to more stress than stay-at-home moms, which could confound the results. Perhaps of greater concern, there is no accepted way to measure or control for the quality of daycare. As a result, even some of the researchers involved in the study have suggested that the observed bullying is produced by mediocre daycare and not by daycare *per se*.

Guideline 6: The researchers used interviews to determine the type and amount of daycare and the level of aggression for each child. One of many potential problems with such interviews is that, as discussed earlier, it is difficult to define aggressiveness. Thus, interviews might produce different results depending on how interviewers ask questions and how respondents define aggression.

Guideline 7: The actual reports produced by the Early Child Care researchers acknowledge potential flaws in the study and emphasize that the study shows only a *possible* link between daycare and bullying. So where does the claim that daycare breeds bullies come from? It comes from the way the findings are presented in the media. Because the actual results do not support this interpretation, the news reports clearly reflect either bias or misunderstanding. For example, journalists who have not studied statistics may not be aware of the difference between finding a link and finding a cause, and their misunderstanding is then reflected in their news reports. Alternatively, the media might deliberately sensationalize the results in order to attract more attention. And in some cases, people who believe that women should stay home with their children might misrepresent the results so as to claim that the study lends scientific support to their views.

Guideline 8: The question of practical significance is very important with this study. On an individual level, many working women don't believe they really have a choice about daycare; they must work for either financial or professional reasons. Thus, they would use daycare even if there were proof that home care was better. On the societal level, the findings might be practically significant if they led to new policies that could improve child care generally. But as long as the results do not yet prove cause and effect, they cannot give clear guidance to policy makers.

In summary, we've found that a provocative headline about daycare and bullying falls apart upon close scrutiny. The research might still be quite valuable, especially because the researchers themselves are aware of potential flaws in the study. The results may lead to future studies with more conclusive results—information that would surely be appreciated by working parents everywhere. But until then, we need to be careful about how we interpret the Early Child Care study results, and even more careful about how we react to any sensational headlines the study may produce.

QUESTIONS FOR DISCUSSION

1. A third potential confounding variable is the amount of time children spend watching television and playing video games, which have also been linked to aggressive behavior. Do you think this variable might affect the Early Child Care study results and, if so, how? What other confounding variables might affect its results? Explain your answers clearly.

2. Suppose you were designing a follow-up study intended to give more conclusive results. Based on what has been learned from the Early Child Care study, what would you do differently?

3. What is your personal opinion about whether it is better for a parent to stay at home or to work? Would your opinion be swayed if the Early Child Care study produced any definitive results about daycare? Why or why not?

4. Some people think the Early Child Care study is extremely valuable, even if it cannot produce definitive results. Others think it is a waste of time and money. What is your opinion of the research? Defend your opinion.

SUGGESTED READING

Peth-Pierce, Robin. "The NICHD Study of Early Child Care," on-line at www.nichd.nih.gov/publications/pubs/early_child_care.htm.

Stolberg, Sheryl Gay. "Science, Studies, and Motherhood." *New York Times*, April 22, 2001.

Sweeney, Jennifer Foote. "The Day-Care Scare, Again." Salon.com, April 20, 2001.

FOCUS ON PUBLIC HEALTH

Is Your Lifestyle Healthy?

Consider the following findings from statistical studies:

- Smoking increases the risk of heart disease.

- Eating margarine can increase the risk of heart disease.

- One or two glasses of wine per day can protect against heart disease, but increase the risk of breast cancer.

You are probably familiar with some of these findings, and perhaps you've even altered your lifestyle as a result of them. But where do they come from? Remarkably, these and hundreds of other important findings on public health come from a huge case-control study that has provided data for hundreds of statistical studies. Known as the Harvard Nurses' Health Study, it is the most enduring public health study ever undertaken. If it has not already changed the way you live, it almost certainly will in the future.

The Harvard Nurses' Health Study began in 1976 when Dr. Frank E. Speizer, a professor at Harvard Medical School, decided to study the long-term effects of oral contraceptives. He mailed questionnaries to approximately 370,000 registered nurses and received more than 120,000 responses. He chose to survey nurses because he believed that their medical training would make their responses more reliable than those of the general public.

As Dr. Speizer and his colleagues sifted through the data in the returned questionnaires, they realized that the study could be expanded to include more than just the effects of contraceptives. Today, this research team continues to follow some 90% of the original 120,000 respondents. Hundreds of other researchers also use the data to study public health.

Annual questionnaires are still a vital part of the study, allowing researchers to gather data about what the nurses eat; what medicines and vitamins they take; whether and how much they exercise, drink, and smoke; and what illnesses they have contracted. Some of the nurses also provide blood samples, which are used to measure such things as cholesterol level, hormone levels, genetic variations, and residues from pesticides and environmental pollutants. Dr. Speizer's faith in nurses has proven justified, as they reliably complete surveys and almost always provide properly drawn and labeled blood samples upon request.

After more than 20 years of correspondence, both the researchers and the nurses say they feel a sense of closeness. Many of the nurses look forward to hearing from the researchers and say that the study has helped them to pay more attention to how they live their lives. Researchers feel deep sorrow when they must record the death of one of the nurses.

The sorrow of death will play an increasing role in the study, as many of the nurses are now entering their 70s and 80s. But this sorrow will also yield a wealth of valuable data about factors that influence longevity and health in old age. Researchers hope that the data will point the way toward definitive understanding about what constitutes a healthy diet, how pollution and chemical exposure influence health, and how we might prevent debilitating diseases like osteoporosis and Alzheimer's. In death, the 120,000 women of the Harvard Nurses' Health Study may give the gift of a better life to future generations.

QUESTIONS FOR DISCUSSION

1. Consider some of the results that are likely to come from the Harvard Nurses' Health Study over the next 10–20 years. What types of results do you think will be most important? Do you think the findings will alter the way you live your life?

2. Explain why the Harvard Nurses' Health Study is essentially an observational, case-control study. Critics sometimes say that the results would be more valid if obtained by experiments rather than such observations. Discuss whether it would be possible to gather similar data by carrying out experiments in a practical and ethical way.

3. In principle, the Harvard Nurses' Health Study is subject to participation bias because only 120,000 of the original 370,000 questionnaires were returned. Should the researchers be concerned about this bias? Why or why not?

4. Another potential pitfall comes from the fact that the questionnaires often deal with sensitive issues of personal health, and researchers have no way to confirm that the nurses answer honestly. Do you think that dishonesty could be leading researchers to incorrect conclusions? Defend your opinion.

5. All of the participants in the Harvard Nurses' Health Study are women. Do you think that the results also are of use to men? Why or why not?

SUGGESTED READING

Brophy, Beth. "Doing It for Science." *U.S. News & World Report*, Vol. 126, March 21, 1999, p. 67.

Yoon, Carol Kaesuk. "In Nurses' Lives, a Treasure Trove of Health Data." *New York Times*, September 15, 1998.

CHAPTER 2

Measurement in Statistics

We all know how to measure quantities such as height, weight, and temperature. However, in statistical studies there are many other kinds of measurements, and we must be sure that they are defined, obtained, and reported carefully. In this chapter, we will discuss a few important concepts associated with measurements and statistics. As you'll soon see, these concepts are crucial to understanding the statistical reports you encounter in your daily life.

LEARNING GOALS

1. Be able to identify data by type (qualitative or quantitative), type of measurement (discrete or continuous), and level of measurement (nominal, ordinal, interval, or ratio).

2. Understand sources of errors and how to report measurement results in a way that indicates possible errors.

3. Understand the different ways in which percentages are used to report statistical results and how percentages are often misused.

4. Understand the concept of an index number, such as the Consumer Price Index.

2.1 Data Types and Levels of Measurement

One of the challenges in statistics is deciding how best to summarize and display data. Different types of data call for different types of summaries. Thus, before we discuss data summaries and displays, it's useful to learn how data are categorized. We can divide this process into two parts: identifying a *data type* and identifying a *level of measurement* for the data.

Data Types

Data come in two basic types: qualitative and quantitative. **Qualitative data** have values that can be placed into nonnumerical *categories*. (For this reason, qualitative data are sometimes called *categorical* data.) For example, eye color data are qualitative because they are categorized by colors such as blue, brown, and hazel. Other examples of qualitative data include flavors of ice cream, names of employers, genders of animals, and "star ratings" of movies or restaurants. Note that star ratings are qualitative even though they involve a *number* of stars (such as three stars or four stars), because the numbers are not necessary; we could rate movies equally well with four nonnumerical categories such as bad, average, good, and excellent.

    Quantitative data have numerical values representing counts or measurements. The times of runners in a race, the incomes of college graduates, and the numbers of students in different classes are all examples of quantitative data.

> ### Data Types
>
> **Qualitative data** consist of values that can be placed into nonnumerical *categories*.
>
> **Quantitative data** consist of values representing counts or measurements.

EXAMPLE 1 Data Types

Classify each of the following sets of data as either qualitative or quantitative.
a. Brand names of shoes in a consumer survey
b. Scores on a multiple-choice exam
c. Letter grades on an essay assignment
d. Numbers on uniforms that identify players on a basketball team

SOLUTION

a. Brand names are categories and therefore represent qualitative data.

b. Scores on a multiple-choice exam are quantitative because they are numerical, representing a count of the number of correct answers.

c. Letter grades on an essay assignment are qualitative because they represent different categories of performance (failing through excellent).

d. Numbers on uniforms are qualitative because they are used solely for identifying players. That is, these numbers don't represent a count or measurement of anything, and they cannot be used in computations. This example shows that, in certain circumstances, numbers can represent qualitative data. ∎

Discrete Versus Continuous Data

Quantitative data can be further classified as continuous or discrete. Data are **continuous** if they can take on *any* value in a given interval. For example, a person's weight can be anything between 0 and a few hundred pounds, so data that consist of weights are continuous. In contrast, data that come from counts are **discrete** because a count can result in only whole numbers and not other values in between. For example, the number of students in your class might be 19 or 27 or 254, but it cannot be 31.5.

> ### Discrete Versus Continuous Data
>
> **Continuous** data can take on *any* value in a given interval.
>
> **Discrete** data can take on only particular values and not other values in between.

EXAMPLE 2   Discrete or Continuous?

For each data set, indicate whether the data are discrete or continuous.

a. Measurements of the time it takes to walk a mile

b. The numbers of calendar years (such as 2002, 2003, 2004)

c. The numbers of dairy cows on different farms

d. The amounts of milk produced by dairy cows on a farm

SOLUTION

a. Time can take on any value, so measurements of time are continuous.

b. The numbers of calendar years are discrete because they cannot have fractional values. For example, on New Year's Eve of 2004, the year will change from 2004 to 2005; we'll never say the year is $2004\frac{1}{2}$.

c. Each farm has a whole number of cows that we can count, so these data are discrete. (You cannot have $1\frac{1}{2}$ cows, for example.)

d. The amount of milk that a cow produces can take on any value in some range, so the milk production data are continuous.   ∎

Levels of Measurement

Another way to classify data is by their *level of measurement*. The simplest level of measurement applies to variables such as eye color, ice cream flavors, or gender of animals. These variables can be described solely by names, labels, or categories. We say that such data are at a **nominal level of measurement**. (The word *nominal* refers to *names* for categories.) The nominal level of measurement does not involve any ranking or ordering of the data. For example, we could not say that blue eyes come before brown eyes or that vanilla ranks higher than chocolate.

When we describe data with a ranking or ordering scheme, such as star ratings of movies or restaurants, we are using an **ordinal level of measurement**. (The word *ordinal* refers to an *order*.) Such data generally cannot be used in any meaningful way for computations. For example, it doesn't make sense to add star ratings—watching three one-star movies is not equivalent to watching one three-star movie.

Time out to think

Consider a survey that asks "What's your favorite flavor of ice cream?" We've said that ice cream flavors represent data at the nominal level of measurement. But suppose that, for convenience, the researchers enter the survey data into a computer by assigning numbers to the different flavors. For example, they assign 1 = vanilla, 2 = chocolate, 3 = cookies and cream, and so on. Does this change the ice cream flavor data from nominal to ordinal? Why or why not?

The ordinal level of measurement provides a ranking system, but it does not allow us to determine precise differences between measurements. For example, there is no way to determine the exact difference between a three-star movie and a two-star movie. In contrast, a temperature of 81°F is hotter than 80°F by the same amount that 28°F is hotter than 27°F. Temperature data are at a higher level of measurement, because the *intervals* (differences) between units on a temperature scale always mean the same definite amount. However, while intervals (which involve subtraction) between Fahrenheit temperatures are meaningful, *ratios* (which involve division) are not. For example, it is *not true* that 20°F is twice as hot as 10°F or that −40°F is twice as cold as −20°F. The reason ratios are meaningless on the Fahrenheit scale is that its *zero point is arbitrary* and does not represent a state of "no heat." If intervals are meaningful but ratios are not, as is the case with Fahrenheit temperatures, we say that the data are at the **interval level of measurement**.

When both intervals and ratios are meaningful, we say that data are at the **ratio level of measurement**. For example, data consisting of distances are at the ratio level of measurement because a distance of 10 kilometers really is twice as far as a distance of 5 kilometers. In general, the ratio level of measurement applies to any scale with a true zero, which is a value that really means *none* of whatever is being measured. In the case of distances, a distance of zero really does mean "no distance." Other examples of data at the ratio level of measurement include weights, speeds, and incomes.

Note that data at the nominal or ordinal level of measurement are always qualitative, while data at the interval or ratio level are always quantitative. Figure 2.1 summarizes the possible data types and levels of measurement.

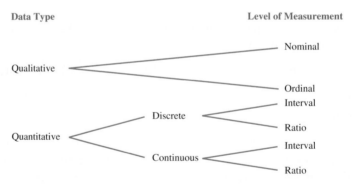

Figure 2.1   Data types and levels of measurement.

Levels of Measurement

The **nominal level of measurement** is characterized by data that consist of names, labels, or categories only. The data are qualitative and cannot be ranked or ordered.

The **ordinal level of measurement** applies to qualitative data that can be arranged in some order (such as low to high). It generally does not make sense to do computations with data at the ordinal level of measurement.

The **interval level of measurement** applies to quantitative data in which intervals are meaningful, but ratios are not. Data at this level have an arbitrary zero point.

The **ratio level of measurement** applies to quantitative data in which both intervals and ratios are meaningful. Data at this level have a true zero point.

EXAMPLE 3 Levels of Measurement

Identify the level of measurement (nominal, ordinal, interval, ratio) for each of the following sets of data.

a. Numbers on uniforms that identify players on a basketball team

b. Student rankings of cafeteria food as excellent, good, fair, or poor

c. Calendar years of historic events, such as 1776, 1945, or 2001

d. Temperatures on the Celsius scale

e. Runners' times in the Boston Marathon

SOLUTION

a. As discussed in Example 1, numbers on uniforms don't count or measure anything. They are at the nominal level of measurement because they are labels and do not imply any kind of ordering.

b. A set of rankings represents data at the ordinal level of measurement because the categories (excellent, good, fair, or poor) have a definite order.

c. An interval of one calendar year always has the same meaning. But ratios of calendar years do not make sense because the choice of the year 0 is arbitrary and does not mean "the beginning of time." Thus, data that consist of calendar years are at the interval level of measurement.

d. Like Fahrenheit temperatures, Celsius temperatures are at the interval level of measurement. An interval of 1°C always has the same meaning, but the zero point (0°C = freezing point of water) is arbitrary and does not mean "no heat."

e. Marathon times have meaningful ratios because they have a true zero point—namely, a time of 0 hours. Thus, for example, a time of 6 hours is twice as long as a time of 3 hours. These data are at the ratio level of measurement. ∎

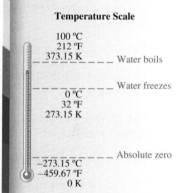

By the Way...

Scientists often measure temperatures on the Kelvin scale. Data on the Kelvin scale are at the ratio level of measurement, because the Kelvin scale has a true zero. A temperature of 0 Kelvin really is the coldest possible temperature. Called *absolute zero*, 0 K is equivalent to about −273.15°C or −459.67°F. (The degree symbol is not used for Kelvin temperatures.) Thus, for example, it is true that 100 K is twice as hot as 50 K.

Temperature Scale

100 °C
212 °F
373.15 K ———— Water boils

———————— Water freezes
0 °C
32 °F
273.15 K

———————— Absolute zero
−273.15 °C
−459.67 °F
0 K

Review Questions

1. What is the distinction between *qualitative data* and *quantitative data*? Give a few examples of each.

2. Distinguish between *discrete* and *continuous* data. Give a few examples of each.

3. Describe each of the four basic levels of measurement. Which levels apply to qualitative data? Which apply to quantitative data?

4. Summarize data types and levels of measurement by describing each of the relationships shown in Figure 2.1.

Exercises

BASIC SKILLS AND CONCEPTS

SENSIBLE STATEMENTS? For Exercises 1–4, determine whether the given statement is sensible and explain why it is or is not.

1. I know these data are not qualitative, so they must be quantitative.

2. I know these qualitative data are not nominal, so they must be ordinal.

3. My commuting times on 25 different days form a set of discrete data.

4. I can make the rankings *bad, average,* and *excellent* quantitative by assigning them values 0, 1, and 2, respectively.

QUALITATIVE VS. QUANTITATIVE DATA. In Exercises 5–18, determine whether the data described are qualitative or quantitative and explain why.

5.   The colors of dogs at a dog show

6.   The blood types of individuals

7.   The average waiting time in checkout lines at a grocery store

8.   The responses of people to potato chips in a taste test where ratings range from 0 = inedible up to 5 = outstanding

9.   The day of the week when survey respondents "feel the best"

10. The high temperature in Anchorage, Alaska, on each day of the year

11. The score that each student receives on a true/false exam

12. The finish positions (first, second, third) of contestants in a running race

13. The responses of people to the question "Will you vote to re-elect your mayor?"

14. The birthdays (not including year) of surveyed people

15. The federal income taxes paid by individuals

16. The state funding per public school pupil in each state

17. The choice of TV shows to watch on a given night

18. The numbers of people watching various TV shows on a given night

DISCRETE OR CONTINUOUS. In Exercises 19–32, state whether the data described are discrete or continuous and explain why.

19. Number of defective computer components that are found each day on an assembly line

20. Taxicab fare in New York City

21. Total number of taxicabs each day in New York City

22. Weights of cars in a safety survey

23. Numbers of exits along highways for various driving routes from Jacksonville to San Diego

24. Distances traveled by airplanes flying from Jacksonville to San Diego

25. Numbers of miles traveled by presidents of major companies in a year

26. Number of cars passing through a busy intersection each hour

27. Average speed of cars passing through a busy intersection each hour

28. Scores on a multiple-choice exam

29. Number of stars in each galaxy in the universe

30. Average length of movies in each of the last 30 years

31. Average theater price for movies for each of the last 30 years

32. Numbers of calories consumed daily by professional basketball players

FURTHER APPLICATIONS

LEVELS OF MEASUREMENT. For the data described in Exercises 33–46, identify the level of measurement (nominal, ordinal, interval, or ratio) and explain your answer.

33. Party affiliation of voters: Democrat, Republican, or independent

34. Classification of voters by income level: low, middle, or high

35. Body temperatures in degrees Celsius of a sample of newborn babies

36. Weights of a sample of newborn babies

37. Types of cars bought by customers at a car dealership in a single month: sedan, stationwagon, sports utility

38. Prices of cars bought by customers at a car dealership in a single month

39. *Consumer Reports* safety ratings of cars: 0 = unsafe up to 3 = safest

40. Years in which Democrats won the U.S. presidential election

41. Final course grades of A, B, C, D, F

42. SAT scores

43. Low temperature on each day of the year in Miami

44. Ratings of people in a survey of TV news anchors: least favorite, acceptable, or favorite

45. Breeds of horses at a horse show

46. Social Security numbers

MEANINGFUL RATIOS? In Exercises 47–54, determine whether the given statement represents a meaningful ratio. Explain.

47. 2 o'clock is twice as late as 1 o'clock

48. 10 km is twice as far as 5 km

49. 500 miles per hour is ten times as fast as 50 miles per hour

50. The new *Star Wars* movie is three times as good as the first one

51. Smith received $1\frac{1}{2}$ times as many votes as Jones

52. The inflation rate in the 1900s was half of what it was in the 1980s

53. A person with an SAT score of 1200 is twice as qualified for college as a person with a score of 600

54. With twice as many hits per month, Web site A is twice as good as Web site B

COMPLETE CLASSIFICATION. In Exercises 55–62, state whether the data described are qualitative or quantitative and give their level of measurement. If the data are quantitative, state whether they are continuous or discrete. Be sure to explain your answers.

55. The number of cars produced by an auto manufacturer each year between 1980 and 2000

56. Ethnicity categories on an application for financial aid

57. A listing of the number of students at your school from each state in the United States (for example, how many from Colorado, California, North Dakota)

58. The collegiate football rankings of the Associated Press

59. Times of day that are on the hour (such as 1 o'clock and 2 o'clock, but not 1:30)

60. Times of runners in a marathon

61. Finishing positions (such as 1st, 2nd, 3rd) of runners in a marathon

62. Net personal income for people in Mississippi

2.2 Dealing with Errors

Now that you understand the different levels and types of measurement, we turn to the issue of how to deal with errors in measurement. First we will consider the various types of errors that can occur, and then we'll discuss how to account for possible errors when we state results. Note that while we will phrase most of this discussion in terms of measurements, it applies equally well to estimates or projections, such as population estimates or projected revenues for a corporation.

Types of Error: Random and Systematic

Broadly speaking, measurement errors fall into two categories: random errors and systematic errors. Let's consider an example to illustrate the difference.

Suppose you work in a pediatric office and use a digital scale to weigh babies. If you've ever worked with babies, you know that they usually aren't very happy about being put on a scale. Their thrashing and crying tends to shake the scale, making the readout jump around. For the case shown in Figure 2.2a, you could equally well record the baby's weight as anything between about 14.5 and 15.0 pounds. We say that the shaking of the scale introduces a **random error** because any particular measurement may be either too high or too low.

Now suppose you have been measuring weights of babies all day long with the scale shown in Figure 2.2b. At the end of the day, you notice that the scale reads 1.2 pounds when there is nothing on it. In that case, every measurement you made was high by 1.2 pounds. This type of error is called a **systematic error** because it is caused by an error in the measurement *system*—an error that consistently (systematically) affects all measurements.

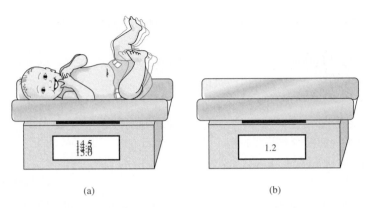

Figure 2.2 (a) The baby's motion introduces random errors. (b) A scale that does not read zero when empty introduces systematic errors.

Two Types of Measurement Error

Random errors occur because of random and inherently unpredictable events in the measurement process.

Systematic errors occur when there is a problem in the measurement system that affects all measurements in the same way.

By the Way...

The fact that urban areas tend to be warmer than they would be in the absence of human activity is often called the *urban heat island effect*. Major causes of this effect include heat released by burning fuel in automobiles, homes, and industry and the fact that pavement and large masonry buildings tend to retain heat from sunlight.

A systematic error affects all measurements in the same way, such as making them all too high or all too low. Thus, if you discover a systematic error, you can go back and adjust the affected measurements. In contrast, the unpredictable nature of random errors makes it impossible to correct for them. However, you can minimize the effects of random errors by making many measurements and averaging them. For example, if you measure the baby's weight ten times, your measurements will probably be too high in some cases and too low in others. You can therefore get a better value by averaging the ten individual measurements.

Time out to think

Call the local phone number that gives the current time. How far off is your clock or watch? Describe possible sources of random and systematic errors in your timekeeping.

EXAMPLE 1   Errors in Global Warming Data

Scientists trying to determine whether human activity is raising the Earth's temperature need to know how the average temperature of the entire Earth, or the *global average temperature*, has changed with time. Consider two difficulties (among many others) in trying to interpret historical temperature data from the early 20th century: (1) temperatures were measured with simple thermometers and the data were recorded by hand; and (2) most temperature measurements were recorded in or near urban areas, which tend to be warmer than surrounding rural areas because of heat released by human activity. Discuss whether each of these two difficulties produces random or systematic errors, and consider the implications of these errors.

SOLUTION   The first difficulty involves *random errors* because people undoubtedly made occasional errors in reading the thermometers, in calibrating the thermometers, and in recording temperature readings. There is no way to predict whether any individual reading is correct, too high, or too low. However, if there are several readings for the same region on the same day, averaging these readings can minimize the effects of the random errors.

The second difficulty involves a *systematic error* because the excess heat in urban areas always causes the temperature reading to be higher than it would be otherwise. If the researchers could determine how much this systematic error affected the temperature readings, they could correct the data for this problem. ∎

The Census

The Constitution of the United States mandates a census of the population every 10 years. The United States Census Bureau conducts the census (and also does many other demographic studies).

In attempting to count the population, the Census Bureau relies largely on a survey that is supposed to be delivered to and returned by every household in the United States. However, many *random errors* occur in this survey process. For example, some people fill out their forms incorrectly and some responses are recorded incorrectly by Census Bureau employees.

The census also is subject to several types of *systematic error*. For example, it is very difficult to deliver surveys to the homeless, and it is difficult to count undocumented aliens, who may be reluctant to reveal their presence in the United States. These systematic errors lead to undercounts in the population. An example of a systematic error leading to an overcount is the double counting of some college students, who are counted by their parents and again at their school residence.

The 2000 census, for example, originally counted about 281.4 million people. However, follow-up statistical studies suggested that the census had missed about 7.6 million people and counted about 4.3 million people twice. Overall, then, the census undercounted the population by over 3 million people. ∎

A little inaccuracy sometimes saves a ton of explanation.

– H. H. Munro (Saki)

Time out to think
The question of whether the Census Bureau should be allowed to adjust its "official" count on the basis of statistical surveys is very controversial. The Constitution calls for an "actual enumeration" of the population (Article 1, Section 2, Subsection 2). Do you believe that this wording precludes or allows the use of statistical surveys in the official count? Defend your opinion. Also discuss reasons why Democrats tend to favor the use of sampling methods while Republicans tend to oppose it.

Size of Errors: Absolute Versus Relative

Besides wanting to know whether an error is random or systematic, we often want to know something about the overall size of the error. For example, is the error big enough to be of concern or small enough to be unimportant? Let's explore the concept of error size.

Suppose you go to the store and buy what you think is 6 pounds of hamburger, but because the store's scale is poorly calibrated you actually get only 4 pounds. You'd probably be upset by this 2-pound error. Now suppose you buy a car which the owner's manual says weighs 3,132 pounds, but you find that it really weighs 3,130 pounds. The error is the same 2 pounds as before, but this error probably doesn't seem very important for the car. This simple example shows that the size of an error can depend on how you look at it.

Speaking more technically, the 2-pound error in both the measurement of the hamburger and the measurement of the car is an **absolute error**—it describes how far the measured value lies from the true value. The **relative error** compares the size of the absolute error to the true value. The relative error for the hamburger is fairly large because the absolute error of 2 pounds is half the true weight of 4 pounds. In this case, we say that the relative error is 2/4, or 50%. In contrast, the relative error for the car is quite small, because 2 pounds is very small compared to the car's true weight of 3,130 pounds. Here, the relative error is 2/3130, which is about 0.0006, or 0.06%.

By the Way...
Speaking of weights of cars, in 2002 you could buy a Hyundai Accent for $3.85 per pound, a Ford Taurus for $5.63 per pound, a BMW 3-Series for $7.99 per pound, a Chevrolet Corvette for $12.44 per pound, or a Jaguar XJ8 for $13.94 per pound.

Errors using inadequate data are much less than those using no data at all.
— Charles Babbage, computer pioneer

Absolute and Relative Error

The **absolute error** describes how far a measured value lies from the true value; it is the *difference* between the measured and true values:

$$\text{absolute error} = \text{measured value} - \text{true value}$$

The **relative error** compares the size of the absolute error to the true value. It is often expressed as a percentage:

$$\text{relative error} = \frac{\text{measured value} - \text{true value}}{\text{true value}} \times 100\%$$

Note that absolute and relative errors are positive when the measured value is greater than the true value, meaning the measurement was on the "too high" side. They are negative when the measured value is less than the true value, meaning the measurement was on the "too low" side.

EXAMPLE 2   Absolute and Relative Error

Find the absolute and relative error in each case.

a. Your true weight is 100 pounds, but a scale says you weigh 105 pounds.

b. The government claims that a program costs $99.0 billion, but an audit shows that the true cost is $100.0 billion.

SOLUTION

a. The measured value is the scale reading of 105 pounds and the true value is 100 pounds.

$$\begin{aligned}\text{absolute error} &= \text{measured value} - \text{true value}\\ &= 105 \text{ lb} - 100 \text{ lb}\\ &= 5 \text{ lb}\end{aligned}$$

$$\begin{aligned}\text{relative error} &= \frac{\text{measured value} - \text{true value}}{\text{true value}} \times 100\%\\ &= \frac{105 \text{ lb} - 100 \text{ lb}}{100 \text{ lb}} \times 100\%\\ &= 5\%\end{aligned}$$

The measured weight is too high by 5 pounds, or by 5%.

A billion here, a billion there; soon you're talking real money.
— Senator Everett Dirksen

b. The measured value is the claimed cost of $99.0 billion and the true value is $100.0 billion.

$$\begin{aligned}\text{absolute error} &= \$99.0 \text{ billion} - \$100.0 \text{ billion}\\ &= -\$1.0 \text{ billion}\end{aligned}$$

$$\begin{aligned}\text{relative error} &= \frac{\$99.0 \text{ billion} - \$100.0 \text{ billion}}{\$100.0 \text{ billion}} \times 100\%\\ &= -1\%\end{aligned}$$

The claimed cost is too low by $1.0 billion, or by 1%. ∎

Describing Results: Accuracy and Precision

Once a measurement is reported, we should evaluate it to see whether it is believable in light of any potential errors. In particular, we should consider two key ideas about any reported value: its *accuracy* and its *precision*. These terms are often used interchangeably in English, but mathematically they refer to two different concepts.

The goal of any measurement is to obtain a value that is as close as possible to the *true value*. The closer the measured value lies to the true value, the greater is the **accuracy** of the measurement. If a census says that the population of your hometown is 72,453 but the true population is 96,000, then the census report is not very accurate. In contrast, if a company projects sales of $7.30 billion and true sales turn out to be $7.32 billion, we would say that projection is quite accurate.

The **precision** of a number describes the amount of detail in a measurement. That is, precision describes how *precisely* a number is measured. For example, a distance given as 2.345 km is more precise than a distance given as 2.3 km, because the first number shows us detail down to the nearest 0.001 km whereas the second number shows us detail only to the nearest 0.1 km. Similarly, an income of $45,678.90 has greater precision than one of $46,000, because the first income is precise to the nearest penny whereas the second is precise only to the nearest thousand dollars.

> **Definitions**
>
> **Accuracy** describes how closely a measurement approximates a true value. An accurate measurement is very close to the true value.
>
> **Precision** describes the amount of detail in a measurement.

We generally assume that the precision with which a number is reported reflects what was actually measured. If you say that you weigh "132 pounds," we assume that you measured your weight only to the nearest pound. In that case, a more precise measurement might find that you weighed, say, 132.3 pounds or 131.6 pounds (both of these would round to 132). In contrast, if you say that you weigh "132.0 pounds," we assume that you measured your weight to the nearest tenth of a pound. This assumption about precision means that numbers reported with unjustified precision are dishonest. For example, if you actually measured your weight only to the nearest pound, it would be dishonest to say that you weighed "132.0 pounds" because that would imply you made a measurement to the nearest tenth of a pound.

EXAMPLE 3   Accuracy and Precision in Your Weight

Suppose that your true weight is 102.4 pounds. The scale at the doctor's office, which can be read only to the nearest quarter pound, says that you weigh $102\frac{1}{4}$ pounds. The scale at the gym, which gives a digital readout to the nearest 0.1 pound, says that you weigh 100.7 pounds. Which scale is more *precise*? Which is more *accurate*?

SOLUTION   The scale at the gym is more *precise* because it gives your weight to the nearest tenth of a pound, whereas the doctor's scale gives your weight only to the nearest quarter pound. However, the scale at the doctor's office is more *accurate* because its value is closer to your true weight. ∎

Time out to think

In Example 3 we needed to know your true weight in order to determine which scale was more accurate. But how would you know your true weight? Can you ever be sure that you know your true weight? Explain.

Mistakes are the portals of discovery.
— James Joyce

By the Way...

The digits in a number that were actually measured are called *significant digits*. All the digits in a number are significant except for zeros that are included so that the decimal point can be properly located. For example, 0.001234 has four significant digits; the zeros are required for proper placement of the decimal point. The number 1,234,000,000 has four significant digits for the same reason. The number 132.0 has four significant digits; the zero is significant because it is not required for the proper placement of the decimal point, and therefore implies an actual measurement of zero tenths.

By the Way...

Projection errors don't always work in the government's favor. In January 2001, the government projected a surplus of $5.6 trillion over the ten years beginning in 2002. By January 2002, the projection had dropped to $1.6 trillion—a loss of $4 trillion in projected revenue. And if you believe the new projection will prove accurate, we have some valuable swampland you might be interested in . . . .

CASE STUDY Government Makes $190 Billion Mistake— in Its Favor!

Like a business, the federal government either takes in more money than it spends, producing a *surplus*, or spends more money than it takes in, producing a *deficit*. Huge deficits plagued the United States government throughout the 1970s, the 1980s, and most of the 1990s, causing the total federal *debt* (the total amount of money the government owes from all previous years combined) to grow to nearly $6 trillion. At the beginning of fiscal year 1998, government economists expected the string of deficits to continue; they projected a $121 billion deficit. Notice that this projection was *precise* to the nearest $1 billion, implying that the actual budget deficit would end up between $120.5 billion and $121.5 billion.

In fact, the final results for 1998 showed a $69 billion *surplus*, making the original projection in error by $190 billion ($121 billion + $69 billion = $190 billion). Thus, the original projection was stated very precisely, but not very accurately! Fortunately for the politicians, this inaccuracy worked in their favor because it meant that they had more money than expected. ∎

By the Way...

The Mars Climate Orbiter was supposed to go into orbit of Mars in late 1999. NASA engineers controlled the spacecraft from Earth by sending precise computer instructions telling it when and for how long to fire its rockets. However, the instructions were sent in English units (pounds) and the spacecraft software interpreted them in metric units (kilograms). Result: The very precise instructions were quite inaccurate, sending the $160 million spacecraft so close to Mars that it burned up in the Martian atmosphere.

CASE STUDY Does the Census Measure the True Population?

Upon completing the 2000 census, the U.S. Census Bureau reported a population of 281,421,906 (on April 1, 2000), thereby implying an exact count of everyone living in the United States. Unfortunately, such a precise count could not possibly be accurate.

Even in principle, the only way to get an exact count of the number of people living in the United States would be to count everyone *instantaneously*. Otherwise, the count would be off because the number of people changes very rapidly. An average of about eight births and four deaths occur every minute in the United States, and a new immigrant enters the United States about every 3 minutes on average. Moreover, random and systematic errors can easily make the census inaccurate by a few million people (see the Case Study on p. 59).

Thus, in reality, the Census Bureau did *not* know the exact population and only found an estimate of the population to within a few million people. Given these facts, it is dishonest to report the population to the nearest person. A more honest report would have used much less precision—for example, stating the population as "about 280 million." In fairness to the Census Bureau, their detailed reports did explain the uncertainties in the population count, but these uncertainties were rarely mentioned by the press. ∎

Summary: Dealing with Errors

The ideas we've covered in this section are a bit technical, but very important to understanding measurements and errors. Let's briefly summarize how the ideas relate to one another.

- Errors can occur in many ways, but generally can be classified into one of two basic types: random errors or systematic errors.

- Whatever the source of an error, its size can be described in two different ways: as an absolute error or as a relative error.

- Once a measurement is reported, we can evaluate it in terms of its accuracy and its precision.

Review Questions

1. Distinguish between *random errors* and *systematic errors*.

2. How can we minimize the effects of random errors on measurements? How can we account for the effects of a systematic error?

3. Distinguish between the *absolute error* and the *relative error* in a measurement. Give an example in which the absolute error is large but the relative error is small and another example in which the absolute error is small but the relative error is large.

4. Distinguish between *accuracy* and *precision*. Give an example of a measurement that is precise but inaccurate and another example that is accurate but imprecise.

5. Why is it dishonest to give measurements with more precision than is justified by the actual measurements?

Exercises

BASIC SKILLS AND CONCEPTS

SENSIBLE STATEMENTS? For Exercises 1–6, determine whether the given statement is sensible and explain why it is or is not.

1. There are 138,232 species of butterflies and moths on the Earth.

2. The measurement taken by the electronic timer must be more accurate than that of the stopwatch.

3. The relative error that a microbiologist makes in measuring a cell must be less than the relative error that an astronomer makes in measuring a galaxy, because cells are smaller than galaxies.

4. The bank teller claims that his errors are random even though they are always to his advantage.

5. The 6 billionth person on Earth was born on October 12, 1999, in Bosnia.

6. I would rather be shortchanged by $1 than by 1%.

7. **TAX AUDIT.** A tax auditor reviewing a tax return looks for several kinds of problems, including these two: (1) mistakes made in entering or calculating numbers on the tax return and (2) places where the taxpayer reported income dishonestly. Discuss whether each problem involves random or systematic errors.

8. **AIDS EPIDEMIC.** Researchers trying to study the progression of the AIDS epidemic need to know how many people are suffering from AIDS, which they can try to determine by studying medical records. Two of the many problems they face in this research are that (1) some people who are suffering from AIDS are misdiagnosed as having other diseases, and vice versa; and (2) some people with AIDS never seek medical help and therefore do not have medical records. Discuss whether each problem involves random or systematic errors.

9. **SAFE AIR TRAVEL.** Before taking off, a pilot is supposed to set the aircraft altimeter to the elevation of the airport. A pilot leaves from Denver (altitude 5,280 feet) with her altimeter set to 2,500 feet. Explain how this affects the altimeter readings throughout the flight. What kind of error is this?

10. **LUMBER YARD.** A lumber yard is cutting boards that are supposed to be 4 feet long. Each board is cut by an employee who first measures and marks the 4-foot length. Later, careful measurements show that the average length of the boards was indeed 4 feet, but that most of the boards were actually slightly longer or shorter than 4 feet. What type of measurement error are we dealing with in the board lengths? Explain.

11. **UNCALIBRATED SCALE.** Imagine that you are a purchaser for a grocery chain, weighing flats of strawberries from local farms. After a long day of weighing, you discover that the scale reads 1.2 pounds even though nothing is on it. Explain how all of the weights that you recorded are affected by this discovery. What kind of error did you introduce into your measurements? How can you account for this error in your earlier measurements?

12. **SUICIDE DATA.** Concerning the collection of data on suicides, the *New York Times Almanac* claims, "Most experts believe that suicide statistics are grimmer than reported. They contend that numerous suicides are categorized as accidents or other deaths to spare families." If this is the case, what kind of error is introduced in suicide data and how does it affect the values of the data?

FURTHER APPLICATIONS

SOURCES OF ERRORS. For each measurement described in Exercises 13–20, discuss any likely sources of random errors and any likely sources of systematic errors.

13. A count of sport utility vehicles passing through a busy intersection during a 20-minute period

14. The population of a county in Mississippi, according to a census count

15. The average income of 25 people, found by checking their tax returns

16. The average income of 25 people, found by asking them their incomes during interviews conducted at a supermarket

17. Weights of sandwich meats, measured in a delicatessen using a well-calibrated scale

18. Weights of adults, measured by weighing them with their clothes on

19. The numbers of popped kernels in "large" boxes of popcorn at a movie theater

20. Times in a swim meet

ABSOLUTE AND RELATIVE ERRORS. In Exercises 21–24, find the absolute and relative errors.

21. The price of a gift you purchase by mail-order is supposed to be $18.50, but instead you are billed for $19.00.

22. The label on a bag of dog food says "20 pounds," but the true weight is only 18 pounds.

23. The bank is supposed to charge you interest of $65 on your loan, but instead charges you only $48.

24. The diameter of a ball-bearing is supposed to be 0.1 centimeter, but instead is 0.2 centimeter.

25. **MINIMIZING ERRORS.** Suppose that 25 people, including yourself, are asked to measure the length of a room to the nearest tenth of a millimeter. Assume that everyone uses the same well-calibrated measuring device, such as a tape measure.
    a. All 25 measurements are not likely to be exactly the same; thus, the measurements will contain some sources of error. Are these errors systematic or random? Explain.
    b. If you want to minimize the effect of random errors in determining the length of the room, which is the better choice: to report your own personal measurement as the length of the room or to report the average of all 25 measurements? Explain.
    c. Describe any possible sources of systematic errors in the measurement of the room length.
    d. Can the process of averaging all 25 measurements help reduce any systematic errors? Why or why not?

26. **MINIMIZING ERRORS.** Twenty children in a fourth-grade classroom are doing a project about the weather. Each has a thermometer outside on the playground, in the same shaded area. At noon each day, each child records the temperature on his or her personal thermometer.
    a. All 20 measurements are not likely to be exactly the same; thus, the measurements will contain some sources of error. Are these errors systematic or random? Explain.
    b. If you want to minimize the effect of random errors in determining the true temperature at noon one day, which is the better choice: to choose one child's measurement at random or to report the average of all 20 measurements? Explain.
    c. Describe any possible sources of systematic errors in the temperature recordings.
    d. Can the process of averaging all 20 measurements help reduce any systematic errors? Why or why not?

27. **ACCURACY AND PRECISION IN HEIGHT.** Suppose your true height is 62.50 inches. A tape measure that can be read to the nearest $\frac{1}{8}$ inch gives your height as $62\frac{3}{8}$ inches. A new laser device at the doctor's office that gives readings to the nearest 0.05 inch gives your height as 62.90 inches. Which measurement is more *precise*? Which is more *accurate*? Explain.

28. **ACCURACY AND PRECISION IN HEIGHT.** Suppose your true height is 62.50 inches. A tape measure that can be read to the nearest $\frac{1}{8}$ inch gives your height as $62\frac{5}{8}$ inches. A new laser device at the doctor's office that gives readings to the nearest 0.05 inch gives your height as 62.50 inches. Which measurement is more *precise*? Which is more *accurate*? Explain.

29. **ACCURACY AND PRECISION IN WEIGHT.** Suppose your weight is 52.55 kilograms. A scale at a health clinic that gives weight measurements to the nearest half kilogram gives your weight as 53 kilograms. A digital scale at the gym that gives readings to the nearest 0.01 kilogram gives your weight as 52.88 kilograms. Which measurement is more *precise*? Which is more *accurate*? Explain.

30. **ACCURACY AND PRECISION IN WEIGHT.** Suppose your weight is 52.55 kilograms. A scale at a health clinic that gives weight measurements to the nearest half kilogram gives your weight as $52\frac{1}{2}$ kilograms. A digital scale at the gym that gives readings to the nearest 0.01 kilogram gives your weight as 51.48 kilograms. Which meas-

urement is more *precise*? Which is more *accurate*? Explain.

BELIEVABLE FACTS? Exercises 31–38 give statements of "fact" coming from statistical measurements. For each statement, briefly discuss possible sources of error in the measurement. Then, in light of the precision with which the measurement is given, discuss whether you think the fact is believable.

31. The population of the United States in 1860 was 31,443,321.

32. The number of deaths in the United States due to coronary heart disease in 2000 was 709,894.

33. The population of Tokyo in 2015 will be 28.7 million.

34. The Petrona Towers in Kuala Lumpur are 1,483 feet tall, making them the world's tallest building as of 2002.

35. The average maximum temperature in January in Lagos, Nigeria, is 88 degrees.

36. The number of deaths due to AIDS in the United States in 2000 was 14,370.

37. The number of cell phone subscribers in the United States in 2000 was 109,478,000.

38. Worldwide in 2000, there were 1,694 endangered or threatened species of animals and plants.

 PROJECTS FOR THE WEB AND BEYOND

For useful links, select "Links for Web Projects" for Chapter 2 at www.aw.com/bbt.

39. **THE 2000 CENSUS.** Go to the Web site for the Census Bureau and find the latest results from the 2000 census. According to the census, what was the population of the United States in 2000? Discuss the accuracy and precision of the claim.

40. **CENSUS CONTROVERSIES.** Use the Library of Congress's "Thomas" Web site to find out about any pending legislation concerning the collection or use of census data. If you find more than one legislative bill pending, choose one to study in depth. Summarize the proposed legislation, and briefly discuss arguments both for and against it.

41. **WRISTWATCH ERRORS.** First, using a wristwatch that is reasonably accurate, set the time to be exact. Use a radio station or telephone time report that states "At the tone, the time is . . . ." If you cannot set the time to the nearest second, record the error for the watch you are using. Now, compare the time on your watch to the times on other watches. Record the errors with positive signs for watches that are ahead of the true time and negative signs for those watches that are behind the true time. Use the concepts of this section to describe the accuracy of the wristwatches in your sample.

IN THE NEWS

1. **RANDOM AND SYSTEMATIC ERRORS.** Find a recent news report that gives a quantity that was measured statistically (for example, a report of population, average income, or the number of homeless people). Write a short description of how the quantity was measured, and briefly describe any likely sources of either random or systematic errors. Overall, do you think that the reported measurement was accurate? Why or why not?

2. **ABSOLUTE AND RELATIVE ERRORS.** Find a recent news report that describes some mistake in a measured, estimated, or projected number (for example, a budget projection that turned out to be incorrect). In words, describe the size of the error in terms of both absolute error and relative error.

3. **ACCURACY AND PRECISION.** Find a recent news article that causes you to question accuracy or precision. For example, the article might report a figure with more precision than you think is justified, or it might cite a figure that you know is inaccurate. Write a one-page summary of the report and explain why you question its accuracy or precision (or both).

2.3 Uses of Percentages in Statistics

Statistical results are often stated with percentages. A percentage is simply a way of expressing a fraction; the words *per cent* literally mean "divided by 100." However, percentages are often used in subtle ways. Consider a statement that appeared in a front-page article in *The New York Times*:

> The rate [of smoking] among 10th graders jumped 45 percent, to 18.3 percent, and the rate for 8th graders is up 44 percent, to 10.4 percent.

All the percentages in this statement are used correctly, but they are very confusing. For example, what does it mean to say "up 44%, to 10.4%"? In this section, we will investigate some of the subtle uses and abuses of percentages. Before we begin, you should review the following basic rules regarding conversions between fractions and percentages.

Conversions Between Fractions and Percentages

To convert a percentage to a common fraction: Replace the % symbol with division by 100; simplify the fraction if necessary.

$$\text{Example: } 25\% = \frac{25}{100} = \frac{1}{4}$$

To convert a percentage to a decimal: Drop the % symbol and move the decimal point two places to the left (that is, divide by 100).

$$\text{Example: } 25\% = 0.25$$

To convert a decimal to a percentage: Move the decimal point two places to the right (that is, multiply by 100) and add the % symbol.

$$\text{Example: } 0.43 = 43\%$$

To convert a common fraction to a percentage: First convert the common fraction to a decimal; then convert the decimal to a percentage.

$$\text{Example: } \frac{1}{5} = 0.2 = 20\%$$

EXAMPLE 1 Newspaper Survey

A newspaper reports that 44% of 1,069 people surveyed said that the President is doing a good job. How many people said that the President is doing a good job?

SOLUTION The 44% represents the fraction of respondents who said the President is doing a good job. Because "of" usually indicates multiplication, we multiply:

$$44\% \times 1{,}069 = 0.44 \times 1{,}069 = 470.36 \approx 470$$

(The symbol $\approx$ means "approximately equal to.") Thus, about 470 people said the President is doing a good job. We rounded the answer to 470 to obtain a whole number of people. ∎

Using Percentages to Describe Change

In statistics, percentages are often used to describe how data change with time. As an example, suppose the population of a town was 10,000 in 1970 and 15,000 in 2000. We can express the change in population in two basic ways:

- Because the population rose by 5,000 people (from 10,000 to 15,000), we say that the **absolute change** in the population was 5,000 people.

- Because the increase of 5,000 people was 50% of the starting population of 10,000, we say that the **relative change** in the population was 50%.

In general, calculating an absolute or relative change always involves two numbers: a starting number, or **reference value**, and a **new value**. Once we identify these two values, we can calculate the absolute and relative change with the following formulas. Note that the absolute and relative changes are positive if the new value is greater than the reference value and negative if the new value is less than the reference value.

Absolute and Relative Change

The **absolute change** describes the actual increase or decrease from a reference value to a new value:

$$\text{absolute change} = \text{new value} - \text{reference value}$$

The **relative change** describes the size of the absolute change in comparison to the reference value and can be expressed as a percentage:

$$\text{relative change} = \frac{\text{new value} - \text{reference value}}{\text{reference value}} \times 100\%$$

Time out to think

Compare the formulas for absolute and relative change to the formulas for absolute and relative error, given in Section 2.2. Briefly describe the similarities you notice.

EXAMPLE 2 World Population Growth

World population in 1950 was 2.6 billion. By the beginning of 2000, it had reached 6.0 billion. Describe the absolute and relative change in world population from 1950 to 2000.

SOLUTION   The reference value is the 1950 population of 2.6 billion and the new value is the 2000 population of 6.0 billion.

$$\begin{aligned}
\text{absolute change} &= \text{new value} - \text{reference value} \\
&= 6.0 \text{ billion} - 2.6 \text{ billion} \\
&= 3.4 \text{ billion}
\end{aligned}$$

$$\begin{aligned}
\text{relative change} &= \frac{\text{new value} - \text{reference value}}{\text{reference value}} \times 100\% \\
&= \frac{6.0 \text{ billion} - 2.6 \text{ billion}}{2.6 \text{ billion}} \times 100\% \\
&\approx 130\%
\end{aligned}$$

World population increased by 3.4 billion people, or by 130%, from 1950 to 2000. ∎

Using Percentages for Comparisons

Percentages are also commonly used to compare two numbers. In this case, the two numbers are the reference value and the compared value.

By the Way...

According to estimates made by the U.S. Census Bureau, world population reached 6 billion on July 19, 1999. According to United Nations estimates, the 6 billion mark was passed on October 12, 1999.

- The **reference value** is the number that we are using as the basis for a comparison.

- The **compared value** is the other number, which we compare to the reference value.

We can then express the absolute or relative difference between these two values with formulas very similar to those for absolute and relative change. Note that the absolute and relative differences are positive if the compared value is greater than the reference value and negative if the compared value is less than the reference value.

Absolute and Relative Difference

The **absolute difference** is the difference between the compared value and the reference value:

$$\text{absolute difference} = \text{compared value} - \text{reference value}$$

The **relative difference** describes the size of the absolute difference in comparison to the reference value and can be expressed as a percentage:

$$\text{relative difference} = \frac{\text{compared value} - \text{reference value}}{\text{reference value}} \times 100\%$$

By the Way...

No one knows all the reasons for the low life expectancy of Russian men, but one contributing factor certainly is alcoholism, which is much more common in Russia than in America.

EXAMPLE 3   Russian and American Life Expectancy

Life expectancy for American men is about 73 years, while life expectancy for Russian men is about 59 years. Compare the life expectancy of American men to that of Russian men in absolute and relative terms. (See Section 10.1 for a discussion of the meaning of life expectancy.)

SOLUTION   We are told to compare the American life expectancy to the Russian life expectancy, which means that we use the Russian life expectancy as the reference value and the American life expectancy as the compared value.

$$\begin{aligned}
\text{absolute difference} &= \text{compared value} - \text{reference value} \\
&= 73 \text{ years} - 59 \text{ years} \\
&= 14 \text{ years}
\end{aligned}$$

$$\begin{aligned}
\text{relative difference} &= \frac{\text{compared value} - \text{reference value}}{\text{reference value}} \\
&= \frac{73 \text{ years} - 59 \text{ years}}{59 \text{ years}} \times 100\% \\
&\approx 24\%
\end{aligned}$$

The life expectancy of American men is 14 years greater in absolute terms and 24% greater in relative terms than the life expectancy of Russian men.

"Of" Versus "More Than"

A subtlety in dealing with percentage statements comes from the way they are worded. Consider a population that *triples* in size from 200 to 600. There are two equivalent ways to state this change with percentages:

- Using *more than*: The new population is 200% *more than* the original population. Here we are looking at the relative change in the population:

$$\text{relative change} = \frac{\text{new value} - \text{reference value}}{\text{reference value}} \times 100\%$$

$$= \frac{600 - 200}{200} \times 100\%$$

$$= 200\%$$

- Using *of*: The new population is 300% *of* the original population, which means it is three times the original population. Here we are looking at the *ratio* of the new population to the original population:

$$\frac{\text{new population}}{\text{original population}} = \frac{600}{200} = 3.00 = 300\%$$

Notice that the percentages in the "more than" and "of" statements are related by 300% = 100% + 200%. This leads to the following general relationship:

> ### *Of* Versus *More Than* (or *Less Than*)
>
> If the new or compared value is *P%* *more than* the **reference value,** then it is (100 + *P*)% *of* the reference value. Similarly, if the new or compared value is *P%* *less than* the reference value, then it is (100 − *P*)% *of* the reference value.

For example, 40% *more than* the reference value is 140% *of* the reference value, and 40% *less than* the reference value is 60% *of* the reference value. When you hear statistics quoted with percentages, it is very important to listen carefully for the key words *of* and *more than* (or *less than*)—and hope that the speaker knows the difference.

EXAMPLE 4   World Population

In Example 2 we found that world population in 2000 was 130% more than world population in 1950. Express this change with an "of" statement.

SOLUTION   World population in 2000 was 130% more than world population in 1950. Because (100 + 130)% = 230%, the 2000 population was 230% *of* the 1950 population. This means that the 2000 population was 2.3 times the 1950 population. ∎

EXAMPLE 5   Sale!

A store is having a "25% off" sale. How does a sale price compare to an original price?

SOLUTION   The "25% off" means that a sale price is 25% *less than* the original price. Thus, the sale price is (100 − 25)% = 75% *of* the original price. For example, if the original price is $100, the sale price is $75. ∎

Time out to think

One store advertises "1/3 off everything!" Another store advertises "Sale prices just 1/3 of original prices!" Which store is having the bigger sale? Explain.

Percentages of Percentages

Percentage changes and percentage differences can be particularly confusing when the values *themselves* are percentages. Suppose your bank increases the interest rate on your savings

By the Way...

World population is currently growing by about 90 million people each year. Thus, in just three years the world adds roughly as many people as the entire population of the United States.

account from 3% to 4%. It's tempting to say that the interest rate increases by 1%, but that statement is ambiguous at best. The interest rate increases by 1 *percentage point*, but the relative change in the interest rate is 33%:

$$\frac{4\% - 3\%}{3\%} \times 100\% = 0.33 \times 100\% = 33\%$$

Thus, you can say that the bank raised your interest rate by 33%, even though the actual rate increased by only 1 percentage point (from 3% to 4%).

Percentage Points Versus %

When you see a change or difference expressed in *percentage points*, you can assume it is an *absolute* change or difference. If it is expressed as a percentage, it is a *relative* change or difference.

EXAMPLE 6   Margin of Error

Based on interviews with a sample of students at your school, you conclude that the percentage of all students who are vegetarians is probably between 20% and 30%. Should you report your result as "25% with a margin of error of 5%" or as "25% with a margin of error of 5 percentage points"? Explain. (See Section 1.1 to review the meaning of margin of error.)

SOLUTION   The range of 20% to 30% comes from subtracting and adding an *absolute difference* of 5 percentage points to 25%. That is,

$$20\% = (25 - 5)\% \qquad \text{and} \qquad 30\% = (25 + 5)\%$$

Thus, the correct statement is "25% with a margin of error of 5 percentage points." It is wrong to say "25% with a margin of error of 5%," because this implies that the error is 5% of 25%, which is only 1.25%. ∎

EXAMPLE 7   Care in Wording

Assume that 40% of the registered voters in Carson City are Republicans. Read the following questions carefully and give the most appropriate answers.

a. The percentage of voters registered as Republicans is 25% higher in Freetown than in Carson City. What percentage of the registered voters in Freetown are Republicans?

b. The percentage of voters registered as Republicans is 25 percentage points higher in Freetown than in Carson City. What percentage of the registered voters in Freetown are Republicans?

SOLUTION

a. We interpret the "25%" as a relative difference, and 25% of 40% is 10% (because $0.25 \times 0.40 = 0.10$). Thus, the percentage of registered Republicans in Freetown is $40\% + 10\% = 50\%$.

b. In this case, we interpret the "25 percentage points" as an absolute difference, so we simply add this value to the percentage of Republicans in Carson City. Thus, the percentage of registered Republicans in Freetown is $40\% + 25\% = 65\%$. ∎

If you can't convince them, confuse them.

— Harry S Truman

Review Questions

1. Briefly describe how percentages represent fractions and how to convert between percentages and common fractions and decimals.

2. What do we mean by a *relative change*? Give an example that illustrates how to calculate a relative change.

3. What do we mean by a *relative difference*? Give an example that illustrates how to calculate a relative difference.

4. Explain the difference between the key words *of* and *more than* when dealing with percentages. How are their meanings related?

5. Explain the difference between the terms *percentage* and *percentage points*.

Exercises

BASIC SKILLS AND CONCEPTS

SENSIBLE STATEMENTS? For Exercises 1–6, determine whether the given statement is sensible and explain why it is or is not.

1. The percentage of households with more than four children decreased by 100,000 households.

2. Brent makes 120% less than Bill each month.

3. Ann is 10% taller than Brenda, so Brenda is 10% shorter than Ann.

4. Fifty percent of the people in the room are men and 50% of the people in the room are single. Therefore, 25% of the people in the room are single men.

5. Pete's prices are 10% more than Paul's prices, so Pete's prices are 110% of Paul's prices.

6. The interest rate at the bank increased by 100%.

7. **FRACTIONS, DECIMALS, PERCENTAGES.** Express the following numbers in all three forms: fraction, decimal, and percentage.
   a. 1/4   c. 0.33333 . . .
   b. 0.45   d. 23%

8. **FRACTIONS, DECIMALS, PERCENTAGES.** Express the following numbers in all three forms: fraction, decimal, and percentage.
   a. 120%   c. 2.34
   b. 0.652 d. −0.98

9. **PERCENTAGE PRACTICE.** Suppose there are 250 women at a convention, 120 of whom are Democrats and 130 of whom are Republicans. There are also 200 men at the convention, 80 of whom are Democrats and 120 of whom are Republicans. What percentage of the people at the convention are
   a. Republicans?   d. female Republicans?
   b. women?   e. male Democrats?
   c. Democrats?

10. **PERCENTAGE PRACTICE.** Suppose there are 75 people on a tour bus. Of the 30 men, 23 are American and 7 are British. Of the 45 women, 22 are American and 23 are British. What percentage of the people on the bus are
    a. American?   d. British women?
    b. men?   e. American men?
    c. British?

PERCENTAGE CHANGE. Exercises 11–16 describe statistical measurements made at two different times. For each pair of measurements, use a percentage to express the relative change.

11. The number of daily newspapers in the United States was 2,226 in 1900 and 1,483 in 1999.

12. The per capita annual consumption of beer in the United States increased from 0.73 gallon in 1940 to 1.24 gallons in 1997.

13. The United States produced 8.0 million cars in 1950 and 12.0 million cars in 1998.

14. Japan produced 5.3 million cars in 1970 and 10.0 million cars in 1998.

15. The winning time in the Indianapolis 500 (a 500-mile automobile race) decreased from 3 hours 36 minutes in 1961 (winner A. J. Foyt) to 2 hours 59 minutes in 2000 (winner Juan Montoya).

16. The total number of libraries in the United States was 28,638 in 1980 and 32,852 in 1999.

PERCENTAGE DIFFERENCE. Exercises 17–23 each state two measurements. For each pair of measurements, use a percentage to express their relative difference. Be sure to be clear about which measurement you use as the reference value and which you use as the compared value.

17. The daily circulation of the *Wall Street Journal* is about 1.75 million (the largest in the country). The daily circulation of the *New York Times* is about 1.09 million (the third largest in the country).

18. In a recent year, Ford sold 371,074 Taurus cars and Honda sold 401,071 Accord cars.

19. As the busiest airport in the world, Atlanta's Hartsfield Airport handled 78 million passengers in a single year. London's Heathrow Airport (fourth busiest) handled 62 million passengers.

20. In a recent year, France ranked as the number one tourist destination with about 71 million international arrivals. The United States ranked third with about 47 million international arrivals.

21. In a recent year, Saudi Arabia produced 8.4 million barrels of crude oil per day, while the United States produced 6.2 million barrels per day.

22. In a recent year, 499,000 males had open-heart surgery in the United States, while 261,000 females had the operation.

23. In a recent year, 99,000 males had total knee replacements in the United States, while 167,000 females had the operation.

FURTHER APPLICATIONS

OF VS. *MORE THAN.* Fill in the blanks in Exercises 24–27. Briefly explain your reasoning in each case.

24. If Jack weighs 40% more than Jill, then Jack's weight is _____% of Jill's weight.

25. If the area of Norway is 24% more than the area of Colorado, then Norway's area is _____% of Colorado's area.

26. If the population of Montana is 20% less than the population of New Hampshire, then Montana's population is _____% of New Hampshire's population.

27. If Henry earns 45% less than Ingrid, then Henry's salary is _____% of Ingrid's salary.

28. MARGIN OF ERROR. A survey found that the percentage of Americans who own answering machines is about 63%. The margin of error was 3 percentage points (International Mass Retail Association, reported in *USA Today*). Would it matter if a newspaper reported the margin of error as "3%"? Explain.

29. MARGIN OF ERROR. Based on a survey of local 10-year-olds, the local newspaper concludes that the percentage of all local 10-year-olds who collect Harry Potter memorabilia is probably between 28% and 40%. How should the newspaper report the margin of error? Explain.

PERCENTAGES OF PERCENTAGES. Exercises 30–33 describe changes in which the measurements themselves are percentages. Express each change in two ways: (1) as an absolute difference in terms of percentage points and (2) as a relative difference in terms of percent.

30. The percentage of high school seniors using alcohol decreased from 68.2% in 1975 to 52.7% now.

31. The percentage of the world's population living in developed countries decreased from 27.1% in 1970 to 19.5% now.

32. The five-year survival rate for Caucasians for all forms of cancer increased from 39% in the 1960s to 61% now.

33. The five-year survival rate for Blacks for all forms of cancer increased from 27% in the 1960s to 48% now.

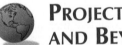

PROJECTS FOR THE WEB AND BEYOND

For useful links, select "Links for Web Projects" for Chapter 2 at www.aw.com/bbt.

34. WORLD POPULATION. Find the current estimate of world population on the U.S. Census Bureau's world population clock. Describe the percentage change in population since the 6 billion mark was passed during 1999. Also find how the population clock estimates are made, and discuss the uncertainties in estimating world population.

35. DRUG USE STATISTICS. Go to the Web site for the National Center on Addiction and Substance Abuse (CASA) and find a recent report giving statistics on substance abuse. Write a one-page summary of the new research, giving at least some of the conclusions in terms of percentages.

36. DEMOGRAPHICS. Consider the following quote (*New York Times*, August 5, 2001): "Hispanics are now the largest minority group in the United States, and 34% of them live in California, 19% in Texas, 10% in New York, and 7% in Florida. Of the Asian population in America, nearly half live in three metropolitan areas: San Francisco, Los Angeles, and New York. And 75% of the black population is in the Southeast and Mississippi Delta." After conducting research, write an essay on how the above figures will change by the year 2020.

IN THE NEWS

1. **PERCENTAGES.** Find three recent news reports in which percentages are used to describe statistical results. In each case, describe the meaning of the percentage.

2. **PERCENTAGE CHANGE.** Find a recent news report in which percentages are used to express the change in a statistical result from one time to another (such as an increase in population or in the number of children who smoke). Describe the meaning of the change. Be sure to watch for key words such as *of* or *more than*.

3. **QUOTE INTERPRETATION.** Consider the quote from the beginning of this section: "The rate [of smoking] among 10th graders jumped 45 percent, to 18.3 percent, and the rate for 8th graders is up 44 percent, to 10.4 percent." Briefly explain the meaning of each of the percentages in this statement.

2.4 Index Numbers

If you listen to the nightly business report, you've probably heard talk about **index numbers**, such as the Consumer Price Index, the Producer Price Index, or the Consumer Confidence Index. Index numbers are very common in statistics because they provide a simple way to compare measurements made at different times or in different places. In this section, we'll investigate the meaning and use of index numbers. Let's start with an example using gasoline prices.

Table 2.1 shows the average price of gasoline in the United States for selected years from 1955 to 2000. (These are real prices from those years; that is, they have *not* been adjusted for inflation.) Suppose that, instead of the prices themselves, we want to know how the price of gasoline in different years compares to the 1975 price. One way to compare the prices would be to express each year's price as a percentage of the 1975 price. For example, by dividing the 1965 price by the 1975 price, we find that the 1965 price was 55.0% of the 1975 price:

$$\frac{1965 \text{ price}}{1975 \text{ price}} = \frac{31.2\cent}{56.7\cent} = 0.550 = 55.0\%$$

Table 2.1 Average Gasoline Prices (per gallon)

Year	Price	Price as a percentage of 1975 price	Price index (1975 = 100)
1955	29.1¢	51.3%	51.3
1965	31.2¢	55.0%	55.0
1975	56.7¢	100.0%	100.0
1985	119.6¢	210.9%	210.9
1995	120.5¢	212.5%	212.5
2000	155.0¢	273.4%	273.4

SOURCE: U.S. Department of Energy.

Proceeding similarly for each of the other years, we can calculate all the prices as percentages of the 1975 price. The results are shown in the third column of Table 2.1. Note that the percentage for 1975 is 100%, because we chose the 1975 price as the reference value.

Now, look at the last column of Table 2.1. It is identical to the third column, except that we dropped the % signs. This simple change converts the numbers from percentages to a *price index*, which is one type of index number. The fact that the 1975 price is our reference value is now shown in the column heading by the statement "1975 = 100." In this case, there's really no difference between stating the comparisons as percentages and as index numbers—it's a matter of choice and convenience. However, as we'll see shortly, it's traditional to use index numbers rather than percentages in cases where many factors are being considered simultaneously.

Index Numbers

An **index number** provides a simple way to compare measurements made at different times or in different places. The value at one particular time (or place) must be chosen as the *reference value* (or *base value*). The index number for any other time (or place) is

$$\text{index number} = \frac{\text{value}}{\text{reference value}} \times 100$$

EXAMPLE 1   Finding an Index Number

Suppose the cost of gasoline today is $1.60 per gallon. Using the 1975 price as the reference value, find the price index number for gasoline today.

SOLUTION   Table 2.1 shows that the price of gas was 56.7¢, or $0.567, per gallon in 1975. If we use the 1975 price as the reference value and the price today is $1.60, the index number for gasoline today is

$$\text{index number} = \frac{\$1.60}{\$0.567} \times 100 = 282.2$$

This index number tells us that the current gasoline price is 282.2% of the 1975 price.   ∎

Time out to think

Find the actual price of gasoline today at a nearby gas station. What is the gasoline price index for today's price, with the 1975 price as the reference value?

Making Comparisons with Index Numbers

The primary purpose of index numbers is to facilitate comparisons. For example, suppose we want to know how much more expensive gas was in 2000 than in 1975. We can get the answer easily from Table 2.1, which uses the 1975 price as the reference value. This table shows that the price index for 2000 was 273.4, which means that the price of gasoline in 2000 was 273.4% of the 1975 price. Equivalently, we can say that the 2000 price was 2.734 times the 1975 price.

Comparisons are also possible when neither value is the reference value. For example, suppose we want to know how much more expensive gas was in 1995 than in 1965. We find the answer by dividing the index numbers for the two years:

$$\frac{\text{index number for 1995}}{\text{index number for 1965}} = \frac{212.5}{55.0} = 3.86$$

The 1995 price was 3.86 times the 1965 price, or 386% of the 1965 price.

EXAMPLE 2   Using the Gas Price Index

Use Table 2.1 to answer the following questions:

a. Suppose that it would have cost $7 to fill your gas tank in 1975. How much would it have cost to fill the same tank in 1995?

b. Suppose that it cost $20 to fill your gas tank in 2000. How much would it have cost to fill the same tank in 1955?

SOLUTION

a. Table 2.1 shows that the price index (1975 = 100) for 1995 was 212.5, which means that the price of gasoline in 1995 was 212.5% of the 1975 price. Thus, the 1995 price of gas that cost $7 in 1975 would have been

$$212.5\% \times \$7 = 2.125 \times \$7 = \$14.875$$

b. Table 2.1 shows that the price index (1975 = 100) for 2000 was 273.4 and the index for 1955 was 51.3. Thus, the cost of gasoline in 1955 compared to the cost in 2000 is

$$\frac{\text{index number for 1955}}{\text{index number for 2000}} = \frac{51.3}{273.4} = 0.1876$$

Gas that cost $20 in 2000 would have cost 0.1876 × $20 = $3.75 in 1955.   ■

The Consumer Price Index

We've seen that the price of gas has risen substantially with time. Most other prices and wages have also risen, a phenomenon called **inflation**. (During the rarer periods when prices or wages decline with time, we call it *deflation*.) Thus, changes in the actual gasoline price are not very meaningful unless we compare them to the overall rate of inflation. The purpose of the **Consumer Price Index (CPI)** is to measure the overall rate of inflation.

The Consumer Price Index is computed and reported monthly by the U.S. Bureau of Labor Statistics. It represents an average of prices in a sample of goods, services, and housing. The monthly sample consists of data from more than 60,000 items. The details of the data collection and index calculation are fairly complex, but the CPI itself is a simple index number. Table 2.2 on page 76 shows the average annual CPI over a 30-year period. Currently, the reference value for the CPI is an average of prices during the period 1982–1984, which is why the table says "1982−1984 = 100."

> **TECHNICAL NOTE**
>
> The government measures two consumer price indices: The CPI-U is based on products thought to reflect the purchasing habits of all urban consumers, whereas the CPI-W is based on the purchasing habits of only wage earners. (Table 2.2 shows the CPI-U.)

The Consumer Price Index

The **Consumer Price Index (CPI),** which is computed and reported monthly, is based on prices in a sample of more than 60,000 goods, services, and housing costs.

Table 2.2 Average Annual Consumer Price Index (1982–1984 = 100)					
Year	CPI	Year	CPI	Year	CPI
1972	41.8	1982	96.5	1992	140.3
1973	44.4	1983	99.6	1993	144.5
1974	49.3	1984	103.9	1994	148.2
1975	53.8	1985	107.6	1995	152.4
1976	56.9	1986	109.6	1996	156.9
1977	60.6	1987	113.6	1997	160.5
1978	65.2	1988	118.3	1998	163.0
1979	72.6	1989	124.0	1999	166.6
1980	82.4	1990	130.7	2000	172.2
1981	90.9	1991	136.2	2001	177.1

By the Way…

Scientists at NASA's Ames Research Center in California ran a speed test in which they pitted an Apple Macintosh G4 computer purchased in 2000 against a Cray supercomputer purchased in 1985. The two machines essentially tied in speed; but at $30 million, the Cray cost 10,000 times as much as the $3,000 Macintosh. That is, the price for equivalent computing power fell by a factor of 10,000 in a 15-year period.

By the Way…

Salaries of professional athletes were once kept low because the players were not allowed to offer their skills in the free market ("free agency"). That changed after star baseball player Curt Flood filed suit against Major League Baseball in 1970. Flood ultimately lost when the Supreme Court ruled in favor of baseball in 1972, but the process he set in motion (toward free agency) was unstoppable.

The CPI allows us to compare overall prices at different times. As an example, let's compare the CPI for 2000 to that for 1985:

$$\frac{\text{CPI for 2000}}{\text{CPI for 1985}} = \frac{172.2}{107.6} = 1.600$$

This tells us that overall prices in 2000 were 1.6 times overall prices in 1985. In other words, products that cost $1,000 in 1985 would, on average, have cost $1,600 in 2000. Of course, individual items may have price changes that are different from the average. For example, computer prices for equivalent computing power actually *fell* from 1985 to 2000. In contrast, the average price of health insurance more than doubled during the same period, so we say that health insurance prices rose faster than the overall rate of inflation.

EXAMPLE 3 Baseball Salaries

In 1987, the mean salary for major league baseball players was $412,000. In 2000, it was $1,988,000. Compare the increase in mean baseball salaries to the overall rate of inflation as measured by the Consumer Price Index.

SOLUTION First, we compare the Consumer Price Indices for 1987 and 2000:

$$\frac{\text{CPI for 2000}}{\text{CPI for 1987}} = \frac{172.2}{113.6} = 1.516$$

That is, average prices in 2000 were about 1.5 times as much as average prices in 1987. During the same period, the mean baseball salary more than quadrupled. Thus, the salaries of major league baseball players rose much more than the overall rate of inflation. ∎

Other Index Numbers

The Consumer Price Index is only one of many index numbers that you'll see in news reports. Some are also price indices, such as the Producer Price Index (PPI), which measures the prices that producers (manufacturers) pay for the goods they purchase (rather than the prices that consumers pay). Other indices attempt to measure more qualitative variables. For example, the Consumer Confidence Index is based on a survey designed to measure consumer attitudes so that businesses can gauge whether people are likely to be spending or saving. New indices are created frequently by groups attempting to provide simple comparisons.

EXAMPLE 4 Health Care Quality Index

Statisticians at Health Risk Management, Inc. invented the Health Care Quality Index to compare the quality of health care in different states. The index represents an average of 46 different measures of health care quality, such as spending on public health, the fraction of children receiving immunizations, and the quality of public drinking water. They chose the average score among all the states as the reference value; thus, the average score is defined to be 100 on this index. Minnesota had the highest index value of 118, while Louisiana had the lowest value at 83. Discuss the meaning of these index values.

SOLUTION   We can start by comparing the Health Care Quality Indices for Minnesota and Louisiana:

$$\frac{\text{Health Care Quality Index Minnesota}}{\text{Health Care Quality Index Louisiana}} = \frac{118}{83} \approx 1.4$$

The Health Care Quality Index for Minnesota is 1.4 times, or 140% of, the index for Louisiana. Remembering how the word *of* relates to the words *more than*, we can also say that the index for Minnesota is 40% more than that for Louisiana. Of course, not everyone will agree that the 46 measures used in this index are the best set of measures on which to rate health care. Thus, while the index provides a useful starting point for discussions of health care quality, it is by no means definitive. Without much more discussion, it is unfair to conclude that health care is 40% better in Minnesota than in Louisiana. ∎

> **By the Way...**
>
> Thinking of becoming a comedian? Then you'll probably want to check the Cost of Laughing Index (there really is one!), which tracks costs of items such as rubber chickens, Groucho Marx glasses, and admission to comedy clubs.

Review Questions

1. What is an index number? What are index numbers used for?

2. Briefly describe how index numbers are calculated and what they mean.

3. Briefly describe how to compare values from two different times or places using index numbers.

4. What is the Consumer Price Index (CPI)? How is it supposed to be related to inflation?

Exercises

BASIC SKILLS AND CONCEPTS

SENSIBLE STATEMENTS? For Exercises 1–5, determine whether the given statement is sensible and explain why it is or is not.

1. The Gas Price Index in 1998 was $1.41 per gallon.

2. If prices increase, then the Consumer Price Index increases.

3. A higher Health Care Quality Index means better health care services on average.

4. If the Consumer Price Index increases, then wages increase, at least on average.

5. If the Consumer Price Index were to double between this year and next year, it would mean that, on average, prices had roughly doubled in a year.

6. **GASOLINE PRICE INDEX.** Suppose the cost of gasoline today is $1.80 per gallon. What is the price index number for gasoline today, with the 1975 price as the reference value? (See the data in Table 2.1.)

7. **GASOLINE PRICE INDEX.** Suppose the cost of gasoline today is $1.45 per gallon. What is the gasoline price index for today's price, with the 1975 price as the reference value? (See the data in Table 2.1.)

8. **RECONSTRUCTING THE GASOLINE PRICE INDEX.** Reconstruct the gasoline price index in Table 2.1 using the price from 1965 as the reference value. (Hint: Create a column for price as a percentage of 1965 price and another column giving the price index with 1965 = 100.)

9. **RECONSTRUCTING THE GASOLINE PRICE INDEX.** Reconstruct the gasoline price index in Table 2.1 using the price from 1985 as the reference value. (Hint: Create a column for price as a percentage of 1985 price and another column giving the price index with 1985 = 100.)

10. **USING THE GAS PRICE INDEX.** Suppose that it would have cost $10 to fill your gas tank in 1975. How much would it have cost to fill the same tank in 1995?

11. **USING THE GAS PRICE INDEX.** Suppose that it would have cost $15 to fill your gas tank in 1985. How much would it have cost to fill the same tank in 1965?

12. **HEALTH CARE SPENDING.** Total spending on health care in the United States rose from $73.2 billion in 1970 to $1,420 billion ($1.42 trillion) in 2001. Compare this rise in health care spending to the overall rate of inflation as measured by the Consumer Price Index.

FURTHER APPLICATIONS

13. **PRIVATE COLLEGE COSTS.** The average annual cost (tuition, fees, and room and board) at four-year private universities rose from $5,900 in 1980 to $23,650 in 2000. Calculate the percentage rise in cost from 1980 to 2000, and compare it to the overall rate of inflation as measured by the Consumer Price Index.

14. **PUBLIC COLLEGE COSTS.** The average annual cost (tuition, fees, and room and board) at four-year public universities rose from $2,490 in 1980 to $10,910 in 2000. Calculate the percentage rise in cost from 1980 to 2000, and compare it to the overall rate of inflation as measured by the Consumer Price Index.

15. **HOME PRICES—SOUTH.** The typical price of a new single-family home in the South (United States) rose from $99,000 in 1990 to $148,000 in 2000. Calculate the percentage rise in cost of a home from 1990 to 2000, and compare it to the overall rate of inflation as measured by the Consumer Price Index.

16. **HOME PRICES—WEST.** The typical price of a new single-family home in the West (United States) rose from $147,500 in 1990 to $196,400 in 2000. Calculate the percentage rise in cost of a home from 1990 to 2000, and compare it to the overall rate of inflation as measured by the Consumer Price Index.

HOUSING PRICE INDEX. Realtors use an index to compare housing prices in major cities throughout the country. The index numbers for several cities are given in the table below. If you know the price of a particular home in your town, you can use the index to find the price of a comparable house in another town:

$$\frac{\text{price}}{\text{(other town)}} = \frac{\text{price}}{\text{(your town)}} \times \frac{\text{index in other town}}{\text{index in your town}}$$

Use the housing price index in Exercises 17–19.

City	Index	City	Index
Juneau, AK	100	Manhattan, NY	495
Palo Alto, CA	365	Tulsa, OK	52
Denver, CO	87	Providence, RI	91
Sioux City, IA	47	Spokane, WA	78
Boston, MA	182	Cheyenne, WY	75

17. Suppose you see a house valued at $250,000 in Denver. Find the price of a comparable house in Palo Alto, Sioux City, and Boston.

18. Suppose you see a house valued at $380,000 in Boston. Find the price of a comparable house in Juneau, Manhattan, and Tulsa.

19. Suppose you see a house valued at $250,000 in Cheyenne. Find the price of a comparable house in Spokane, Denver, and Juneau.

 ## PROJECTS FOR THE WEB AND BEYOND

For useful links, select "Links for Web Projects" for Chapter 2 at www.aw.com/bbt.

20. **CONSUMER PRICE INDEX.** Go to the Consumer Price Index home page and find the latest news release with updated figures for the CPI. Summarize the news release and any important ongoing trends in the CPI.

21. **PRODUCER PRICE INDEX.** Go to the Producer Price Index (PPI) home page. Read the overview and recent news releases. Write a short summary describing the purpose of the PPI and how it is different from the CPI. Also summarize any important recent trends in the PPI.

22. **CONSUMER CONFIDENCE INDEX.** Use a search engine to find recent news about the Consumer Confidence Index. After studying the news, write a short summary of what the Consumer Confidence Index is trying to measure and describe any recent trends in the Consumer Confidence Index.

23. **PROJECT: HUMAN DEVELOPMENT INDEX.** The United Nations Development Programme regularly releases its Human Development Report. A closely watched finding of this report is the Human Development Index (HDI), which measures the overall achievements in a country in three basic dimensions of human development: life expectancy, educational attainment, and adjusted income. Find the most recent copy of this report and investigate exactly how the HDI is defined and computed.

24. **CONVENIENCE STORE INDEX.** Go to a local supermarket and find the prices of a few staples, such as bread, milk, juice, and coffee. Compute the total cost of those items. Then go to a few smaller convenience stores and find the prices of the same items. Using the supermarket total as the reference value, compute the index numbers for the convenience stores.

25. **GASOLINE PRICES.** Collect data, make a convincing graph, and write a persuasive argument to either defend or refute the statement that gasoline prices have *not* increased over the past 30 years relative to the cost of living.

IN THE NEWS

1. **CONSUMER PRICE INDEX.** Find a recent news report that includes a reference to the Consumer Price Index. Briefly describe how the Consumer Price Index is important in the story.

2. **INDEX NUMBERS.** Find a recent news report that includes an index number other than the Consumer Price Index. Describe the index number and its meaning, and discuss how the index is important in the story.

Chapter Review Exercises

1. Of the 2,223 people who were aboard the *Titanic*, 31.76% survived when the ship sank on Monday, April 15, 1912.

   a. Find the true number of people who survived when the *Titanic* sank.
   b. Is the number of survivors a value from a discrete data set or a continuous data set? Explain.
   c. Of the people who were aboard the *Titanic*, 531 were women or children. What percentage of *Titanic* passengers were women or children?
   d. There were 45 girls aboard the *Titanic*, and there were 42% more boys than girls. How many boys were aboard?
   e. If we compile the ages of all passengers aboard the *Titanic*, what is the level of measurement (nominal, ordinal, interval, ratio) of those ages?
   f. If we list the passengers according to the categories of men, women, boys, and girls, what is the level of measurement (nominal, ordinal, interval, ratio) of this data set?

2. The College of Newport conducts a study of graduates and, based on a sample of 250 graduates, reports that "graduates earn an average of $43,782.64 per year."

   a. Because the salary is reported to the nearest penny, would you say that it is very accurate or very precise?
   b. In what way is the reported value of $43,782.64 misleading?
   c. If college officials estimate that next year's income level will be 4% more than the current level, what is next year's level projected to be?
   d. If the reported income level is suspected of being erroneous because graduates with low incomes are less likely to respond, is the suspected error random or systematic? Explain.

   e. A subsequent graduate study reveals that the true annual income is $39,376.69. If the larger initial value is used as the reference value, what is the amount of the absolute error and what is the relative error?

3. The following table lists U.S. barley production (in millions of bushels) for different years (based on data from the U.S. Department of Agriculture).

Year	Barley
1990	422
1991	464
1992	455
1993	398
1994	375
1995	360
1996	396
1997	360
1998	352
1999	284
2000	317
2001	297

   a. Using the 1994 value as the reference value, find the index number for 1996.
   b. Using the 1994 value as the reference value, find the index number for 2001.
   c. When we compare the 2000 and 2001 barley production amounts, we see a decrease. What is the absolute change from 2000 to 2001? What is the relative change?
   d. The first column of the table lists years. What is the level of measurement (nominal, ordinal, interval, ratio) of such data?
   e. The second column of the table lists amounts of barley production. What is the level of measurement (nominal, ordinal, interval, ratio) of such data?

FOCUS ON POLITICS

Who Benefits from a Tax Cut?

Politicians have a remarkable capacity to cast numbers in whatever light best supports their beliefs. Consider the two charts shown in Figure 2.3. Both purport to show the effects of the *same* proposed tax cut, yet they appear to support radically different conclusions. (This particular tax cut did not become law.)

The chart in Figure 2.3a reflects numbers supplied by Congressional Republicans, who supported the tax cut. It suggests that the tax cut would benefit families of all incomes similarly in percentage terms, with slightly greater benefits going to middle-income families than to the poor or the rich. Figure 2.3b reflects numbers supplied by Democrats, who opposed the tax cut. This chart suggests that the benefits would go disproportionately to the rich. How could the Republicans and the Democrats make such remarkably different claims?

Both sides used the same data, but chose to show it very differently. The most important difference arose from the fact that the two sides calculated the tax cut "benefits" in different ways. The Republicans calculated the *average* tax cut that would be received by families in each group. For example, the last bar in Figure 2.3a shows that families with incomes over

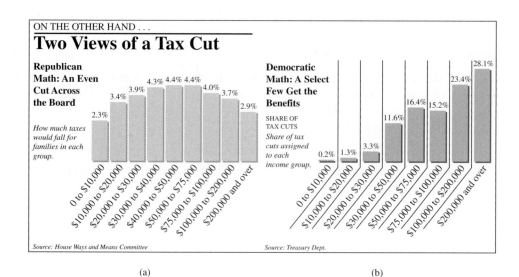

ON THE OTHER HAND . . .

Two Views of a Tax Cut

Republican Math: An Even Cut Across the Board

How much taxes would fall for families in each group.

2.3% 3.4% 3.9% 4.3% 4.4% 4.4% 4.0% 3.7% 2.9%

0 to $10,000 / $10,000 to $20,000 / $20,000 to $30,000 / $30,000 to $40,000 / $40,000 to $50,000 / $50,000 to $75,000 / $75,000 to $100,000 / $100,000 to $200,000 / $200,000 and over

Source: House Ways and Means Committee

Democratic Math: A Select Few Get the Benefits

SHARE OF TAX CUTS
Share of tax cuts assigned to each income group.

0.2% 1.3% 3.3% 11.6% 16.4% 15.2% 23.4% 28.1%

0 to $10,000 / $10,000 to $20,000 / $20,000 to $30,000 / $30,000 to $50,000 / $50,000 to $75,000 / $75,000 to $100,000 / $100,000 to $200,000 / $200,000 and over

Source: Treasury Dept.

(a)   (b)

Figure 2.3 SOURCE: *New York Times.*

Politicians and government officials usually abuse numbers and logic in the most elementary ways. They simply cook figures to suit their purpose, use obscure measures of economic performance, and indulge in horrendous examples of chart abuse, all in the name of disguising unpalatable truths.

—A. K. Dewdney,
200% of Nothing

$200,000 would get an average tax cut of 2.9%. Of course, a 2.9% tax cut means many more actual dollars to someone paying high taxes than to someone paying a little. For example, someone who pays $100,000 in taxes would save $2,900 with a 2.9% tax cut, while someone paying only $1,000 in taxes would save only $29.

The Democrats calculated the percentage of *total* benefits that would go to each group. For example, the last bar in Figure 2.3b shows that families with incomes over $200,000 would receive 28.1% of the total benefits from the tax cut. But this does *not* mean that these families are getting a 28.1% tax cut. To see why, consider the effects of an across-the-board tax cut of the same percentage for everyone. Because families with incomes above $200,000 pay more than one-fourth of the total income taxes collected by the U.S. government, these families would receive about one-fourth of the total benefit of *any* across-the-board cut. For example, if the government collected $1 trillion in taxes, a 10% across-the-board cut would mean savings of $100 billion for taxpayers as a whole. In that case, because one-fourth of the $1 trillion was paid by families with incomes over $200,000, one-fourth of the $100 billion in savings would go to these same families. Thus, these families would get 25% of the *benefits* of the tax cut, even though their actual tax cut would be only 10%.

By using percentages in different ways, each side was able to support its own position regarding the proposed tax cut. Neither side was fundamentally dishonest, but neither told the whole truth either.

QUESTIONS FOR DISCUSSION

1. The percentages in Figure 2.3a represent the *relative* tax savings to each income group. How do these relative savings compare to the *absolute* savings for each group? (Hint: In estimating the absolute tax savings, remember that the amount of tax that any family pays is some percentage of its income, and that lower-income families generally pay a smaller percentage of their income in taxes.)

2. A secondary reason for the differences in the two graphs is that the two sides defined *income* differently. For example, the Democrats chose to allocate corporate profits to the individual incomes of stockholders, which means that a person holding stock was judged to have a higher income by the Democrats than by the Republicans. How did this difference in defining income affect the two graphs?

3. Do you think that either of the charts in Figure 2.3 accurately portrays the overall "fairness" of the proposed tax cut? If so, which one and why? If not, how do you think the numbers could be portrayed more fairly?

4. Have any tax cuts or tax increases been proposed by the U.S. Congress or the President this year? Discuss the fairness of current proposals.

SUGGESTED READING

Branegan, Jay and Zagorin. "Nation: Who Needs a Tax Cut?" *Time*, August 2, 1999.

Mitchell, Alison. "Sometimes a Tax Cut Can Be a Hard Sell." *New York Times*, April 11, 1999.

"Two Views of a Tax Cut." *New York Times*, April 7, 1995.

FOCUS ON ECONOMICS

Does the Consumer Price Index Really Measure Inflation?

Suppose you earn $30,000 this year. How much will you need to earn to maintain the *same* standard of living next year? The answer to this question depends on inflation, which has traditionally been tracked by the Consumer Price Index (CPI). For example, Table 2.2 shows that from 1995 to 1996 the CPI rose by

$$\frac{\text{CPI}_{1996} - \text{CPI}_{1995}}{\text{CPI}_{1995}} \times 100\% = \frac{156.9 - 152.4}{152.4} \approx 3.0\%$$

If this change truly reflects the underlying rate of inflation, a person would have needed to earn 3.0% more money in 1996 than in 1995 to maintain the same standard of living.

However, many economists argue that the CPI overstates the true increase in the cost of living. The thrust of the argument is that the sampling methods used to measure the CPI create several forms of bias that tend to make the CPI overstate inflation. For example, in tracking the CPI, the Bureau of Labor Statistics compares monthly changes in prices of certain items at certain stores. Thus, when an item's price rises at one store, it tends to increase the CPI. However, consumers may not experience an increase in their cost of living if they can find the same item for a lower price at a different store or if they can substitute a similar but lower-priced item. Another bias occurs because the CPI tracks prices of items purchased by "typical" consumers. Thus, today's CPI includes items that were not available just a couple of decades ago, such as DVD players, cable television, and computers. The CPI does not factor in any improvement in the standard of living that these items represent.

An economic commission, the Boskin Commission, concluded that the CPI overstates the actual rate of inflation by between 0.8 and 1.6 percentage points per year. In that case, the actual inflation rate from 1995 to 1996 was between 1.4% and 2.2%, rather than the 3.0% suggested by the rise in the CPI.

These small percentages may not sound very important, but they have a tremendous impact on national finances. For example:

- Annual increases in individual Social Security benefits are supposed to match the rate of inflation so that recipients can maintain a constant standard of living. But benefits actually have *risen* in value if the CPI really does overstate inflation (Figure 2.4).

- The levels of income at which different income tax rates take effect (tax brackets) are supposed to rise with the rate of inflation. Otherwise, people with no change in their living standards would gradually move to higher tax brackets. But if inflation is overstated, tax rates have effectively come down.

If I had to populate an asylum with people certified insane, I'd just pick 'em from all those who claim to understand inflation.

—Will Rogers

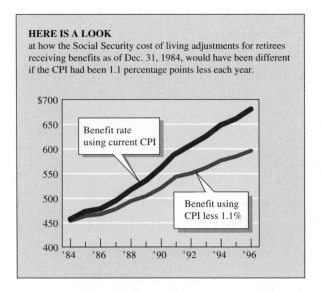

HERE IS A LOOK
at how the Social Security cost of living adjustments for retirees receiving benefits as of Dec. 31, 1984, would have been different if the CPI had been 1.1 percentage points less each year.

Benefit rate using current CPI

Benefit using CPI less 1.1%

Figure 2.4   SOURCE: *Washington Post.*

Because Social Security is a large fraction of the federal budget and income taxes are the federal government's primary source of revenue, these effects add up to huge amounts of money. The Boskin Commission recommended that the government adjust Social Security benefits and tax rates annually by an amount that is 1.1 percentage points lower than the change in the CPI. Incredibly, this adjustment could save the federal government over *$1 trillion* during the next decade.

The CPI also has important implications for economic data. For example, it has been widely reported that average wages have remained relatively stagnant (unchanged) over the past two to three decades. Such reports are based on the fact that wages have risen at approximately the same rate as the CPI. But if the CPI overstates inflation by 1% per year, then wages have actually risen faster than inflation by 1% per year—which adds up to a substantial gain in real wages over two or three decades. Clearly, knowing the truth about the CPI and inflation is very important to understanding economics. Unfortunately, the truth in this case remains a topic of hot debate.

QUESTIONS FOR DISCUSSION

1. Using the data in Table 2.2, calculate the total percentage increase in the CPI between 1990 and 2000. List a few things for which this percentage increase in price seems realistic, too high, and too low. Based on your lists, do you think the CPI accurately measures changes in the cost of living?

2. Suppose that you had the resources to sample the prices from 100,000 sources each month. Can you think of a way to choose the sample that would avoid the existing biases in the CPI?

3. Discuss some of the political ramifications of the Boskin Commission recommendations. For example, how would they affect individual Social Security recipients? How would they affect tax rates? If you were a member of Congress, would you vote to accept the Boskin Commission recommendations? Why or why not?

SUGGESTED READING

Boskin, Michael J. "The CPI Commission." *Business Economics*, Vol. 32, April 1, 1997, p. 60.

Boskin, Michael J., Ellen R. Dulberger, Robert J. Gordon, Zvi Griliches, and Dale W. Jorgenson. "Toward a More Accurate Measure of the Cost of Living." Final Report to the Senate Finance Committee, from the Advisory Commission to Study the Consumer Price Index, Washington, DC, Senate Finance Committee, 1996.

Edmondson, Brad. "Inflation Infatuation, or Is the Price Right?" *American Demographics*, Vol. 18, December 1, 1996, p. 4.

Stevenson, Richard W. "Economists Readjust Estimate of Overstatement of Inflation." *New York Times*, March 1, 2000.

CHAPTER 3

The greatest value of a picture is when it forces us to notice what we never expected to see.

— John Tukey

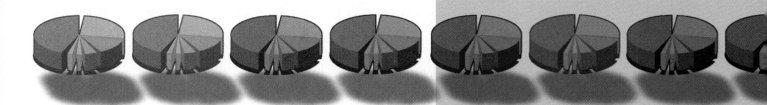

Visual Displays of Data

Whether you look at a newspaper, a corporate annual report, or a government study, you are almost sure to see tables and graphs of statistical data. Some of these tables and graphs are very simple; others can be quite complex. Some make it easy to understand the data; others may be confusing or even misleading. In this chapter, we'll study some of the many ways in which statistical data are commonly displayed in tables and graphs. Because the ability to convey concepts through graphs is so valuable in today's data-driven society, the skills developed in this chapter are crucial for success in nearly every profession.

LEARNING GOALS

1. Create and interpret frequency tables.

2. Create and interpret basic types of graphs, including bar graphs, dotplots, pie charts, histograms, stem-and-leaf plots, line charts, and time-series diagrams.

3. Interpret the many types of more complex graphics that are commonly found in news media.

4. Critically evaluate graphics and identify some of the most common ways in which graphics can be misleading.

3.1 Frequency Tables

Professor Delaney makes the following list of the grades she gave to her 25 students on a set of essays:

A C C B C D C C F D C C C B B A B D B A A B F C B

This list contains all of the grade data, but it is not very easy to read. A much easier way to display these data is by making a table in which we record the number of times, or **frequency,** that each grade appears. The result, shown in Table 3.1, is called a **frequency table.** The five possible grades (A, B, C, D, F) are called the **categories** (or classes) for the table.

Table 3.1 Frequency Table for a Set of Essay Grades

Grade	Frequency
A	4
B	7
C	9
D	3
F	2
Total	**25**

Definition

A basic **frequency table** has two columns:

- One column lists all the **categories** of data.
- The other column lists the **frequency** of each category, which is the number of data values in the category.

EXAMPLE 1   Taste Test

The Rocky Mountain Beverage Company wants feedback on its new product, Coral Cola, and sets up a taste test with 20 people. Each individual is asked to rate the taste of the cola on a 5-point scale:

(bad taste) 1 2 3 4 5 (excellent taste)

The 20 ratings are as follows:

1 3 3 2 3 3 4 3 2 4 2 3 5 3 4 5 3 4 3 1

Construct a frequency table for these data.

SOLUTION   The variable of interest is taste, and this variable can take on five values: the taste categories 1 through 5. (Note that the data are qualitative and at the ordinal level of measurement.) We make a table with these five categories in the left column and their frequencies in the right column, as shown in Table 3.2.

Table 3.2 Taste Test Ratings

Taste scale	Frequency
1	2
2	3
3	9
4	4
5	2
Total	**20**

Binning Data

Consider the data in Table 3.3, showing the average annual energy use per person, in millions of BTUs, in each of the 50 states. The 50 numbers in this data set (not counting the U.S. average) range from 215 million BTUs per person (New York) to 1,139 million BTUs per person (Alaska), and most of the numbers appear only once. How can we efficiently make a frequency table of these data?

The answer is to create categories that each span some range of data values. For example, we could create a category for all data values between 200 and 299 million BTUs, then create a second category for data values between 300 and 399 million BTUs, and so on. We then count the frequency (number of data values) in each category, generating the frequency table in Table 3.4. This process is called **binning** the data because each category acts like a separate bin into which we can pour some of the data values.

> ## By the Way...
>
> A BTU, or British thermal unit, is the energy required to raise the temperature of one pound of water by 1°F. It is equivalent to 252 calories or 1,055 joules.

Definition

When it is impossible or impractical to have a category for every value in a data set, we **bin** (or group) the data into categories (*bins*), each covering a range of the possible data values.

Table 3.3 Average Annual Energy Use per Person, by State, in Millions of BTUs

State	Millions of BTUs per person	State	Millions of BTUs per person	State	Millions of BTUs per person
Alabama	455	Louisiana	87	Ohio	363
Alaska	1,139	Maine	414	Oklahoma	415
Arizona	246	Maryland	260	Oregon	333
Arkansas	402	Massachusetts	246	Pennsylvania	322
California	240	Michigan	331	Rhode Island	237
Colorado	287	Minnesota	352	South Carolina	382
Connecticut	240	Mississippi	393	South Dakota	323
Delaware	368	Missouri	313	Tennessee	377
Florida	248	Montana	435	Texas	559
Georgia	349	Nebraska	354	Utah	326
Hawaii	216	Nevada	350	Vermont	256
Idaho	391	New Hampshire	248	Virginia	311
Illinois	323	New Jersey	320	Washington	396
Indiana	447	New Mexico	340	West Virginia	449
Iowa	375	New York	215	Wisconsin	342
Kansas	406	North Carolina	323	Wyoming	846
Kentucky	459	North Dakota	546	**U.S. Average**	**344**

NOTE: Data include all energy uses, including residential, commercial, industrial, and transportation.
SOURCE: *Statistical Abstract of the United States*

Table 3.4 Frequency Table for the Energy Use Data Listed in Table 3.3

Annual energy use per person (millions of BTUs)	Frequency (number of states)
200–299	12
300–399	24
400–499	9
500–599	2
600–699	0
700–799	0
800–899	2
900–999	0
1,000–1,099	0
1,100–1,199	1
Total	**50**

Table 3.5 Data for Stocks in the Dow Jones Industrial Average, 2000

Company	Revenue (billions)	Return	Rank	Company	Revenue (billions)	Return	Rank
Exxon Mobil Corp.	210.4	10.3%	1	Intel Corp.	33.7	−26.9%	41
Wal-Mart Stores Inc.	193.3	−22.8%	2	DuPont Co.	29.2	−24.4%	56
General Motors Corp.	184.6	−27.8%	3	Johnson & Johnson	29.1	14.3%	57
General Electric Co.	129.9	−6.1%	5	International Paper Co.	28.2	−25.5%	61
Citigroup Inc.	111.8	23.6%	6	United Technologies	26.6	22.6%	64
IBM	88.4	−20.8%	8	Walt Disney Co.	25.4	−0.3%	67
AT&T Corp.	66.0	N/A	9	Honeywell	25.0	−16.6%	71
Philip Morris Co.	63.3	106.4%	11	American Express Co.	23.7	−0.3%	74
J.P. Morgan & Co.	60.1	−10.1%	12	ALCOA	23.1	−18.0%	77
SBC Communications	51.5	0.1%	14	Microsoft Corp.	23.0	−62.9%	79
Boeing Co.	51.3	61.2%	15	Coca-Cola Co.	20.5	5.9%	93
Hewlett-Packard Co.	48.8	−31.7%	19	Caterpillar, Inc.	20.2	4.1%	99
Home Depot Inc.	45.7	−33.3%	23	3M	16.7	26.4%	118
Merck & Co.	40.4	41.7%	30	McDonald's Corp.	14.2	−15.1%	138
Procter & Gamble	40.0	−27.1%	31	Eastman Kodak Co.	14.0	−38.6%	141

NOTE: "Return" data reflect Total Return to Investors in 2000.
SOURCE: djindexes.com, Fortune.com.

By the Way...

The 30 stocks that make up the Dow list are chosen by the editors of the *Wall Street Journal*. On occasion, the stocks on the list are changed. For example, in October 1999, Chevron, Sears, Goodyear, and Union Carbide were removed from the Dow and replaced by Intel, Microsoft, Home Depot, and SBC Communications. The new stocks are thought to be more representative of today's overall economy than those that they replaced.

EXAMPLE 2 The Dow Stocks

For the 30 stocks of the Dow Jones Industrial Average, Table 3.5 shows the annual revenue (in billions of dollars), the one-year total return, and the rank on the Fortune 500 list of largest U.S. companies. Create a frequency table for the revenue. Discuss the pros and cons of the binning choices.

SOLUTION The revenue data range from $14.0 billion (Eastman Kodak) to $210.4 billion (Exxon Mobil). There are many possible ways to bin data for this range; here's one good way and the reasons for it:

- We create bins spanning a range from $0 to $220 billion. This covers the full range of the data, with a little extra room below the lowest data value and above the highest data value.

- We give each bin a width of $20 billion so that we can span the $0 to $220 billion range with eleven bins. Also, the width of $20 billion is a convenient number that helps make the table easy to read.

- Because the data values are given to the nearest tenth (of a billion dollars), we also define the bins to the nearest tenth so that they do not overlap. That is, bins go from $0 to $19.9 billion, from $20.0 to $39.9 billion, and so on.

With these choices, we get the frequency table shown in Table 3.6. This table works fairly well because it is easy to study. If it had more bins, it would be difficult to read. But note that more than a third of the companies have revenues in the first bin. Thus, if the table had fewer bins, even more companies would fall into the first bin, making it difficult to see the distribution of the data over the range of values. ∎

Time out to think

Consider three other possible ways of binning the data for the frequency table in Table 3.6: (1) 3 bins spanning the range $0 to $300 billion; (2) 10 bins spanning the range $0 to $220 billion; (3) 22 bins spanning the range $0 to $220 billion. Briefly discuss the pros and cons of each of these choices.

Table 3.6 Frequency Table for the Annual Revenue Data in Table 3.5

Annual revenue (billions of dollars)	Frequency (number of companies)
0–19.9	3
20–39.9	12
40–59.9	6
60–79.9	3
80–99.9	1
100–119.9	1
120–139.9	1
140–159.9	0
160–179.9	0
180–199.9	2
200–219.9	1
Total	**30**

Relative Frequency

Let's reconsider the essay grades listed in Table 3.1. We might want to know not only the number of students who received each grade, but also the fraction or percentage of students who received those grades. We call these fractions (or proportions or percentages) for each category the **relative frequencies**. For example, 4 of the 25 students received A grades, so the relative frequency of A grades is 4/25, or 0.16, or 16%. Table 3.7 repeats the data from Table 3.1, but this time with an added column for the relative frequency.

Table 3.7 Relative Frequency Table

Grade	Frequency	Relative frequency
A	4	4/25 = 0.16
B	7	7/25 = 0.28
C	9	9/25 = 0.36
D	3	3/25 = 0.12
F	2	2/25 = 0.08
Total	**25**	**1**

The sum of the relative frequencies must equal 1 (or 100%), because each individual relative frequency is a fraction of the total frequency. (Sometimes, rounding causes the total to be slightly different from 1.)

"Data! Data! Data!" he cried impatiently. "I can't make bricks without clay."

—Sherlock Holmes in Sir Arthur Conan Doyle's The Adventure of the Copper Beeches

Definition

The **relative frequency** of any category is the proportion or percentage of the data values that fall in that category:

$$\text{relative frequency} = \frac{\text{frequency in category}}{\text{total frequency}}$$

Cumulative Frequency

Look one more time at the essay grades in Table 3.1. What if we want to know how many students got a grade of C or better? We could, of course, add the frequencies of A, B, and C grades to find that 20 students got a C or better. Often, however, tables do this arithmetic for us by showing the **cumulative frequencies**, or the number of data values in a particular category *and all preceding* categories. Table 3.8 repeats the data from Table 3.1, but this time with an added column for the cumulative frequency.

Note that the cumulative frequency for the last category must always equal the total number of data values, which is the total frequency.

TECHNICAL NOTE

Most frequency tables start with the lowest category, but tables of grades commonly start with the highest category (A). Category order does not affect frequencies or relative frequencies, but *does* affect cumulative frequency. For example, if the table to the right had the categories in reverse order, the cumulative frequency for C would be the number of grades of C or worse (rather than C or better).

Table 3.8 Cumulative Frequency Table

Grade	Frequency	Cumulative frequency
A	4	4
B	7	7 + 4 = 11
C	9	9 + 7 + 4 = 20
D	3	3 + 9 + 7 + 4 = 23
F	2	2 + 3 + 9 + 7 + 4 = 25
Total	**25**	**25**

Definition

The **cumulative frequency** of any category is the number of data values in that category *and all preceding* categories.

It's important to keep in mind that cumulative frequencies make sense only for data categories that have a clear order. That is, we can use cumulative frequencies for data at the ordinal, interval, and ratio levels of measurement, but not for data at the nominal level of measurement.

EXAMPLE 3 More on the Taste Test

Using the taste test data from Example 1, create a frequency table with columns for the relative and cumulative frequencies. What percentage of the respondents gave the cola the highest rating? What percentage gave the cola one of the three lowest ratings?

SOLUTION We find the relative frequencies by dividing each category frequency by the total frequency of 20. We find the cumulative frequencies by adding the frequency in each category to the sum of the frequencies in all preceding categories. Table 3.9 shows the results. The relative frequency column shows that 0.10, or 10%, of the respondents gave the cola the highest rating. The cumulative frequency column shows that 14 out of 20 people, or 70%, gave the cola a rating of 3 or lower.

TECHNICAL NOTE

A cumulative frequency divided by the total frequency is called a *relative cumulative frequency*. For example, in Table 3.9 the relative cumulative frequency of 3 or lower is 14/20 = 0.70.

Table 3.9 Relative and Cumulative Frequencies

Taste scale	Frequency	Relative frequency	Cumulative frequency
1	2	2/20 = 0.10	2
2	3	3/20 = 0.15	3 + 2 = 5
3	9	9/20 = 0.45	9 + 3 + 2 = 14
4	4	4/20 = 0.20	4 + 9 + 3 + 2 = 18
5	2	2/20 = 0.10	2 + 4 + 9 + 3 + 2 = 20
Total	**20**	**1**	**20**

EXAMPLE 4   Energy Data

To the frequency table for energy data in Table 3.4, add columns for the relative and cumulative frequencies. Discuss any trends that seem particularly revealing or surprising.

SOLUTION   We find the relative frequencies by dividing each category frequency by the total frequency, which is 50 in this case. We find the cumulative frequencies by adding the frequency in each category to the sum of the frequencies in all preceding categories. Table 3.10 shows the results.

The table reveals many interesting facts about annual energy use per person in the different states. For example, nearly half the states (0.48, or 48%) fall into the single category of 300–399 million BTUs per person per year. Moreover, the cumulative frequency column shows that 45 of the 50 states fall into one of the first three categories, or energy use between 200 and 499 million BTUs per person per year. Two more states are only slightly above this range, falling into the category of 500–599 million BTUs. The three remaining states have annual energy use per person far above that of all the others. (Using terminology that we'll discuss in Chapter 4, we say that these three states are *outliers* because their values differ so much from those of other states.)

> ## By the Way...
>
> The portions of total energy going to residential, commercial, industrial, and transportation uses vary widely among the states. However, industrial and transportation uses are particularly high in the three states with the highest total energy use (Alaska, Louisiana, and Wyoming).

Table 3.10   Binned Energy Data

Annual energy use per person (millions of BTUs)	Frequency	Relative frequency	Cumulative frequency
200–299	12	0.24	12
300–399	24	0.48	36
400–499	9	0.18	45
500–599	2	0.04	47
600–699	0	0.00	47
700–799	0	0.00	47
800–899	2	0.04	49
900–999	0	0.00	49
1,000–1,099	0	0.00	49
1,100–1,199	1	0.02	50
Total	**50**	**1**	**50**

Time out to think

Be careful in interpreting Table 3.10 in Example 4. For example, the 0.24 relative frequency for 200–299 million BTUs means that 24% of the 50 states have per capita energy use in this range. Does this also mean that 24% of all Americans use between 200 and 299 million BTUs of energy each year? Why or why not?

Review Questions

1. What is a *frequency table*? Explain what we mean by the *categories* and *frequencies*. Make your own example of a simple frequency table.

2. What is the purpose of *binning*? Give an example of a data set for which we would want to use bins when making a frequency table. How should you choose the bins for your example? Why?

3. What do we mean by *relative frequency*? Why should the sum of the relative frequencies be 1 (or 100%)?

4. What do we mean by *cumulative frequency*? How is the cumulative frequency useful?

Exercises

BASIC SKILLS AND CONCEPTS

SENSIBLE STATEMENTS? For Exercises 1–6, determine whether the given statement is sensible and explain why it is or is not.

1. A friend tells you that her frequency table has two columns labeled *State* and *Median Income*.

2. A friend tells you that after binning her table with two columns labeled *State* and *Median Income*, she produced a frequency table.

3. The relative frequency of category A in a table is 1.22.

4. The cumulative frequency of category X in a table is 1.04.

5. For a given data set, as the width of the bins decreases, the number of bins increases.

6. The frequency of a category in a frequency table is greater than the cumulative frequency of that category.

7. FREQUENCY TABLE PRACTICE. Suppose Professor Diaz records the following final grades in one of her courses.

A A A A B B B B B B B C C
C C C C C C D D D F F

Make a frequency table for these grades. Include columns for relative frequency and cumulative frequency. Briefly explain the meaning of each column.

8. FREQUENCY TABLE PRACTICE. A guide book for New York City lists 5 five-star restaurants (the highest rating), 10 four-star restaurants, 20 three-star restaurants, 15 two-star restaurants, and 5 one-star restaurants. Make a frequency table for these ratings. Include columns for relative frequency and cumulative frequency. Briefly explain the meaning of each column.

MILLENNIUM CHAMPIONSHIPS. The last world track and field championships of the second millennium were held in Seville, Spain, from August 21 to 29, 1999. Exercises 9 and 10 give some of the results. In each case, do the following.

a. Make a frequency table for the data, using bins with a width of one foot. Include columns for relative frequency and cumulative frequency.

b. Make a frequency table for the data, using bins with a width of two feet. Include columns for relative frequency and cumulative frequency.

9. The following are the results for the women's long jump.

Entrant (country)	Distance (feet-inches)
Niurka Montalvo (Spain)	23-2
Fiona May (Italy)	22-9.25
Marion Jones (USA)	22-5
Lyudmila Galkina (Russia)	22-4.5
Joanne Wise (Great Britain)	22-1.75
Dawn Burrell (USA)	22-1.5
Susen Tiedtke (Germany)	21-11
Maurren Maggi (Brazil)	21-11
Nicole Boegman (Australia)	21-9
Erica Johansson (Sweden)	21-9
Olga Rublyova (Russia)	21-6.25
Shana Williams (USA)	21-4.75

10. The following are the results for the men's pole vault.

Entrant (country)	Height (feet-inches)
Maksim Tarasov (Russia)	19-9
Dmitriy Markov (Australia)	19-4.25
Aleksandr Averbukh (Israel)	19-0.25
Danny Ecker (Germany)	18-8.25
Nick Hysong (USA)	18-8.25
Tim Lobinger (Germany)	18-8.25
Igor Potapovich (Kazakhstan)	18-8.25
Michael Stolle (Germany)	18-8.25
Danny Krasnov (Israel)	18-0.5
Okkert Brits (South Africa)	18-0.5

11. WEIGHTS OF COKE. Construct a frequency table for the weights (in pounds) given below of 36 cans of regular Coke. Start the first bin at 0.7900 lb and use a bin width of 0.0050 lb. Discuss your findings.

0.8192	0.8150	0.8163	0.8211	0.8181	0.8247
0.8062	0.8128	0.8172	0.8110	0.8251	0.8264
0.7901	0.8244	0.8073	0.8079	0.8044	0.8170
0.8161	0.8194	0.8189	0.8194	0.8176	0.8284
0.8165	0.8143	0.8229	0.8150	0.8152	0.8244
0.8207	0.8152	0.8126	0.8295	0.8161	0.8192

12. **WEIGHTS OF DIET COKE.** Construct a frequency table for the weights (in pounds) given below of 36 cans of Diet Coke. Start the first bin at 0.7750 lb and use a bin width of 0.0050 lb. Discuss your findings.

0.7773	0.7758	0.7896	0.7868	0.7844	0.7861
0.7806	0.7830	0.7852	0.7879	0.7881	0.7826
0.7923	0.7852	0.7872	0.7813	0.7885	0.7760
0.7822	0.7874	0.7822	0.7839	0.7802	0.7892
0.7874	0.7907	0.7771	0.7870	0.7833	0.7822
0.7837	0.7910	0.7879	0.7923	0.7859	0.7811

13. **MISSING INFORMATION.** The following table shows grades for a term paper in an English class. The table is incomplete. Use the information given to fill in the missing entries and complete the table.

Category	Frequency	Relative frequency
A	?	?
B	?	18%
C	?	24%
D	11	?
F	6	?
Total	50	?

14. **MISSING INFORMATION.** The following table shows grades for performances in a drama class. The table is incomplete. Use the information given to fill in the missing entries and complete the table.

Category	Frequency	Cumulative frequency
A	?	1
B	6	?
C	7	?
D	?	23
F	?	25
Total	?	?

15. **DOW STOCK RETURNS.** Make a frequency table for the return data in Table 3.5. How do the returns for these companies compare to the average return on a bank account, which was about 3% for the same period?

16. **INTERPRETING FAMILY DATA.** Consider the following frequency table for the number of children in American families.
    a. According to the data, how many families are there in America?
    b. How many families have two or fewer children?
    c. What percent of American families have no children?
    d. What percent of American families have three or more children?

Number of children	Number of families (millions)
0	35.54
1	14.32
2	13.28
3	5.13
4 or more	1.97

17. **OSCAR-WINNING ACTORS.** The following data show the ages of 34 recent Academy Award–winning male actors at the time when they won their award. Make a frequency table for the data, using bins of 20–29, 30–39, and so on. Do actors appear to be more likely to win Oscars when they are younger, older, or neither? (See Example 3 in Section 3.2 for a comparison with the ages of Oscar-winning actresses.)

32	37	36	32	51	53	33	61	35
45	55	39	76	37	42	40	32	60
38	56	48	48	40	43	62	43	42
44	41	56	39	46	31	47		

18. **STATE PER CAPITA INCOME.** Create a frequency table for the state per capita income data in Appendix A. Use bins with a width of $4,000. Include a column for relative frequency. After you complete the table, study it carefully and write one or two paragraphs discussing the trends you find to be of greatest interest.

19. **STATE TAXES.** Create a frequency table for the state tax data in Appendix A. Use bins with a width of $500. Include a column for relative frequency. After you complete the table, study it carefully and write one or two paragraphs discussing the trends you find to be of greatest interest.

20. **STATE EDUCATION EXPENDITURES.** Create a frequency table for the state education expenditures (funding per pupil) data in Appendix A. Use bins with a width of $500. Include a column for relative frequency. After you complete the table, study it carefully and write one or two paragraphs discussing the trends you find to be of greatest interest.

21. **STATE TEACHER SALARIES.** Create a frequency table for the state teacher salary data in Appendix A. Use bins with a width of $4,000. Include a column for relative frequency. After you complete the table, study it carefully and write one or two paragraphs discussing the trends you find to be of greatest interest.

FURTHER APPLICATIONS

22. COMPUTER KEYBOARDS. The traditional keyboard configuration is called a *Qwerty* keyboard because of the positioning of the letters QWERTY on the top row of letters. Developed in 1872, the Qwerty configuration supposedly forced people to type slower so that the early typewriters would not jam. Developed in 1936, the Dvorak keyboard supposedly provides a more efficient arrangement by positioning the most used keys on the middle row (or "home" row), where they are more accessible.

A *Discover* magazine article suggested that you can measure the ease of typing by using this point rating system: Count each letter on the home row as 0, count each letter on the top row as 1, and count each letter on the bottom row as 2. For example, the word *statistics* would result in a rating of 7 on the Qwerty keyboard and 1 on the Dvorak keyboard, as shown below.

S T A T I S T I C S	
Qwerty keyboard 0 1 0 1 1 0 1 1 2 0	(sum = 7)
Dvorak keyboard 0 0 0 0 0 0 0 0 1 0	(sum = 1)

Using this rating system with each of the 52 words in the Preamble to the Constitution, we get the rating values below.

Qwerty Keyboard Word Ratings:

2	2	5	1	2	6	3	3	4	2
4	0	5	7	7	5	6	6	8	10
7	2	2	10	5	8	2	5	4	2
6	2	6	1	7	2	7	2	3	8
1	5	2	5	2	14	2	2	6	3
1	7								

Dvorak Keyboard Word Ratings:

2	0	3	1	0	0	0	0	2	0
4	0	3	4	0	3	3	1	3	5
4	2	0	5	1	4	0	3	5	0
2	0	4	1	5	0	4	0	1	3
0	1	0	3	0	1	2	0	0	0
1	4								

a. Create a frequency table for the Qwerty word ratings data. Use bins of 0–2, 3–5, 6–8, 9–11, and 12–14. Include a column for relative frequency.

b. Create a frequency table for the Dvorak word ratings data, using the same bins as in part a. Include a column for relative frequency.

c. Based on your results from parts a and b, which keyboard arrangement is easier for typing? Explain.

23. DOUBLE BINNING. The students in a statistics class conduct a transportation survey of students in their high school. Among other data, they record the age and mode of transportation between home and school for each student. The following table gives some of the data that were collected. For age: 1 = 14 years, 2 = 15 years, 3 = 16 years, 4 = 17 years, 5 = 18 years. For transportation: 1 = walk, 2 = school bus, 3 = public bus, 4 = drive, 5 = other.

Student	Age	Transportation	Student	Age	Transportation
1	1	1	11	3	5
2	5	1	12	5	5
3	2	2	13	1	2
4	3	5	14	5	5
5	4	3	15	5	5
6	1	1	16	4	4
7	5	2	17	2	2
8	2	1	18	3	1
9	3	4	19	3	3
10	1	3	20	1	4

a. Classify the two variables, age and transportation, as qualitative or quantitative, and give the level of measurement for each.

b. In order to be analyzed or displayed, the data must be binned with respect to both variables. Count the number of students in each of the 25 age/transportation categories and fill in the blank cells in the following table.

		Transportation				
		1	2	3	4	5
Age	1	2	1	1	1	
	2	1	2			
	3	1		1	1	2
	4			1	1	
	5	1	1			3

PROJECTS FOR THE WEB AND BEYOND

For useful links, select "Links for Web Projects" for Chapter 3 at www.aw.com/bbt.

24. **ENERGY TABLE.** The U.S. Energy Information Administration (EIA) Web site offers dozens of tables relating to energy use, energy prices, and pollution. Explore the selection of tables. Find a table of raw data (that is, a table similar to Table 3.3) that is of interest to you and convert it to an appropriate frequency table. Briefly discuss what you can learn from the frequency table that is less obvious in the raw data table.

25. **ENDANGERED SPECIES.** The Web site for the World Conservation Monitoring Centre in Great Britain provides data on extinct, endangered, and threatened animal species. Explore these data and summarize some of your more interesting findings with frequency tables.

26. **NAVEL DATA.** The *navel ratio* is defined to be a person's height divided by the height (from the floor) of his or her navel. An old theory says that, on average, the navel ratio of humans is the golden ratio: $(1 + \sqrt{5})/2$. Measure the navel ratio of each person in your class. What percentage of students have a navel ratio within 5% of the golden ratio? What percentage of students have a navel ratio within 10% of the golden ratio? Does the old theory seem reliable?

27. **YOUR OWN FREQUENCY TABLE (UNBINNED).** Collect your own frequency data for some set of categories that will *not* require binning. (For example, you might collect data by asking friends to do a taste test on some brand of cookie.) State how you collected your data, and make a list of all your raw data. Then summarize the data in a frequency table. Include a column for relative frequency, and also include a column for cumulative frequency if it is appropriate. (Cumulative frequency is appropriate for all data except those at the nominal level of measurement.)

28. **YOUR OWN FREQUENCY TABLE (BINNED).** Collect your own frequency data for some set of categories that *will* require binning (for example, weights of your friends or scores on a recent exam). State how you collected your data, and make a list of all your raw data. Then summarize the data in a frequency table. Include columns for relative frequency and cumulative frequency.

IN THE NEWS

1. **FREQUENCY TABLES.** Find a recent news article that includes some type of frequency table. Briefly describe the table and how it is useful to the news report. Do you think the table was constructed in the best possible way for the article? If so, why? If not, what would you have done differently?

2. **RELATIVE FREQUENCIES.** Find a recent news article that gives at least some data in the form of relative frequencies. Briefly describe the data, and discuss why relative frequencies were useful in this case.

3. **CUMULATIVE FREQUENCIES.** Find a recent news article that gives at least some data in the form of cumulative frequencies. Briefly describe the data, and discuss why cumulative frequencies were useful in this case.

4. **TEMPERATURE DATA.** Look in a newspaper for a weather report that lists the expected high temperatures in many American cities. (Make a photocopy of the data so that your instructor can see them.) Choosing appropriate bins, make a frequency table for the high temperature data. Include columns for relative frequency and cumulative frequency. Briefly describe how and why you chose your bins.

3.2 Picturing Distributions of Data

A frequency table shows us how a variable is distributed over chosen categories. We say that a frequency table summarizes the **distribution** of data. While tables can be extraordinarily useful, we often gain deeper insight into a distribution by seeing it displayed in a picture or graph. In this section, we'll study some of the most common methods for pictorially displaying distributions of data.

Definition

The **distribution** of a variable refers to the way its values are spread over all possible values. We can summarize a distribution in a table or show a distribution visually with a graph.

Bar Graphs, Dotplots, and Pareto Charts

A **bar graph** is one of the simplest ways to picture a distribution. Bar graphs generally are used for qualitative data. Each bar represents the frequency (or relative frequency) for a particular category: the higher the frequency, the longer the bar. The bars can be either vertical or horizontal.

Let's create a vertical bar graph from the essay grade data in Table 3.1. We need five bars, one for each of the five categories (the grades A, B, C, D, F). The height of each bar should correspond to the frequency for its category. Figure 3.1 shows the result. Note the following key features of the graph's construction:

- Because the highest frequency is 9 (the frequency for C grades), we chose to make the vertical scale run from 0 to 10. This ensures that even the tallest bar does not quite touch the top of the graph.

- The graph should not be too short or too tall. In this case, it looks about right to choose a total height of 5 cm, which is convenient because it means that each 1 cm of height corresponds to a frequency of 2.

Table 3.1 (repeated)	
Grade	Frequency
A	4
B	7
C	9
D	3
F	2
Total	**25**

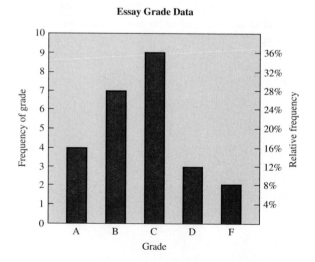

Figure 3.1 Bar graph for the essay grade data in Table 3.1.

- The height of each bar should be proportional to its frequency. For example, because each 1 cm of height corresponds to a frequency of 2, the bar representing a frequency of 4 should have a height of 2 cm.

- Because the data are qualitative, the widths of the bars have no special meaning and there is no reason for them to touch one another. We therefore draw them with uniform width.

Labeling graphs is extremely important. Without proper labels, a graph is meaningless. The following summary gives the important labels for almost any graph. Notice how these rules were applied in Figure 3.1.

Important Labels for Graphs

Title/caption: The graph should have a title or caption (or both) that explains what is being shown and, if applicable, lists the source of the data.

Vertical scale and title: Numbers along the vertical axis should clearly indicate the scale. The numbers should line up with the *tick marks*—the marks along the axis that precisely locate the numerical values. Include a label that describes the variable shown on the vertical axis.

Horizontal scale and title: The categories should be clearly indicated along the horizontal axis. (Tick marks may not be necessary for qualitative data, but should be included for quantitative data.) Include a label that describes the variable shown on the horizontal axis.

Legend: If multiple data sets are displayed on a single graph, include a legend or key to identify the individual data sets.

A **dotplot** is a variation on a bar graph in which we use dots rather than bars to represent the frequencies. Each dot represents one data value; for example, a stack of 4 dots means a frequency of 4. Figure 3.2 shows a dotplot for the essay data set. Dotplots are especially useful when making graphs of raw data, because you can tally the data by making a dot for each data value.

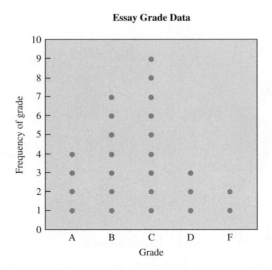

Figure 3.2 Dotplot for the essay grade data in Table 3.1.

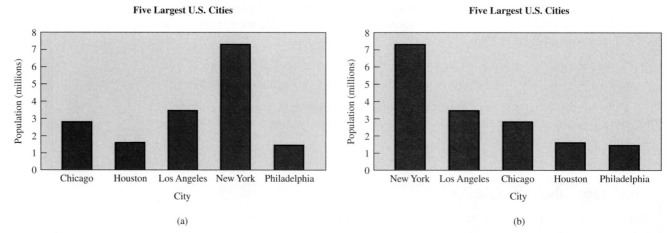

Figure 3.3 (a) Bar graph showing populations for the five largest cities in the United States. (b) Pareto chart for the same data. SOURCE: U.S. Census Bureau.

By the Way...

Pareto charts were invented by Italian economist Vilfredo Pareto (1848–1923). Pareto is best known for developing methods of analyzing income distributions, but his most important contributions probably were in developing new ways of applying mathematics and statistics to economic analysis.

Figure 3.3a shows a bar graph of populations for the five largest cities in the United States. In this case, the five cities are the bins and their populations are the frequencies; the cities are listed in alphabetical order. Figure 3.3b shows the same data, but with the bars arranged in descending order. A bar graph in which the bars are arranged in frequency order is often called a **Pareto chart**. Note that rearranging the order of the bars makes sense only if the data are at the nominal level of measurement, as they are when the categories are cities. For example, it would not make sense to make a Pareto chart for the ordinal-level grade data in Figure 3.1, because putting the bars in frequency order would put the grades in the order C, B, A, D, F.

Time out to think

Would it be practical to make a dotplot for the population data in Figure 3.3? Would it make sense to make a Pareto chart for data concerning SAT scores? Explain.

Definitions

A **bar graph** consists of bars representing frequencies (or relative frequencies) for particular categories. The bar lengths are proportional to the frequencies.

A **dotplot** is similar to a bar graph, except each individual data value is represented with a dot.

A **Pareto chart** is a bar graph with the bars arranged in frequency order. Pareto charts make sense only for data at the nominal level of measurement.

EXAMPLE 1 Carbon Dioxide Emissions

Carbon dioxide is released into the atmosphere by the combustion of fossil fuels (oil, coal, natural gas). Because carbon dioxide in the atmosphere can potentially lead to global warm-

Table 3.11 The World's Eight Leading Emitters of Carbon Dioxide			
Country/region	Total carbon dioxide emissions (millions of metric tons of carbon)	Population (millions)	Per person carbon dioxide emissions (metric tons of carbon)
United States	1,495	271	5.5
China	740	1,244	0.6
Russia	405	147	2.8
Japan	288	126	2.3
India	253	955	0.3
Germany	227	82	2.8
United Kingdom	147	59	2.5
Canada	138	30	4.6

SOURCE: U.S. Department of Energy, based on 1998 emissions.

TECHNICAL NOTE

Table 3.11 gives emissions in terms of the weight of carbon contained in the emitted carbon dioxide; thus, it does not include the weight of the oxygen in the carbon dioxide.

ing, the U.S. Department of Energy tracks carbon dioxide emissions from countries around the world. Table 3.11 gives data for the eight countries that emit the most carbon dioxide each year. Make Pareto charts for the total emissions and for the average emissions per person. Discuss why the two charts look so different.

SOLUTION   The categories are the countries, and the frequencies are the data values. Because the range of data values for total carbon dioxide emissions goes from 138 to 1,495, a range of 0 to 1,500 makes a good choice for the vertical scale. To make the Pareto chart, we put the bars in descending order of size. Each bar's height corresponds to its data value, and we label the category (country) under the bar. Because the categories are nominal, the bars should not touch each other. Figure 3.4a shows the graph for total emissions. To make the Pareto chart for the per person carbon dioxide emissions, we use a vertical scale range of 0 to 6 to encompass all the data (Figure 3.4b).

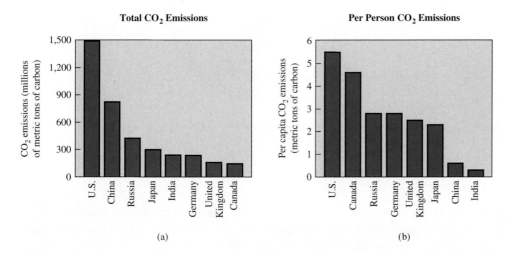

Figure 3.4   Pareto charts for (a) carbon dioxide emissions by country and (b) per person carbon dioxide emissions.

The two Pareto charts have the countries in different orders. This tells us that the biggest total emitters of carbon dioxide are not necessarily the biggest per person emitters. For example, China ranks as the second largest total emitter, but its per person emissions are far below those of the United States or any of the European countries listed. ∎

Time out to think

The combined population of China and India is more than eight times the U.S. population, yet U.S. carbon dioxide emissions are larger than those of China and India combined. Why do you think this is the case? What consequences might there be for the world if China and India had the same per person carbon dioxide emissions as the United States?

Pie Charts

Pie charts are usually used to show relative frequency distributions. A circular pie represents the total relative frequency of 100%, and the sizes of the individual slices, or wedges, represent the relative frequencies of different categories. Pie charts are used almost exclusively for qualitative data.

As a simple example, consider the registered voters in Rochester County: 25% are Democrats, 25% are Republicans, and 50% are independents. We can show these party affiliations in a pie chart. Because Democrats and Republicans each represent 25% of the voters, the wedges for Republicans and Democrats each occupy 25%, or one-fourth, of the pie. Independents represent half of the voters, so their wedge occupies the remaining half of the pie. Figure 3.5 shows the result. As usual, note the importance of clear labeling.

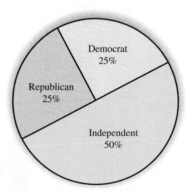

Figure 3.5   Party affiliations of registered voters in Rochester County.

Definition

A **pie chart** is a circle divided so that each wedge represents the *relative frequency* of a particular category. The wedge size is proportional to the relative frequency. The entire pie represents the total relative frequency of 100%.

A pie chart like the one in Figure 3.5 is easy to create because the wedge sizes represent very simple fractions. For more complex pie charts, you must either make careful angle measurements or use software that makes the measurements. And while pie charts can be very useful for simple data sets, the following example shows that complex pie charts may not always be the best way to present data.

EXAMPLE 2   Student Majors

Figure 3.6 is a pie chart showing planned major areas for first-year college students. Make a Pareto chart showing the same data. What are the three most popular major areas? Comment on the relative ease with which this question can be answered with the pie chart and the Pareto chart.

SOLUTION   Figure 3.7 shows the Pareto chart for the data. This chart makes it immediately obvious that the three most popular major areas are business (16.7%), arts and humanities (12.1%), and professional (11.6%). ("Professional" includes students majoring in the fields with professional licensing, such as architecture, nursing, and pharmacy.) In contrast, it takes a fair amount of study of the pie chart before we can easily list the three most popular major areas.   ▮

Time out to think
Example 2 discussed an advantage of a Pareto chart over a pie chart for showing the data concerning major areas. Do you think the pie chart has any advantages over the Pareto chart? If so, what?

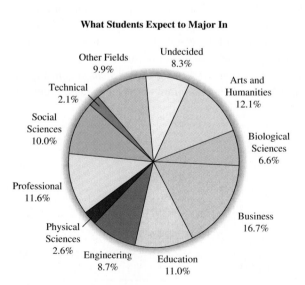

Figure 3.6   Planned major areas for first-year college students, 2001. SOURCE: *The Chronicle of Higher Education.*

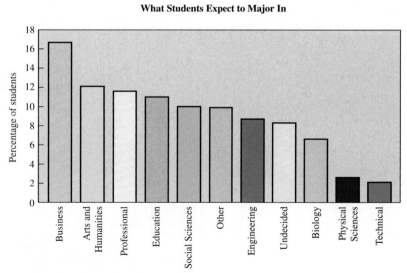

Figure 3.7   Pareto chart for the data in Figure 3.6.

Histograms and Stem-and-Leaf Plots

Figure 3.8 shows a bar graph for the energy use data of Table 3.4. The horizontal axis is marked with the categories of energy use per person per year (in millions of BTUs), and the vertical axis is marked with the frequencies (number of states) that correspond to each category. As in all bar graphs, the lengths of the bars are proportional to the frequencies. However, unlike the bar graphs we made earlier with qualitative data, the bars on this graph fall into a natural order based on the category values. In addition, the widths of the bars in Figure 3.8 have a specific meaning—in this case, telling us the range of values in each of the energy use categories. This type of bar graph, in which the bars have a natural order and the bar widths have specific meaning, is called a **histogram**. The bars in a histogram touch each other because there are no gaps between the categories.

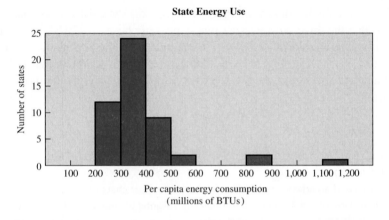

Figure 3.8 Histogram for the data in Table 3.4.

The energy use histogram clearly reveals the trends listed in a frequency table like Table 3.4. However, neither the frequency table nor the histogram provides all the details of the original data set in Table 3.3. For example, the histogram tells us that two states have energy use in the category 800–899 million BTUs per person, but it does not tell us which two states or the precise energy use values for these states. The **stem-and-leaf plot** (or *stemplot*) in Figure 3.9 gives a more detailed look at the data. It looks like a histogram turned sideways, except in

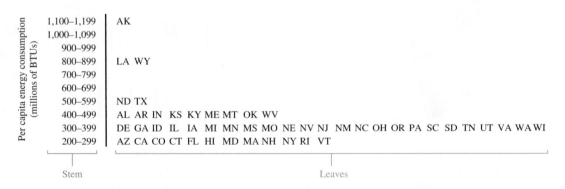

Figure 3.9 Stem-and-leaf plot for the energy use data from Table 3.3.

place of bars we see a listing of specific data sources or values for each category. In this case, the stem-and-leaf plot shows the individual states with data values in each category. We can now see, for example, that the two states in the category 800–899 million BTUs per person are Louisiana and Wyoming.

Another type of stem-and-leaf plot lists the individual data *values* rather than the data sources. For example, Figure 3.10 shows a stem-and-leaf plot for the per person carbon dioxide emissions data in Table 3.11. In this case, each data value is represented with a stem consisting of the first digit (and the decimal point) and a leaf consisting of the remaining digit. We can read the data values directly from this plot. For example, the first row shows the data values 0.3 and 0.6.

Stem	Leaves
0.	3 6
1.	
2.	3 5 8 8
3.	
4.	6
5.	5

Figure 3.10 Stem-and-leaf plot showing numerical data—in this case, the per person carbon dioxide emissions from Table 3.11.

Definitions

A **histogram** is a bar graph showing a distribution for quantitative data (at the interval or ratio level of measurement); the bars have a natural order and the bar widths have specific meaning.

A **stem-and-leaf plot** (or *stemplot*) is much like a histogram turned sideways, except in place of bars we see a listing of the individual data sources or values.

EXAMPLE 3   Oscar-Winning Actresses

Table 3.12 shows the ages of 34 recent Academy Award–winning actresses at the time when they won their award. Make a histogram to display these data. Discuss the results.

SOLUTION   The largest frequency is 15 (actresses), so we choose 20 as the height of the histogram. As in any bar graph, the height of each bar corresponds to the frequency for the category. The width of each bar spans a full 10 years, so the bars touch one another. Figure 3.11 shows the result. We see that actresses are most likely to win Oscars when they are fairly young. In contrast, male actors tend to win Oscars at older ages (see Exercise 17 in Section 3.1). Many actresses believe these facts reflect a subtle form of discrimination, in which Hollywood producers rarely make movies that feature older women in strong character roles.

Table 3.12 Ages of Actresses at Time of Academy Award	
Age	Number of actresses
20–29	7
30–39	15
40–49	6
50–59	1
60–69	3
70–79	1
80–89	1

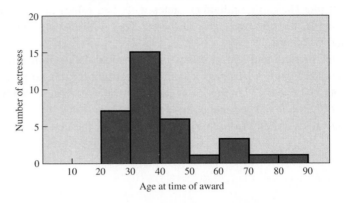

Figure 3.11   Histogram for ages of Academy Award–winning actresses.

Time out to think

What additional information would you need to create a stem-and-leaf plot for the ages of actresses when they won Academy Awards? What would the stem-and-leaf plot look like?

Line Charts

Like a histogram, a **line chart** shows a distribution of quantitative data. However, instead of using bars, a line chart connects a series of dots. The vertical position of each dot represents a data value; the dot goes where the top of a bar would go on a histogram. Figure 3.12 shows a line chart for the energy data in Table 3.4. For comparison, it is overlaid by the histogram shown earlier (see Figure 3.8).

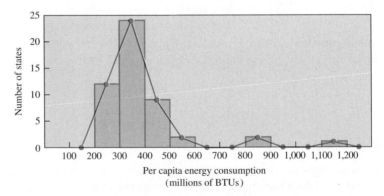

Figure 3.12   Line chart for the energy use data, with a histogram overlaid for comparison.

Note one important subtlety in interpreting a line chart: The horizontal positions of the dots correspond to the *centers* of the bins. For example, the dot representing the bin for 300–399 million BTUs is located at 349.5 million BTUs. Therefore, if you didn't know better, you might think the dot meant that 24 states had an energy use of *exactly* 349.5 million BTUs per person, when actually 24 states are in the *range* of 300–399 million BTUs per person.

Definition

A **line chart** shows a distribution of quantitative data as a series of dots connected by lines. For each dot, the horizontal position is the *center* of the bin it represents and the vertical position is the frequency value for the bin.

EXAMPLE 4   Age Distribution

Table 3.13 shows the distribution of the U.S. population according to age categories. Make a histogram and a line chart for these data. If you made a similar graph for age data 20 years from now, how would you expect it to be different?

Table 3.13 Age Distribution of U.S. Population

Age category	Population (millions)	Age category	Population (millions)	Age category	Population (millions)
0–4	19.0	35–39	22.6	70–74	8.8
5–9	19.9	40–44	21.8	75–79	7.2
10–14	19.2	45–49	18.8	80–84	4.7
15–19	19.4	50–54	15.8	85–89	2.5
20–24	17.7	55–59	12.3	90–94	1.1
25–29	18.6	60–64	10.2	95–99	0.3
30–34	20.2	65–69	9.6	100 or more	0.06

SOURCE: U.S. Census Bureau.

SOLUTION   The data are already binned in 5-year intervals (except for the last category of "100 or more"). We make the histogram by representing the frequency (population) in each category with a bar. Figure 3.13 shows the result. Superimposed on this histogram is a line chart in which the dots are placed at the center of each category. For example, the dot representing 25–29-year-olds is located at 27.5 on the horizontal axis. For consistency, we show the "100 or more" category as if it were 100–104, but the label on the horizontal axis shows what

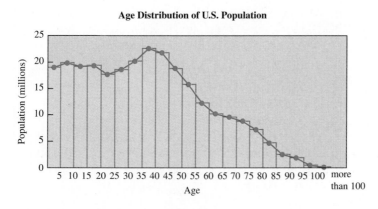

Figure 3.13   Histogram and line chart for the age data in Table 3.13.

it really means. Many frequency tables include open-ended categories such as "100 or more" when a small number of extreme data values are spread over a wide range.

If we made a similar graph in 20 years, we would expect to see two major differences. First, all the frequencies would be higher because of overall population growth. Second, there would be more growth in the higher age categories than in the lower age categories because advances in medical technology are allowing more people to live longer. ■

Time-Series Diagrams

Table 3.14 shows how the homicide rate in the United States has changed with time. The categories are the years and the data are the homicide rates (measured in deaths per 100,000 people). We can represent these data with either a histogram or a line chart; Figure 3.14 shows a line chart. Because the horizontal axis represents time in this case, we say that this graph is a **time-series diagram**.

Table 3.14 U.S. Homicide Rate per 100,000 People

Year	Homicides (per 100,000 people)	Year	Homicides (per 100,000 people)	Year	Homicides (per 100,000 people)
1960	5.1	1974	9.8	1988	8.4
1961	4.8	1975	9.6	1989	8.7
1962	4.6	1976	8.8	1990	9.4
1963	4.6	1977	8.8	1991	9.8
1964	4.9	1978	9.0	1992	9.3
1965	5.1	1979	9.7	1993	9.5
1966	5.6	1980	10.2	1994	9.0
1967	6.2	1981	9.8	1995	8.2
1968	6.9	1982	9.1	1996	7.4
1969	7.3	1983	8.3	1997	6.8
1970	7.9	1984	7.9	1998	6.3
1971	8.6	1985	7.9	1999	5.7
1972	9.0	1986	8.6	2000	5.5
1973	9.4	1987	8.3		

SOURCE: *Statistical Abstract of the United States.*

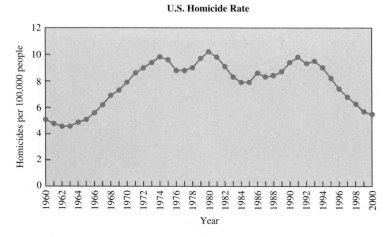

Figure 3.14 Time-series diagram for the homicide rate data of Table 3.14.

Definition

A histogram or line chart in which the horizontal axis represents *time* is called a **time-series diagram.**

EXAMPLE 5   Declining Death Rates

Figure 3.15 shows a time-series diagram for the death rate (deaths per 1,000 people) in the United States since 1900. (For example, the 1905 death rate of 15 means that, for each 1,000 people living at the beginning of 1905, 15 people died during the year.) Discuss the general trend and reasons for the trend. Also consider the spike in 1919: If someone told you that this spike was due to battlefield deaths in World War I, would you believe it? Explain.

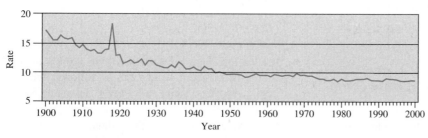

Death Rates per 1,000 Population

Source: National Center for Health Statistics

Figure 3.15   Historical U.S. death rates per 1,000 people.

SOLUTION   The general trend in death rates is clearly downward, presumably due to improvements in medical science. For example, bacterial diseases such as pneumonia were major killers in the early 1900s, but are largely curable with antibiotics today. The spike in 1919 *does* coincide with the end of World War I. However, if the spike were due to battlefield casualties, we might expect it to span the several years of World War I and we might expect to see a similar spike during World War II. The absence of these features suggests that the 1919 spike may not be due to World War I. In fact, the reason for the spike was a deadly epidemic of influenza.

By the Way...

The influenza epidemic of 1919 killed 850,000 people in the United States and an estimated 20 million people worldwide. Most deaths were caused by secondary bacterial infections that would be curable today with antibiotics.

Review Questions

1. What do we mean by a *distribution* of data?

2. What is a bar graph? How is it similar to and different from a dotplot? What is a Pareto chart?

3. Describe the importance of labeling on a graph, and briefly discuss the kinds of labels that should be included on graphs.

4. Describe the design and construction of a pie chart. When are pie charts useful?

5. What is a histogram? In what ways is it a special type of bar graph? How is a stem-and-leaf plot similar to a histogram? How is it different?

6. What is a line chart? Describe how you can make a line chart from a histogram, and vice versa.

7. What is a time-series diagram? Give an example.

Exercises

BASIC SKILLS AND CONCEPTS

SENSIBLE STATEMENTS? For Exercises 1–6, determine whether the given statement is sensible and explain why it is or is not.

1. A bar graph could be used to display the number of people in Chicago in each of 10 age categories.

2. Jack decided to convert his histogram to a line chart.

3. A time-series diagram would be used to show the net sales of the 10 largest restaurant chains in 2002.

4. Jill decided to convert her stem-and-leaf plot to a histogram.

5. A pie chart could be used to display the sales of pies at a large bakery each month for a year.

6. By rearranging a histogram, one can create a Pareto chart.

7. **NCAA BASKETBALL CHAMPIONS.** The table below shows the number of colleges or universities that won more than one NCAA basketball championship from 1960 to 2002.

Team	Number of championships
Cincinnati	2
Duke	3
Indiana	3
Kentucky	3
Louisville	2
Michigan State University	2
North Carolina	2
North Carolina State	2
UCLA	11

a. Make a bar graph for these data.
b. Make a dotplot for these data.
c. Make a Pareto chart for these data.

8. **TOP RETAILERS.** The following table gives the top eight retail companies in the United States and their total sales volume in 2000.
a. Make a bar graph for these data.
b. Make a Pareto chart for these data.

Company	Sales (billions of dollars)
Albertson's	36.8
Home Depot	45.7
JC Penney	33.0
Kmart	37.0
Kroger	49.0
Sears, Roebuck	40.9
Target	36.9
Wal-Mart	193.3

9. **HIGH SCHOOL GRADUATION RATES.** The high school graduation rates for various ethnic groups in the United States in a recent year were as follows: white, 83.0%; black, 74.9%; Asian and Pacific Islander, 84.9%; Hispanic, 54.7%.
a. Make a bar graph for these data.
b. Make a Pareto chart for these data.
c. Suppose that you were in charge of creating a program to increase high school graduation rates. In one to two paragraphs, explain how these data might influence the way in which you designed your program.

10. **WHAT PEOPLE ARE READING.** The pie chart in Figure 3.16 shows the results of a survey about what people are reading.
a. Summarize these data in a table of relative frequencies.
b. Make a Pareto chart for these data.
c. Which do you think is a better representation of the data: the pie chart or the Pareto chart? Why?

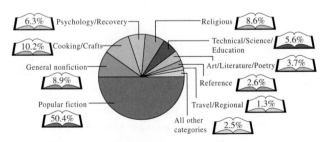

Source: Book Industry Study Group

Figure 3.16 SOURCE: *Wall Street Journal Almanac.*

11. **WORLD POPULATION.** The following table shows world population data by continent (2000).

Continent	Population (billions)
Africa	0.78
Asia	3.68
Europe	0.73
North America	0.31
Oceania	0.03
South/Central America	0.52

a. Draw a pie chart and a Pareto chart to display these data. Which display is more effective? Why?

b. Would a line chart be appropriate for this display? Why or why not?

12. **WORLD LAND MASS.** The following table shows areas of the world's land masses.

Continent	Area (millions of sq. miles)
Asia	17.2
Africa	11.6
North America	9.3
South America	6.9
Australia	3.0
Europe	3.8
Antarctica	5.1
All others	2.1

a. Draw a pie chart and a Pareto chart to display these data. Which display is more effective? Why?

b. Would a line chart be appropriate for this display? Why or why not?

13. **DOW JONES HISTOGRAM.** Make a histogram for the binned revenue data shown in Table 3.6. Be sure to include appropriate labels.

14. **DOW JONES LINE CHART.** Make a line chart for the binned revenue data shown in Table 3.6. Be sure to include appropriate labels.

15. **STATE PER CAPITA INCOME.** Using the data in Appendix A and the frequency table from Exercise 18 in Section 3.1, do the following:

a. Create a histogram for the state per capita income data.

b. Create a stem-and-leaf plot for these data.

c. Write one or two paragraphs discussing any interesting trends that are evident in the graphics.

16. **STATE TAXES.** Repeat Exercise 15, but this time for the state tax data in Appendix A and the frequency table from Exercise 19 in Section 3.1.

17. **STATE EDUCATION EXPENDITURES.** Repeat Exercise 15, but this time for the state education expenditures data in Appendix A and the frequency table from Exercise 20 in Section 3.1.

18. **STATE TEACHER SALARIES.** Repeat Exercise 15, but this time for the state teacher salary data in Appendix A and the frequency table from Exercise 21 in Section 3.1.

SURVEY OF FIRST-YEAR STUDENTS. Exercises 19–21 are taken from the annual survey of first-year students conducted by the Higher Education Research Institute at UCLA. The data are for the 2000–2001 academic year.

19. The following table gives the ages of first-year college students. Make a histogram and a line chart for these data. For the "16 or less" and "21 or more" categories, position the bars and dots at 16 and 21, respectively, but label the horizontal axis clearly with their actual meanings.

Age	Percent of sample
16 or less	0.1
17	1.9
18	68.9
19	27.4
20	1.0
21 or more	0.7

20. The following table gives the stated religion of first-year college students. Make a Pareto chart for these data. (Note: The "other religions" category consists of religions that were stated by less than 1% of the students in the sample.)

Religion	Percent of sample
Baptist	11.6
Catholic	30.5
Episcopal	1.7
Jewish	2.8
Lutheran	5.8
Methodist	6.4
Mormon	1.5
Presbyterian	4.0
United Church of Christ	1.5
Other religions	19.3
No religion	14.9

21. The following table gives the number of "other colleges" (besides the one they attend) to which first-year college students applied. Make a histogram and line chart for these data.

Other colleges	Percent of sample
0	20.4
1	13.0
2	16.1
3	17.4
4	12.1
5	7.9
6	5.3
7–10	6.5
11 or more	1.3

FURTHER APPLICATIONS

22. **AVERAGE FAMILY SIZE.** The following table (U.S. Census Bureau) gives the average family size in the United States from 1940 to 2000.

Year	Family size	Year	Family size
1940	3.76	1980	3.29
1950	3.54	1985	3.23
1960	3.67	1990	3.17
1965	3.70	1995	3.19
1970	3.58	2000	3.17
1975	3.42		

a. Make a time-series histogram for these data.
b. Make a time-series line chart for these data.
c. Can you give plausible historical and sociological explanations for any of the patterns in these data? Explain in one or two paragraphs.

23. **DRUNK DRIVING DEATHS.** Figure 3.17 shows the number of automobile fatalities in the United States in which alcohol was involved for each year from 1982 to 2000.

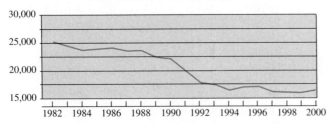

Total Fatalities in Alcohol-Related Crashes

Source: National Highway Traffic Safety Administration

Figure 3.17

a. How many alcohol-related fatalities were there in 1982? in 2000? Comment on the overall trend over this period.
b. What is the percent change in alcohol-related fatalities over this period?
c. The total numbers of automobile fatalities in 1982 and 2000 were 43,945 and 41,821, respectively. What percentage of all fatalities in these two years involved alcohol?
d. In view of your answer to part c, can you offer explanations for the trend in these data?

MOST APPROPRIATE DISPLAY. Exercises 24–27 describe data sets, but do not give actual data. For each data set, state the type of graphic that you feel would be most appropriate for displaying the data, if they were available. Explain your choice.

24. IQ scores of college statistics students

25. Colors of new cars

26. Causes of accidental deaths

27. Mean home prices in the United States over the past 50 years

PROJECTS FOR THE WEB AND BEYOND

For useful links, select "Links for Web Projects" for Chapter 3 at www.aw.com/bbt.

28. **CO_2 Emissions.** Look for updated data on international carbon dioxide emissions at the Web site for the *International Energy Annual*, published by the U.S. Energy Information Administration (EIA). Create an updated or expanded version of Figure 3.4a. Discuss any new features of your updated graph.

29. **ENERGY TABLE.** Explore the full set of energy tables at the U.S. Energy Information Administration (EIA) Web site. Choose a table that you find interesting and make a graph of its data. You may choose any of the graph types discussed in this section. Explain how you made your graph, and briefly discuss what can be learned from it.

30. **STATISTICAL ABSTRACT.** Go to the Web site for the *Statistical Abstract of the United States*. Explore the selection of "frequently requested tables." Choose one table of interest to you and make a graph from its data. You may choose any of the graph types discussed in this section. Explain how you made your graph and briefly discuss what can be learned from it.

31. NAVEL DATA. The *navel ratio* is defined to be a person's height divided by the height (from the floor) of his or her navel. An old theory says that, on average, the navel ratio of humans is the golden ratio: $(1 + \sqrt{5})/2$. Measure the navel ratio of each person in your class. Create an appropriate display of the navel data. Discuss any special properties of this distribution.

IN THE NEWS

1. BAR GRAPHS. Find a recent news article that includes a bar graph with qualitative data categories.
   a. Briefly explain what the bar graph shows, and discuss whether it helps make the point of the news article. Are the labels clear?
   b. Briefly discuss whether this bar graph could be recast as a dotplot.
   c. Is this bar graph already a Pareto chart? If so, explain why you think it was drawn this way. If not, do you think it would be clearer if the bars were rearranged to make a Pareto chart? Explain.

2. PIE CHARTS. Find a recent news article that includes a pie chart. Briefly discuss the effectiveness of the pie chart. For example, would it be better if the data were displayed in a bar graph rather than a pie chart? Could the pie chart be improved in other ways?

3. HISTOGRAMS. Find a recent news article that includes a histogram. Briefly explain what the histogram shows, and discuss whether it helps make the point of the news article. Are the labels clear? Is the histogram a time-series diagram? Explain.

4. LINE CHARTS. Find a recent news article that includes a line chart. Briefly explain what the line chart shows, and discuss whether it helps make the point of the news article. Are the labels clear? Is the line chart a time-series diagram? Explain.

3.3 Graphics in the Media

The basic graphs we have studied so far are only the beginning of the many ways to depict data visually. In this section, we will explore some of the more complex types of graphics that are common in the media.

Multiple Bar Graphs and Line Charts

A **multiple bar graph** is a simple extension of a regular bar graph: It has two or more sets of bars that allow comparison between two or more data sets. All the data sets must involve the same categories so that they can be displayed on the same graph. Figure 3.18 is a multiple bar graph showing the amount of money that children receive from their parents. The categories are age ranges of children. The three sets of bars represent weekly income from three different parental sources: allowance, handouts, and payments for chores. (Note: The median, which we will discuss in Section 4.1, is the middle of a set of values; for example, the median allowance of $3.00 for children ages 9–10 means that half these children received more than $3.00 and half received less than $3.00.)

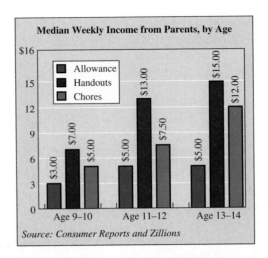

Figure 3.18 A multiple bar graph. SOURCE: *Wall Street Journal Almanac.*

A **multiple line chart** follows the same basic idea as a multiple bar chart, but shows the related data sets with lines rather than bars. Figure 3.19 shows a multiple line chart of stock, bond, and gold prices over a 12-week period. This particular chart is also a time-series diagram, because it shows data over time.

EXAMPLE 1   Reading the Investment Graph

Consider Figure 3.19. Suppose that, on July 7, you had invested $100 in a stock fund that tracks the S&P 500, $100 in a bond fund that follows the Lehman Index, and $100 in gold. If you sold all three funds on August 4, how much would you have gained or lost?

SOLUTION   The graph shows that the $100 in the stock fund would have been worth about $101 on August 4. The $100 bond investment would have declined in value to about $96. The gold investment would have held its initial value of $100. Thus, on August 4, your complete portfolio would have been worth

$$\$101 + \$96 + \$100 = \$297$$

You would have lost $3 on your total investment of $300.

By the Way...

Gold was once considered to be a solid investment and an important part of any investment portfolio. However, gold prices have languished in recent decades. By 2000, gold was worth only about $300 per ounce—much less than half its peak price of $850 per ounce in 1980.

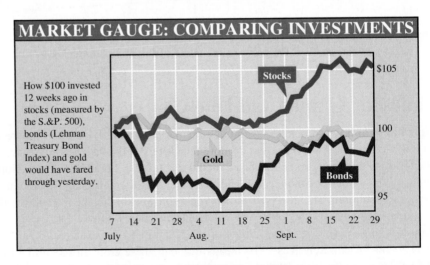

Figure 3.19 A multiple line chart. SOURCE: *New York Times.*

EXAMPLE 2 Graphic Conversion

Figure 3.20 is a multiple bar graph of the numbers of U.S. homes with computers, computers with modems, and computers with Internet access over a five-year period. Redraw this graph as a multiple line chart. Briefly discuss the trends shown on the graphs.

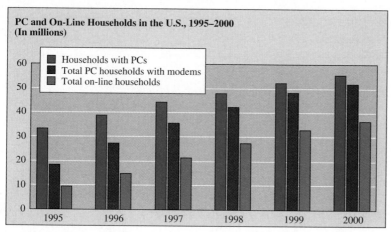

Source: Jupiter Communications

Figure 3.20 A multiple bar graph of computer usage. SOURCE: *Wall Street Journal Almanac.*

SOLUTION To convert the graphic to a multiple line chart, we must change the sets of three bars to a set of three lines. We place a dot corresponding to the height of each bar in the *center* of each category (year). Whereas the three bars were side by side on the multiple bar graph, the three corresponding dots will be aligned vertically. We then connect the dots of the same color with a color-coded line, as shown in Figure 3.21.

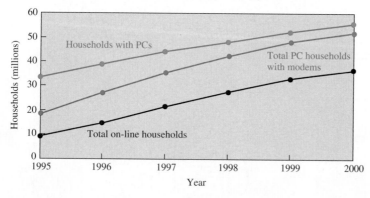

Figure 3.21 Multiple line chart showing the data from Figure 3.20.

All three data sets show an increase with time. However, the number of on-line households is increasing more rapidly than the number of households with computers. This suggests that not just new computer users are going on-line; many people who already had computers are also going on-line. If we project the trends into the future, it seems likely that the number of on-line households will approach the number of households with computers. ∎

Stack Plots

Another way to show two or more related data sets simultaneously is with a **stack plot**, which shows different data sets in a vertical stack. Figure 3.22 shows immigration data with stacked

By the Way...

In 1993, only 3 million people worldwide were connected to the Internet. By 2001, 160 million Americans were connected and approximately 510 million people were connected worldwide.

Country of Origin

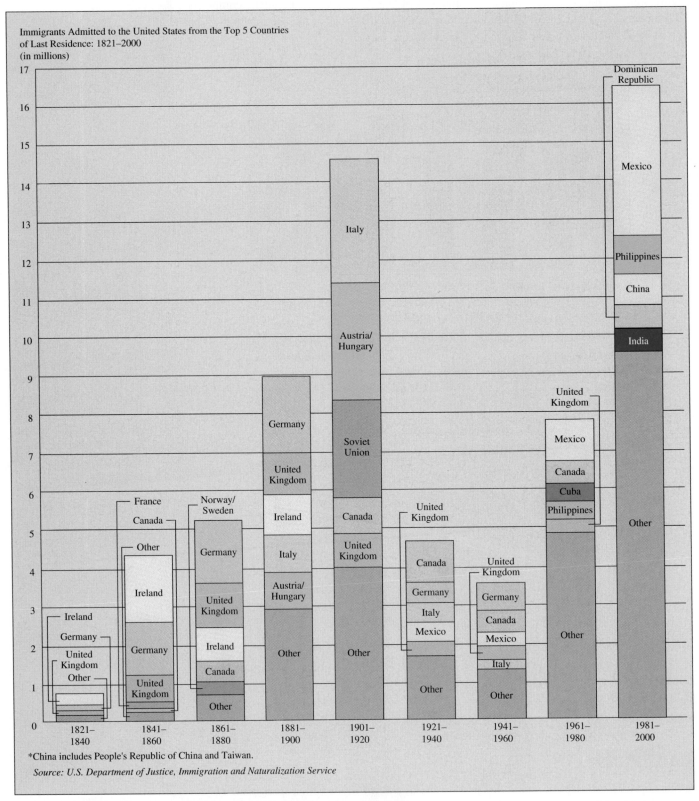

Immigrants Admitted to the United States from the Top 5 Countries
of Last Residence: 1821–2000
(in millions)

*China includes People's Republic of China and Taiwan.

Source: U.S. Department of Justice, Immigration and Naturalization Service

Figure 3.22   A stack plot using stacked bars. SOURCE: *Wall Street Journal Almanac.*

bars. Each individual bar consists of several separate bars stacked on top of one another. The slices in each bar represent numbers of immigrants from particular countries. The stacked bars make it easy to see some general trends. For example, we see that total immigration dropped dramatically after 1920, but returned to near-record levels by the 1980s and 1990s. Other trends are more difficult to see with this type of graphic. For example, it takes some effort to detect that immigration from Ireland peaked during the period 1841–1860.

Time out to think

Describe how the data in Figure 3.22 would appear if shown on a multiple bar graph rather than a stack plot. What trends would be easier to see? What trends would be more difficult to see?

EXAMPLE 3   Stacked Line Chart

Figure 3.23 shows death rates (deaths per 100,000 people) for four diseases over the period 1900 to 2000. Based on this graph, what was the death rate for pneumonia in 1980? Discuss the general trends visible on this graph.

By the Way...

Since the mid-1980s, there has been a small but noticeable resurgence of tuberculosis in the United States. Part of the resurgence is due to new strains of the disease that are resistant to most common drug treatments.

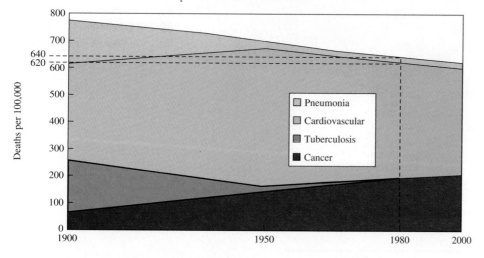

Death Rates for Various Diseases: 1900–2000

Figure 3.23   A stack plot using stacked wedges.

SOLUTION   Each disease has its own region, or wedge, shown by color-coding. The *thickness* of a wedge at a particular time tells us the death rate for the corresponding disease at that time. For 1980, the wedge for pneumonia extends from about 620 to 640 on the vertical axis, giving it a thickness of 20. Thus, the death rate for pneumonia in 1980 was 20 deaths per 100,000 people. The graph shows several important trends. First, the downward slope of the top wedge shows that the overall death rate from these four diseases decreased substantially, from nearly 800 deaths per 100,000 in 1900 to about 600 in 2000. The drastic decline in the thickness of the tuberculosis wedge shows that this disease was once a major killer, but has been nearly wiped out since 1950. The thickening of the cancer wedge shows that the death rate from cancer has gradually increased.   ∎

Geographical Data

The energy use data in Table 3.3 are an example of **geographical data**, because the raw data correspond to different geographical locations. We used these data earlier to make a frequency table (Table 3.4) and a histogram (Figure 3.8). However, these displays do not give us a sense of any geographical patterns in the data. For example, we are unable to see whether states in the northeast all have similar levels of energy use. Figure 3.24 shows one way to display the geographic trends. The categories are color-coded according to the key, and each state is colored appropriately.

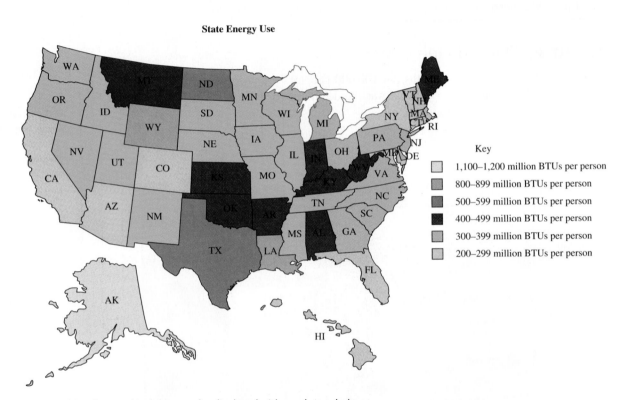

State Energy Use

Key
- 1,100–1,200 million BTUs per person
- 800–899 million BTUs per person
- 500–599 million BTUs per person
- 400–499 million BTUs per person
- 300–399 million BTUs per person
- 200–299 million BTUs per person

Figure 3.24 Geographical data can be displayed with a color-coded map.

Time out to think
What can you learn from the histogram in Figure 3.8 that you cannot learn easily from the geographic display in Figure 3.24, and vice versa? Do you see any surprising geographic trends in Figure 3.24? Explain.

Figure 3.24 works well for this data set because each state is associated with a unique number. When data vary continuously across geographical areas, we can use a **contour map.** Figure 3.25 shows a contour map of temperature over the United States at a particular time. Each **contour** connects locations with the same temperature. For example, the temperature is 50°F everywhere along the contour labeled 50°, and it is 60°F everywhere along the contour labeled 60°. Between these two contours, the temperature is between 50°F and 60°F. The closely packed curves in the northeast indicate that the temperature varies substantially over short distances. To make the graph easier to read, the regions between adjacent contours are color-coded.

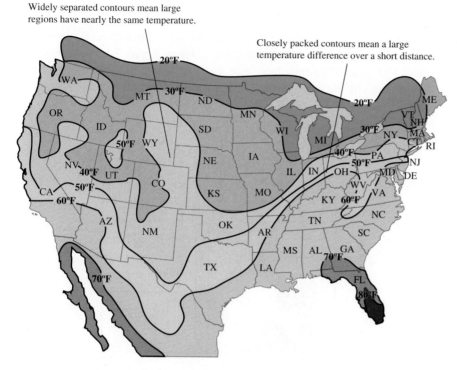

Widely separated contours mean large
regions have nearly the same temperature.

Closely packed contours mean a large
temperature difference over a short distance.

Figure 3.25   Geographical data that vary continuously, such as temperatures, can be
displayed with a contour map.

EXAMPLE 4   A Contour Elevation Map

Contour plots are also used to show geographical elevations. Figure 3.26 shows elevation contours around Boulder, Colorado. Discuss a few of the key features shown on the map.

SOLUTION   The labels on the figure indicate key features. The contours are widely spaced in the east, where the terrain is relatively flat and the elevations are fairly constant. Westward, the contours become closely spaced where the mountains rise up from the plains. The concentric closed contours in the center of the map surround peaks.

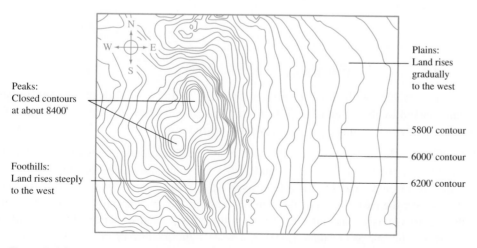

Peaks:
Closed contours
at about 8400'

Foothills:
Land rises steeply
to the west

Plains:
Land rises
gradually
to the west

5800' contour

6000' contour

6200' contour

Figure 3.26   A contour elevation map for the region around Boulder, CO.

Three-Dimensional Graphics

Today, computer software makes it easy to give almost any graph a three-dimensional appearance. For example, Figure 3.27 shows the same bar graph as Figure 3.1, but "dressed up" with a three-dimensional look. They may look nice, but the three-dimensional effects are purely cosmetic; they do not provide any information that wasn't already shown in Figure 3.1.

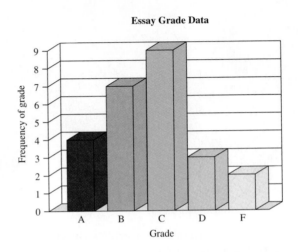

Figure 3.27   This graph has a three-dimensional appearance, but it shows only two-dimensional data.

In contrast, each of the three axes in Figure 3.28 carries distinct information, making the graph a true three-dimensional graph. Researchers studying migration patterns of a bird species (the *Bobolink*) counted the number of birds flying over seven New York cities throughout the night. As shown on the inset map, the cities were aligned east-west so that the researchers would learn what parts of the state the birds flew over, and at what times of night, as they headed south for the winter. Thus, the three axes measure *number of birds, time of night,* and *east-west location.*

EXAMPLE 5   Bird Migration

Based on Figure 3.28, at about what time was the largest number of birds flying over the east-west line marked by the seven cities? Over what part of New York did most of the birds fly? More specifically, approximately how many birds passed over Oneonta at about 12:30 A.M.?

SOLUTION   The number of birds detected in all the cities peaked between 4 and 6 hours after 8:30 P.M., or between about 12:30 and 2:30 A.M. More birds flew over the two easternmost cities of Oneonta and Jefferson than over cities farther west. Thus, most of the birds were flying over the eastern part of the state. To answer the specific question about Oneonta, note that 12:30 A.M. is 4 hours after 8:30 P.M. On the graph, this time appears to align with the lower peak on the line at Oneonta. Looking across to the *number of birds* axis, we see that about 30 to 40 birds were flying over Oneonta at that time.

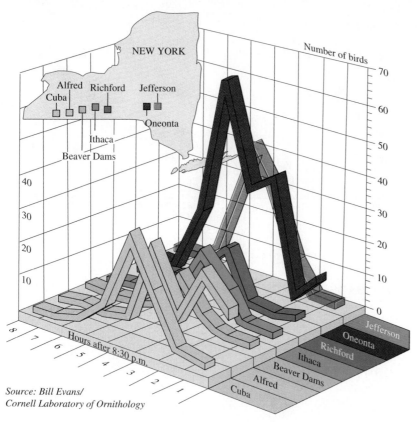

SONIC MAPPING TRACES BIRD MIGRATION

Sensors across New York State counted each occurrence of the nocturnal flight call of the bobolink to trace the fall migration on the night of Aug. 28–29, 1993. Computerized, the data showed the heaviest swath passing over the eastern part of the state.

*Source: Bill Evans/
Cornell Laboratory of Ornithology*

Figure 3.28 This graph shows true three-dimensional data. SOURCE: *New York Times.* ∎

Complex Graphics

All of the graphic types we have studied so far are common and fairly easy to create. But the media today are often filled with many varieties of even more complex graphics. For example, Figure 3.29 on page 122 shows a graphic concerning the participation of women in the Olympics. This single graphic combines a line chart, many pie charts, and numerical data. It is certainly a case of a picture being worth far more than a thousand words.

EXAMPLE 6 Olympic Women

Describe three trends shown in Figure 3.29.

SOLUTION The line chart shows that the total number of women competing in the Olympics has risen fairly steadily, especially since the 1960s, reaching nearly 4,500 in the 2000 games. The pie charts show that the percentage of women among all competitors has also increased, reaching 42% in the 2000 games. The bold numbers at the bottom show that the number of events for women has also increased dramatically, reaching 121 in the 2000 games. ∎

Time out to think
Which of the trends shown in Figure 3.29 are likely to continue over the next few Olympic games? Which are not? Explain.

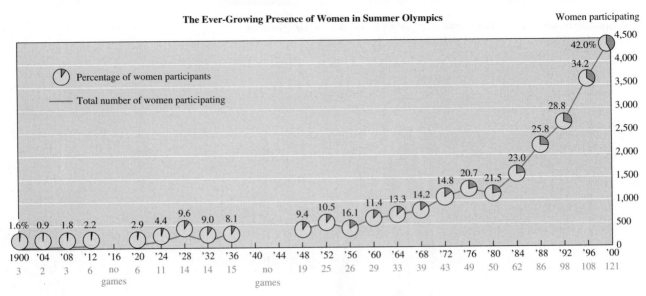

The Ever-Growing Presence of Women in Summer Olympics

Source: International Olympic Committee

Figure 3.29   Women in the Olympics. SOURCE: *New York Times*, August 20, 2000.

Review Questions

1. Briefly describe the construction and use of multiple bar charts and multiple line charts.

2. Briefly explain how to read and interpret a stack plot. In what ways are stack plots good for comparisons? In what ways are stack plots difficult to read?

3. What are geographical data? Briefly describe at least two ways to display geographical data.

4. What is a contour on a contour map? What does it mean when contours are close together? What does it mean when they are far apart?

5. What are three-dimensional graphics? Explain the difference between graphics that only *appear* three-dimensional and those that show truly three-dimensional data.

Exercises

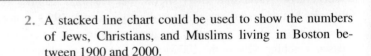

BASIC SKILLS AND CONCEPTS

SENSIBLE STATEMENTS? For Exercises 1–4, determine whether the given statement is sensible and explain why it is or is not.

1. Sue claims that she needs a three-dimensional graph to display the high temperature in Sacramento on each day of 2002.

2. A stacked line chart could be used to show the numbers of Jews, Christians, and Muslims living in Boston between 1900 and 2000.

3. A contour map could be used to display the number of Wal-Mart stores in each of the 50 states.

4. Chloe is drawing a multiple line chart to display the numbers of male and female Ph.D.s in 10 different disciplines in 2002.

5. **TEENAGE SMOKING.** The graphic in Figure 3.30 shows data regarding smoking by teenage students.
   a. In words, discuss the trends revealed on this graphic.
   b. Redraw the graph as a multiple line chart. Briefly discuss the advantages and disadvantages of the two different representations for this particular data set.

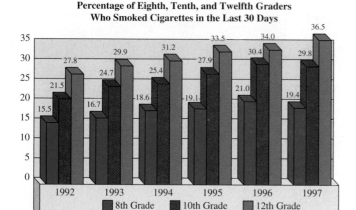

Source: The Monitoring the Future Study, University of Michigan

Figure 3.30 SOURCE: *Wall Street Journal Almanac.*

New Home Prices

Median Sales Prices of New Privately Owned One-Family Houses Sold, by Region (in thousands)

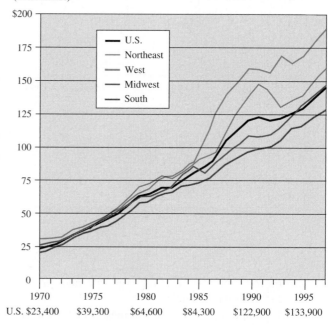

| U.S. | $23,400 | $39,300 | $64,600 | $84,300 | $122,900 | $133,900 |

Source: U.S. Census Bureau and U.S. Department of Housing and Urban Development

Figure 3.31 SOURCE: *Wall Street Journal Almanac.*

6. **HOME PRICES BY REGION.** The graph in Figure 3.31 shows home prices in different regions of the United States. Note that the data have *not* been adjusted for the effects of inflation.
   a. In words, describe the general trends that apply to the home price data for all regions.
   b. In words, describe any differences that you notice among the different regions.
   c. In words, describe how this graph would look different if the data were adjusted for the effects of inflation.

7. **EDUCATION AND EARNINGS.** Consider the display in Figure 3.32 of mean earnings in three different years according to levels of education (U.S. Census Bureau, *Current Population Survey*).
   a. What is the purpose of this graph? Briefly explain the meaning of each of the three sets of bars on the graph.
   b. The graph has a three-dimensional appearance. Is it showing true three-dimensional data, or is the appearance purely cosmetic? Do you think the three-dimensional appearance helps or hinders the display?
   c. Compare the change in earnings between 1975 and 1995 for people with bachelor's degrees to the change for people who did not graduate from high school. What do these data say about the value of a college education?

Mean Earnings of Workers 18 Years and Over by Educational Attainment, 1975 to 1995

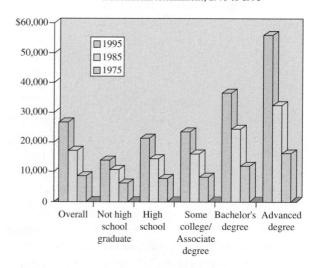

Source: Census Bureau, Current Population Survey

Figure 3.32 SOURCE: *TIME Almanac.*

8. **MARRIAGE AND DIVORCE RATES.** The following table gives the marriage and divorce rates in this country for selected years since 1900. Both rates are given in units of marriages/divorces per 1,000 people in the population (Department of Health and Human Services).

Year	Marriage	Divorce	Year	Marriage	Divorce
1900	9.3	0.7	1965	9.3	2.5
1910	10.3	0.9	1970	10.6	3.5
1920	12.0	1.6	1975	10.1	4.9
1930	9.2	1.6	1980	10.6	5.2
1940	12.1	2.0	1985	10.2	5.0
1950	11.1	2.6	1990	9.8	4.7
1960	8.5	2.2	1995	7.6	4.1

a. Make a multiple bar graph for these data, in which one bar represents the marriage rate and one represents the divorce rate.
b. Make a multiple line chart for these data, in which one line represents the marriage rate and one represents the divorce rate.
c. Why do these data consist of marriage and divorce *rates* rather than total numbers of marriages and divorces? Comment on any trends that you observe in these rates, and give plausible historical and sociological explanations for these trends.

9. **IMMIGRATION STACK PLOT.** Answer the following questions by studying Figure 3.22 on page 116.
a. How many Irish immigrants arrived in this country in the period 1841–1860?
b. How many Canadian immigrants arrived in this country in the period 1901–1920?
c. How many Mexican immigrants arrived in this country in the period 1981–2000?
d. Briefly discuss the historical trends in immigration, and offer plausible historical explanations for some of these trends.

10. **FEDERAL SPENDING.** The stack plot in Figure 3.33 shows the changes in major spending categories of the federal budget. Note that the "net interest" category represents interest payments on the national debt, and the "all other" category includes spending on such things as education, environmental cleanup, and scientific research. Interpret the stack plot and discuss some of the trends it reveals.

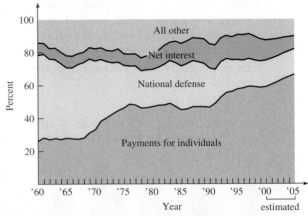

Percentage Composition of Federal Government Outlays

Figure 3.33   SOURCE: Budget for Fiscal Year 2001.

11. **COLLEGE DEGREES.** The stack plot in Figure 3.34 shows the numbers of college degrees awarded to men and women over time.

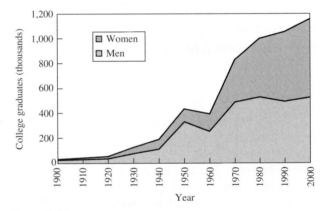

College Degrees Awarded

Figure 3.34

a. Estimate the numbers of college degrees awarded to men and to women (separately) in 1930 and in 2000.
b. Compare the numbers of degrees awarded to men and to women (separately) in 1980 and 2000.
c. During what decade did the *total* number of degrees awarded increase the most?
d. Compare the *total* numbers of degrees awarded in 1950 and 2000.
e. Do you think the stack plot is an effective way to display these data? Briefly discuss other ways that might have been used instead.

12. **STUDENT ENROLLMENT.** The stack plot in Figure 3.35 shows the enrollments in mathematics courses at two-year colleges over a 25-year period. The courses are divided into three levels.

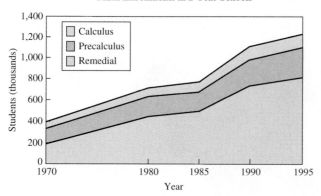

Math Enrollments in 2-Year Schools

Figure 3.35 SOURCE: National Science Foundation.

a. Which category of courses experienced the greatest increase in size between 1970 and 1995?
b. Estimate the numbers of students in remedial courses in 1970 and 1995. What is the percent change?
c. Estimate the numbers of students in precalculus courses in 1970 and 1995. What is the percent change?
d. How many more students took remedial courses than calculus courses in 1995?
e. Do you think the stack plot is an effective way to display these data? Briefly discuss other ways that might have been used instead.

FURTHER APPLICATIONS

13. **CREATING A STACK PLOT.** Redraw Figure 3.18 using stacked bars rather than side-by-side bars. Briefly discuss the advantages and disadvantages of the two different representations for this particular data set.

14. **MELANOMA MORTALITY.** Figure 3.36 shows how the mortality from *melanoma* (a form of skin cancer) varies on a county-by-county basis across the United States. The legend shows that the darker the shading in a county, the higher the mortality rate. Discuss a few of the trends revealed in the figure. If you were researching skin cancer, which regions might warrant special study? Why?

15. **SCHOOL SEGREGATION.** One way of measuring segregation is the likelihood that a black student will have classmates who are white. A recent *New York Times* study found that, by this measure, segregation increased significantly in the 1990s. Figure 3.37 (next page) shows the probability that a black student would have white classmates, by county, during a recent academic year. Do there appear to be any significant regional differences? Can you pick out any differences between urban and rural areas? Discuss possible explanations for a few of the trends that you see in the figure.

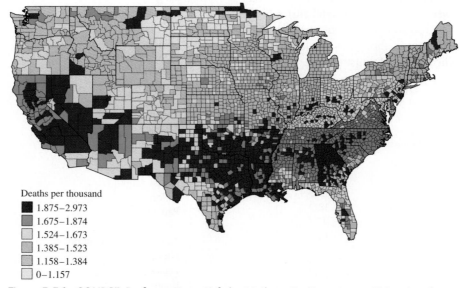

Female Melanoma Mortality Rates by County

Figure 3.36 SOURCE: Professor Karen Kafadar, Mathematics Department, University of Colorado at Denver.

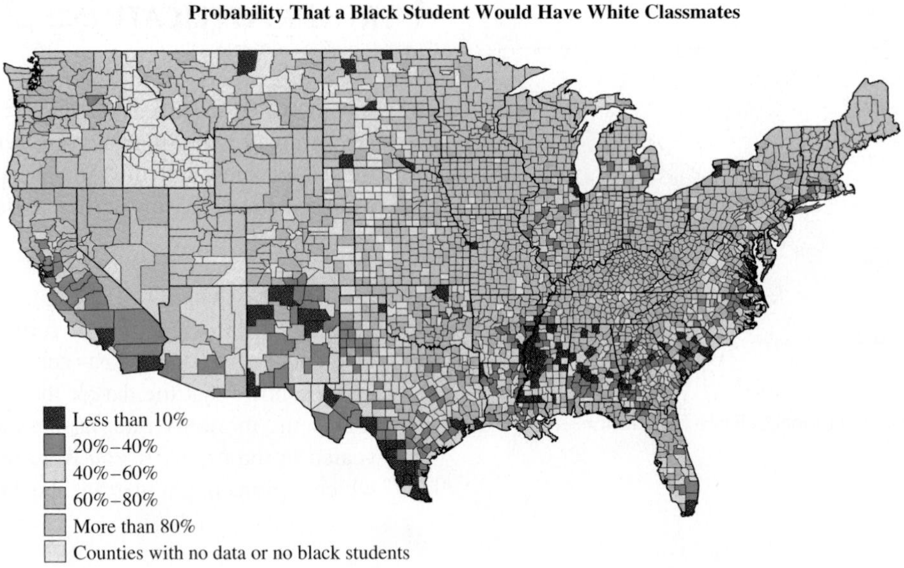

Probability That a Black Student Would Have White Classmates

■ Less than 10%
■ 20%–40%
□ 40%–60%
■ 60%–80%
■ More than 80%
□ Counties with no data or no black students

Figure 3.37 SOURCE: *New York Times*.

CREATING GRAPHICS. Exercises 16–20 give tables of real data. For each table, make a graphical display of the data. You may choose any graphic type that you feel is appropriate to the data set. In addition to making the display, write a few sentences explaining why you chose this type of display and a few sentences describing interesting patterns in the data.

16. ALCOHOL ON THE ROAD. The following table gives the number of automobile fatalities in which (i) no alcohol was involved, (ii) moderate alcohol was involved (blood alcohol content between 0.01 and 0.09), and (iii) a high level of alcohol was involved (blood alcohol content above 0.1). All figures are in thousands of deaths (National Highway Traffic Safety Administration).

Year	No alcohol	Moderate alcohol	High alcohol
1982	18.7	4.8	20.4
1984	20.5	4.8	19.0
1986	22.0	5.1	18.9
1988	23.5	4.9	18.7
1990	22.5	4.4	17.7
1992	21.4	3.6	14.2
1994	24.1	3.5	13.1
1996	24.8	3.8	13.4
1998	25.5	3.7	12.2
2000 (est.)	25.1	3.9	12.8

17. CANCER TYPES. The following table lists deaths in 2000 due to various types of cancer, for men and women (American Cancer Society).

Type of cancer	Deaths (thousands)
Men	
Lung	93.1
Prostate	31.9
Colon	28.0
Urinary	15.7
Non-Hodgkin's lymphoma	14.4
Pancreas	13.7
Leukemia	12.1
Liver	9.7
Esophagus	9.2
Stomach	7.6
Women	
Lung	68.8
Breast	40.8
Colon	28.0
Pancreas	14.5
Ovary	14.0
Non-Hodgkin's lymphoma	13.1
Uterus	11.1
Leukemia	9.6
Brain	5.9
Stomach	5.4

18. PERCENT NEVER MARRIED. The following table shows the percentages, for 1970 and 1998, of men and women in various age categories who were never married (U.S. Census Bureau).

Women	1970	1998	Men	1970	1998
20–24	35.8	70.3	20–24	54.7	83.4
25–29	10.5	38.6	25–29	19.1	51.0
30–34	6.2	21.6	30–34	9.4	29.2
35–39	5.4	14.3	35–39	7.2	21.6
40–44	4.9	9.9	40–44	6.3	15.6

19. DAILY NEWSPAPERS. The following table gives the number of daily newspapers and their total circulation (in millions) for selected years since 1920 (*Editor & Publisher*).

Year	Number of daily newspapers	Circulation (millions)
1920	2,042	27.8
1930	1,942	39.6
1940	1,878	41.1
1950	1,772	53.9
1960	1,763	58.8
1970	1,748	62.1
1980	1,747	62.2
1990	1,611	62.3
2000	1,485	56.1

20. FIREARM FATALITIES. The following table summarizes deaths due to firearms in different nations in a recent year (Coalition to Stop Gun Violence).

Country	Total firearms deaths	Homicides by firearms	Suicides by firearms	Fatal accidents by firearms
United States	35,563	15,835	18,503	1,225
Germany	1,197	168	1,004	25
Canada	1,189	176	975	38
Australia	536	96	420	20
Spain	396	76	219	101
United Kingdom	277	72	193	12
Sweden	200	27	169	4
Vietnam	131	85	16	30
Japan	93	34	49	10

PROJECTS FOR THE WEB AND BEYOND

For useful links, select "Links for Web Projects" for Chapter 3 at www.aw.com/bbt.

21. WEATHER MAPS. Many Web sites offer contour maps with current weather data. For example, you can use the Yahoo Weather site to generate many different contour weather maps. Generate at least two contour weather maps and discuss what they show.

22. THE FEDERAL BUDGET. The U.S. Office of Management and Budget (OMB) publishes an annual *Citizen's Guide to the Federal Budget*. Go to the Web site for this guide and click on the list of charts and tables. Explore some of the charts. Pick two charts of particular interest to you and discuss the data they show.

23. CANCER CURE. As shown in Figure 3.23, cancer is one of the leading causes of death today. Nevertheless, scientists have made great progress in treating many forms of cancer. Go to the American Cancer Society Web site and investigate research into cancer cures. Read about one or two recent studies, and write a short report on what you learn. Be sure to include graphics in your report.

24. GRAPHIC DECEPTION. Refer to the *USA Today* illustration in Figure 3.38, in which percentages are represented by volumes of portions of someone's head. Are the data presented in a format that makes them easy to understand and compare? Are the data presented in a way that does not mislead? Could the same information be presented in a better way? If so, construct your own graph that better depicts the given information.

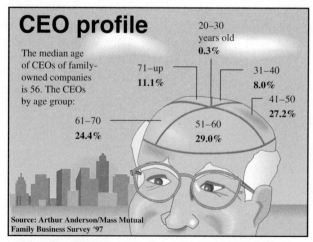

Figure 3.38 SOURCE: Anne R. Carey and Grant Jarding, "A Look at Statistics That Shape Your Finances," *USA Today*.

IN THE NEWS

1. **MULTIPLE BAR GRAPHS.** Find an example of a multiple bar graph or multiple line chart in a recent news report. Comment on the effectiveness of the display. Could another display have been used to depict the same data?

2. **STACK PLOTS.** Find an example of a stack plot in a recent news report. Comment on the effectiveness of the display. Could another display have been used to depict the same data?

3. **GEOGRAPHICAL DATA.** Find an example of a graph of geographical data in a recent news report. Comment on the effectiveness of the display. Could another display have been used to depict the same data?

4. **THREE-DIMENSIONAL DISPLAYS.** Find an example of a three-dimensional display in a recent news report. Are three dimensions needed, or are they included for cosmetic reasons? Comment on the effectiveness of the display. Could another display have been used to depict the same data?

5. **FANCY NEWS GRAPHICS.** Find an example in the news of a graphic that combines two or more of the basic graphic types. Briefly explain what the graphic is showing, and discuss the effectiveness of the graphic.

3.4 A Few Cautions About Graphics

As we have seen, graphics can offer clear and meaningful summaries of statistical data. However, even well-made graphics can be misleading if we are not careful in interpreting them, and poorly made graphics are almost always misleading. Moreover, some people use graphics in deliberately misleading ways. In this section, we discuss a few of the more common ways in which graphics can lead us astray.

Perceptual Distortions

Many graphics are drawn in a way that distorts our perception of them. Figure 3.39 shows one of the most common types of distortion. Dollar-shaped bars are used to show the declining value of the dollar over time. The *lengths* of the bars represent the data, but our eyes tend to focus on the *areas* of the bars. For example, the right bar is supposed to show that a dollar in 2000 was worth 48% as much as a dollar in 1980. Its length is indeed 48% of that of the left bar, but its area is much smaller in comparison (about 23% of the area of the left bar). This gives the perception that the value of the dollar shrank even more than it really did.

By the Way...

German researchers in the latter part of the 19th century studied many types of graphics. The type of distortion shown in Figures 3.39 and 3.40 was so common that they gave it its own name, which translates roughly as "the old goosing up the effect by squaring the eyeball trick."

Figure 3.39 The lengths of the dollars are proportional to their spending power, but our eyes are drawn to the areas, which decline more than the lengths.

Homes with Cable TV

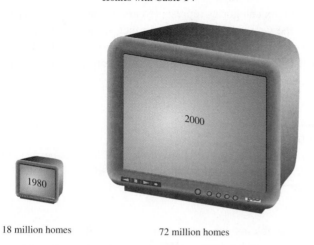

18 million homes   72 million homes

Figure 3.40   The heights of the TVs are the important measure in this figure, but our eyes are drawn to their volumes.

Even greater distortion occurs when a graphic shows volumes where length is the important measure. Figure 3.40 uses television sets to represent the number of houses with cable in 1980 and 2000. This number increased by a factor of about 4 during that period, from 18 million homes to 72 million homes, as shown by the *heights* of the TVs in the figure. However, our eyes are drawn to the *volumes* of the TVs, which differ by the much greater factor of $4^3 = 64$. Thus, the figure makes the increase look much larger than it really was.

The easiest person to deceive is one's own self.

—Edward Bulwer-Lytton

Watch the Scales

Figure 3.41 shows a multiple bar graph concerning home ownership in the United States. At first glance, it appears that the percentage of people owning their own homes rose much faster

Home, Sweet Home

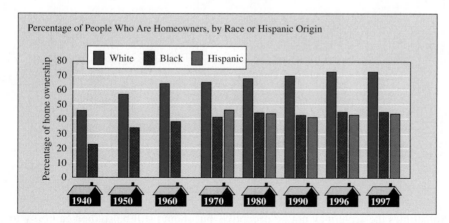

Source: U.S. Census Bureau

Figure 3.41   Home ownership, by race or Hispanic origin. Note the nonuniform horizontal scale.
SOURCE: *Wall Street Journal Almanac.*

during the period 1940–1990 than it did during the 1990s. But look at the horizontal scale more closely: The first six categories represent years that are a decade apart, while the last two categories represent the years 1996 and 1997. In fact, if the small increase from 1996 to 1997 were repeated each year for an entire decade, the increase would not look much different from the trends in earlier decades. This graph is misleading on first impression because it does not use a uniform scale for the horizontal axis.

Time out to think

Based on Figure 3.41 and your own intuition, predict U.S. home ownership rates (by race or Hispanic origin) in 2020. Explain the reasoning behind your predictions.

Similar scaling problems can occur with the vertical axis. Figure 3.42a shows the percentage of college students between 1910 and 1990 who were female. At first glance, it appears that there was a huge increase in this percentage between 1950 and 1990. But note that the vertical axis scale does not begin at zero and does not end at 100%. The increase looks far less dramatic if we redraw the graph with the vertical axis covering the full range of 0 to 100% (Figure 3.42b). From a mathematical point of view, leaving out the zero point on a scale is perfectly honest and can make it easier to see small-scale trends in data. Nevertheless, as this example shows, it can be visually deceptive if you don't study the scale carefully.

Women as a Percentage of All College Students

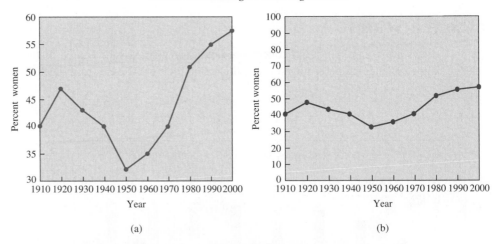

Figure 3.42 Both graphs show the same data, but they look very different because their vertical scales have different ranges.

Sometimes the scale may not be deceptive, but should still be studied carefully to avoid misinterpretation. Consider Figure 3.43a, which shows the change in computer speed from 1950 to 2000. At first glance, it may appear that the speeds have been increasing linearly; for example, it might look as if the speed increased by the same amount from 1990 to 2000 as it

Computer Speed

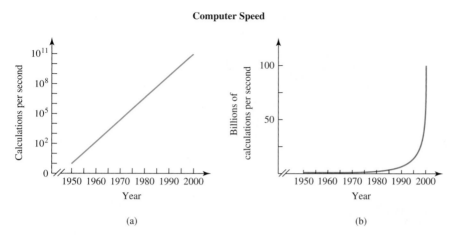

(a)   (b)

Figure 3.43 Both graphs show the same data, but the graph on the left uses an exponential scale.

did from 1950 to 1960. However, if we look closely, we see that each tick mark represents a *tenfold* increase in speed. Now we see that computer speed grew from about 1 to 100 calculations per second from 1950 to 1960 and from about 100 million to 10 billion calculations per second between 1990 and 2000. This type of scale is called an *exponential scale* because it grows by powers of 10 and powers of 10 are *exponents.* (For example, 3 is the exponent in 10^3.) In general, exponential scales are useful for displaying data that vary over a huge range of values. Recasting the computer data with an ordinary scale, as in Figure 3.43b, makes it impossible to see any detail in the early years shown on the graph, because the speeds have grown so rapidly.

Time out to think
Based on Figure 3.43a, can you predict the speed of the fastest computers in 2010? Could you make the same prediction with Figure 3.43b? Explain.

CASE STUDY Asteroid Threat

Asteroids and comets occasionally hit the Earth. Small ones tend to burn up in the atmosphere or create small craters on impact. But larger ones can cause substantial devastation. About 65 million years ago, an asteroid about 10 kilometers in diameter hit the Earth, leaving a 200-kilometer-wide crater on the coast of the Yucatan peninsula in Mexico. Many scientists believe this impact caused the extinction of about three-quarters of all species living on Earth at the time, including all the dinosaurs.

Clearly, a similar impact would be bad news for our civilization. Thus, we might want to understand the likelihood of such an event. Figure 3.44 shows a graph relating the size of impacting asteroids and comets to the frequency with which such objects hit the Earth. Because of the wide range of sizes and time scales involved, *both* axes on this graph are exponential. The horizontal axis shows impactor (asteroid or comet) sizes, with each tick representing a power of 10. The vertical axis shows the frequency of impact; moving up on the vertical axis corresponds to more frequent events. With this double exponential graph, we can see trends clearly. For example, small objects of about 1 meter in size strike the Earth every day, but cause little damage. At the other extreme, objects large enough to cause a mass extinction hit only about once every hundred million years.

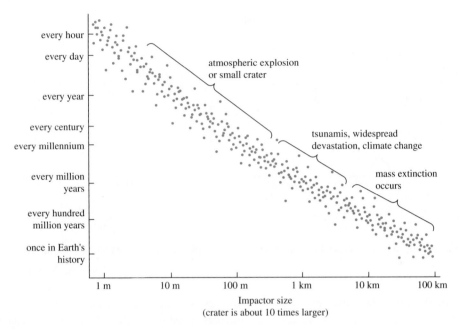

Figure 3.44 This graph shows how the frequency of impacts—and the magnitude of their effects—depends on the size of the impactor. Note that smaller impacts are much more frequent than larger ones.

The intermediate cases are probably the most worrisome. The graph indicates that objects that could cause "widespread devastation"—such as wiping out the population of a small state—can be expected as often as once every thousand years. This is often enough to warrant at least some preventive action. Currently, astronomers are trying to make more precise predictions about when an object might hit the Earth. If they discover an object that will hit the Earth, then we will need to find a way to deflect it to prevent the impact. ∎

By the Way...

Gasoline prices in Europe and Japan are typically $4 to $5 dollars per gallon, primarily because gasoline taxes are much higher than in the United States and Canada.

Unadjusted Economic Data

How much are you paying for gasoline? The red curve in Figure 3.45 shows actual U.S. gasoline prices over many decades. This curve gives the impression that gasoline is much more expensive than it was a few decades ago. However, the actual prices do not take into account the effects of inflation. The blue curve gives a much more honest depiction of gasoline prices because these prices have been adjusted for inflation (as measured by the CPI [see Section 2.4]). Economists say that this curve reflects the *real cost* of gasoline. We now see, for example, that the *real cost* of gasoline in the late 1990s was almost as low as it had ever been. This example points out the fact that price comparisons over time are virtually meaningless unless the prices are adjusted for inflation.

Time out to think

How does the current price of gasoline compare to the current price of bottled water? Briefly discuss any implications of this price comparison.

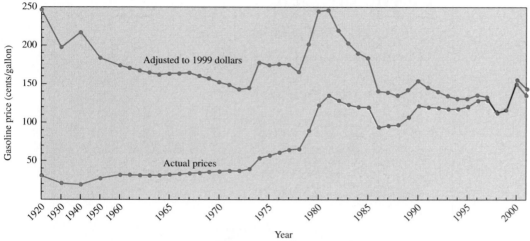

Graph shows data only at 10-year intervals prior to 1960.

Figure 3.45   Trends in gasoline prices look very different when actual prices are adjusted for inflation.
SOURCE: American Petroleum Institute.

Percentage Change Graphs

Is college getting more or less expensive? A quick look at the downward-sloping lines in the lower graph in Figure 3.46 might give the impression that college costs fell steadily in the 1990s. But on closer examination, we find that this is not the case at all. The vertical axis on this graph represents the *percentage increase* in costs. Thus, the downward-sloping lines show only that the percentage increase in college costs slowed, *not* that the actual costs went down. In fact, the actual costs have risen substantially, as shown in the upper graph. Moreover, because the rate of inflation (as measured by the CPI) was less than the rate of increase in college costs, the *real cost* of public colleges steadily rose. Graphs that show percentage change are very common, particularly with economic data. Although they are perfectly honest, you can be easily misled unless you interpret them with great care.

Pictographs

Pictographs are graphs embellished with additional art work. The art work may make the graph more appealing, but it can also distract or mislead. Figure 3.47 is a pictograph showing the rise in world population from 1804 to 2054 (numbers for future years are based on United Nations projections). The lengths of the bars correctly correspond to world population for the years listed. However, the artistic embellishments of this graph can be deceptive in several ways. For example, your eye may be drawn to the figures of people lining the globe. Because this line of people rises from the left side of the pictograph to the center and then falls, it might give the impression that future world population will decline. In fact, the line of people is purely decorative and carries no information.

Perhaps the most serious problem with this pictograph is that it makes it appear that world population has been rising linearly. However, notice that the time intervals on the horizontal axis are not the same in each case. For example, the interval between the bars for

Get your facts first, and then you can distort them as much as you please.
—Mark Twain

By the Way...

Demographers often characterize population growth by a doubling time—the time it takes the population to double. During the late 20th century, the doubling time for human population was about 40 years. If population continued to double at this rate, world population would reach 34 billion by 2100 and 192 billion by 2200. By about 2650, human population would be so large that it would not fit on the Earth, even if everyone stood elbow-to-elbow everywhere.

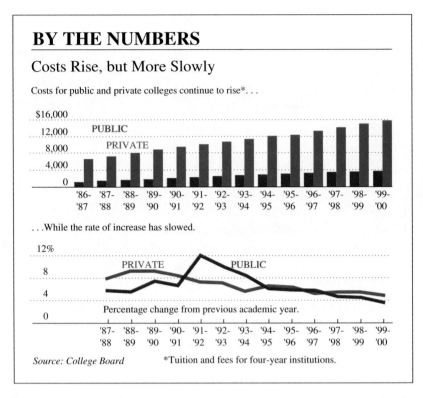

Figure 3.46 The lower graph shows percentage change, which must be interpreted with great care. SOURCE: *New York Times*.

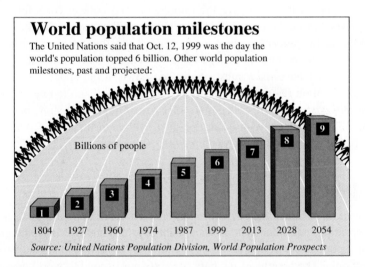

Figure 3.47 A pictograph. SOURCE: James Abundis and Bob Laird, *USA TODAY*.

1 billion and 2 billion people is 123 years (from 1804 to 1927), but the interval between the bars for 5 billion and 6 billion people is only 12 years (from 1987 to 1999).

Pictographs are very common. As this example shows, however, you have to study them carefully to extract the essential information and not be distracted by the cosmetic effects.

Review Questions

1. Describe several ways in which perceptual distortions can arise in graphics and how they can be misleading.

2. How can graphics be misleading when the scales do not go all the way to zero?

3. What is an exponential scale? When is it useful to use an exponential scale on a graphic?

4. In displaying economic data, why does it matter whether the data are adjusted for inflation?

5. Explain how a graph that shows percentage change can show descending bars (or a descending line) even when the variable of interest is increasing.

6. What is a pictograph? Briefly explain how a pictograph can enhance a graph and how a pictograph can also make a graph misleading.

Exercises

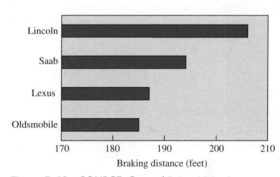

BASIC SKILLS AND CONCEPTS

1. **THREE-DIMENSIONAL PIES.** The pie charts in Figure 3.48 give the percentage of Americans in three age categories in 1990 and 2050 (projected).
   a. Consider the 1990 age distribution. The actual percentages for the three categories for 1990 were 87.5% (others), 11.3% (60–84), and 1.2% (85+). Does the pie chart show these values accurately? Explain.
   b. Consider the 2050 age distribution. The actual percentages for the three categories for 2050 were 80.0% (others), 15.4% (60–84), and 4.6% (85+). Does the pie chart show these values accurately? Explain.
   c. Using the actual percentages given in parts a and b, draw flat (two-dimensional) pie charts to display these data. Explain why these pie charts give a more accurate picture than the three-dimensional pies.
   d. Comment on the general trends shown in the two pie charts.

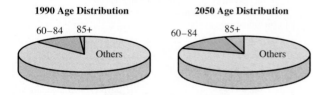

Figure 3.48   SOURCE: U.S. Census Bureau.

2. **BRAKING DISTANCES.** Figure 3.49 shows the braking distance for four different cars. Discuss the ways in which this display might be deceptive. How much greater is the braking distance of Lincolns than the braking distance of Oldsmobiles? Draw the display in a fairer way.

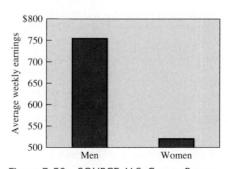

Figure 3.49   SOURCE: *Car and Driver Magazine.*

3. **COMPARING EARNINGS.** Consider the bar graph in Figure 3.50 comparing the average weekly earnings of men and women. Identify any misleading aspects of the display. Draw the display in a fairer way.

Figure 3.50   SOURCE: U.S. Census Bureau.

4. **COMPUTER SALES.** The following table shows the numbers of worldwide shipments of personal computers since 1980.

Year	Worldwide shipments
1981	1,630,000
1983	11,120,000
1985	14,700,000
1987	16,680,000
1989	21,330,000
1991	26,970,000
1993	38,850,000
1995	60,170,000
1997	82,400,000
1999	112,000,000
2000	131,700,000
2001 (est.)	129,600,000

a. Make a time-series graph of these data, using a uniform scale on both axes.
b. Make an exponential graph of these data in which the subdivisions on the vertical axis are 1 million, 10 million, and 100 million.
c. Discuss the advantages and disadvantages of the graph in part b compared to the graph in part a.

5. **WORLD POPULATION.** Recast Figure 3.47 as a graph with a proper horizontal axis. What trends are clear in your new graph that are not clear in the original?

6. **CIGARETTE PICTOGRAPH.** Consider the pictograph in Figure 3.51. Briefly describe what the graph is showing. Discuss whether it is effective in its purpose and whether it is deceptive in any way.

FURTHER APPLICATIONS

7. **SEASONAL EFFECTS ON SCHIZOPHRENIA?** The graph in Figure 3.52 shows data regarding the relative risk of schizophrenia among people born in different months.

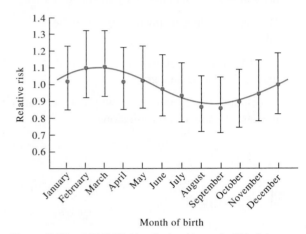

Figure 3.52   SOURCE: *New England Journal of Medicine.*

a. Note that the scale of the vertical axis does not include zero. Sketch the same risk curve using an axis that includes zero. Comment on the effect of this change.
b. Each value of the relative risk is shown with a dot at its most likely value and with an "error bar" indicating the range in which the data value probably lies. The study concludes that "the risk was also significantly associated with the season of birth." Given the size of the error bars, does this claim appear justified? (Is it possible to draw a flat line that passes through all of the error bars?)

8. **CHARITABLE DONATIONS.** The graph in Figure 3.53 shows the total amount of money given to charity in the United States from 1967 to 1997.

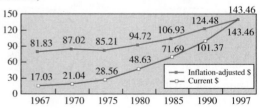

Source: American Association of Fund-Raising Counsel Trust for Philanthropy

Figure 3.53   SOURCE: *Wall Street Journal Almanac.*

Adults with Reported Tobacco Use and Reported Exposure to Environmental Tobacco Smoke, 1988–1991

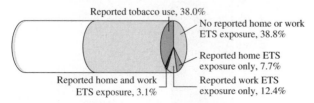

Source: Centers for Disease Control and Prevention

Figure 3.51   SOURCE: *Wall Street Journal Almanac.*

a. Briefly discuss the general trend revealed on this graph.

b. Suppose someone claimed that Americans were several times more generous in 1997 than they were in the 1960s. Would this statement be accurate? Why or why not?

c. Can we conclude from this graph that individual Americans gave more to charity in the 1990s than in the 1960s? If so, explain how. If not, explain what additional data would be needed to answer this question.

9. **PERCENT CHANGE IN THE CPI.** The graph in Figure 3.54 shows the percent change in the CPI over 14 years. In what year (of the years displayed) was the change in the CPI the greatest? In what year was the change in the CPI the least? Based on this graph, what can you conclude about changes in prices during the period shown?

Percent Change in U.S. CPI

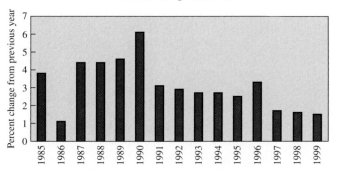

Figure 3.54   SOURCE: U.S. Bureau of Labor Statistics.

10. **CONSTANT DOLLARS.** The graph in Figure 3.55 shows the minimum wage in the United States over 60 years, together with its value in "constant dollars." In two or three paragraphs, summarize what the graph shows.

Tracking the Minimum Wage

The hourly federal minimum wage and its purchasing power in 1997 dollars

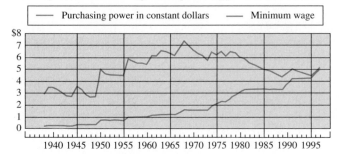

Figure 3.55   SOURCE: *Wall Street Journal Almanac.*

11. **DOUBLE HORIZONTAL SCALE.** The graph in Figure 3.56 shows *simultaneously* the number of births in this country during two time periods: 1946–1964 and 1977–1994. When did the first baby boom peak? When did the second baby boom peak? Why do you think the designers of this display chose to superimpose the two time intervals, rather than use a single time scale from 1946 through 1994?

Baby Boomers and Their Babies

The baby-boom generation, born between 1946 and 1964, produced their own smaller boom between 1977 and 1994.

Number of U.S. Births, 1946 through 1964 and 1977 through 1994 (in millions)

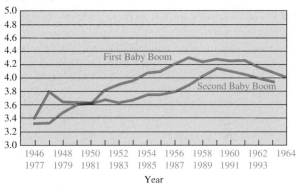

Source: National Center for Health Statistics

Figure 3.56   SOURCE: *Wall Street Journal Almanac.*

PROJECTS FOR THE WEB AND BEYOND

For useful links, select "Links for Web Projects" for Chapter 3 at www.aw.com/bbt.

12. **USA SNAPSHOT.** *USA Today* offers a daily pictograph for its "USA Snapshot." Study today's snapshot. Briefly discuss its purpose and effectiveness.

13. **CREATING DISPLAYS.** Currently, women earn about 74 cents for each dollar earned by men doing the same job. Draw a graph that depicts this information objectively, then draw another graph that exaggerates the difference.

IN THE NEWS

1. **DISTORTIONS IN THE NEWS.** Find an example in a recent news report of a graph that involves some type of perceptual distortion. Explain the effects of the distortion, and describe how the graph could have been drawn more honestly.

2. **SCALE PROBLEMS IN THE NEWS.** Find an example in a recent news report of a graph in which the vertical scale does not start at zero. Suggest why the graph was drawn that way and also discuss any ways in which the graphic might be misleading as a result.

3. **ECONOMIC GRAPH IN THE NEWS.** Find an example in a recent news report of a graph that shows economic data over time. Are the data adjusted for inflation? Discuss the meaning of the graph and any ways in which it might be deceptive.

4. **PICTOGRAPH IN THE NEWS.** Find an example of a pictograph in a recent news report. Discuss what the pictograph attempts to show, and discuss whether the artistic embellishments help or hinder this purpose.

5. **OUTSTANDING NEWS GRAPH.** Find a graph from a recent news report that, in your opinion, is truly outstanding in displaying data visually. Discuss what the graph shows, and explain why you think it is so outstanding.

6. **NOT-SO-OUTSTANDING NEWS GRAPH.** Find a graph from a recent news report that, in your opinion, fails in its attempt to display data visually in a meaningful way. Discuss what the graph was trying to show, explain why it failed, and explain how it could have been done better.

Chapter Review Exercises

Various media reports have claimed that it rains more on weekends. Listed below are measured amounts of rainfall (in inches) in Boston for all of the Wednesdays and Sundays in a recent year. Use these data for Exercises 1–4.

Sunday Rain:

0.05	0.00	0.00	0.00	0.00	0.00	0.00
0.00	0.09	0.00	0.00	0.00	0.00	0.00
0.28	0.00	0.00	0.00	0.00	0.00	0.27
0.01	0.00	0.00	0.33	0.00	0.00	0.00
0.00	0.00	0.44	0.00	0.01	0.00	0.00
0.00	0.00	0.00	0.00	0.23	0.00	0.00
0.00	0.21	1.28	0.01	0.02	0.28	0.00
0.00	0.00	0.01				

Wednesday Rain:

0.00	0.00	0.00	0.14	0.00	0.64	0.00
0.01	0.01	0.18	0.00	0.00	0.31	0.00
0.00	0.00	0.00	0.00	0.00	0.02	0.06
0.00	0.00	0.00	0.27	0.00	0.08	0.06
0.00	0.00	0.00	0.00	0.64	0.08	0.00
0.12	0.01	0.00	0.00	0.00	0.00	0.00
0.00	0.00	0.00	0.00	0.00	0.06	0.00
0.00	0.00	0.00				

1. **a.** Construct a frequency table for the Sunday rainfall amounts. Use bins of 0.00–0.09, 0.10–0.19, 0.20–0.29, and so on.
   b. Construct a frequency table for Wednesday rainfall amounts. Use bins of 0.00–0.09, 0.10–0.19, 0.20–0.29, and so on.
   c. Compare the frequency tables from parts a and b. What notable differences are there?

2. **a.** Construct a relative frequency table for the Sunday rainfall amounts. Use bins of 0.00–0.09, 0.10–0.19, 0.20–0.29, and so on.
   b. Using the frequency table from part a, construct a cumulative frequency table for the Sunday rainfall amounts.

3. **a.** Use the result from Exercise 1a to construct a histogram for the Sunday rainfall amounts.
   b. Use the result from Exercise 1b to construct a histogram for the Wednesday rainfall amounts.
   c. Compare the histograms from parts a and b. How are they similar and how are they different?

4. On the same set of axes, construct line charts for the Sunday rainfall amounts and the Wednesday rainfall amounts. Compare the rainfall amounts on those two days. Based on these two sets of measurements, does it appear to rain more on weekends, as media reports have claimed?

5. A study was conducted to determine how people get jobs. The table below lists data from 400 randomly selected subjects. The data are based on results from the National Center for Career Strategies. Construct a Pareto chart that corresponds to the given data. What seems to be the most effective approach for someone who would like to get a job?

Job sources of survey respondents	Frequency
Help-wanted ads	56
Executive search firms	44
Networking	280
Mass mailing	20

6. Construct a pie chart from the data given in Exercise 5. Compare the pie chart to the Pareto chart. Can you determine which graph is more effective in showing the relative importance of job sources?

FOCUS ON HISTORY

Can War Be Described with a Graph?

Can a war be described with a graph? Figure 3.57, created by Charles Joseph Minard in 1869, does so remarkably well. This graph tells the story of Napoleon's ill-fated Russian campaign of 1812, sometimes called Napoleon's death march.

The underlying map on Minard's graph shows a roughly 500-mile strip of land extending from the Niemen River on the Polish-Russian border to Moscow. The blue strip depicts the march of Napoleon's army. On Minard's original drawing, each millimeter of width represented 6,000 men; this reproduction is shown at a smaller size than the original. The march begins at the far left, where the strip is widest. Here, an army of 422,000 men triumphantly began a march toward Moscow on June 24, 1812. At the time, it was the largest army ever mobilized.

The narrowing of the strip as it approaches Moscow represents the unfolding decimation of the army. (The offshoots represent battalions that were sent off in other directions along the way.) Napoleon had brought only minimal food supplies, and hot summer weather accompanied by heavy rains brought rampant disease. Starvation, disease, and combat losses killed thousands of men each day. By the time the army entered Moscow on September 14, it had shrunk to 100,000 men. The worst was yet to come.

To Napoleon's dismay, the Russians evacuated Moscow prior to the French army's arrival. Deprived of the opportunity to engage the Russian troops and feeling that his army's condition was too poor to continue on to the Russian capital of St. Petersburg, Napoleon took his troops southward out of Moscow. The lower part of the strip on the graph represents the retreat, and the black line near the bottom of the figure shows the nighttime temperatures as winter approached. We see that freezing temperatures had already set in by October 18.

Temperatures plunged below 0°F in late November. The sudden narrowing of the lower strip around November 28 shows where 22,000 men perished on the banks of the Berezina River. Three-fourths of the survivors froze to death over the next few days, many on the bitter cold night of December 6. By the time the army reached Poland on December 14, only 10,000 of the original 422,000 remained.

In a famous analysis of graphical techniques, author Edward Tufte described Minard's graph as possibly "the best statistical graphic ever drawn." But a more dramatic statement came from a contemporary of Minard, E. J. Marey, who wrote that this graphic "brought tears to the eyes of all France."

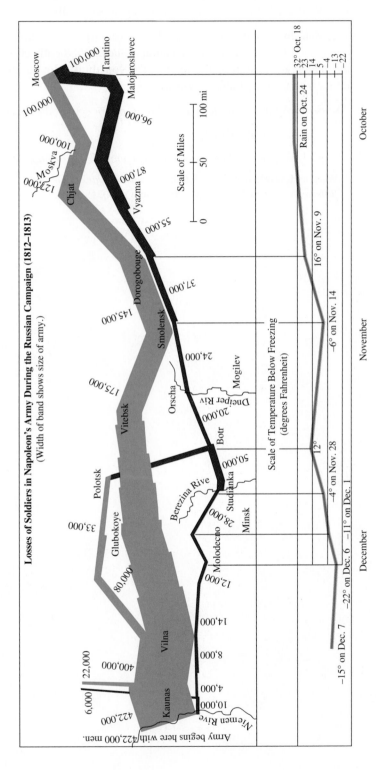

Figure 3.57   SOURCE: Edward R. Tufte, *The Visual Display of Quantitative Information* (Cheshire, CT: Graphics Press, 1983). Reprinted with permission.

QUESTIONS FOR DISCUSSION

1. Discuss how this graph helps to overcome the impersonal nature of the many deaths in a war. What kind of impact does it have on you personally?

2. Note that this graph plots six variables: two variables of direction (north-south and east-west), the size of the army, the location of the army, the direction of the army's movement, and temperatures during the retreat. Do you think Minard could have gotten the point across with fewer variables? Why or why not?

3. Discuss how you might make a similar graph for some other historical or political event.

SUGGESTED READING

Tufte, Edward R. *The Visual Display of Quantitative Information*. Graphics Press, 1992.

Wainer, Howard. *Visual Revelations: Graphical Tales of Fate and Deception from Napoleon Bonaparte to Ross Perot*. Copernicus (New York), 1997.

FOCUS ON ENVIRONMENT

How Much Carbon Dioxide Is in the Atmosphere?

You've undoubtedly heard about *global warming*, which refers to the idea that human activity (such as burning fossil fuels and clearing forests) may be raising the average temperature of the Earth. The questions of how much global warming will occur and how much damage it will cause are quite controversial. But most scientists believe they understand the primary cause of global warming quite well: increased amounts of carbon dioxide in the atmosphere. Carbon dioxide is the most common of several *greenhouse gases* that trap heat from the Sun. In theory, increasing the amount of these greenhouse gases in our atmosphere will cause the Earth to warm up. As usual, good graphics can make a complex situation much clearer.

Figure 3.58 shows data collected on Mauna Loa (Hawaii) regarding the atmospheric concentration of carbon dioxide. The carbon dioxide concentration is measured in *parts per million* (ppm); for example, 320 ppm means that there are 320 carbon dioxide molecules in every million molecules of air. Note that this is much less than 1%, since 1% means 10,000 parts per million. The yearly wiggles on the graph show that the carbon dioxide concentration varies with the seasons. Despite these wiggles, the long-term trend is clearly upward.

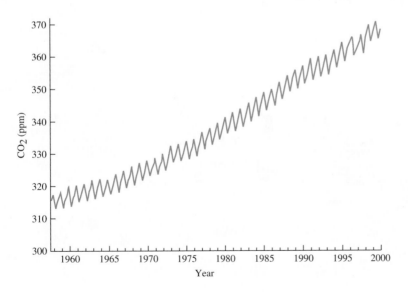

Figure 3.58 Carbon dioxide concentration data collected on Mauna Loa (Hawaii).

The Mauna Loa data have enabled scientists to calibrate other techniques for estimating the carbon dioxide concentration in the past. Scientists can then examine trends in energy use and carbon dioxide emissions to predict how the carbon dioxide concentration will continue to rise in the future. Figure 3.59 shows past estimates going back to the beginning of the Indus-

Global Warming

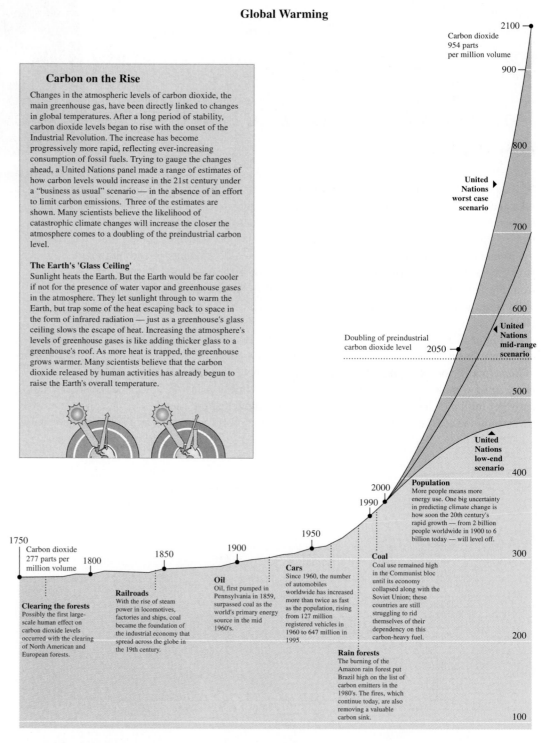

Carbon on the Rise

Changes in the atmospheric levels of carbon dioxide, the main greenhouse gas, have been directly linked to changes in global temperatures. After a long period of stability, carbon dioxide levels began to rise with the onset of the Industrial Revolution. The increase has become progressively more rapid, reflecting ever-increasing consumption of fossil fuels. Trying to gauge the changes ahead, a United Nations panel made a range of estimates of how carbon levels would increase in the 21st century under a "business as usual" scenario — in the absence of an effort to limit carbon emissions. Three of the estimates are shown. Many scientists believe the likelihood of catastrophic climate changes will increase the closer the atmosphere comes to a doubling of the preindustrial carbon level.

The Earth's 'Glass Ceiling'

Sunlight heats the Earth. But the Earth would be far cooler if not for the presence of water vapor and greenhouse gases in the atmosphere. They let sunlight through to warm the Earth, but trap some of the heat escaping back to space in the form of infrared radiation — just as a greenhouse's glass ceiling slows the escape of heat. Increasing the atmosphere's levels of greenhouse gases is like adding thicker glass to a greenhouse's roof. As more heat is trapped, the greenhouse grows warmer. Many scientists believe that the carbon dioxide released by human activities has already begun to raise the Earth's overall temperature.

Carbon dioxide 954 parts per million volume

United Nations worst case scenario

Doubling of preindustrial carbon dioxide level

United Nations mid-range scenario

United Nations low-end scenario

Population
More people means more energy use. One big uncertainty in predicting climate change is how soon the 20th century's rapid growth — from 2 billion people worldwide in 1900 to 6 billion today — will level off.

Carbon dioxide 277 parts per million volume

Clearing the forests
Possibly the first large-scale human effect on carbon dioxide levels occurred with the clearing of North American and European forests.

Railroads
With the rise of steam power in locomotives, factories and ships, coal became the foundation of the industrial economy that spread across the globe in the 19th century.

Oil
Oil, first pumped in Pennsylvania in 1859, surpassed coal as the world's primary energy source in the mid 1960's.

Cars
Since 1960, the number of automobiles worldwide has increased more than twice as fast as the population, rising from 127 million registered vehicles in 1960 to 647 million in 1995.

Coal
Coal use remained high in the Communist bloc until its economy collapsed along with the Soviet Union; these countries are still struggling to rid themselves of their dependency on this carbon-heavy fuel.

Rain forests
The burning of the Amazon rain forest put Brazil high on the list of carbon emitters in the 1980's. The fires, which continue today, are also removing a valuable carbon sink.

Figure 3.59   SOURCE: *New York Times*.

trial Revolution. Note that the carbon dioxide concentration has increased by about 30%, from 277 ppm in 1750 to about 360 ppm in 2000. Going past 2000, the graph shows three possible scenarios from United Nations scientists; presumably, the actual future carbon dioxide concentration will be somewhere between the low-end and high-end scenarios.

Note that Figure 3.59 not only shows the trends in carbon dioxide concentration, but also provides a wealth of other information. With words and pictures, it explains why this concentration is rising and how it is thought to cause global warming. Thus, it provides an outstanding example of the power of graphics to help us make sense of our world.

QUESTIONS FOR DISCUSSION

1. Study Figure 3.59 carefully. Based on what you learn, what do you think are the major sources of the additional carbon dioxide going into the atmosphere today? Which countries are the major sources of carbon dioxide emissions? Explain.

2. Discuss some of the factors that will affect the future concentration of carbon dioxide in the atmosphere. What kinds of assumptions about these factors do you think went into the low-end and high-end scenarios, respectively?

3. Calculate the percent increase in carbon dioxide concentration from 1750 to 2100 under each of the three future scenarios shown. Based on what you have learned about global warming here and elsewhere, do you think that we should be taking steps to reduce carbon dioxide emissions? If so, what steps would you advocate?

4. Note that the carbon dioxide data in Figure 3.58 were collected on the tall mountain of Mauna Loa on the big island of Hawaii. Why is this a good place to collect these data? Do you think the data could have been collected equally well near a major city like Los Angeles? Explain.

SUGGESTED READING

Bennett, Jeffrey; Donahue, Megan; Schneider, Nicholas; and Voit, Mark. *The Cosmic Perspective*, 2nd ed. Addison Wesley, 2002 (see Chaper 13).

Revkin, Andrew C. "Who Cares About a Few Degrees," *New York Times*, December 1, 1997.

Schneider, Stephen H. *Laboratory Earth*. Basic Books, 1998.

CHAPTER 4

Describing Data

In Chapter 3, we discussed methods for displaying data distributions with tables and graphs. Now we are ready to study how we can summarize the characteristics of a distribution in terms of just a few properties and numbers. In particular, we'll discuss common methods for describing the center, shape, and variation of a collection of data. These methods are central to data analysis and, as you'll see, have applications to nearly every statistical study you encounter in the news.

LEARNING GOALS

1. Understand and calculate common measures of center, including mean, median, and mode.

2. Describe the general shape of a distribution in terms of modes, skewness, and variation.

3. Understand and calculate common measures of variation, including range, the five-number summary, variance, and standard deviation.

4.1 What Is Average?

By now, you probably suspect that the term *average* can have multiple meanings. Measures of average are really ways of numerically describing the center of a data set. As you will see in this section, the most appropriate definition of *average* depends on the situation.

Mean, Median, and Mode

Let's begin by discussing three common measures of center: mean, median, and mode. Consider the science fiction movie series in Table 4.1, each consisting of several movies (that is, an original movie and its sequels or prequels). We find the **mean** of this data set by dividing the total number of movies by the five titles:

$$\text{mean} = \frac{4 + 3 + 5 + 9 + 5}{5} = \frac{26}{5} = 5.2$$

In other words, these five series have a mean of 5.2 movies among them. More generally, we find the mean of any data set by dividing the sum of all the data values by the number of data values. The mean is what most people think of as the average. In essence, it represents the balance point for a quantitative data distribution, as shown in Figure 4.1.

The **median** is the middle value in any quantitative data set. To find a median, we arrange the data values in ascending (or descending) order, repeating data values that appear more than once. If the number of values is odd, there is exactly one value in the middle of the list, and this value is the median. If the number of values is even, there are two values in the middle of the list, and the median is the number that lies halfway between them. For the data in Table 4.1, putting the frequencies in ascending order results in this list: 3, 4, 5, 5, 9. The median number of movies is 5 because 5 is the middle number in the list.

The **mode** is the most common value or group of values in a data set. In the case of the movies, the mode is 5 because this value occurs twice in the data set, while the other values occur once. A data set may have one mode, more than one mode, or no mode. Sometimes the mode refers to a group of closely spaced values rather than a single value. The mode is used

Table 4.1 Five Science Fiction Movie Series

Series	Number of movies
Alien	4
Back to the Future	3
Planet of the Apes*	5
Star Trek	9
Star Wars	5

*Not including remake.

By the Way...

A recent study at the University of Chicago discovered that the median allowance for teenagers in the United States is $50 per week. The study estimates that teenagers receive over $1 billion per week in allowances.

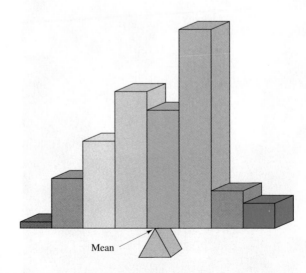

Figure 4.1 A histogram made from blocks would balance at the position of its mean.

more commonly for qualitative data than for quantitative data, as neither the mean nor the median can be used with qualitative data.

> ## Definitions
>
> The **mean** is what we most commonly call the average value. It is found as follows:
>
> $$\text{mean} = \frac{\text{sum of all values}}{\text{total number of values}}$$
>
> The **median** is the middle value in the sorted data set (or halfway between the two middle values if the number of values is even).
>
> The **mode** is the most common value (or group of values) in a data set.

When rounding, we will use the following rule for all the calculations discussed in this chapter.

> ## Rounding Rule for Statistical Calculations
>
> State your answers with *one* more decimal place of precision than is found in the raw data. Example: The mean of 2, 3, and 5 is 3.3333..., which we round to 3.3. Because the raw data are whole numbers, we round to the nearest tenth. *As always, round only the final answer and not any intermediate values used in your calculations.*

> **TECHNICAL NOTE**
>
> If the measure of center has the same number of significant digits as the original data, you can either include an extra zero or use the exact result without the extra decimal place. For example, the mean of 2 and 4 can be expressed as 3 or 3.0.

Note that we applied this rule in our example of the movies. The data in Table 4.1 consist of whole numbers, but we stated the mean as 5.2.

EXAMPLE 1   Price Data

Eight grocery stores in a small town sell the PR energy bar for the following prices:

$$\$1.09 \quad \$1.29 \quad \$1.29 \quad \$1.35 \quad \$1.39 \quad \$1.49 \quad \$1.59 \quad \$1.79$$

For these prices, find the mean, median, and mode.

SOLUTION   The *mean* price is $1.41:

$$\text{mean} = \frac{(\$1.09 + \$1.29 + \$1.29 + \$1.35 + \$1.39 + \$1.49 + \$1.59 + \$1.79)}{8}$$

$$= \$1.41$$

To find the *median*, we first sort the data in ascending order:

$$\underbrace{\$1.09, \$1.29, \$1.29,}_{\text{3 values below}} \quad \underbrace{\$1.35, \$1.39,}_{\text{2 middle values}} \quad \underbrace{\$1.49, \$1.59, \$1.79}_{\text{3 values above}}$$

Because there are eight prices (an even number), there are two values in the middle of the list: $1.35 and $1.39. Therefore the median lies halfway between these two values, which we calculate by adding them and dividing by 2:

$$\text{median} = \frac{\$1.35 + \$1.39}{2} = \$1.37$$

Using the rounding rule, we could express the mean and median as $1.410 and $1.370, respectively.

The *mode* is $1.29 because this price occurs more times than any other price. ∎

EXAMPLE 2   Oceans and Seas

Table 4.2 lists areas, in square kilometers, of the world's oceans and seas. For these areas, find the mean, median, and mode.

Table 4.2 Areas of the World's Oceans and Seas

Ocean/Sea	Area (sq. km)	Ocean/Sea	Area (sq. km)	Ocean/Sea	Area (sq. km)
Pacific Ocean	165,760,000	South China Sea	2,319,000	Japan Sea	1,007,800
Atlantic Ocean	82,400,000	Bering Sea	2,291,900	Andaman Sea	797,700
Indian Ocean	65,527,000	Gulf of Mexico	1,592,800	North Sea	575,200
Arctic Ocean	14,090,000	Okhotsk Sea	1,589,700	Red Sea	438,000
Mediterranean Sea	2,965,800	East China Sea	1,249,200	Baltic Sea	422,200
Caribbean Sea	2,718,200	Hudson Bay	1,232,300		

SOURCE: *TIME Almanac.*

SOLUTION   The sum of the areas for the 17 oceans and seas listed is 346,976,800 square kilometers. (You should confirm this sum for yourself.) Thus, we find the *mean* by dividing this sum by 17:

$$\text{mean} = \frac{346,976,800 \text{ sq. km}}{17} = 20,410,400 \text{ sq. km}$$

Table 4.2 is already arranged in descending order, which makes it easy to find the median. With 17 items on the list, the 9th value is the middle value because 8 values lie above it and 8 values lie below it. The 9th item on the list is the Gulf of Mexico, so the *median* is its area of 1,592,800 square kilometers.

These data have no *mode* because no value occurs more than once. However, if we look at groups of values, 5 of the 17 oceans and seas have areas between 1 and 2 million square kilometers. Thus, we might say that the most common areas are those between about 1 and 2 million square kilometers.   ∎

Effects of Outliers

The five graduating seniors on a college basketball team receive the following first-year contract offers to play in the National Basketball Association (zero indicates that the player did not receive a contract offer):

$$0 \quad 0 \quad 0 \quad 0 \quad \$3,500,000$$

The mean contract offer is

$$\text{mean} = \frac{0 + 0 + 0 + 0 + \$3,500,000}{5} = \$700,000$$

Is it therefore fair to say that the *average* senior on this basketball team received a $700,000 contract offer?

Not really. The problem is that the single player receiving the large offer makes the mean much larger than it would be otherwise. In fact, if we ignore this one player and look only at the other four, the mean contract offer is zero! Because this one value of $3,500,000 is so extreme compared to the others, we say that it is an **outlier**. As our example shows, an outlier can pull the mean significantly upward (or downward), thereby making the mean unrepresentative of the data set as a whole.

Definition

An **outlier** in a data set is a value that is much higher or much lower than almost all other values.

While the outlier pulls the mean contract offer upward, it has no effect on the median contract offer, which remains zero for the five players. In general, the value of an outlier has no effect on the median, because outliers don't lie in the middle of a data set. Outliers do not affect the mode either. (If we delete an outlier, the median can change because we are changing the number of values in the data set.)

Time out to think

Is it fair to use the median as the average contract offer for the five players? Why or why not?

Deciding how to deal with outliers is one of the more important issues in statistics. There are no general rules, so you must rely on your judgment. Sometimes, as in our basketball example, the outlier is a legitimate value that must be understood in order to interpret the mean and median properly. Other times, outliers may indicate mistakes in a data set. Table 4.3 summarizes the characteristics of the mean, median, and mode, including the effects of outliers on each measure.

By the Way...

A survey once found that geography majors from the University of North Carolina had a far higher mean starting salary than geography majors from other schools. The reason for the high mean turned out to be a single outlier—the basketball superstar and geography major named Michael Jordan.

Table 4.3 Comparison of Mean, Median, and Mode

Measure	Definition	How common?	Existence	Takes every value into account?	Affected by outliers?	Advantages
Mean	$\dfrac{\text{sum of all values}}{\text{total number of values}}$	most familiar "average"	always exists	yes	yes	commonly understood; works well with many statistical methods
Median	middle value	common	always exists	no (aside from counting the total number of values)	no	when there are outliers, may be more representative of an "average" than the mean
Mode	most frequent value	sometimes used	may be no mode, one mode, or more than one mode	no	no	most appropriate for data at the nominal level of measurement (see Section 2.1)

EXAMPLE 3   Mistake?

A track coach wants to determine an appropriate pulse range for her athletes during their workouts. She chooses five of her best runners and asks them to wear heart monitors during a workout. In the middle of the workout, she reads the following pulse rates for the five athletes: 130, 135, 140, 145, 325. Which is a better measure of the average in this case—the mean or the median? Why?

SOLUTION   Four of the five values are fairly close together and seem reasonable for mid-workout pulse rates. The high value of 325 is an outlier. This outlier seems unreasonable (perhaps caused by a faulty heart monitor), because no one can have such a high pulse rate without being in cardiac arrest. If the coach uses the mean as the average, the outlier will affect it. If she uses the median, she'll have a more reasonable value, because it won't be affected by the suspicious data point. ∎

I am only an average man, but, by George, I work harder at it than the average man.

—Theodore Roosevelt

Confusion About "Average"

The different meanings of "average" can lead to confusion. Sometimes this confusion arises because we are not told whether the "average" is the mean or the median, and other times because we are not given enough information about how the average was computed. The following examples illustrate a few such situations.

EXAMPLE 4   Wage Dispute

A newspaper surveys wages for workers in regional high-tech companies and reports an average of $22 per hour. The workers at one large firm immediately request a pay raise, claiming that they work as hard as employees at other companies but their average wage is only $19. The management rejects their request, telling them that they are *overpaid* because their average wage, in fact, is $23. Can both sides be right? Explain.

SOLUTION   Both sides can be right if they are using different definitions of average. In this case, the workers may be using the median while the management uses the mean. For example, imagine that there are only five workers at the company and their wages are $19, $19, $19, $19, and $39. The median of these five wages is $19 (as the workers claimed), but the mean is $23 (as management claimed). ∎

EXAMPLE 5   Which Mean?

All 100 first-year students at a small college take three courses in the Core Studies program. Two courses are taught in large lectures, with all 100 students in a single class. The third course is taught in 10 classes of 10 students each. Students and administrators get into an argument about whether classes are too large. The students claim that the mean size of their Core Studies classes is 70. The administrators claim that the mean class size is only 25. Can both sides be right? Explain.

Figures won't lie, but liars will figure.

—Charles H. Grosvenor

SOLUTION   Both sides are right, but they are talking about different means. The students calculated the mean size of the classes in which each student is personally enrolled. Each student is taking two classes with enrollments of 100 each and one class with an enrollment of 10, so the mean size of each student's classes is

$$\frac{\text{total enrollment in student's classes}}{\text{number of classes student is taking}} = \frac{100 + 100 + 10}{3} = 70$$

The administrators calculated the mean enrollment in all classes. There are two classes with 100 students each and 10 classes with 10 students each, making a total enrollment of 300 students in 12 classes. Thus, the mean enrollment per class is

$$\frac{\text{total enrollment}}{\text{number of classes}} = \frac{300}{12} = 25$$

The two claims about the mean are both correct, but very different because the students and administrators are talking about different means. The students calculated the mean class size *per student*, while the administrators calculated the mean number of students *per class*. ∎

Time out to think

In Example 5, could the administrators redistribute faculty assignments so that all classes have 25 students each? How? Discuss the advantages and disadvantages of such a change.

Weighted Mean

Suppose your course grade is based on four quizzes and one final exam. Each quiz counts as 15% of your final grade, and the final counts as 40%. Your quiz scores are 75, 80, 84, and 88, and your final exam score is 96. What is your overall score?

　　Because the final exam counts more than the quizzes, a simple mean of the five scores does not give your final score. Instead, we must assign a *weight* (indicating the relative importance) to each score. In this case, we assign weights of 15 (for the 15%) to each of the quizzes and 40 (for the 40%) to the final. We then find the **weighted mean** by adding the products of each score and its weight and then dividing by the sum of the weights:

$$\text{weighted mean} = \frac{(75 \times 15) + (80 \times 15) + (84 \times 15) + (88 \times 15) + (96 \times 40)}{15 + 15 + 15 + 15 + 40}$$

$$= \frac{8745}{100} = 87.45$$

The weighted mean of 87.45 properly accounts for the different weights of the quizzes and the exam. Following the rounding rule, we round this score to 87.5.

　　Weighted means are appropriate whenever the data values vary in their degree of importance. You can always find a weighted mean using the following formula.

> *The average family exists only on paper and its average budget is fiction, invented by statisticians for the convenience of statisticians.*
>
> —Sylvia Porter, financial columnist

Definition

A **weighted mean** accounts for variations in the relative importance of data values. Each data value is assigned a weight, and the weighted mean is

$$\text{weighted mean} = \frac{\text{sum of (each data value} \times \text{its weight)}}{\text{sum of all weights}}$$

Time out to think

Because the weights are percentages in the course grade example, we could think of the weights as 0.15 and 0.40 rather than 15 and 40. Calculate the weighted mean by using the weights of 0.15 and 0.40. Do you still find the same answer? Why or why not?

EXAMPLE 6 GPA

Randall has 38 credits with a grade of A, 22 credits with a grade of B, and 7 credits with a grade of C. What is his grade point average (GPA)? Base the GPA on values of 4.0 points for an A, 3.0 points for a B, and 2.0 points for a C.

SOLUTION The grades of A, B, and C represent data values of 4.0, 3.0, and 2.0, respectively. The numbers of credits are the weights. Thus, the As represent a data value of 4 with a weight of 38, the Bs represent a data value of 3 with a weight of 22, and the Cs represent a data value of 2 with a weight of 7. The weighted mean is

$$\text{weighted mean} = \frac{(4 \times 38) + (3 \times 22) + (2 \times 7)}{38 + 22 + 7} = \frac{232}{67} = 3.46$$

Following our rounding rule, we round Randall's GPA from 3.46 to 3.5. ∎

EXAMPLE 7 Stock Voting

Voting in corporate elections is usually weighted by the amount of stock owned by each voter. Suppose a company has five stockholders who vote in an election to determine whether the company will embark on a new advertising campaign. The votes (Y = yes, N = no) are as follows:

Stockholder	Shares owned	Vote
A	225	Y
B	170	Y
C	275	Y
D	500	N
E	90	N

According to the company's bylaws, the measure needs 60% of the vote to pass. Does it pass?

SOLUTION We can regard a yes vote as a value of 1 and a no vote as a value of 0. The number of shares is the weight for the vote of each stockholder. Thus, stockholder A's vote represents a value of 1 with a weight of 225, stockholder B's vote represents a value of 1 with a weight of 170, and so on. The weighted mean vote is

$$\text{weighted mean} = \frac{(1 \times 225) + (1 \times 170) + (1 \times 275) + (0 \times 500) + (0 \times 90)}{225 + 170 + 275 + 500 + 90}$$

$$= \frac{670}{1260} = 0.53$$

The weighted vote is 0.53, or 53%, in favor. This is short of the required 60%, so the measure does not pass. ∎

Means with Summation Notation (Optional Section)

Many statistical formulas, including the formula for the mean, can be written compactly with a mathematical notation called *summation notation*. The symbol Σ (the Greek capital letter *sigma*) is called the *summation sign* and indicates that a set of numbers should be added. We use the symbol x to represent *each* value in a data set, so we write the sum of all the data values as

$$\text{sum of all values} = \sum x$$

For example, if a sample consists of 25 exam scores, $\sum x$ represents the sum of all 25 scores. Similarly, if a sample consists of the incomes of 10,000 families, $\sum x$ represents the total dollar value of all 10,000 incomes.

We use n to represent the total number of values in the sample. Thus, the general formula for the mean is

$$\bar{x} = \text{sample mean} = \frac{\text{sum of all values}}{\text{total number of values}} = \frac{\sum x}{n}$$

The symbol $\bar{x}$ is the standard symbol for the mean of a sample. When dealing with the mean of a population rather than a sample, statisticians instead use the Greek letter μ *(mu)*.

Summation notation also makes it easy to express a general formula for the weighted mean. Again we use the symbol x to represent each data value, and we let w represent the weight of each data value. The sum of the products of each data value and its corresponding weight is $\sum (x \times w)$. The sum of the weights is $\sum w$. Therefore, the formula for the weighted mean is

$$\text{weighted mean} = \frac{\sum (x \times w)}{\sum w}$$

TECHNICAL NOTE

Summations are often written with the use of an *index* that specifies how to step through the sum. For example, the symbol x_i indicates the *i*th data value in the set; the letter *i* is the index. We then write the sum of all values as

$$\sum_{i=1}^{n} x_i$$

We read this expression as "the sum of the x_i values, starting with $i = 1$ and continuing to $i = n$, where n is the total number of data values in the set." With this notation, the mean is written

$$\bar{x} = \frac{1}{n} \sum_{i=1}^{n} x_i$$

Means and Medians with Binned Data (Optional Section)

The ideas of this section can be extended to binned data simply by assuming that the middle value in the bin represents all the data values in the bin. For example, consider the following table of 50 binned data values:

Bin	Frequency
0–6	10
7–13	10
14–20	10
21–27	20

The middle value of the first bin is 3, so we assume that the value of 3 occurs 10 times. Continuing this way, we have for the total of the 50 values in the table

$$(3 \times 10) + (10 \times 10) + (17 \times 10) + (24 \times 20) = 780$$

Thus, the mean is 780/50 = 15.6. With 50 values, the median is between the 25th and 26th values. These values fall within the bin 14–20, so we call this bin the **median class** for the data. The mode is the bin with the highest frequency—the bin 21–27 in this case.

Review Questions

1. Define and distinguish among *mean, median,* and *mode*.

2. What are outliers? Describe the effects of outliers on the mean, median, and mode.

3. Briefly explain the cause of the confusion about the "average" in each of Examples 4 and 5.

4. What is a weighted mean? Give a few examples in which a weighted mean is useful.

Exercises

BASIC SKILLS AND CONCEPTS

SENSIBLE STATEMENTS? For Exercises 1–6, determine whether the given statement is sensible and explain why it is or is not.

1. A data set should be discarded if the mean exceeds the mode.

2. A student with an average of 65 computes her new average after earning a 70 on the last exam. Her new average is 72.

3. Observing that the mean weight of a group of patients is 154 pounds and the median weight is 145 pounds, the doctor concludes that there must be an outlier on the heavy side.

4. Noting that there are three modes in his data set, Rob assumes that there is an error in his data gathering.

5. The two means in the data lie at 102 and 201.

6. The two medians in the data set lie at 23 and 28.

MEAN, MEDIAN, AND MODE. Exercises 7–14 each list a set of numbers. In each case, find the mean, median, and mode of the listed numbers.

7. Weights (in milligrams) of Bufferin aspirin tablets:

    672.2   679.2   669.8   672.6   672.2   662.2

8. Body temperatures (in degrees Fahrenheit) of randomly selected normal and healthy adults:

    98.6   98.6   98.0   98.0   99.0

    98.4   98.4   98.4   98.4   98.6

9. Blood alcohol concentrations of drivers involved in fatal crashes and then given jail sentences (based on data from the U.S. Department of Justice):

    0.27   0.17   0.17   0.16   0.13   0.24

    0.29   0.24   0.14   0.16   0.12   0.16

10. Golf scores from a recent WPGA tournament:

    69   70   70   71   73   73   75   76   76   78

    80   81   81   82   83   84   85   86   87   88

11. High temperatures (in °F) during a 15-day period in Alaska in March:

    15 11 10 9 0 2 4 5 5 7 10 12 15 18 19

12. Ages of selected U.S. Presidents at the time of inauguration:

    57   61   57   57   58   57   61   54

    68   51   49   64   50   48   65

13. Weights (in grams) of randomly selected M&M plain candies:

    0.957   0.912   0.842   0.925   0.939   0.886

    0.914   0.913   0.958   0.947   0.920

14. Weights (in grams) of quarters in circulation:

    5.60 5.63 5.58 5.56 5.66 5.58 5.57 5.59

    5.67 5.61 5.84 5.73 5.53 5.58 5.52 5.65

    5.57 5.71 5.59 5.53 5.63 5.68

15. ALPHABETIC STATES. The following table gives the total area in square miles (land and water) of the seven states whose names begin with the letters A through C.

State	Area
Alabama	52,200
Alaska	615,200
Arizona	114,000
Arkansas	53,200
California	158,900
Colorado	104,100
Connecticut	5,500

a. Find the mean area and median area for these states.

b. Which state is an outlier on the high end? If you eliminate this state, what are the new mean and median areas for this data set?

c. Which state is an outlier on the low end? If you eliminate this state, what are the new mean and median areas for this data set?

16. **OUTLIER COKE.** Cans of Coca-Cola vary slightly in weight. Here are the measured weights of seven cans, in pounds:

0.8161 0.8194 0.8165 0.8176 0.7901 0.8143 0.8126

a. Find the mean and median of these weights.
b. Which, if any, of these weights would you consider to be an outlier? Explain.
c. What are the mean and median weights if the outlier is excluded?

FURTHER APPLICATIONS

17. **RAISING YOUR GRADE.** Suppose you have scores of 70, 75, 80, and 70 on quizzes in a mathematics class.
    a. What is the mean of these scores?
    b. What score would you need on the next quiz to have an overall mean of 75?
    c. If the maximum score on the quizzes is 100, is it possible to have a mean of 80 after the fifth quiz? Explain.

18. **RAISING YOUR GRADE.** Suppose you have scores of 60, 70, 65, 85, and 85 on exams in a sociology class.
    a. What is the mean of these scores?
    b. What score would you need on the next exam to have an overall mean of 75?
    c. If the maximum score on the exams is 100, what is the maximum mean score that you could possibly have after the next exam? Explain.

19. **NEW MEAN.** Suppose that after six quizzes you have a mean score of 80 (out of 100). If you get a score of 90 on the next quiz, what is your new mean? What is the maximum mean score that you could have after the next quiz? What is the minimum mean score that you could have after the next quiz?

20. **NEW MEAN.** Suppose that after six quizzes you have a mean score of 75 (out of 100). If you get a score of 95 on the next quiz, what is your new mean? What is the maximum mean score that you could have after the next quiz? What is the minimum mean score that you could have after the next quiz?

21. **NEW BATTING AVERAGE.** Suppose that after 30 at-bats, a baseball player has a batting average of 0.300. If she gets a hit on the next at-bat, what will her new batting average be? (Recall that the batting average is the number of hits divided by the number of at-bats.)

22. **COMPARING AVERAGES.** Suppose that school district officials claim that the average reading score for fourth graders in the district is 73 (out of 100). As a principal, you know that your fourth graders had the following scores: 55, 60, 68, 70, 87, 88, 95. Would you be justified in claiming that your students scored above the district average? Explain.

23. **COMPARING AVERAGES.** Suppose the National Basketball Association (NBA) reports the average height of basketball players to be 6′8″. As a coach you know that the players on your starting lineup have heights of 6′5″, 6′6″, 6′6″, 7′0″, and 7′2″. Would you be justified in claiming that your starting lineup has above average height for the NBA? Explain.

APPROPRIATE AVERAGE. Exercises 24–29 list "averages" that someone might want to know. In each case, state whether the mean, median, or some other measure would give a better description of the "average." Explain your reasoning.

24. The average income of all adults in a large city

25. The average weight of the oranges in a large box

26. The average number of times that people change jobs during their careers

27. The average number of pieces of lost luggage per flight for each airline company

28. The average daily high temperature over a period of a month

29. The average waiting time in the checkout lines of a supermarket

30. **AVERAGE PEACHES.** Suppose a grocer has three baskets of peaches. One holds 50 peaches and weighs 18 pounds, one holds 55 peaches and weighs 22 pounds, and the third holds 60 peaches and weighs 24 pounds. What is the mean weight of the individual peaches? Explain.

31. **AVERAGE CONFUSION.** Imagine you are a teacher. Your first-period class with 25 students had a mean score of 86% on the midterm exam. Your second-period class with 30 students had a mean score of 84% on the same exam. Does it follow that the mean score for both classes is 85%? Explain.

32. **WHICH MEAN?** Suppose 300 students at a high school all take the same four courses. Three of the courses are taught in 15 classes of 20 students each. The fourth course is taught in three classes of 100 students each. Explain how the school's principal could claim that the mean class size is 25 students, while upset parents could claim that the mean class size is 40 students. Which mean do you think is a fairer description of class sizes?

33. **FINAL GRADE.** Your course grade is based on one midterm that counts as 15% of your final grade, one class project that counts as 20% of your final grade, a set of homework assignments that counts as 40% of your final grade, and a final exam that counts as 25% of your final grade. Your midterm score is 75, your project score is 90, your homework score is 85, and your final exam score is 72. What is your overall final score?

34. **BATTING AVERAGE.** Recall that a batting average in baseball is determined by dividing the total number of hits by the total number of at-bats (neglecting walks, errors, and a few other special cases). A player goes 2 for 4 (2 hits in 4 at-bats) in the first game, 0 for 3 in the second game, and 3 for 5 in the third game. What is his batting average? In what way is this number an "average"?

35. **AVERAGING AVERAGES.** Suppose a player has a batting average over many games of 0.400. In his next game, he goes 2 for 4, which is a batting average of 0.500 for the game. Does it follow that his new batting average is $(0.400 + 0.500)/2 = 0.450$? Explain.

36. **EGG PERCENTAGES.** Suppose a farm inspector finds that 8% of the eggs tested at one farm contain salmonella and 12% of the eggs at another farm contain salmonella. Does it follow that between the two farms, a total of 10% of the eggs tested contain salmonella? Why or why not?

37. **SLUGGING AVERAGE.** Another measure of hitting performance in baseball is called the slugging average. In finding a slugging average, a single is worth 1 point, a double is worth 2 points, a triple is worth 3 points, and a home run is worth 4 points. A player's slugging average is the total number of points divided by the total number of at-bats (neglecting walks, errors, and a few other special cases). Suppose a player has three singles in five at-bats in the first game, a triple and a single in four at-bats in the second game, and a double and a home run in five at-bats in the third game.
    a. What is his batting average?
    b. What is his slugging average?
    c. Is it possible for a slugging average to be more than 1? Explain.

38. **STOCKHOLDER VOTING.** Imagine that a small company has four stockholders: A holds 400 shares, B holds 300 shares, C holds 200 shares, and D holds 100 shares. In a vote on a new advertising campaign, A votes yes and B, C, and D vote no. Explain how the outcome of the vote can be expressed as a weighted mean. What is the outcome of the vote?

39. **GPA.** One system for computing a grade point average (GPA) assigns 4 points to an A, 3 points to a B, 2 points to a C, and 1 point to a D. What is the GPA of a student who gets an A in a 5-credit course, and a B, a C, and a D in each of three 3-credit courses?

40. **U.S. POPULATION CENTER.** Imagine taking a huge, flat map of the United States and placing weights on it to represent where people live. The point at which the map would balance is called the mean center of population. Figure 4.2 shows how the location of the mean center of population has shifted from 1790 to 1990. Briefly explain the pattern shown on this map.

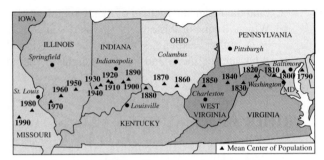

Figure 4.2   Mean center of population.
SOURCE: *Statistical Abstract of the United States.*

PROJECTS FOR THE WEB AND BEYOND

For useful links, select "Links for Web Projects" for Chapter 4 at www.aw.com/bbt.

41. **SALARY DATA.** Many Web sites offer data on salaries in different careers. Find salary data for a career you are considering. What are the mean and median salaries for this career? How do these salaries compare to those of other careers that interest you?

42. **IS THE MEDIAN THE MESSAGE?** Read the article "The Median Isn't the Message," by Stephen Jay Gould, which is posted on the Web. Write a few paragraphs in which you describe the message that Gould is trying to get across. How is this message important to other patients diagnosed with cancer?

43. **NAVEL DATA.** The *navel ratio* is defined to be a person's height divided by the height (from the floor) of his or her navel. Measure the navel ratio of each person in your class. Then, bin the data, make a frequency table, and draw a histogram of the distribution. What is the mean the distribution? What is the median of the distribution? An old theory says that, on average, the navel ratio of humans is the golden ratio: $(1 + \sqrt{5})/2$. Does this theory seem accurate based on your observations?

IN THE NEWS

1. DAILY AVERAGES. Cite three examples of averages that you deal with in your own life (such as grade point average or batting average). In each case, explain whether the average is a mean, a median, or some other type of average. Briefly describe how the average is useful to you.

2. AVERAGES IN THE NEWS. Find three recent news articles that refer to some type of average. In each case, explain whether the average is a mean, a median, or some other type of average.

4.2 Shapes of Distributions

So far in this chapter we have discussed how to describe the center of a quantitative data distribution with measures such as the mean and median. We now turn our attention to describing the overall *shape* of a distribution. We can *see* the complete shape of a distribution on a graph. In essence, our current goal is to describe the general shape in words. Although such descriptions carry less information than the complete graph, they are still useful. We will focus on three characteristics of a distribution: its number of modes, its symmetry or skewness, and its variation.

Because we are interested primarily in the *general* shapes of distributions, it's often easier to examine graphs that show smooth curves rather than actual data sets. Figure 4.3 shows three examples of this idea, two in which the actual distributions are shown as histograms and one in which the actual distribution is shown as a line chart. The smooth curves approximate the actual distributions, but do not show all of their details.

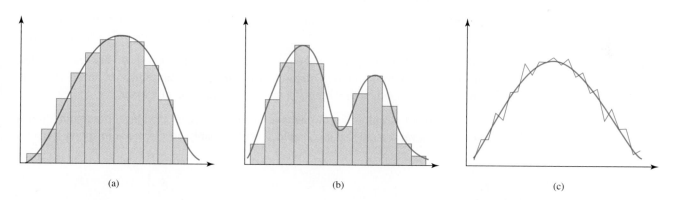

(a) (b) (c)

Figure 4.3 The smooth curves approximate the actual shapes of the distributions.

Number of Modes

One simple way of describing the shape of a distribution is by its number of peaks, or modes. Figure 4.4a shows a distribution, called a **uniform distribution**, that has no mode because all data values have the same frequency. Figure 4.4b shows a distribution with a single peak at its mode. It is called a **single-peaked**, or **unimodal**, distribution. By convention, any peak in a

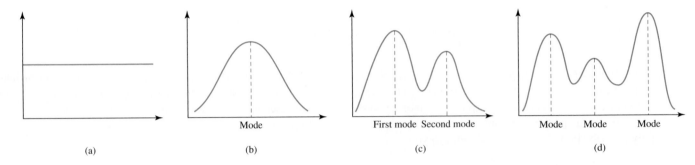

Figure 4.4 (a) A uniform distribution has no mode. (b) A single-peaked distribution has one mode. (c) A bimodal distribution has two modes. (d) A trimodal distribution has three modes.

distribution is considered a mode, even if not all peaks have the same height. For example, the distribution in Figure 4.4c is said to have two modes—even though the second peak is lower than the first; it is a *bimodal* distribution. Similarly, the distribution in Figure 4.4d is said to have three modes; it is a *trimodal* distribution.

EXAMPLE 1 Number of Modes

How many modes would you expect for each of the following distributions? Why? Make a rough sketch for each distribution, with clearly labeled axes.

a. Heights of 1,000 randomly selected adult women

b. Hours spent watching football on TV in January for 1,000 randomly selected adult Americans

c. Weekly sales throughout the year at a retail clothing store for children

d. The number of people with particular last digits (0 through 9) in their Social Security numbers

SOLUTION   Figure 4.5 shows sketches of the distributions.

a. The distribution of heights of women is single-peaked because many women are at or near the mean height, with fewer and fewer women at heights much greater or less than the mean.

b. The distribution of times spent watching football on TV for 1,000 randomly selected adult Americans is likely to be bimodal (two modes). One mode represents the mean watching time of men, and the other represents the mean watching time of women.

c. The distribution of weekly sales throughout the year at a retail clothing store for children is likely to have several modes. For example, it will probably have a mode in spring for sales

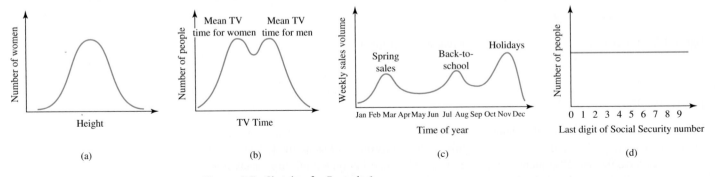

Figure 4.5 Sketches for Example 1.

of summer clothing, a mode in late summer for back-to-school sales, and another mode in winter for holiday sales.

d. The last digits of Social Security numbers are essentially random, so the number of people with each different last digit (0 through 9) should be about the same. That is, about 10% of all Social Security numbers end in 0, 10% end in 1, and so on. It is therefore a uniform distribution with no mode. ∎

Symmetry or Skewness

A second simple way to describe the shape of a distribution is in terms of its symmetry or skewness. A distribution is **symmetric** if its left half is a mirror image of its right half. The distributions in Figure 4.6 are all symmetric.

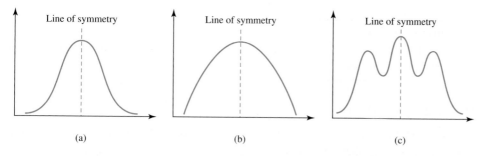

(a)   (b)   (c)

Figure 4.6   These distributions are all symmetric because their left halves are mirror images of their right halves. Note that (a) and (b) are single-peaked (unimodal), whereas (c) is triple-peaked (trimodal).

A distribution that is not symmetric must have values that tend to be more spread out on one side than on the other. In this case, we say that the distribution is **skewed**. Figure 4.7a shows a distribution in which the values are more spread out on the left, meaning that some values are outliers at low values. We say that such a distribution is **left-skewed**. It is helpful to think of such a distribution as having a tail that has been pulled toward the left. Figure 4.7b shows a distribution in which the values are more spread out on the right, making it **right-skewed**. It looks as if it has a tail pulled toward the right.

Figure 4.7 also shows how skewness affects the relative positions of the mean, median, and mode. By definition, the mode is at the peak in a single-peaked distribution. A left-skewed

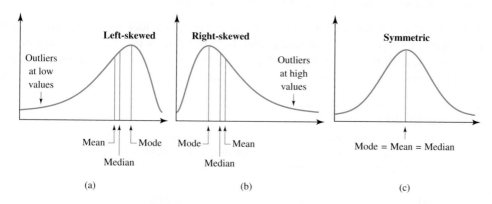

(a)   (b)   (c)

Figure 4.7   (a) Skewed to the left (left-skewed): The mean and median are less than the mode.
(b) Skewed to the right (right-skewed): The mean and median are greater than the mode.
(c) Symmetric distribution: The mean, median, and mode are the same.

distribution pulls the mean and median to the left—that is, to values less than the mode. Moreover, outliers at the low end of the data set tend to make the mean less than the median (see Table 4.3 on page 151). Similarly, a right-skewed distribution pulls the mean and median to the right—that is, to values greater than the mode. In such cases, the outliers at the high end of the data set tend to make the mean greater than the median. When the distribution is symmetric and single-peaked, both the mean and the median are equal to the mode.

Definitions

A distribution is **symmetric** if its left half is a mirror image of its right half.

A distribution is **left-skewed** if its values are more spread out on the left side.

A distribution is **right-skewed** if its values are more spread out on the right side.

Time out to think

In a skewed distribution, the median is a better measure of the center of the distribution than the mean. Why do you think this is true?

EXAMPLE 2   Skewness

For each of the following situations, state whether you expect the distribution to be symmetric, left-skewed, or right-skewed. Explain.

a. Heights of a sample of 100 women

b. Family income in the United States

c. Speeds of cars on a road where a visible patrol car is using radar to detect speeders.

SOLUTION

a. The distribution of heights of women is symmetric because roughly equal numbers of women are shorter and taller than the mean, and extremes of height are rare on either side of the mean.

b. The distribution of family income is right-skewed. Most families are middle-class, so the mode of this distribution is a middle-class income (somewhere around $40,000 in the United States). But a few very high-income families pull the mean to a considerably higher value, stretching the distribution to the right (high-income) side.

c. Drivers usually slow down when they are aware of a patrol car looking for speeders. Few if any drivers will be exceeding the speed limit, but some drivers tend to slow to well below the speed limit. Thus, the distribution of speeds will be left-skewed, with a mode near the speed limit but a few cars going well below the speed limit. ∎

Time out to think

In ordinary English, the term *skewed* is often used to mean something that is distorted or depicted in an unfair way. How is this use of *skew* related to its meaning in statistics?

Variation

A third way to describe a distribution is by its **variation**, which is a measure of how much the data values are spread out. A distribution in which most data are clustered together has a low variation. As shown in Figure 4.8a, such a distribution has a fairly sharp peak. The variation is higher when the data are distributed more widely around the center, which makes the peak broader. Figure 4.8b shows a distribution with a moderately high variation and Figure 4.8c shows a distribution with an even higher variation. In the next section, we'll discuss methods for describing the variation quantitatively.

By the Way...

In 1997, doctors told 25-year-old Lance Armstrong that, according to the averages, he had less than a 50-50 chance of surviving his testicular cancer (it had already spread to his brain). In 2001, besides being in complete remission, Armstrong won his third consecutive *Tour de France* (bicycle race), considered by many to be the most grueling of all professional athletic competitions. His story shows that, on a personal level, variation can be much more important than any average.

> ### Definition
> **Variation** describes how widely data are spread out about the center of a data set.

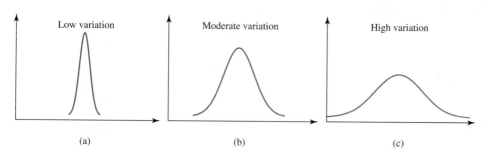

Figure 4.8 From left to right, these three distributions have increasing variation.

EXAMPLE 3   Variation in Marathon Times

How would you expect the variation to differ between times in the Olympic marathon and times in the New York marathon? Explain.

SOLUTION   The Olympic marathon invites only elite runners, whose times are likely to be clustered relatively near world-record times. The New York marathon allows runners of all abilities, whose times are spread over a very wide range (from near the world record to many hours). Therefore, the variation among the times should be greater in the New York marathon than in the Olympic marathon. ■

Review Questions

1. Briefly explain how we can use smooth curves to approximate a distribution in a way that helps us see its general features. What kind of information is lost when we replace a complete distribution with a smoothed approximation of it?

2. Give simple examples of a uniform distribution, a single-peaked distribution, a bimodal distribution, and a distribution with more than two modes.

3. What do we mean when we say that a distribution is symmetric? Give simple real examples of a symmetric distribution, a left-skewed distribution, and a right-skewed distribution.

4. What do we mean by the variation of a distribution? Give simple examples of distributions with different amounts of variation.

Exercises

BASIC SKILLS AND CONCEPTS

SENSIBLE STATEMENTS? For Exercises 1–5, determine whether the given statement is sensible and explain why it is or is not.

1. Because this data set has two modes, it cannot be symmetric.

2. This distribution is left-skewed because it has outliers to the left.

3. Josh works at a veterinary clinic and weighs dogs. He claims that there is less variation in the weights of 10 Rottweilers than in the weights of 10 dogs of different breeds.

4. It's Josh again (see Exercise 3). This time he makes a single distribution of the weights of 10 toy poodles, 10 setters, and 10 St. Bernards. He observes that the distribution has one mode.

5. Jean concludes that the mean of her symmetric distribution is greater than the median.

6. OLD FAITHFUL. The histogram in Figure 4.9 shows the times between eruptions of Old Faithful geyser in Yellowstone National Park for a sample of 300 eruptions (which means 299 times between eruptions). Draw a smooth curve over the histogram that captures its general features. Then classify the distribution according to its number of modes and symmetry or skewness. In words, summarize the meaning of your results.

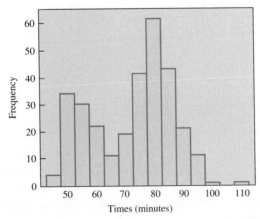

Times Between Eruptions of Old Faithful

Figure 4.9
SOURCE: Hand et al., *Handbook of Small Data Sets*.

7. CHIP FAILURES. The histogram in Figure 4.10 shows the time until failure for a sample of 108 computer chips.

Draw a smooth curve over the histogram that captures its general features. Then classify the distribution according to its number of modes and symmetry or skewness. In words, summarize the meaning of your results.

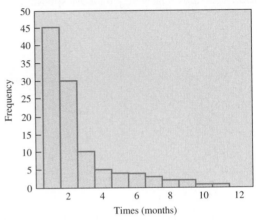

Failure Time of Computer Chips

Figure 4.10
SOURCE: Hand et al., *Handbook of Small Data Sets*.

8. RUGBY WEIGHTS. The histogram in Figure 4.11 shows the weights of a sample of 391 rugby players. Draw a smooth curve over the histogram that captures its general features. Then classify the distribution according to its number of modes and symmetry or skewness. In words, summarize the meaning of your results.

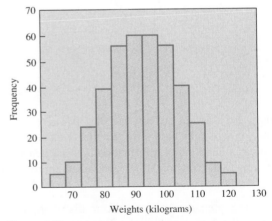

Weight of Rugby Players

Figure 4.11
SOURCE: Hand et al., *Handbook of Small Data Sets*.

FURTHER APPLICATIONS

9. **FAMILY INCOME.** Suppose you study family income in a random sample of 300 families. Your results can be summarized as follows:

   • The mean family income was $41,000.
   • The median family income was $35,000.
   • The highest and lowest incomes were $250,000 and $2,400, respectively.

   a. Draw a rough sketch of the income distribution, with clearly labeled axes. Describe the distribution as symmetric, left-skewed, or right-skewed.
   b. How many families in the sample earned less than $35,000? Explain how you know.
   c. Based on the given data, can you determine how many families earned more than $41,000? Why or why not?

DESCRIBING DISTRIBUTIONS. For each distribution described in Exercises 10–29, answer the following:

A. How many modes would you expect for the distribution? Explain.
b. Would you expect the distribution to be symmetric, left-skewed, or right-skewed? Explain.

10. The exam scores of 50 students who received grades of A on the exam

11. The exam scores of all 200 students who took a fair exam with a mean score of 75%

12. The weights of 100 eighth-grade students

13. The weights of 100 professional football players

14. The weights of people who use an ice rink that is open only to professional figure skaters in the morning and to professional hockey players in the afternoon and evening

15. The weights of cars in the new car lot at a car dealership in which about half the inventory consists of compact cars and the other half consists of sport utility vehicles

16. The number of buses that leave a busy bus terminal each hour over a 24-hour period

17. The times between successive buses at a bus stop when all buses are running on time

18. The delays in minutes of scheduled flights from a large airport

19. The speeds of drivers as they pass through a school zone

20. The ages of people who visit an art museum

21. The ages of people who visit an amusement park

22. The numbers of people with particular last digits (0 through 9) in their phone numbers

23. The number of players in each doubles match of a women's tennis tournament

24. The number of cars that pass through a busy intersection per minute over a 24-hour period

25. The monthly sales of swimming suits over a one-year period at a store in San Diego

26. The incomes of people sitting in luxury boxes at the Super Bowl

27. The incomes of people watching the Super Bowl on TV

28. The salaries of major league baseball players

29. The batting averages of major league baseball players

PROJECTS FOR THE WEB AND BEYOND

For useful links, select "Links for Web Projects" for Chapter 4 at www.aw.com/bbt.

30. **NEW YORK MARATHON.** The Web site for the New York marathon gives frequency data for finish times in the most recent marathon. Study the data, make a rough sketch of the distribution, and describe the shape of the distribution in words.

31. **TAX STATS.** The IRS Web site provides statistics collected from tax returns on income, refunds, and much more. Choose a set of statistics from this Web site and study the distribution. Describe the distribution in words, and discuss anything you learn that is relevant to national tax policies.

32. **SOCIAL SECURITY DATA.** Survey a sample of fellow students, asking each to indicate the last digit of her or his Social Security number. Also ask each participant to indicate the fifth digit. Draw one graph showing the distribution of the last digits and another graph showing the distribution of the fifth digits. Compare the two graphs. What notable difference becomes apparent?

4.3 Measures of Variation

In Section 4.2, we saw how to describe variation qualitatively. The idea of variation is so important that there are several common ways to describe it quantitatively.

Why Variation Matters

We mortals cross the ocean of this world,
Each in his average cabin of a life.
—Robert Browning

Imagine that you observe customers waiting in line for tellers at two different banks. Customers at Big Bank can enter any one of three different lines leading to three different tellers. Best Bank also has three tellers, but all customers wait in a single line and are called to the next available teller. The following values are waiting times (in minutes) for 11 customers at each bank. The times are arranged in ascending order.

Big Bank (three lines):

| 4.1 | 5.2 | 5.6 | 6.2 | 6.7 | 7.2 | 7.7 | 7.7 | 8.5 | 9.3 | 11.0 |

Best Bank (one line):

| 6.6 | 6.7 | 6.7 | 6.9 | 7.1 | 7.2 | 7.3 | 7.4 | 7.7 | 7.8 | 7.8 |

You'll probably find more unhappy customers at Big Bank than at Best Bank, but this is *not* because the average wait is any longer. In fact, you should verify for yourself that the mean and median waiting times are 7.2 minutes at both banks. The difference in customer satisfaction comes from the *variation* at the two banks. The waiting times at Big Bank vary over a fairly wide range, so a few customers have long waits and are likely to become annoyed. In contrast, the variation of the waiting times at Best Bank is small, so all customers feel they are being treated roughly equally.

By the Way...

The idea of waiting in line (or *queuing*) is important not only for people but also for data, particularly for data streaming through the Internet. Major corporations often employ statisticians to help them make sure that data move smoothly and without bottlenecks through their servers and Web pages.

Time out to think

Explain why Big Bank, with three separate lines, should have a greater variation in waiting times than Best Bank. Then consider several places where you commonly wait in lines, such as the grocery store, a bank, a concert ticket outlet, or a fast food restaurant. Does each place use a single customer line or multiple lines? If the place uses multiple lines, do you think a single line would be better? Explain.

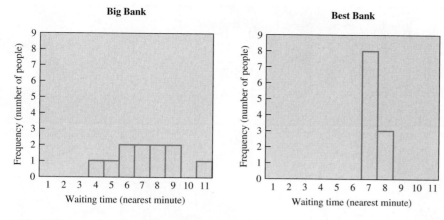

Figure 4.12   Histograms for the waiting times at Big Bank and Best Bank, shown with data binned to the nearest minute.

We can depict the difference in the two variations by making histograms in which we bin the data values to the nearest minute (Figure 4.12). However, for the purposes of statistical comparisons, we'd like to have a way to describe each variation with numbers. In the rest of this section, we will discuss several methods, including range, the five-number summary, and the standard deviation.

Range

The simplest way to describe the variation of a data set is to compute its **range**, defined as the difference between the lowest (minimum) and highest (maximum) values. For the example of the two banks, the waiting times for Big Bank vary from 4.1 to 11.0 minutes, so the range is $11.0 - 4.1 = 6.9$ minutes. The waiting times for Best Bank vary from 6.6 to 7.8 minutes, so the range is $7.8 - 6.6 = 1.2$ minutes. The range for Big Bank is much larger, reflecting its greater variation.

> ## Definition
> The **range** of a distribution is the difference between its highest and lowest data values:
>
> $$\text{range} = \text{highest value (max)} - \text{lowest value (min)}$$

Although the range is easy to compute and can be useful, it occasionally can be misleading, as the next example shows.

EXAMPLE 1   Misleading Range

Consider the following two sets of quiz scores for nine students. Which set has the greater range? Would you also say that this set has the greater variation?

Quiz 1:	1	10	10	10	10	10	10	10	10
Quiz 2:	2	3	4	5	6	7	8	9	10

SOLUTION   The range for Quiz 1 is $10 - 1 = 9$ points and the range for Quiz 2 is $10 - 2 = 8$ points. Thus, the range is greater for Quiz 1. However, aside from a single low score (an outlier), Quiz 1 has no variation at all because every other student got a 10. In contrast, no two students got the same score on Quiz 2, and the scores are spread throughout the list of possible scores. Thus, Quiz 2 has greater variation even though Quiz 1 has greater range.   ∎

DILBERT reprinted by permission of United Feature Syndicate, Inc.

Quartiles and the Five-Number Summary

A better way to describe variation is to consider a few intermediate data values in addition to the high and low values. The most common way involves looking at the **quartiles**, or values that divide the data distribution into quarters. The following list repeats the waiting times at the two banks, with the quartiles shown in bold. Note that the middle quartile, which divides the data set in half, is simply the median.

			Lower quartile ↓			Median ↓			Upper quartile ↓		
Big Bank:	4.1	5.2	**5.6**	6.2	6.7	**7.2**	7.7	7.7	**8.5**	9.3	11.0
Best Bank:	6.6	6.7	**6.7**	6.9	7.1	**7.2**	7.3	7.4	**7.7**	7.8	7.8

TECHNICAL NOTE

Statisticians do not universally agree on this procedure for calculating quartiles, and different procedures can result in different values.

Definitions

The **lower quartile** (or *first quartile*) divides the lowest fourth of a data set from the upper three-fourths. It is the median of the data values in the *lower half* of a data set. (Exclude the middle value in the data set if the number of data points is odd.)

The **middle quartile** (or *second quartile*) is the overall median.

The **upper quartile** (or *third quartile*) divides the lowest three-fourths of a data set from the upper fourth. It is the median of the data values in the *upper half* of a data set. (Exclude the middle value in the data set if the number of data points is odd.)

Once we know the quartiles, we can describe a distribution with a **five-number summary**, consisting of the low value, the lower quartile, the median, the upper quartile, and the high value. For the waiting times at the two banks, the five-number summaries are as follows:

Big Bank:		*Best Bank:*	
low	= 4.1	low	= 6.6
lower quartile	= 5.6	lower quartile	= 6.7
median	= 7.2	median	= 7.2
upper quartile	= 8.5	upper quartile	= 7.7
high	= 11.0	high	= 7.8

The Five-Number Summary

The **five-number summary** for a data distribution consists of the following five numbers:

low value   lower quartile   median   upper quartile   high value

---

We can display the five-number summary with a graph called a **boxplot** (sometimes called a *box-and-whisker* plot). Using a number line for reference, we enclose the values from the lower to the upper quartiles in a box. We then draw a line through the box at the median and add two "whiskers," extending from the box to the low and high values. Figure 4.13 shows boxplots for the bank waiting times. Both the box and the whiskers for Big Bank are broader than those for Best Bank, indicating that the waiting times have greater variation at Big Bank.

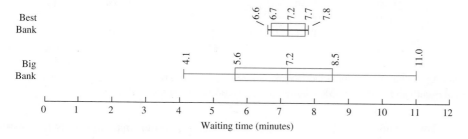

Figure 4.13   Boxplots show that the variation of the waiting times is greater at Big Bank than at Best Bank.

Drawing a Boxplot

Step 1. Draw a number line that spans all the values in the data set.

Step 2. Enclose the values from the lower to the upper quartile in a box. (The thickness of the box has no meaning.)

Step 3. Draw a line through the box at the median.

Step 4. Add "whiskers" extending to the low and high values.

TECHNICAL NOTE

Some boxplots are drawn with outliers marked by an asterisk (∗) and the whiskers extending only to the smallest and largest *nonoutliers*.

EXAMPLE 2   Passive and Active Smoke

One way to study exposure to cigarette smoke is by measuring blood levels of *serum cotinine*, a metabolic product of nicotine that the body absorbs from cigarette smoke. Table 4.4 lists serum cotinine levels from samples of 50 smokers ("active smoke") and 50 nonsmokers who are exposed to cigarette smoke at home or at work ("passive smoke"). Compare the two data sets (smokers and nonsmokers) with five-number summaries and boxplots, and discuss your results.

Table 4.4 Serum Cotinine Levels (nanograms per milliliter of blood) in Samples of 50 Smokers and 50 Nonsmokers Exposed to Passive Smoke, with Data Values Listed in Ascending Order

Order number	Smokers	Nonsmokers	Order number	Smokers	Nonsmokers
1	0.08	0.03	26	34.21	0.82
2	0.14	0.07	27	36.73	0.97
3	0.27	0.08	28	37.73	1.12
4	0.44	0.08	29	39.48	1.23
5	0.51	0.09	30	48.58	1.37
6	1.78	0.09	31	51.21	1.40
7	2.55	0.10	32	56.74	1.67
8	3.03	0.11	33	58.69	1.98
9	3.44	0.12	34	72.37	2.33
10	4.98	0.12	35	104.54	2.42
11	6.87	0.14	36	114.49	2.66
12	11.12	0.17	37	145.43	2.87
13	12.58	0.20	38	187.34	3.13
14	13.73	0.23	39	226.82	3.54
15	14.42	0.27	40	267.83	3.76
16	18.22	0.28	41	328.46	4.58
17	19.28	0.30	42	388.74	5.31
18	20.16	0.33	43	405.28	6.20
19	23.67	0.37	44	415.38	7.14
20	25.00	0.38	45	417.82	7.25
21	25.39	0.44	46	539.62	10.23
22	29.41	0.49	47	592.79	10.83
23	30.71	0.51	48	688.36	17.11
24	32.54	0.51	49	692.51	37.44
25	32.56	0.68	50	983.41	61.33

NOTE: The column "Order number" is included to make it easier to read the table.
SOURCE: National Health and Nutrition Examination Survey, National Institutes of Health.

SOLUTION The two data sets are already in ascending order, making it easy to construct the five-number summary. Each has 50 data points, so the median lies halfway between the 25th and 26th values. For the smokers, the 25th and 26th values are 32.56 and 34.21, respectively, so the median is

$$\frac{32.56 + 34.21}{2} = 33.385$$

For the nonsmokers, the 25th and 26th values are 0.68 and 0.82, respectively, so the median is

$$\frac{0.68 + 0.82}{2} = 0.75$$

The lower quartile is the median of the *lower half* of the values, which is the 13th value in each set. The upper quartile is the median of the *upper half* of the values, which is the 38th value in each set. Thus, the five-number summaries for the two data sets are as follows.

Active Smoke:

low value	= 0.08 ng/ml
lower quartile	= 12.58 ng/ml
median	= 33.385 ng/ml
upper quartile	= 187.34 ng/ml
high	= 983.41 ng/ml

Passive Smoke:

low value	= 0.03 ng/ml
lower quartile	= 0.20 ng/ml
median	= 0.75 ng/ml
upper quartile	= 3.13 ng/ml
high	= 61.33 ng/ml

By the Way…

Passive smoke appears to be particularly harmful to young children. Apparently, the toxins in cigarette smoke have a greater effect on developing bodies than on full-grown adults. A similar effect is found for most other toxins, which is why it is especially important to limit children's exposure to toxic chemicals.

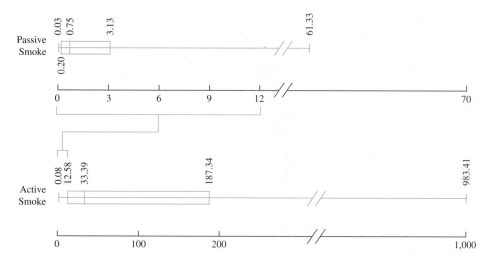

Figure 4.14   Boxplots for the data in Table 4.4.

Figure 4.14 shows boxplots for the two data sets. The boxplots make it easy to see some key features of the data sets. For example, it is immediately clear that the active smokers have a higher median level of serum cotinine, as well as a greater variation in levels. We conclude that smokers absorb considerably more nicotine than do nonsmokers exposed to passive smoke. Nevertheless, the levels in the passive smokers are more than we would expect in people who had no exposure to cigarette smoke. Indeed, the nonsmoker with the high value for passive smoke has absorbed more nicotine than the median smoker. We conclude that passive smoke can expose nonsmokers to significant amounts of nicotine. Given the known dangers of cigarette smoke, these results give us reason to be concerned about possible health effects from passive smoke.   ∎

Percentiles

Quartiles divide a data set into four segments. It is possible to divide a data set even more. For example, *quintiles* divide a data set into 5 segments, and *deciles* divide a data set into 10 segments. It is particularly common to divide data sets into 100 segments using **percentiles**. Roughly speaking, for example, the 35th percentile is a value that separates the bottom 35% of the data values from the top 65%. (To be more precise, the 35th percentile is greater than or equal to at least 35% of the data values, and it is less than or equal to at least 65% of the data values.)

If a data value lies between two percentiles, it is common to say that the data value lies *in* the lower percentile. For example, if you score higher than 84.7% of all people taking a college entrance examination, we say that your score is in the 84th percentile.

> **TECHNICAL NOTE**
>
> Like quartiles, percentiles may be calculated using slightly different procedures.

Definition

The *n*th percentile of a data set divides the bottom *n*% of data values from the top (100 − *n*)%. A data value that lies between two percentiles is often said to lie *in* the lower percentile. You can approximate the percentile of any data value with the following formula:

$$\text{percentile of data value} = \frac{\text{number of values less than this data value}}{\text{total number of values in data set}} \times 100$$

There are different procedures for finding a data value corresponding to a given percentile, but one approximate approach is to find the Lth value, where L is the product of the percentile (in decimal form) and the sample size. For example, with 50 sample values, the 12th percentile is around the $0.12 \times 50 = 6$th value.

EXAMPLE 3   Smoke Exposure Percentiles

Answer the following questions concerning the data in Table 4.4.

a. What is the percentile for the data value of 104.54 ng/ml for smokers?

b. What is the percentile for the data value of 61.33 ng/ml for nonsmokers?

c. What data value marks the 36th percentile for the smokers? For the nonsmokers?

SOLUTION   The following results are approximate.

a. The data value of 104.54 ng/ml for smokers is the 35th data value in the set, which means that 34 data values lie below it. Thus, its percentile is

$$\frac{\text{number of values less than 104.54 ng/ml}}{\text{total number of values in data set}} \times 100 = \frac{34}{50} \times 100 = 68$$

In other words, the 35th data value marks the 68th percentile.

b. The data value of 61.33 ng/ml for nonsmokers is the 50th and highest data value in the set, which means that 49 data values lie below it. Thus, its percentile is

$$\frac{\text{number of values less than 61.33 ng/ml}}{\text{total number of values in data set}} \times 100 = \frac{49}{50} \times 100 = 98$$

In other words, the highest data value in this set lies in the 98th percentile.

c. Because there are 50 data values in the set, the 36th percentile is around the $0.36 \times 50 = 18$th value. For smokers this value is 20.16 ng/ml, and for nonsmokers it is 0.33 ng/ml.   ∎

Standard Deviation

Although the five-number summary gives an excellent characterization of variation, statisticians often prefer to describe variation with a single number. The single number most commonly used to describe variation is called the **standard deviation**. The standard deviation is not difficult to understand, but we'll discuss how it is calculated before we explore its meaning.

First, we must understand the idea of a **deviation from the mean** (or just *deviation*). The deviation from the mean for a particular data value tells us how far that value lies from the mean. In our earlier example, the mean waiting time was 7.2 minutes for both Big Bank and Best Bank. A waiting time of 8.2 minutes therefore has a deviation from the mean of 1 minute, since it is 1 minute greater than the mean of 7.2 minutes. A waiting time of 5.2 minutes has a deviation from the mean of −2 minutes (*negative* 2 minutes), since it is 2 minutes less than the mean of 7.2 minutes.

In essence, the standard deviation is a measure of the average of all the deviations from the mean. However, the actual mean of the deviations is always zero, because the positive deviations exactly balance the negative deviations. Therefore, the standard deviation calculation involves first finding a mean of the *squares* of the deviations (as the squares will all be positive) and taking a square root in the end. (For technical reasons, we divide the sum of the squares by the total number of data values *minus* 1.)

TECHNICAL NOTE

In finding the standard deviation, we divide the sum of the squared deviations by the total number of data values minus 1 when dealing with data from a *sample*. When dealing with an entire population, we do not subtract the 1. In this book, we will use only the sample formula.

Calculating the Standard Deviation

To calculate the standard deviation for any data set:

Step 1. Compute the mean of the data set. Then find the deviation from the mean for every data value by subtracting the mean from the data value. That is, for every data value:

$$\text{deviation from mean} = \text{data value} - \text{mean}$$

Step 2. Find the squares (second power) of all the deviations from the mean.

Step 3. Add all the squares of the deviations from the mean.

Step 4. Divide this sum by the total number of data values *minus* 1.

Step 5. The standard deviation is the square root of this quotient.

Overall, these steps produce the standard deviation formula:

$$\text{standard deviation} = \sqrt{\frac{\text{sum of (deviations from the mean)}^2}{\text{total number of data values} - 1}}$$

(This formula is shown in summation notation on p. 176.)

> **TECHNICAL NOTE**
>
> The result of Step 4 is called the *variance* of the distribution. In other words, the standard deviation is the square root of the variance. Although the variance is used in many advanced statistical computations, we will not use it in this book.

Note that, because we square the deviations in Step 3 and then take the square root in Step 5, the units of the standard deviation are the same as the units of the data values. For example, if the data values have units of minutes, the standard deviation also has units of minutes. Although the standard deviation formula may look complex, it is easy to use for small data sets; but it is tedious to use for large data sets. Fortunately, many calculators and computers automate these calculations.

EXAMPLE 4   Calculating Standard Deviation

Calculate the standard deviations for the waiting times at Big Bank and Best Bank.

SOLUTION   We follow the five steps to calculate the standard deviations. Table 4.5 shows how to organize the work in the first three steps. The first column for each bank lists the waiting

Table 4.5 Calculating Standard Deviation

	Big Bank			Best Bank	
Time	Deviation (Time − Mean)	(Deviation)^2	Time	Deviation (Time − Mean)	(Deviation)^2
4.1	4.1 − 7.2 = −3.1	$(-3.1)^2 = 9.61$	6.6	6.6 − 7.2 = −0.6	$(-0.6)^2 = 0.36$
5.2	5.2 − 7.2 = −2.0	$(-2.0)^2 = 4.00$	6.7	6.7 − 7.2 = −0.5	$(-0.5)^2 = 0.25$
5.6	5.6 − 7.2 = −1.6	$(-1.6)^2 = 2.56$	6.7	6.7 − 7.2 = −0.5	$(-0.5)^2 = 0.25$
6.2	6.2 − 7.2 = −1.0	$(-1.0)^2 = 1.00$	6.9	6.9 − 7.2 = −0.3	$(-0.3)^2 = 0.09$
6.7	6.7 − 7.2 = −0.5	$(-0.5)^2 = 0.25$	7.1	7.1 − 7.2 = −0.1	$(-0.1)^2 = 0.01$
7.2	7.2 − 7.2 = 0.0	$(0.0)^2 = 0.0$	7.2	7.2 − 7.2 = 0.0	$(0.0)^2 = 0.0$
7.7	7.7 − 7.2 = 0.5	$(0.5)^2 = 0.25$	7.3	7.3 − 7.2 = 0.1	$(0.1)^2 = 0.01$
7.7	7.7 − 7.2 = 0.5	$(0.5)^2 = 0.25$	7.4	7.4 − 7.2 = 0.2	$(0.2)^2 = 0.04$
8.5	8.5 − 7.2 = 1.3	$(1.3)^2 = 1.69$	7.7	7.7 − 7.2 = 0.5	$(0.5)^2 = 0.25$
9.3	9.3 − 7.2 = 2.1	$(2.1)^2 = 4.41$	7.8	7.8 − 7.2 = 0.6	$(0.6)^2 = 0.36$
11.0	11.0 − 7.2 = 3.8	$(3.8)^2 = 14.44$	7.8	7.8 − 7.2 = 0.6	$(0.6)^2 = 0.36$
		Sum = 38.46			**Sum = 1.98**

times (in minutes), the second column lists the deviations from the mean (Step 1), and the third column lists the squares of the deviations (Step 2). We add all the squared deviations to find the sum at the bottom of the third column (Step 3). Note that we can calculate the deviations because we already know that the mean waiting time is 7.2 minutes for both banks. Step 4 tells us to divide this sum (of the squared deviations) by the total number of data values *minus* 1. Because there are 11 data values, we divide by 10:

$$Big\ Bank: \quad \frac{38.46}{10} = 3.846$$

$$Best\ Bank: \quad \frac{1.98}{10} = 0.198$$

Finally, Step 5 tells us that the standard deviation is the square root of the number from Step 4. Thus, the standard deviations are

$$Big\ Bank: \quad standard\ deviation = \sqrt{3.846} = 1.96\ minutes$$

$$Best\ Bank: \quad standard\ deviation = \sqrt{0.198} = 0.44\ minute$$

In other words, while both banks had the same mean waiting time of 7.2 minutes, typical waiting times tended to be about 1.96 minutes away from this mean at Big Bank but only 0.44 minute away from this mean at Best Bank. Again, we see that the waiting times showed greater variation at Big Bank, explaining why the lines at Big Bank annoyed more customers than did those at Best Bank. ∎

Time out to think

Look closely at the individual deviations in Table 4.5 in Example 4. Do the standard deviations for the two data sets seem like reasonable "averages" for the deviations? Explain.

Interpreting the Standard Deviation

We've seen that the standard deviation is a good way to describe variation, but it would be nice to understand the numerical value of standard deviations in a more precise way. The following simple rule allows us to interpret the standard deviation in many cases. Although it provides only very rough approximations and breaks down on occasion, this rule can help us understand the standard deviation.

TECHNICAL NOTE

Another way of interpreting the standard deviation uses a mathematical rule called *Chebyshev's Theorem*. It states that, for any data distribution, at least 75% of all data values lie within two standard deviations of the mean, and at least 89% of all data values lie within three deviations of the mean. Although we will not use this theorem in this book, you may encounter it if you take a more advanced statistics course.

The Range Rule of Thumb

The standard deviation is *approximately* related to the range of a distribution by the **range rule of thumb**:

$$standard\ deviation \approx \frac{range}{4}$$

If we know the range of a distribution (range = high − low), we can use this rule to estimate the standard deviation. Alternatively, if we know the standard deviation, we can use this rule to estimate the low and high values as follows:

$$low\ value \approx mean - (2 \times standard\ deviation)$$
$$high\ value \approx mean + (2 \times standard\ deviation)$$

The range rule of thumb does not work well when the high or low values are outliers.

Note that while the range rule of thumb often provides a simple way of interpreting the standard deviation, it is not the only way. In particular, the standard deviation has a more precise interpretation for distributions that have bell-shaped graphs (called normal distributions). We will study this use of the standard deviation in Chapter 5.

EXAMPLE 5   Using the Range Rule of Thumb

Use the range rule of thumb to estimate the standard deviations for the waiting times at Big Bank and Best Bank. Compare the estimates to the actual values found in Example 4.

SOLUTION   The waiting times for Big Bank vary from 4.1 to 11.0 minutes, which means a range of $11.0 - 4.1 = 6.9$ minutes. The waiting times for Best Bank vary from 6.6 to 7.8 minutes, for a range of $7.8 - 6.6 = 1.2$ minutes. Thus, the range rule of thumb gives the following estimates for the standard deviations:

$$\text{Big Bank:} \quad \text{standard deviation} \approx \frac{6.9}{4} = 1.7$$

$$\text{Best Bank:} \quad \text{standard deviation} \approx \frac{1.2}{4} = 0.3$$

The actual standard deviations calculated in Example 4 are 1.96 and 0.44, respectively. For these two cases, the estimates from the range rule of thumb slightly underestimate the actual standard deviations. Nevertheless, the estimates put us in the right ballpark, showing that the rule is useful. ∎

EXAMPLE 6   Estimating a Range

Studies of the gas mileage of a BMW under varying driving conditions show that it gets a mean of 22 miles per gallon with a standard deviation of 3 miles per gallon. Estimate the minimum and maximum typical gas mileage amounts that you can expect under ordinary driving conditions.

SOLUTION   From the range rule of thumb, the low and high values for gas mileage are approximately

$$\text{low value} \approx \text{mean} - 2 \times \text{standard deviation} = 22 - (2 \times 3) = 16$$

$$\text{high value} \approx \text{mean} + 2 \times \text{standard deviation} = 22 + (2 \times 3) = 28$$

The range of gas mileage for the car is roughly from a minimum of 16 miles per gallon to a maximum of 28 miles per gallon. ∎

EXAMPLE 7   SAT Range

The mean score on the mathematics SAT for women is 496, and the standard deviation is 108. Use the range rule of thumb to estimate the minimum and maximum typical scores for women on the mathematics SAT. Discuss your results.

SOLUTION   From the range rule of thumb, the low and high values for the SAT scores are approximately

$$\text{low value} \approx \text{mean} - 2 \times \text{standard deviation}$$
$$= 496 - (2 \times 108) = 280$$
$$\text{high value} \approx \text{mean} + 2 \times \text{standard deviation}$$
$$= 496 + (2 \times 108) = 712$$

By the Way...

Technologies such as catalytic converters have helped reduce the amounts of many pollutants emitted by cars (per mile driven), but increasing gas mileage is the only way to reduce the carbon dioxide emissions implicated in global warming. This is a major reason why auto manufacturers are trying to develop high-mileage "hybrid" vehicles and zero-emission vehicles that run on electricity or fuel cells.

In fact, the actual minimum and maximum possible scores are 200 and 800, respectively. The range rule of thumb tells us that *most* women score between 280 and 712. Scores below 280 or above 712 are unusual. ∎

Standard Deviation with Summation Notation (Optional Section)

The summation notation introduced earlier makes it easy to write the standard deviation formula in a compact form. Recall that x represents the individual values in a data set and $\bar{x}$ represents the mean of the data set. We can therefore write the deviation from the mean for any data value as

$$\text{deviation} = \text{data value} - \text{mean} = x - \bar{x}$$

We can now write the sum of all squared deviations as

$$\text{sum of all squared deviations} = \Sigma(x - \bar{x})^2$$

The remaining steps in the calculation of the standard deviation are to divide this sum by $n - 1$ and then take the square root. You should confirm for yourself that the following formula summarizes the five steps in the earlier box:

$$s = \text{standard deviation} = \sqrt{\frac{\Sigma(x - \bar{x})^2}{n - 1}}$$

The symbol s is the conventional symbol for the standard deviation of a sample. For the standard deviation of a population, statisticians use the Greek letter σ (*sigma*), and the term $n - 1$ in the formula is replaced by n. Consequently, you will get slightly different results for the standard deviation depending on whether you assume the data represent a sample or a population.

TECHNICAL NOTE

The formula for the *variance* is

$$s^2 = \frac{\Sigma(x - \bar{x})^2}{n - 1}$$

The standard symbol for the variance, s^2, reflects the fact that it is the square of the standard deviation.

Review Questions

1. Describe how we define and calculate the range of a distribution.

2. What are the quartiles of a distribution? How do we find them?

3. Define the five-number summary, and explain how we can depict it visually with a boxplot.

4. What are the percentiles of a distribution? How do we calculate the percentile of a particular data value?

5. Explain what is meant by the deviation from the mean of each data point in a distribution.

6. Describe the process by which we calculate a standard deviation. Give a simple example of its calculation (such as calculating the standard deviation of the numbers 2, 3, 4, 4, and 6). What is the value of the standard deviation if all of the sample values are the same?

7. Briefly describe the use of the range rule of thumb for interpreting the standard deviation. What are its limitations?

Exercises

BASIC SKILLS AND CONCEPTS

SENSIBLE STATEMENTS?   For Exercises 1–6, determine whether the given statement is sensible and explain why it is or is not.

1. Sue's exam score was at the class median and was in the 62nd percentile.

2. Roger's exam score was at the class mean and was in the 43rd percentile.

3. June is a bird watcher. She claims that because hawks are large birds, the boxplot for the weights of 10 hawks is wider than the boxplot for the weights of 10 birds of different species.

4. The standard deviation of the hourly wages of 10 lawyers is less than the standard deviation of the hourly wages of 10 fast food servers.

5. The gas mileages of 10 randomly selected cars from a used car lot have a greater range than the gas mileages of 10 Honda Accords.

6. Carl notices that the ages in his data set are between 10 and 70, while the standard deviation is 50.

7. BIG BANK VERIFICATION. Find the mean and median for the waiting times at Big Bank given in the beginning of this section. Show your work clearly, and verify that both are 7.2 minutes.

8. BEST BANK VERIFICATION. Find the mean and median for the waiting times at Best Bank given in the beginning of this section. Show your work clearly, and verify that both are 7.2 minutes.

9. CALCULATING PERCENTILES. Suppose there are 1,000 students at your school.
   a. Imagine that your GPA is higher than that of 835 students. What is the percentile of your GPA?
   b. Imagine that your cardiovascular fitness is better than that of 921 students. What is your percentile for fitness?
   c. Imagine that you are taller than 125 of the students. What is your percentile for height?

10. CALCULATING PERCENTILES. Suppose you live in a town with 35,000 residents.
    a. Imagine that you are older than 7,800 residents. What is your percentile for age in your town?
    b. Imagine that the amount of garbage you generate is more than that of 22,500 residents. What is your percentile for garbage generation?
    c. Imagine that your average commute time to school or work is greater than that of 14,200 residents. What is your percentile for commute time?

11. PERCENTILES. Consider the data set

$$\{1, 2, 3, \ldots 199, 200\}$$

   a. Which data value is the 30th percentile?
   b. Which data value is the 60th percentile?
   c. Which data value is the 83rd percentile?

12. UNDERSTANDING STANDARD DEVIATION. The following four sets of 7 numbers all have a mean of 9.

   $\{9, 9, 9, 9, 9, 9, 9\}$, $\{8, 8, 9, 9, 9, 10, 10\}$,
   $\{8, 8, 8, 9, 10, 10, 10\}$, $\{6, 6, 6, 9, 12, 12, 12\}$

   a. Make a histogram for each set.
   b. Give the five-number summary and draw a boxplot for each set.
   c. Compute the standard deviation for each set.
   d. Based on your results, briefly explain how the standard deviation provides a useful, single-number summary of the variation in these data sets.

13. UNDERSTANDING STANDARD DEVIATION. The following four sets of 7 numbers all have a mean of 6.

   $\{6, 6, 6, 6, 6, 6, 6\}$, $\{5, 5, 6, 6, 6, 7, 7\}$,
   $\{5, 5, 5, 6, 7, 7, 7\}$, $\{3, 3, 3, 6, 9, 9, 9\}$

   a. Make a histogram for each set.
   b. Give the five-number summary and draw a boxplot for each set.
   c. Compute the standard deviation for each set.
   d. Based on your results, briefly explain how the standard deviation provides a useful, single-number summary of the variation in these data sets.

COMPARING VARIATIONS. For each of Exercises 14–19, do the following:

a. Find the mean, median, and range for each of the two data sets.

b. Give the five-number summary and draw a boxplot for each of the two data sets.

c. Find the standard deviation for each of the two data sets.

d. Apply the range rule of thumb to estimate the standard deviation of each of the two data sets. How well does the rule work in each case? Briefly discuss why it does or does not work well.

e. Based on all your results, compare and discuss the two data sets in terms of their center and variation.

14. The following data sets give waiting times in minutes of 11 customers at bus terminals in Atlanta and Boston.

   Atlanta:

   5.5 6.0 4.5 5.0 7.0   6.5 5.0 7.5 5.5   4.0 8.0

   Boston:

   5.5 8.0 2.0 5.0 8.5 12.0 1.5 6.5 9.5 10.0 6.0

15. The following data sets give the ages in years of a sample of cars in a faculty parking lot and a student parking lot at the College of Portland.

   Faculty: 2 3 1 0 1 2 4 3 3 2 1

   Student: 5 6 8 2 7 10 1 4 6 10 9

16. The following data sets give the driving speeds in miles per hour of the first nine cars to pass through a school zone and the first nine cars to pass through a downtown intersection.

   School: 20 18 23 21 19 18 17 24 25

   Downtown: 29 31 35 24 31 26 36 31 28

17. The following data sets give the numbers of games played in the Stanley Cup finals (hockey) from 1992 to 2001 and in the NBA finals from 1992 to 2001; seven games is the maximum and four games is the minimum in each final series.

   Stanley Cup: 4 5 7 4 4 4 4 6 6 7

   NBA Finals: 6 6 7 4 6 6 6 5 6 5

18. The following data sets show the ages of the first seven U.S. Presidents (Washington through Jackson) and seven recent U.S. Presidents (Nixon through G. W. Bush) at the time of inauguration.

   First 7: 57 61 57 57 58 57 61

   Last 7:   56 61 52 69 64 46 54

19. The following data sets give the approximate lengths of Beethoven's nine symphonies and Mahler's nine symphonies (in minutes).

   Beethoven:

   28 36 50 33 30 40 38 26 68

   Mahler:

   52 85 94 50 72 72 80 90 80

FURTHER APPLICATIONS

20. TEACHER SALARY PERCENTILES. According to the National Education Association, Alaska ranks first among the 50 states in salaries for public school teachers, Delaware ranks 12th, Ohio ranks 17th, South Carolina ranks 37th, and Wyoming ranks 42nd. In what percentile is each of these states?

21. SMALL BABY. A friend has just returned from taking her baby to the doctor's office for his 18-month checkup. She tells you that the baby is very healthy, but very small—the doctor claimed the baby was in the *negative* third percentile! Is this possible? Explain.

22. GREAT VISION. After returning from the eye doctor, your friend tells you that he has great vision—his vision is in the 105th percentile. Is this possible? Explain.

23. PIZZA DELIVERIES. After recording pizza delivery times for two different pizza shops, you conclude that one pizza shop has a mean delivery time of 45 minutes with a standard deviation of 3 minutes and the other shop has a mean delivery time of 42 minutes with a standard deviation of 20 minutes. Interpret these figures. If you liked the pizzas from both shops equally well, which one would you order from? Why?

24. MANAGING COMPLAINTS. You manage a small ice cream shop in which your employees scoop the ice cream by hand. Each night, you total your sales and the total volume of ice cream sold. You find that on nights when an employee named Sam is working, the mean price of the ice cream sold is $1.75 per pint with a standard deviation of $0.05. On nights when an employee named Kevin is working, the mean price of the ice cream sold is $1.70 per pint with a standard deviation of $0.35. Which employee is more likely to be generating complaints of "too small" servings? Explain.

25. PORTFOLIO STANDARD DEVIATION. The book *Investments* by Zvi Bodie, Alex Kane, and Alan Marcus claims that the returns for investment portfolios with a single

stock have a standard deviation of 0.55, while the returns for portfolios with 32 stocks have a standard deviation of 0.325. Explain how the standard deviation measures the risk in these two types of portfolios.

26. CHOOSING A FUND. Your job allows you to contribute to a retirement plan in which you have to choose one of two investment funds. In the past, Fund A has grown by a mean of 5% per year with a standard deviation of 1%. Fund B has grown by a mean of 9% per year with a standard deviation of 7%. Which fund is a better choice for long-term investments? Which fund is a better choice if you will be withdrawing your money next year? Explain.

27. BATTING STANDARD DEVIATION. For the last 100 years, the mean batting average in the major leagues has remained fairly constant at about 0.260. However, the standard deviation of batting averages has decreased from about 0.049 in the 1870s to 0.031 in the present. What does this tell us about the batting averages of players? Based on these facts, would you expect batting averages above 0.350 to be more or less common today than in the past? Explain.

PROJECTS FOR THE WEB AND BEYOND

For useful links, select "Links for Web Projects" for Chapter 4 at www.aw.com/bbt.

28. SECONDHAND SMOKE. At the Web sites of the American Lung Association and the U.S. Environmental Protection Agency, find statistical data concerning the health effects of secondhand (passive) smoke. Write a short summary of your findings and your opinions about whether and how this health issue should be addressed by government.

29. KIDS AND THE MEDIA. A recent study by the Kaiser Family Foundation looked at the role of media (for example, television, books, computers) in the lives of children. The report, which is on the Kaiser Family Foundation Web site, gives many data distributions concerning, for example, how much time children spend daily with each medium. Study at least three of the distributions in the report that you find particularly interesting. Summarize each distribution in words, and discuss your opinions of the social consequences of the findings.

30. MEASURING VARIANCE. The range and standard deviation use different approaches to measure variation in a data set. Construct two different data sets configured so that the range of the first set is *greater than* the range of the second set (suggesting that the first set has more variation) but the standard deviation of the first set is *less than* the standard deviation of the second set (suggesting that the first set has less variation).

IN THE NEWS

1. RANGES IN THE NEWS. Find two examples of data distributions in recent news reports; they may be given either as tables or as graphs. In each case, state the range of the distribution and explain its meaning in the context of the news report. Estimate the standard deviation by applying the range rule of thumb.

2. SUMMARIZING A NEWS DATA SET. Find an example of a data distribution given in the form of a table in a recent news report. Make a five-number summary and a boxplot for the distribution.

Chapter Review Exercises

1. Refer to the weights (in pounds) of cans of regular Coke and diet Coke given below.

Regular Coke:

0.8192	0.8150	0.8163	0.8211	0.8181
0.8247	0.8062	0.8128	0.8172	0.8110
0.8251	0.8264	0.7901	0.8244	0.8073
0.8079				

Diet Coke:

0.7773	0.7758	0.7896	0.7868	0.7844
0.7861	0.7806	0.7830	0.7852	0.7879
0.7881	0.7826	0.7923	0.7852	0.7872
0.7813				

a. Find the mean, median, and range for each of the two data sets.

b. Give the five-number summary and draw a boxplot for each of the two data sets.

c. Find the standard deviation for each of the two data sets.

d. Apply the range rule of thumb to estimate the standard deviation of each of the two data sets. How well does the rule work in each case? Briefly discuss why it does or does not work well.

e. Based on all your results, compare and discuss the two data sets in terms of their center and variation.

f. Identify any notable differences between the two data sets. How might such differences be explained?

2. a. What is the standard deviation for a sample of 50 values, all of which are the same?

b. Which of the following two car batteries would you prefer to buy, and why?

 • One taken from a population with a mean life of 48 months and a standard deviation of 2 months

 • One taken from a population with a mean life of 48 months and a standard deviation of 6 months

c. If an outlier is included with a sample of 50 values, what is the effect of the outlier on the mean?

d. If an outlier is included with a sample of 50 values, what is the effect of the outlier on the median?

e. If an outlier is included with a sample of 50 values, what is the effect of the outlier on the range?

f. If an outlier is included with a sample of 50 values, what is the effect of the outlier on the standard deviation?

FOCUS ON THE STOCK MARKET

What's Average About the Dow?

As "averages" go, this one is extraordinary. You can't watch the news without hearing what happened to it, and many people spend hours tracking it each day. It is by far the most famous indicator of stock market performance. We are talking, of course, about the Dow Jones Industrial Average, or DJIA for short. But what exactly is it?

The easiest way to understand the DJIA is by looking at its history. As the modern industrial era got under way in the late 19th century, most people considered stocks to be dangerous and highly speculative investments. One reason was a lack of regulation that made it easy for wealthy speculators, unscrupulous managers, and corporate raiders to manipulate stock prices. But another was that, given the complexities of daily stock trading, even Wall Street professionals had a hard time figuring out whether stocks in general were going up (a "bull market") or down (a "bear market"). Charles H. Dow, the founder (along with Edward D. Jones) and first editor of the *Wall Street Journal,* believed he could rectify this problem by creating an "average" for the stock market as a whole. If the average was up, the market was up, and if the average was down, the market was down.

To keep the average simple, Dow chose 12 large corporations to include in his average. On May 26, 1896, he added the stock prices of these 12 companies and divided by 12, finding a *mean* stock price of $40.94. This was the first value for the DJIA. As Dow had hoped, it suddenly became easy for the public to follow the market's direction just by comparing his average from day to day, month to month, or year to year.

The basic idea behind the DJIA is still the same, although the list now includes 30 stocks rather than 12 (see Table 3.5). However, the DJIA is no longer the mean price of its 30 stocks. Instead, it is calculated by adding the prices of its 30 stocks and dividing by a special divisor. Because of this divisor, we now think of the DJIA as an *index* that helps us keep track of stock values, rather than as an actual average of stock prices.

The divisor is designed to preserve continuity in the underlying value represented by the DJIA, and it therefore must change whenever the list of 30 stocks changes and whenever a company on the list has a stock split. A simple example shows why the divisor must change when the list changes. Suppose the DJIA consisted of only 2 stocks (rather than 30): Stock A with a price of $100 and Stock B with a price of $50. The mean price of these two stocks is ($100 + $50)/2 = $75. Now, suppose we change the list by replacing Stock B with Stock C and that Stock C's price is $200. The new mean is ($100 + $200)/2 = $150. Thus, merely changing a stock on the list would raise the mean price from $75 to $150. Therefore, to keep the "value" of the DJIA constant when we change this list, we must divide the new mean of $150 by 2. In this way, the DJIA remains "75" both before and after the list change, but we can no longer think of this 75 as a mean price in dollars.

To see why a stock split changes the divisor, again suppose the index consists of just two stocks: Stock X at $100 and Stock Y at $50, for a mean price of $75. Now, suppose Stock X undergoes a 2-for-1 stock split, so that its new price is $50. With both stocks now priced at $50, the mean price after the stock split would also be $50. Thus, even though a stock split does not affect a company's total value (it only changes the number and prices of its shares), we'd find a drop in the mean price from $75 to $50. In this case, we can preserve continuity by dividing the new mean of 50 by 2/3 (which is equivalent to multiplying by 3/2) so that the DJIA holds at 75 both before and after the stock split.

Just as in our simple examples, the real divisor changes with every list change or stock split, so it has changed many times since Charles Dow first calculated the DJIA as an actual mean. The current value of the divisor is published daily in the *Wall Street Journal*.

Given that there are now well over 10,000 actively traded stocks, it might seem remarkable that a sample of only 30 could reflect overall market activity. But today, when computers make it easy to calculate stock market "averages" in many other ways, we can look at historical data and see that the DJIA has indeed been a reliable indicator of overall market performance. Figure 4.15 shows the historical performance of the DJIA.

If you study Figure 4.15 carefully, you may be tempted to think that you can see patterns that would allow you to forecast precise values of the market in the future. Unfortunately, no one has ever found a way to make reliable forecasts, and most economists now believe that such forecasts are impossible.

The futility of trying to forecast the market is illustrated by the story of the esteemed Professor Benjamin Graham, often called the father of "value investing." In the spring of 1951, one of his students came to him for some investment advice. Professor Graham noted that the DJIA then stood at 250, but that it had fallen below 200 at least once during every year since its inception in 1896. Because it had not yet fallen below 200 in 1951, Professor Graham advised his student to hold off on buying until it did. Professor Graham presumably followed his own advice, but the student did not. Instead, the student invested his "about 10

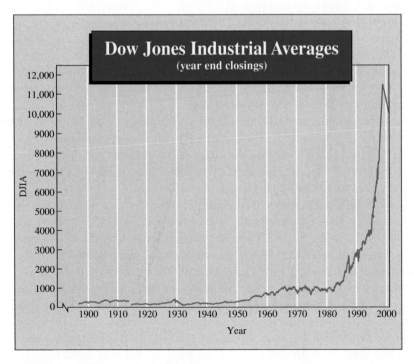

Figure 4.15 Historical values of the DJIA.

thousand bucks" in the market right away. As it turned out, the market never did fall below 200 in 1951 or any time thereafter. And the student, named Warren Buffet, became a billionaire many times over.

QUESTIONS FOR DISCUSSION

1. The stock market is still considered a riskier investment than, say, bank savings accounts or bonds. Nevertheless, financial advisors almost universally recommend holding at least some stocks, which is quite different from the situation that prevailed a century ago. What role do you think the DJIA played in building investors' confidence in the stock market?

2. The DJIA is only one of many different stock market indices in wide use today. Briefly look up a few other indices, such as the S&P 500, the Russell 2000, and the NASDAQ. How do these indices differ from the DJIA? Do you think that any of them should be considered more reliable indicators of the overall market than the DJIA? Why or why not?

3. The 30 stocks in the DJIA represent a *sample* of the more than 10,000 actively traded stocks. However, it is not a random sample because it is chosen by particular editors for particular reasons that sometimes include personal biases. Suppose that you chose a random sample of 30 stocks and tracked their prices. Do you think that such a random sample would track the market as well as the stocks in the DJIA? Why or why not?

4. Create your own "portfolio" of 10 stocks that you'd like to own, and assume you own 100 shares of each. Calculate the total value of your portfolio today, and track price changes over the next month. At the end of the month, calculate the percent change in the value of your portfolio. How did the performance of your portfolio compare to the performance of the DJIA during the month? If you really owned these stocks, would you continue to hold them or would you sell? Explain.

SUGGESTED READING

Morris, K. M., and Siegel, A. M. *The Wall Street Journal Guide to Personal Finance.* Light Bulb Press (New York), 1997.

Prestbo, John (Editor). *The Market's Measure: An Illustrated History of America Told Through the Dow Jones Industrial Average.* Dow Jones & Company, 1999.

You can also find information about the DJIA regularly in the *Wall Street Journal, New York Times, Money, Business Week,* and numerous other periodicals.

FOCUS ON ECONOMICS

Are the Rich Getting Richer?

In April 1999, Bill Gates became the world's first person with a net worth of over $100 billion. His wealth thereby exceeded the gross national products of all but the 18 richest countries in the world. If his wealth had continued to grow at the same rate, he would have been on his way to being a *trillionaire* by 2005. It's easy to look at this story and lament that the rich keep getting richer, while the rest of us are left behind. But is it true?

We can't draw general conclusions about whether the rich are getting richer from the wealth of a single person. Instead, we must look at the overall income distribution. Economists have developed a number, called the *Gini Index,* that is used to describe the level of equality or inequality in the income distribution. The Gini Index is defined so that it can range only between 0 and 1. A Gini Index of 0 indicates perfect income equality, in which every person has precisely the same income. A Gini Index of 1 indicates perfect inequality, in which a single person has all the income and no one else has anything. Figure 4.16 shows the Gini Index in the United States since 1947. Note that the Gini Index fell from 1947 to 1968, indicating that the income distribution became more uniform during this period. The Gini Index has generally risen ever since, indicating that the rich are, indeed, getting richer.

Although the Gini Index provides a simple, single-number summary of income inequality, the number itself is fairly difficult to interpret (and to calculate). An alternative way to look at the income distribution is to study income *quintiles,* which divide the population into fifths by income. Often, the highest quintile is further broken down to show how the top 5% of income earners compares to others.

Figure 4.17 shows the share of total income received by each quintile and the top 5% in the United States in four different decades. The height of each bar (the number on top of it) represents the share of total income. For example, the 3.6 on the bar for the lowest quintile in 1997 means that the poorest 20% of the population received only 3.6% of the total income in the United States. Similarly, the 49.4 on the bar for the top quintile in 1997 means that the richest 20% of the population received 49.4% of the income. Note also that the richest 5% received 21.7% of the income. If you study this graph carefully, you'll see that the share of income received by the first four quintiles—which means all but the richest 20% of the population—dropped since 1967. Meanwhile, the share received by the richest 20% rose substantially, from 43.8% in 1967 to 49.4% in 1997, as did the share of the top 5%. Thus, this graph also confirms that the rich have been getting richer compared to most of the population.

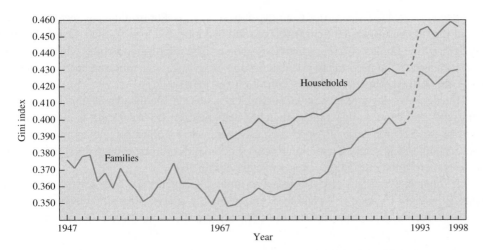

Figure 4.16 Gini Index for families and households, 1947–1997. Household data, which include single people and households in which the members are not part of the same family, have been taken only since 1967. The dashed segments in 1993 indicate a change in the methodology for data collection, so the corresponding rise in the Gini Index may be partially or wholly due to this change rather than a real change in income inequality. SOURCE: U.S. Census Bureau.

Now that we've established that the rich are getting richer, the next question is whether it matters. Most people, including most economists, have traditionally assumed that rising income inequality is bad for democracies. But a few economists from both the left and the right of the political spectrum argue that the change in recent decades is different. For one thing, the change meets a widely accepted ethical criterion called the *Pareto criterion,* after the Italian economist Vilfredo Pareto (for whom Pareto charts are also named): Any change is good if it makes someone better off without making anyone else worse off. The Pareto criterion appears to be satisfied because overall growth in the economy has been helping nearly everyone. In other words, most people may have a smaller percentage of total income than they had in the past, but they still have more absolute income and therefore are living better than they did in the past.

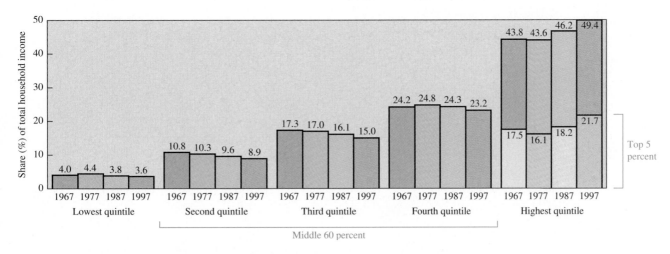

Figure 4.17 Share of total household income by quintile (and top 5%): 1967, 1977, 1987, and 1997. SOURCE: U.S. Bureau of the Census.

Secondly, today's rich differ from the rich in the past. For example, as recently as 1980, 60% of the "Forbes 400" (the richest 400 people) had inherited most of their wealth. By 1997, only 20% of the Forbes 400 represented old money. The implication is that, whereas you had to be born rich in the past, today you can *become* rich by getting educated and working hard. Surely, it is a good thing to encourage education and hard work.

Finally, while overall income inequality has increased, the income inequality among different races and between men and women has decreased. In other words, it is now easier than it was in the past for African Americans, Hispanics, and women to earn as much as white males. Again, this is surely a good thing for our democratic values, even if we still have a long way to go before the inequalities are completely eliminated.

Of course, even if the recent increase in income inequality has been good for the United States, it may not be good if it continues. Unfortunately, no one knows precisely what causes income inequality to increase, and therefore no one knows how to reverse the current trend.

QUESTIONS FOR DISCUSSION

1. Compare several different ways of looking at the data shown in Figure 4.16 and Figure 4.17. For example, does one seem to indicate a larger change in income inequality than the other? Can you think of other possible ways to display income data that might give a different picture than those shown here?

2. Do you agree that the Pareto criterion is a good way to evaluate the ethics of economic change? Why or why not?

3. Overall, do you think the increase in income inequality has been a good or bad thing for the United States? Will it be good if the trend continues? Defend your opinion.

4. Although economic data suggest that the vast majority of Americans are better off today than they were a few decades ago, the poorest Americans still live in very bad economic conditions. What do you think can or should be done to help improve the lives of the poor? Can your suggestion be implemented without harming the overall economy? Explain.

SUGGESTED READING

Nasar, Sylvia. "Is the U.S. Income Gap Really a Big Problem?" *New York Times,* April 4, 1999.

Weinberg, Daniel H. A *Brief Look at Postwar U.S. Income Inequality*, Current Population Report, P60–191. U.S. Census Bureau, June 1996.

CHAPTER 5

Nothing in life is to be feared. It is only to be understood.

—Marie Curie

A Normal World

When you walk into a store, how do you know if a sale price is really a good price? When you exercise and your heart rate rises, how do you know if it has risen enough, but not too much, for a good workout? If your 12-year-old daughter runs a mile in 5 minutes, is she a future Olympic hopeful? These questions seem very different, but from a statistical standpoint they are very similar: Each one asks whether a particular number (price, heart rate, running time) is somehow unusual. In this chapter, we will discuss how we can answer such questions with the aid of the bell-shaped distribution called the *normal distribution*. In the process, we will extend the ideas about distributions that we have encountered in previous chapters.

LEARNING GOALS

1. Understand what is meant by a normal distribution and be able to identify situations in which a normal distribution is likely to arise.

2. Know how to interpret the normal distribution in terms of the 68-95-99.7 rule and standard scores.

3. Learn about the important role of the Central Limit Theorem in statistics.

5.1   What Is Normal?

Suppose a friend is pregnant and due to give birth on June 30. Would you advise her to schedule an important business meeting for June 16, two weeks before the due date? Because she won't be able to attend the meeting if she is in labor or already on maternity leave, answering this question requires knowing whether the baby is likely to arrive 14 or more days before the due date. For that, we need to examine data concerning due dates and actual birth dates.

Figure 5.1 is a histogram for a distribution of 300 natural births at Providence Memorial Hospital. The horizontal axis shows how many days before or after the due date a baby was born: Negative numbers represent births *prior* to the due date, zero represents a birth on the due date, and positive numbers represent births *after* the due date. The left vertical axis shows the number of births for each four-day bin. For example, the frequency of 35 for the highest bar corresponds to the bin from −2 days to 2 days; it shows that out of the 300 total births in the sample, 35 births occurred within two days of the due date.

To answer our question about whether a birth is likely to occur 14 or more days early, it is more useful to look at the *relative frequencies*. Recall that the relative frequency of any data value is its frequency divided by the total number of data values (see Section 3.1). Figure 5.1 shows relative frequencies on the right vertical axis. For example, the bin for −14 days to −10 days has a relative frequency of about 0.07, or 7%. That is, about 7% of the 300 births occurred between 14 days and 10 days before the due date.

Now we can find the proportion of births that occurred 14 or more days before the due date: We simply add the relative frequencies for the bins that correspond to 14 or more days before the due date. Adding the heights of the bars from −14 to the left, we get a total of about 0.21, which says that about 21% of the births in this data set occurred 14 or more days before the due date. We could say that your friend has about a 0.21, or 21%, chance of her baby being born on or before the date of the business meeting. If the meeting is important, it might be good to schedule it earlier.

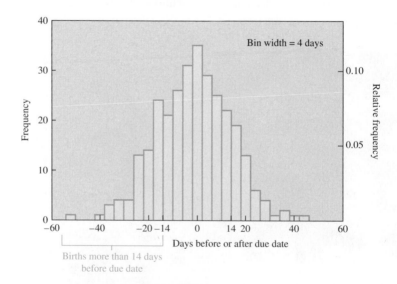

Figure 5.1   Histogram of frequencies (left axis) and relative frequencies (right axis) for birth dates relative to due date. Negative numbers refer to births before the due date; positive numbers refer to births after the due date. The width of each bin is four days.

Time out to think

Suppose the friend wants a three-month maternity leave after the birth. Based on the data in Figure 5.1 and assuming a due date of June 30, should she promise to be at work on October 10?

The Normal Shape

The distribution of the birth data has a fairly distinctive shape, which is easier to see if we overlay the histogram with a smooth curve, as shown in Figure 5.2. For our present purposes, the shape of this smooth distribution has three very important characteristics (see Section 4.2):

- The distribution is *single peaked*. Its mode, or most common birth date, is the due date.

- The distribution is *symmetric* around its single peak; therefore, its median and mean are the same as its mode. The median is the due date because equal numbers of births occur before and after this date. The mean is also the due date because, for every birth before the due date, there is a birth the same number of days after the due date.

- The distribution is spread out in a way that makes it resemble the shape of a bell, so we call it a "bell-shaped" distribution.

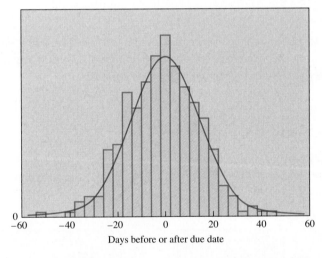

Figure 5.2   A smooth normal distribution curve is drawn over the histogram of Figure 5.1.

Time out to think

The histogram in Figure 5.2, which includes only natural births, is fairly symmetric. How would the shape of the histogram change if it included induced labor births?

The smooth distribution in Figure 5.2, with these three characteristics, is called a **normal distribution**. (Note that the actual birth data are not exactly normal.) All normal distributions have the same characteristic bell shape, but they can differ in their mean and in their variation. Figure 5.3 shows two different normal distributions. Both have the same mean, but distribution (a) has greater variation. As we'll discuss in the next section, knowing the *standard deviation* (see Section 4.3) of a normal distribution tells us everything we need to know

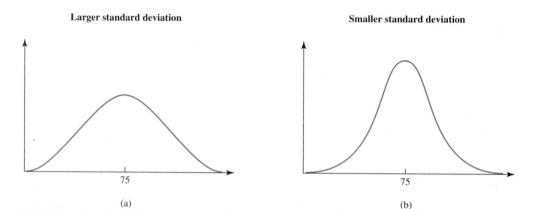

Larger standard deviation   **Smaller standard deviation**

(a)   (b)

Figure 5.3   Both distributions are normal and have the same mean of 75, but the distribution on the left has a larger standard deviation.

TECHNICAL NOTE

A normal distribution with mean μ and standard deviation σ is given by the formula

$$y = \frac{e^{-\frac{1}{2}[(x-\mu)/\sigma]^2}}{\sigma\sqrt{2\pi}}$$

This formula is not used in this book, but it does algebraically describe the shape of the normal distribution.

about its variation. Thus, a normal distribution can be fully described with just two numbers: its mean and its standard deviation.

Definition

The **normal distribution** is a symmetric, bell-shaped distribution with a single peak. Its peak corresponds to the mean, median, and mode of the distribution. Its variation can be characterized by the standard deviation of the distribution.

EXAMPLE 1   Normal Distributions?

Consider the two distributions in Figure 5.4: (a) a famous data set of the chest sizes of 5,738 Scottish militiamen collected in about 1846 and (b) the distribution of the popula-

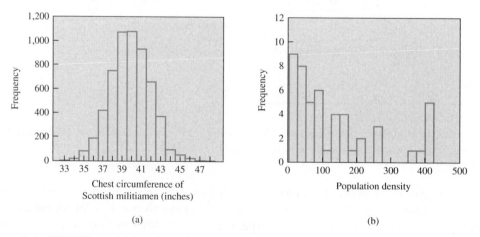

(a)   (b)

Figure 5.4   SOURCE: (a) Adolphe Quetelet, *Lettres à S. A. R. le Duc Régnant de Saxe-Cobourg et Gotha*, 1846.

tion densities of the 50 states. Which distribution appears to be a normal distribution? Explain.

SOLUTION   The distribution in Figure 5.4a is nearly symmetric, with a mean between 39 and 40 inches; it is nearly normal. The distribution in Figure 5.4b shows that most states have low population densities, but a few have much higher densities. This fact makes the distribution right-skewed, so it is not a normal distribution. ∎

The Normal Distribution and Relative Frequencies

Recall that the total relative frequency for any data set must be 1 (see Section 3.1). Now consider the smooth curve for the normal distribution in Figure 5.2. Although we no longer have individual bars, we can still associate the height of the normal curve with the relative frequency. The fact that the relative frequencies must sum to 1 becomes the condition that the area under the normal curve must be 1. We just need to be sure that we interpret the curve as shown in Figure 5.5. The shaded region to the left of −14 days comprises about 18% of the total area under the curve (which differs from the 21% found earlier because the actual data are not exactly normal). Thus, the relative frequency of data values less than −14 days is about 0.18. This means that about 18% of births are more than 14 days early. Similarly, the shaded region to the right of 18 days comprises about 12% of the total area under the curve. Thus, the relative frequency of data values greater than 18 days is about 0.12. This means that about 12% of births are more than 18 days late. Altogether, we see that 18% + 12% = 30% of all births are either more than 14 days early or more than 18 days late.

Now that we see how the area under the curve relates to relative frequency, we can make one further generalization: *The relative frequency for any range of data values is the area under the curve covering that range of values.* For example, consider the region in Figure 5.5 between −14 days and 18 days. The area of this region comprises about 100% − 30% = 70% of the total area under the curve. Therefore, the relative frequency of data values between −14 days and 18 days is about 0.70.

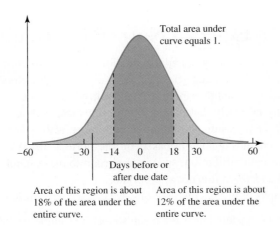

Figure 5.5   The area under the curve to the left of −14 days is the relative frequency of values less than −14 days, which is about 0.18. The area under the curve to the right of 18 days is the relative frequency of values greater than 18 days, which is about 0.12.

Relative Frequencies and the Normal Distribution

- The area that lies under the normal distribution curve corresponding to a range of values on the horizontal axis is the relative frequency of those values.

- Because the total relative frequency must be 1, the *total area under the normal distribution* curve must equal 1.

Time out to think

Suppose 7% of births occur more than 22 days late. What percentage of births occur between 14 days early and 22 days late? Explain.

EXAMPLE 2 Estimating Areas

Look again at the normal distribution in Figure 5.5.

a. Estimate the percentage of births occurring between 0 and 60 days after the due date.

b. Estimate the percentage of births occurring between 14 days before and 14 days after the due date.

SOLUTION

a. About half of the total area under the curve lies in the region between 0 days and 60 days. Thus, the area under this portion of the curve is very close to 0.5. This means that about 50% of the births in the sample occur between 0 and 60 days after the due date.

b. From Figure 5.5, we found that about 18% of the births occur more than 14 days before the due date. Because the graph is symmetric, about 18% must also occur more than 14 days after the due date. Thus, about 18% + 18% = 36% of births occur more than 14 days before or after the due date, leaving about 100% − 36% = 64% of the births in the sample to occur between 14 days before the due date and 14 days after the due date. ∎

When Can We Expect a Normal Distribution?

The normal distribution is a very good approximation to the distribution of many variables of practical interest. Physical characteristics such as weight, height, blood pressure, and reflex times generally follow a normal distribution. Standardized test scores, such as those for SATs or IQ tests, are usually normally distributed. Sports statistics, such as batting averages, times in a swimming event, or high jump results in a track meet, also tend to follow a normal distribution. Indeed, much of statistics is based on the normal distribution. However, not all variables follow a normal distribution, so it is important to understand when the normal distribution can be used and when it is not appropriate.

By studying graphs of normal distributions, we can see what makes a distribution normal:

- It must have values clustered near the mean so that it is single-peaked.

- The values must be spread evenly around the mean so that it is symmetric.

- Large deviations from the mean must be increasingly rare so that it has the characteristic bell shape.

By the Way...

The Scottish politician John Sinclair (1754–1835) was one of the first collectors of economic, demographic, and agricultural data. He is credited with introducing the words *statistics* and *statistical* into the English language, having heard them used in Germany to refer to matters of state.

On a deeper level, any quantity that is influenced by many different factors tends to be normally distributed. Physical traits are influenced by many different genetic and environmental effects. Standardized test scores reflect the performance of many individuals working on many different test questions. Sports statistics involve many players with different skills performing under different conditions.

Conditions for a Normal Distribution

A data set that satisfies the following four criteria is likely to have a nearly normal distribution:

1. Most data values are clustered near the mean, giving the distribution a well-defined single peak.
2. Data values are spread evenly around the mean, making the distribution symmetric.
3. Larger deviations from the mean become increasingly rare, producing the tapering tails of the distribution.
4. Individual data values result from a combination of many different factors, such as genetic and environmental factors.

By the Way...

The normal distribution curve is often called a Gaussian curve in honor of the 19th-century German mathematician Carl Friedrich Gauss. The American logician Charles Peirce introduced the term *normal distribution* in about 1870.

EXAMPLE 3   Is It a Normal Distribution?

Which of the following variables would you expect to have a normal or nearly normal distribution?

a. Scores on a very easy test

b. Heights of a random sample of adult women

c. The number of Macintosh apples in each of 100 full bushel baskets

SOLUTION

a. Tests have a maximum possible score (100%) that limits the size of data values. If the test is easy, the mean will be high and many scores will be close to the maximum possible. The few lower scores may be spread out well below the mean. We therefore expect the distribution of scores to be left-skewed and non-normal.

b. Height is determined by a combination of many factors (the genetic makeup of both parents and possibly environmental or nutritional factors). We expect the mean height for the sample to be close to the mode (most common height). We also expect there to be roughly equal numbers of women above and below the mean, and extremely large and small heights should be rare. Thus, height is nearly normally distributed.

c. The number of apples in a bushel basket varies with the size of the apples. We expect that in the distribution there will be a single mode that should be close to the mean number of apples per basket. The number of baskets with more than the mean number of apples should be close to the number of baskets with fewer than the mean number of apples. Thus, the number of apples per basket has a nearly normal distribution. ∎

Time out to think

Would you expect scores on a moderately difficult exam to have a normal distribution? Suggest two more quantities that you would expect to be normally distributed.

Review Questions

1. Describe the shape of a normal distribution curve.

2. Explain in words how the mean and standard deviation define a normal distribution.

3. Explain the relationship between the relative frequency of a set of data values and the area under the relative frequency histogram.

4. What is the total area under any relative frequency histogram? Why?

5. Give three properties of a data set that has a normal distribution.

6. Describe four examples of variables that have normal or nearly normal distributions.

Exercises

BASIC SKILLS AND CONCEPTS

SENSIBLE STATEMENTS? For Exercises 1–6, determine whether the given statement is sensible and explain why it is or is not.

1. Kate's data set of ages has a normal distribution with a mean of 45 and a median of 54.

2. Lori's distribution of exam scores is normal and has two modes.

3. Rose's baby was three weeks overdue, which is not unusual because about 40% of all babies are three or more weeks overdue.

4. The mean of a normally distributed set of weights is 123 pounds. It follows that 60% of the weights are over 130 pounds.

5. Ted measures the lengths of 100 eight-foot beams at the local lumberyard. He predicts in advance that the distribution of lengths will be normally distributed.

6. Wendy tosses a fair die 100 times and notices that the frequencies of 1, 2, 3, 4, 5, and 6 are normally distributed.

7. WHAT IS NORMAL? Identify the distribution in Figure 5.6 that is not normal. Of the two normal distributions, which has the larger standard deviation?

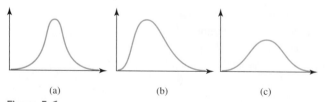

(a)   (b)   (c)

Figure 5.6

8. WHAT IS NORMAL? Identify the distribution in Figure 5.7 that is not normal. Of the two normal distributions, which has the larger standard deviation?

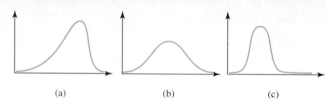

(a)   (b)   (c)

Figure 5.7

NORMAL VARIABLES. For each of the data sets in Exercises 9–18, state whether you would expect it to be normally distributed. Explain your reasoning.

9. Verbal SAT scores

10. Rental rates for office space in large cities

11. The weights of candy bars that say "4 ounces" on the wrapper

12. The numbers of spades, hearts, diamonds, and clubs that are dealt during an evening of card playing

13. The time it takes you to bicycle to school using the same route over the course of a school year

14. The areas of lakes around the world

15. The weights of bags of a particular brand of potato chips

16. The waiting times at a bus stop if the bus comes once every ten minutes and you arrive randomly

17. The prices of a particular model car at different dealers

18. The times of swimmers in the 100-meter butterfly at a swim meet

19. MOVIE LENGTHS. Figure 5.8 shows a histogram for the lengths of 60 movies. The mean movie length is 110.5 minutes. Is this distribution close to normal? Should this variable have a normal distribution? Why or why not?

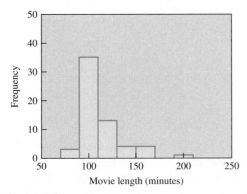

Figure 5.8

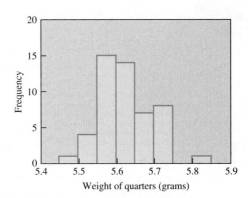

Figure 5.10

20. **HEART RATES.** Figure 5.9 shows a histogram for the heart rates of 98 students. The mean heart rate is 71.2 beats per minute. Is this distribution close to normal? Should this variable have a normal distribution? Why or why not?

22. **ASPIRIN WEIGHTS.** Figure 5.11 shows a histogram for the weights of 30 randomly selected aspirin tablets. The mean weight is 665.4 milligrams. Is this distribution close to normal? Should this variable have a normal distribution? Why or why not?

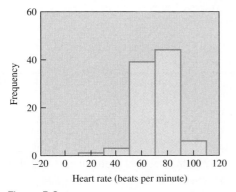

Figure 5.9

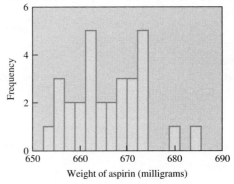

Figure 5.11

21. **QUARTER WEIGHTS.** Figure 5.10 shows a histogram for the weights of 50 randomly selected quarters. The mean weight is 5.62 grams. Is this distribution close to normal? Should this variable have a normal distribution? Why or why not?

FURTHER APPLICATIONS

23. **DATA FROM THE STATES.** The histograms in Figure 5.12 show the distribution of three different variables for the 50 states (the data table is given in Appendix A).

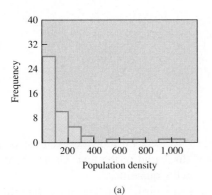

(a)

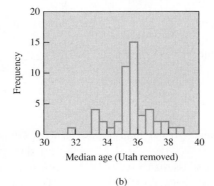

(b)

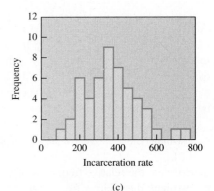

(c)

Figure 5.12

a. While none of these distributions is perfectly normal, some are closer than others. Which distributions appear to be most nearly normal? Which distribution appears to be least like a normal distribution?

b. Choose at least one of the other variables in the data table in Appendix A. Make a histogram and discuss why you think this variable is distributed as it is. How do you explain the extreme values of the variable and why do they occur in the states they do? Do some states have unexpected values of the variable? Discuss whether the variable has a near normal distribution and why.

24. **AREAS AND RELATIVE FREQUENCIES.** Consider the normal curve in Figure 5.13, giving relative frequencies in a distribution of men's heights. The distribution has a mean of 69.6 inches and a standard deviation of 2.8 inches.

a. What is the total area under the curve?

b. Estimate (using area) the relative frequency of values less than 67.

c. Estimate the relative frequency of values greater than 67.

d. Estimate the relative frequency of values between 67 and 70.

e. Estimate the relative frequency of values greater than 70.

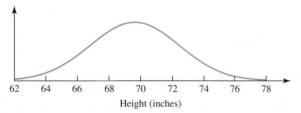

Figure 5.13

25. **AREAS AND RELATIVE FREQUENCIES.** Consider the normal curve in Figure 5.14, giving the relative frequencies in a distribution of IQ scores. The distribution has a mean of 100 and a standard deviation of 16.

a. What is the total area under the curve?

b. Estimate (using area) the relative frequency of values less than 100.

c. Estimate the relative frequency of values greater than 110.

d. Estimate the relative frequency of values less than 110.

e. Estimate the relative frequency of values between 100 and 110.

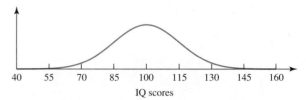

Figure 5.14

26. **ESTIMATING AREAS.** Consider the normal curve in Figure 5.15, giving the relative frequencies in a distribution of systolic blood pressures for a sample of female students. The distribution has a standard deviation of 14.

a. What is the mean of the distribution?

b. Estimate (using area) the percentage of students whose blood pressure is less than 100.

c. Estimate the percentage of students whose blood pressure is between 110 and 130.

d. Estimate the percentage of students whose blood pressure is greater than 130.

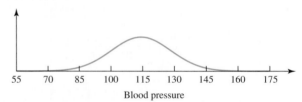

Figure 5.15

27. **ESTIMATING AREAS.** Consider the normal curve in Figure 5.16, giving the relative frequencies in a distribution of body weights for a sample of male students.

a. What is the mean of the distribution?

b. Estimate (using area) the percentage of students whose weight is less than 140.

c. Estimate the percentage of students whose weight is greater than 170.

d. Estimate the percentage of students whose weight is between 140 and 160.

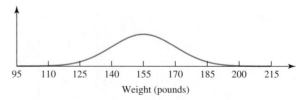

Figure 5.16

28. **BIRTH DISTRIBUTIONS.** Consider the example of pregnancy due dates discussed in the text. Using Figure 5.2, estimate the following quantities.

a. The percentage of births that are more than 20 days overdue

b. The percentage of births that are overdue by any amount

c. The percentage of births that are less than 10 days overdue

d. The percentage of births that are more than 10 days premature

29. **CHEST DATA FOR SCOTTISH MILITIAMEN.** Consider the distribution shown in Figure 5.4a. Estimate the following quantities.

    a. The percentage of men in the sample with chest sizes less than 38 inches

    b. The relative frequency of chest sizes greater than 43 inches

    c. The percentage of men in the sample with chest sizes greater than or equal to 38 inches

    d. The relative frequency of chest sizes less than or equal to 39 inches

30. **POPULATION DENSITIES.** Consider the distribution shown in Figure 5.4b. Estimate the following quantities.

    a. The percentage of states in the sample with a density less than 200 people per square mile

    b. The relative frequency of states in the sample with a density greater than 300 people per square mile

    c. The percentage of states in the sample with a density greater than 100 people per square mile

    d. The relative frequency of states in the sample with a density less than 50 people per square mile

PROJECTS FOR THE WEB AND BEYOND

For useful links, select "Links for Web Projects" for Chapter 5 at www.aw.com/bbt.

31. **SAT SCORE DISTRIBUTIONS.** The College Board Web site gives the distribution of scores for the verbal and mathematics SAT (in 50-point bins). Collect these data and make histograms for both tests. Discuss the truth of the claim that SAT scores are normally distributed.

32. **USING FERRET.** One of the most powerful data-gathering systems for federal data is called FERRET (Federal Electronic Research and Review Extraction Tool). It is available to the public on the Web and provides access to all or parts of economic, demographic, and health data sets. Go to the FERRET site and choose a specific variable of interest (for example, adjusted gross income or computer ownership). Following the step-by-step instructions given at the site, view a data set related to your variable. (Most data sets are very large, but it is possible to view small parts of them.) Experiment with downloading the data set or a part of it. Make a histogram of the data set and discuss whether you have found a normally distributed variable.

33. **FINDING NORMAL DISTRIBUTIONS.** Using the guidelines given in the text, choose a variable that you think should be nearly normally distributed. Collect at least 30 data values for the variable and make a histogram. Comment on how closely the distribution fits a normal distribution. In what ways does it differ from a normal distribution? Try to explain these differences.

34. **MOVIE LENGTHS.** Collect data to support or refute the claim that movies have gotten shorter over the decades. Specifically, make histograms of movie lengths for the 1940s through the 1990s, find the mean movie length for each sample, and comment on whether these distributions are normal. Discuss your results and give plausible reasons for any trends that you observe.

IN THE NEWS

1. **NORMAL DISTRIBUTIONS.** Rarely does a news article refer to the actual distribution of a variable or state that a variable is normally distributed. Nevertheless, variables mentioned in news reports must have some distribution. Find two variables in news reports that you suspect have nearly normal distributions. Explain your reasoning.

2. **NON-NORMAL DISTRIBUTIONS.** Find two variables in news reports that you suspect *do not* have nearly normal distributions. Explain your reasoning.

5.2 Properties of the Normal Distribution

Consider a *Consumer Reports* survey in which participants were asked how long they owned their last TV set before they replaced it. The variable of interest in this survey is *replacement time for television sets*. Based on the survey, the distribution of replacement times has a mean of about 8.2 years, which we denote μ (the Greek letter *mu*). The standard deviation of the distribution is about 1.1 years, which we denote σ (the Greek letter *sigma*). Making the reasonable assumption that the distribution of TV replacement times is approximately normal, we can picture it as shown in Figure 5.17.

Time out to think

Apply the four criteria for a normal distribution (see Section 5.1) to explain why the distribution of TV replacement times should be approximately normal.

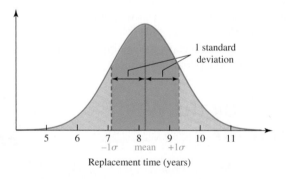

Figure 5.17   Normal distribution for replacement times for TV sets with a mean of $\mu = 8.2$ years and a standard deviation of $\sigma = 1.1$ years.

TECHNICAL NOTE

A normal distribution can have any value for the mean and any positive value for the standard deviation. The term *standard normal distribution* specifically refers to a normal distribution with a mean of 0 and a standard deviation of 1. There are some symmetric, bell-shaped distributions that are not normal.

Because all normal distributions have the same bell shape, knowing the mean and standard deviation of a distribution allows us to say a lot about where the data values lie. For example, if we measure areas under the curve in Figure 5.17, we find that about two-thirds of the area lies within 1 standard deviation of the mean—that is, between $8.2 - 1.1 = 7.1$ years and $8.2 + 1.1 = 9.3$ years. Thus, the TV replacement time is between 7.1 and 9.3 years for about two-thirds of the people. Similarly, about 95% of the area lies within 2 standard deviations of the mean—that is, between $8.2 - 2.2 = 6.0$ years and $8.2 + 2.2 = 10.4$ years. Thus, the TV replacement time is between 6.0 and 10.4 years for about 95% of the people. A simple rule, called the **68-95-99.7 rule**, gives precise guidelines for the percentage of data values that lie within 1, 2, and 3 standard deviations of the mean for any normal distribution. The following box states the rule in words, and Figure 5.18 shows it visually.

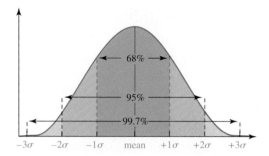

Figure 5.18   Normal distribution illustrating the 68-95-99.7 rule.

The 68-95-99.7 Rule for a Normal Distribution

- About 68% (actually 68.3%) of the data points fall within 1 standard deviation of the mean. (Note that 68% is slightly over two-thirds.)
- About 95% (actually 95.4%) of the data points fall within 2 standard deviations of the mean.
- About 99.7% of the data points fall within 3 standard deviations of the mean.

EXAMPLE 1   SAT Scores

The tests that make up the verbal and math SAT (and the GRE, LSAT, and GMAT) are designed so that their scores are normally distributed with a mean of $\mu = 500$ and a standard deviation of $\sigma = 100$. Interpret this statement.

SOLUTION   By the 68-95-99.7 rule, about 68% of students have scores within 1 standard deviation, or 100 points, of the mean of 500 points. That is, about 68% of students score between 400 and 600. About 95% of students score within 2 standard deviations (200 points) of the mean, or between 300 and 700. And about 99.7% of students score within 3 standard deviations (300 points) of the mean, or between 200 and 800. Figure 5.19 shows this interpretation graphically; the horizontal axis shows both actual scores and distances from the mean in standard deviations.

TECHNICAL NOTE

As discussed in the focus section at the end of the chapter, the mean score on a particular SAT may differ from 500 depending on the test (math or verbal) and the year it is given.

Distribution of SAT Scores

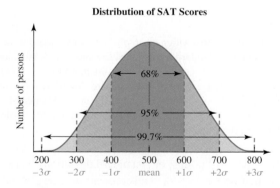

Figure 5.19   Normal distribution for SAT scores, showing the percentages associated with 1, 2, and 3 standard deviations.

EXAMPLE 2   Detecting Counterfeits

Vending machines can be adjusted to reject coins above and below certain weights. The weights of legal U.S. quarters have a normal distribution with a mean of 5.67 grams and a standard deviation of 0.0700 gram. If a vending machine is adjusted to reject quarters that weigh more than 5.81 grams and less than 5.53 grams, what percentage of legal quarters will be rejected by the machine?

SOLUTION   A weight of 5.81 is 0.14 gram, or 2 standard deviations, above the mean. A weight of 5.53 is 0.14 gram, or 2 standard deviations, below the mean. Thus, the machine accepts quarters that are within 2 standard deviations of the mean. Conversely, the machine rejects quarters that are more than 2 standard deviations from the mean. By the 68-95-99.7 rule, 95% of legal quarters will be accepted and 5% of legal quarters will be rejected.   ∎

Applying the 68-95-99.7 Rule

We can apply the 68-95-99.7 rule to determine when data values lie 1, 2, or 3 standard deviations from the mean. For example, suppose that 1,000 students take an exam and the scores are normally distributed with a mean of $\mu = 75$ and a standard deviation of $\sigma = 7$.

A score of 82 is 7 points, or 1 standard deviation, above the mean of 75. The 68-95-99.7 rule tells us that about 68% of the scores are *within* 1 standard deviation of the mean. It follows that about $100\% - 68\% = 32\%$ of the scores are *more than* 1 standard deviation from the mean. Half of this 32%, or 16%, of the scores are more than 1 standard deviation *below* the mean; the other 16% of the scores are more than 1 standard deviation *above* the mean (Figure 5.20a). Thus, about 16% of 1,000 students, or 160 students, scored above 82.

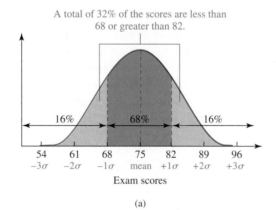

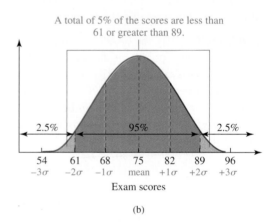

(a)   (b)

Figure 5.20   A normal distribution of test scores with a mean of 75 and a standard deviation of 7. (a) Because 68% of the scores lie *within* 1 standard deviation of the mean, 32% must lie *more than* 1 standard deviation from the mean. Half of these 32%, or 16%, are more than 1 standard deviation below the mean, while the other 16% are more than 1 standard deviation above the mean. (b) Because 95% of the scores lie *within* 2 standard deviations of the mean, 5% must lie *more than* 2 standard deviations from the mean; < 2.5% are more than 1 standard deviation below the mean, and 2.5% are more than 1 standard deviation above the mean.

Similarly, a score of 61 is 14 points, or 2 standard deviations, below the mean of 75. The 68-95-99.7 rule tells us that about 95% of the scores are *within* 2 standard deviations of the mean, so about 5% of the scores are *more than* 2 standard deviations from the mean. Half of this 5%, or 2.5%, of the scores are more than 2 standard deviations *below* the mean (Figure 5.20b). Thus, about 2.5% of 1,000 students, or 25 students, scored below 61. Because 95% of the scores fall between 61 and 89, we sometimes say the scores outside this range are unusual, since they are relatively rare.

Identifying Unusual Results

In statistics, we often need to distinguish values that are typical, or "usual," from values that are unusual. By applying the 68-95-99.7 rule, we find that about 95% of all values from a normal distribution lie within 2 standard deviations of the mean. This implies that, among all values, 5% lie more than 2 standard deviations away from the mean. We can use this property to identify values that are relatively "unusual": **Unusual values** are values that are more than 2 standard deviations away from the mean.

EXAMPLE 3   Traveling and Pregnancy

Consider again the question of whether you should advise a pregnant friend to schedule an important business meeting 14 days before her due date. Actual data suggest that the number of days between the birth date and the due date is normally distributed with a mean of $\mu = 0$ days and a standard deviation of $\sigma = 15$ days. How would you help your friend make the decision? Would a birth 14 days before the due date be considered "unusual"?

SOLUTION   Your friend is assuming that she will not have given birth 14 days, or roughly 1 standard deviation, before her due date. But because this outcome is well within 2 standard deviations of the mean, it is not unusual. By the 68-95-99.7 rule, the day of birth for 68% of pregnancies is within 1 standard deviation of the mean, or between -15 days and 15 days from the due date. This means that about $100\% - 68\% = 32\%$ of pregnancies occur either more than 15 days early or more than 15 days late. Therefore, half of this 32%, or 16%, of all pregnancies are more than 15 days early (see Figure 5.21). You should tell your friend that about 16% of all births occur more than 15 days before the due date. If your friend likes to think in terms of probability, you could say that there is a 0.16 (about 1 in 6) chance that she will give birth on or before her meeting date.

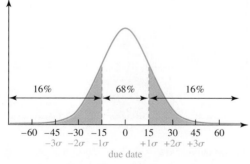

Figure 5.21   About 16% of births occur more than 15 days before the due date.

Time out to think

For the sample data in Figure 5.1, roughly 21% of births occurred at least 14 days early. Now, assuming a normal distribution, we have found that roughly 16% of births occur at least 14 days early. Give a possible reason for the discrepancy.

EXAMPLE 4   Normal Heart Rate

You measure your resting heart rate at noon every day for a year and record the data. You discover that the data have a normal distribution with a mean of 66 and a standard deviation of 4. On how many days was your heart rate below 58 beats per minute?

SOLUTION   A heart rate of 58 is 8 (or 2 standard deviations) below the mean. According to the 68-95-99.7 rule, about 95% of the data points are within 2 standard deviations of the mean. Thus, 2.5% of the data points are more than 2 standard deviations *below* the mean, and 2.5% of the data points are more than 2 standard deviations *above* the mean. On 2.5% of 365 days, or about 9 days, your measured heart rate was below 58 beats per minute.   ∎

Time out to think

As Example 4 suggests, many measurements of the resting heart rate of a *single* individual are normally distributed. Would you expect the average resting heart rates of *many* individuals to be normally distributed? Which distribution would you expect to have the larger standard deviation? Why?

EXAMPLE 5   Finding a Percentile

On a visit to the doctor's office, your fourth-grade daughter is told that her height is 1 standard deviation above the mean for her age and sex. What is her percentile for height? Assume that heights of fourth-grade girls are normally distributed.

SOLUTION   Recall that a data value lies in the *n*th percentile if *n*% of the data values are *less than or equal to* it (see Section 4.3). According to the 68-95-99.7 rule, 68% of the heights are within 1 standard deviation of the mean. Therefore, 34% of the heights (half of 68%) are between 0 and 1 standard deviation *above* the mean. We also know that, because the distribution is symmetric, 50% of all heights are below the mean. Therefore, 50% + 34% = 84% of all heights are less than 1 standard deviation above the mean (Figure 5.22). Thus, your daughter is in the 84th percentile for heights among fourth-grade girls.

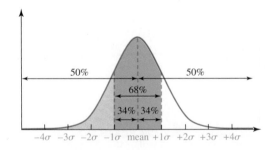

Figure 5.22   Normal distribution curve showing 84% of scores less than 1 standard deviation above the mean.   ∎

Standard Scores

The 68-95-99.7 rule applies only to data values that are 1, 2, or 3 standard deviations from the mean. We can generalize this rule if we know precisely how many standard deviations from the mean a particular data value lies. The number of standard deviations a data value lies above or below the mean is called its **standard score** (or *z-score*), often abbreviated by the letter *z*. For example:

- The standard score of the mean is $z = 0$, because it is 0 standard deviations from the mean.

- The standard score of a data value 1.5 standard deviations *above* the mean is $z = 1.5$.

- The standard score of a data value 2.4 standard deviations *below* the mean is $z = -2.4$.

The following box summarizes the computation of standard scores.

Computing Standard Scores

The number of standard deviations a data value lies above or below the mean is called its standard score (or z-score), defined by

$$z = \text{standard score} = \frac{\text{data value} - \text{mean}}{\text{standard deviation}}$$

The standard score is positive for data values above the mean and negative for data values below the mean.

EXAMPLE 6   Finding Standard Scores

The Stanford-Binet IQ test is scaled so that scores have a mean of 100 and a standard deviation of 16. Find the standard scores for IQs of 85, 100, and 125.

SOLUTION   We calculate the standard scores for these IQs by using the standard score formula with a mean of 100 and standard deviation of 16.

$$\text{Standard score for 85:} \quad z = \frac{85 - 100}{16} = -0.94$$

$$\text{Standard score for 100:} \quad z = \frac{100 - 100}{16} = 0.0$$

$$\text{Standard score for 125:} \quad z = \frac{125 - 100}{16} = 1.56$$

We can interpret these standard scores as follows: 85 is 0.94 standard deviation *below* the mean, 100 is equal to the mean, and 125 is 1.56 standard deviations *above* the mean. Figure 5.23 shows the distribution of IQ scores and their standard scores.

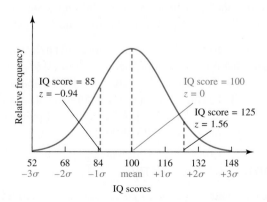

Figure 5.23   Standard scores for IQ scores of 85, 100, and 125.

Standard Scores and Percentiles

Once we know the standard score of a data value, the properties of the normal distribution allow us to find its *percentile* in the distribution. This is usually done with a *standard score table* such as Table 5.1. For each of many standard scores in a normal distribution, the table gives the percentage of values in the distribution less than that value. For example, the table shows that 55.96% of the values in a normal distribution have a standard score less than 0.15. In other words, a data value with a standard score of 0.15 lies in the 55th percentile.

EXAMPLE 7   Cholesterol Levels

Cholesterol levels in men 18 to 24 years of age are normally distributed with a mean of 178 and a standard deviation of 41.

a. What is the percentile for a 20-year-old man with a cholesterol level of 190?

b. What cholesterol level corresponds to the 90th percentile, the level at which treatment may be necessary?

SOLUTION

a. The *standard score* for a cholesterol level of 190 is

$$z = \text{standard score} = \frac{\text{data value} - \text{mean}}{\text{standard deviation}} = \frac{190 - 178}{41} \approx 0.29$$

Table 5.1 shows that a standard score of 0.29 corresponds to about the 61st percentile.

b. Table 5.1 shows that 90.32% of all data values have a standard score less than 1.3. Thus, the 90th percentile is about 1.3 standard deviations above the mean. Given the mean cholesterol level of 178 and the standard deviation of 41, a cholesterol level 1.3 standard deviations above the mean is

$$\underbrace{178}_{\text{mean}} + \underbrace{(1.3 \times 41)}_{\substack{\text{1.3 standard} \\ \text{deviations}}} = 231.3$$

A cholesterol level of about 231 corresponds to the 90th percentile. ∎

EXAMPLE 8   IQ Scores

IQ scores are normally distributed with a mean of 100 and a standard deviation of 16 (see Example 6). What are the IQ scores for people in the 75th and 40th percentiles on IQ tests?

SOLUTION   Table 5.1 shows that the 75th percentile falls *between* standard scores of 0.65 and 0.70; we can estimate that it has a standard score of about 0.67. This corresponds to an IQ that is 0.67 standard deviation, or about $0.67 \times 16 = 11$ points, above the mean of 100. Thus, a person in the 75th percentile has an IQ of 111. The 40th percentile corresponds to a standard score of approximately -0.25, or a score that is $0.25 \times 16 = 4$ points *below* the mean of 100. Thus, a person in the 40th percentile has an IQ of 96. ∎

EXAMPLE 9   Women in the Army

The heights of American women aged 18 to 24 are normally distributed with a mean of 65 inches and a standard deviation of 2.5 inches. In order to serve in the U.S. Army, women must

Table 5.1 Standard Scores and Percentiles for a Normal Distribution (cumulative values from the *left*)							
Standard score	%	Standard score	%	Standard score	%	Standard score	%
−3.5	0.02	−1.0	15.87	0.0	50.00	1.1	86.43
−3.0	0.13	−0.95	17.11	0.05	51.99	1.2	88.49
−2.9	0.19	−0.90	18.41	0.10	53.98	1.3	90.32
−2.8	0.26	−0.85	19.77	0.15	55.96	1.4	91.92
−2.7	0.35	−0.80	21.19	0.20	57.93	1.5	93.32
−2.6	0.47	−0.75	22.66	0.25	59.87	1.6	94.52
−2.5	0.62	−0.70	24.20	0.30	61.79	1.7	95.54
−2.4	0.82	−0.65	25.78	0.35	63.68	1.8	96.41
−2.3	1.07	−0.60	27.43	0.40	65.54	1.9	97.13
−2.2	1.39	−0.55	29.12	0.45	67.36	2.0	97.72
−2.1	1.79	−0.50	30.85	0.50	69.15	2.1	98.21
−2.0	2.28	−0.45	32.64	0.55	70.88	2.2	98.61
−1.9	2.87	−0.40	34.46	0.60	72.57	2.3	98.93
−1.8	3.59	−0.35	36.32	0.65	74.22	2.4	99.18
−1.7	4.46	−0.30	38.21	0.70	75.80	2.5	99.38
−1.6	5.48	−0.25	40.13	0.75	77.34	2.6	99.53
−1.5	6.68	−0.20	42.07	0.80	78.81	2.7	99.65
−1.4	8.08	−0.15	44.04	0.85	80.23	2.8	99.74
−1.3	9.68	−0.10	46.02	0.90	81.59	2.9	99.81
−1.2	11.51	−0.05	48.01	0.95	82.89	3.0	99.87
−1.1	13.57	0.0	50.00	1.0	84.13	3.5	99.98

NOTE: The % column gives the percentage of values in the distribution less than the corresponding standard score.

be between 58 inches and 80 inches tall. What percentage of women are ineligible to serve based on their height?

SOLUTION The standard scores for the army's minimum and maximum heights of 58 inches and 80 inches are

$$For\ 58\ inches: \quad z = \frac{58 - 65}{2.5} = -2.8$$

$$For\ 80\ inches: \quad z = \frac{80 - 65}{2.5} = 6.0$$

Table 5.1 shows that a standard score of −2.8 corresponds to the 0.26 percentile. A standard score of 6.0 does not appear in Table 5.1, which means it is above the 99.98th percentile (the highest percentile shown in the table). Thus, 0.26% of all women are too short to serve in the army and fewer than 0.02% of all women are too tall to serve in the army. Altogether, fewer than about 0.28% of all women, or about 1 out of every 400 women, are ineligible to serve in the army based on their height. ∎

Toward Probability

Suppose you pick a baby at random and ask whether the baby was born more than 15 days prior to his or her due date. Because births are normally distributed around the due date with a standard deviation of 15 days, we know that 16% of all births occur more than 15 days prior to the due date (see Example 3). Thus, for an individual baby chosen at random, we can say that there's a 16% = 0.16 chance (about 1 in 6) that the baby was born more than 15 days early. In other words, the properties of the normal distribution allow us to make a *probability statement* about an individual. In this case, our statement is that the probability of a birth occurring more than 15 days early is 0.16.

This example shows that the properties of the normal distribution can be restated in terms of ideas of probability. In fact, much of the work we will do throughout the rest of this text is closely tied to ideas of probability. For this reason, we will devote the next chapter to studying fundamental ideas of probability. For now, we will use the basic ideas we have discussed so far to introduce one of the most important concepts in statistics in the next section.

Review Questions

1. What are the basic characteristics of a normal distribution?

2. State the 68-95-99.7 rule and explain its meaning.

3. Briefly discuss how you can apply the 68-95-99.7 rule to make statements about the number of data values above or below 1, 2, or 3 standard deviations from the mean. Give an example of such an application.

4. What is a standard score? How is it related to the standard deviation? What is the standard score of a data value that is 1.7 standard deviations below the mean?

5. Briefly explain how to read the standard score table (Table 5.1).

6. Explain how you can find the percentile of a data value from its standard score. Give a few examples.

Exercises

BASIC SKILLS AND CONCEPTS

SENSIBLE STATEMENTS? For Exercises 1–6, determine whether the given statement is sensible and explain why it is or is not.

1. A normally distributed set of test scores has a mean of 40 and a standard deviation of 64.

2. The weights of babies born at Belmont Hospital in April are normally distributed with a mean of 9.6 pounds and a standard deviation of 2.3 pounds.

3. Stan's SAT score was in the 64th percentile, which was 1 standard deviation above the mean.

4. Howard's IQ has a *z*-score of 0.5, which puts him in the 45th percentile.

5. Megan's IQ has a *z*-score of −1.2, which puts her 1 standard deviation above the mean.

6. Striving for the lowest possible score in the golf tournament, Helen was pleased to finish with a score 2 standard deviations below the mean.

7. USING THE 68-95-99.7 RULE. Assume that a set of test scores is normally distributed with a mean of 100 and a standard deviation of 20. Use the 68-95-99.7 rule to find the following quantities.
   a. Percentage of scores less than 100
   b. Relative frequency of scores less than 120
   c. Percentage of scores less than 140
   d. Percentage of scores less than 80
   e. Relative frequency of scores less than 60
   f. Percentage of scores greater than 120
   g. Percentage of scores greater than 140
   h. Relative frequency of scores greater than 80
   i. Percentage of scores between 80 and 120
   j. Percentage of scores between 80 and 140

8. **USING THE 68-95-99.7 RULE.** Assume the resting heart rates for a sample of individuals are normally distributed with a mean of 70 and a standard deviation of 15. Use the 68-95-99.7 rule to find the following quantities.
   a. Percentage of rates less than 70
   b. Percentage of rates less than 55
   c. Relative frequency of rates less than 40
   d. Percentage of rates less than 85
   e. Relative frequency of rates less than 100
   f. Percentage of rates greater than 85
   g. Percentage of rates greater than 55
   h. Relative frequency of rates greater than 40
   i. Percentage of rates between 55 and 85
   j. Percentage of rates between 70 and 100

9. **COIN WEIGHTS.** Consider the following table, showing the official mean weight and estimated standard deviation for five U.S. coins. Suppose a vending machine is designed to reject all coins with weights more than 2 standard deviations above or below the mean. For each coin, find the range of weights that are acceptable to the vending machine. In each case, what percentage of legal coins are rejected by the machine?

Coin	Weight (grams)	Estimated standard deviation (grams)
Cent	2.500	0.03
Nickel	5.000	0.06
Dime	2.268	0.03
Quarter	5.670	0.07
Half dollar	11.340	0.14

GRAPHING THE NORMAL DISTRIBUTION. For Exercises 10 and 11, make a rough sketch by hand of a normal distribution with the given mean and standard deviation (the exact scale is not important here).
a. On this graph, indicate (with vertical lines or shading) the region under the curve that is within 1 standard deviation on either side of the mean. What percentage of the observations lie in this region?
b. Indicate the region under the curve that is within 2 standard deviations on either side of the mean. What percentage of the observations lie in this region?

10. The mean is 100 and the standard deviation is 10.

11. The mean is 50 and the standard deviation is 15.

12. **STANDARD SCORES AND PERCENTILES.** Use Table 5.1 to answer each question.
    a. What is the standard score of a data value that is 1 standard deviation above the mean? In what percentile is that data value?
    b. What is the standard score of a data value that is 1.5 standard deviations below the mean? In what percentile is that data value?
    c. What is the standard score of a data value that is 2 standard deviations below the mean? In what percentile is that data value?
    d. What is the standard score of a data value that is 1.5 standard deviations above the mean? In what percentile is that data value?

13. **STANDARD SCORES AND PERCENTILES.** Use Table 5.1 to answer each question.
    a. What is the standard score of a data value that is 0.5 standard deviation above the mean? In what percentile is that data value?
    b. What is the standard score of a data value that is 1 standard deviation below the mean? In what percentile is that data value?
    c. What is the standard score of a data value that is 2 standard deviations above the mean? In what percentile is that data value?
    d. What is the standard score of a data value that is 1.9 standard deviations above the mean? In what percentile is that data value?

14. **STANDARD SCORES AND PERCENTILES.** Using Table 5.1, find how many standard deviations each data value is above or below the mean.
    a. A data value in the 94th percentile
    b. A data value in the 6th percentile
    c. A data value in the 14th percentile
    d. A data value in the 86th percentile

15. **STANDARD SCORES AND PERCENTILES.** Using Table 5.1, find how many standard deviations each data value is above or below the mean.
    a. A data value in the 85th percentile
    b. A data value in the 10th percentile
    c. A data value in the 54th percentile
    d. A data value in the 23rd percentile

16. **HEIGHTS TO STANDARD SCORES AND PERCENTILES.** According to data from the National Health Survey, the heights of all adult women are normally distributed with a mean of 63.6 inches and a standard deviation of 2.5 inches. Give the standard score and percentile of women with each of the following heights.
    a. 65 inches   c. 62.5 inches
    b. 63 inches   d. 63.8 inches

17. **PREGNANCY LENGTHS TO STANDARD SCORES AND PERCENTILES.** Lengths of pregnancies are normally distributed with a mean of 268 days and a standard deviation of 15 days. Give the standard score and percentile of pregnancies of the following lengths.

a. 260 days   c. 255 days
b. 270 days   d. 265 days

FURTHER APPLICATIONS

18. **SAT SCORES.** According to the College Board, the mean score for the verbal SAT for women is 502 with a standard deviation of 109. The mean score for the verbal SAT for men is 509 with a standard deviation of 112.
    a. What percentage of women score above 611?
    b. What percentage of women score between 493 and 611?
    c. What percentage of men score above 621?
    d. What percentage of men score between 497 and 621?

19. **GRE SCORES.** Suppose that the scores on the Graduate Record Exam (GRE) are normally distributed with a mean of 497 and a standard deviation of 115.
    a. If a graduate school requires a GRE score of 650 for admission, to what percentile does this correspond?
    b. If a graduate school requires a GRE score in the 95th percentile for admission, to what actual score does this correspond?

20. **TEXTBOOK SURVEY.** Suppose that, at a university with 25,000 students, the amount of money spent by students on books is normally distributed with a mean of $150 and a standard deviation of $25. About how many of the 25,000 students spend less than $100 on books each semester? Explain.

21. **INCOME DISTRIBUTION.** Assume that the income distribution for assembly line workers at a large manufacturing plant is nearly normal with a mean of $30,000 and a standard deviation of $6,000. Estimate the percentage of workers who earn between $27,000 and $33,000.

22. **STANFORD-BINET IQ TEST SCORES.** Assume that IQ test scores are normally distributed with a mean of 100 and a standard deviation of 16. Determine the range of IQ scores for the middle 50% of the population (that is, the scores for people in percentiles 25 through 75).

23. **CALIBRATING BAROMETERS.** Researchers for a manufacturer of barometers (devices to measure atmospheric pressure) read each of 50 barometers at the same time of day. The mean of the readings is 30.4 (inches of mercury) with a standard deviation of 0.23 inch.
    a. What percentage of the barometers read over 31?
    b. What percentage of the barometers read less than 30?
    c. The company decides to reject barometers that read more than 1.5 standard deviations above or below the mean. What is the critical reading below which barometers will be rejected? What is the critical reading above which barometers will be rejected?
    d. What would you take as the actual atmospheric pressure at the time the barometers were read? Explain.

24. **SPELLING BEE SCORES.** At the district spelling bee, the 60 girls have a mean score of 71 points with a standard deviation of 6, while the 50 boys have a mean score of 66 points with a standard deviation of 5 points. Those students with a score greater than 75 are eligible to go to the state spelling bee. What percent of those going to the state bee will be girls?

25. **BEING A MARINE.** According to data from the National Health Survey, the heights of adult men are normally distributed with a mean of 69.0 inches and a standard deviation of 2.8 inches. The U.S. Marine Corps requires that men have heights between 64 inches and 78 inches. What percentage of American men are eligible for the Marines based on height?

26. **MOVIE LENGTHS.** According to the data in Exercise 19 of Section 5.1, the mean length of the movies in the sample is 110.5 minutes with a standard deviation of 22.4 minutes. Assume that the movie lengths are normally distributed.
    a. What fraction of movies are more than 2 hours long?
    b. What fraction of movies are less than $1\frac{1}{2}$ hours long?
    c. What is the probability that a randomly selected movie will be less than $2\frac{1}{2}$ hours long?

27. **SCOTTISH ARMY CHEST SIZES.** The following table gives the actual data for Figure 5.4a. The mean of the distribution is $\mu = 39.85$ inches, the standard deviation is $\sigma = 2.07$ inches, and $n = 5,738$.

Chest size	33	34	35	36	37	38	39	40
Frequency	3	18	81	185	420	749	1,073	1,079

Chest size	41	42	43	44	45	46	47	48
Frequency	934	658	370	92	50	21	4	1

   a. If the distribution were exactly normal, what percentage of the data values would be within 1 standard deviation of the mean?
   b. For the actual distribution, what percentage of the data values are within 1 standard deviation of the mean?
   c. If the distribution were exactly normal, what percentage of the data values would be within 2 standard deviations of the mean?
   d. For the actual distribution, what percentage of the data values are within 2 standard deviations of the mean?
   e. Comment on how closely this distribution approximates a normal distribution.

28. **STATES DATA.** Consider the states data given in Appendix A. Choose one of the variables and do the following. First, make a histogram of the variable, choosing the bin size so that the histogram has 10–15 bars. Find the mean and standard deviation (using the range rule of thumb from Section 4.3) of the distribution and mark the mean clearly on the histogram. Determine the percentage of data

values that lie within 1 standard deviation of the mean. Based on these observations, discuss whether the variable is normally or nearly normally distributed. Comment on the reasons for any deviations from a normal distribution.

PROJECTS FOR THE WEB AND BEYOND

For useful links, select "Links for Web Projects" for Chapter 5 at www.aw.com/bbt.

29. DATA AND STORY LIBRARY. Visit the Data and Story Library Web site and find one example (of many) of a normal distribution. Write a one-page account of your findings. Include a careful description of the variable under consideration, details of the distribution, and an explanation of why you expect this variable to have an approximately normal distribution.

30. STATE VITAL STATISTICS. The National Center for Health Statistics provides vital statistics by state (and county) under State Tabulated Data. Find a variable (for example, births, deaths, or infant deaths). Make a histogram showing this variable for each state. Comment on whether the variable is normally distributed across the states.

31. NORMAL DISTRIBUTION DEMONSTRATIONS ON THE WEB. Do a Web search on the keywords "normal distribution" and find an animated demonstration of the normal distribution. Describe how the demonstration works and the useful features that you observed.

32. ESTIMATING A MINUTE. Ask survey subjects to estimate 1 minute without looking at a watch or clock. Each subject should say "go" for the beginning of the minute and then "stop" when he or she thinks that 1 minute has passed. (Alternatively, you could repeatedly time yourself by looking away from any watch or clock and then noting the correct time when you think that 1 minute has passed.) Use a watch to record the actual times. Construct a graph of the estimates. Is the graph approximately normal? Does the mean appear to be close to 1 minute?

5.3   The Central Limit Theorem

A high school English teacher has 100 seniors taking a college placement test. The test is designed to have a mean score of 500 and a standard deviation of 100. Using the methods of this chapter, she can determine the percentage of individual scores that are likely to be above, say, 600. But can she predict anything about the performance of her *group* of 100 students? For example, what is the likelihood that the mean score of the group will be above 600? This type of question, in which we ask about the mean score for a group or sample drawn from a much larger population, can be answered with the *Central Limit Theorem.*

Before we get to the theorem itself, we can develop some useful insight by thinking about dice rolling. Suppose we roll *one* die 1,000 times and record the outcome of each roll, which can be the number 1, 2, 3, 4, 5, or 6. Figure 5.24 shows a histogram of outcomes. All

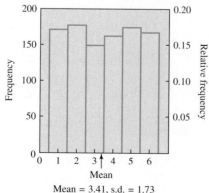

Mean = 3.41, s.d. = 1.73

Figure 5.24   Frequency and relative frequency distribution of outcomes from rolling one die 1,000 times.

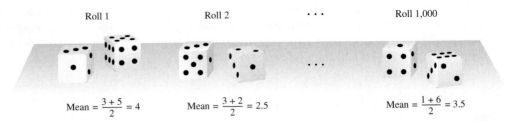

Roll 1   Roll 2   · · ·   Roll 1,000

Mean = $\frac{3+5}{2} = 4$ Mean = $\frac{3+2}{2} = 2.5$ Mean = $\frac{1+6}{2} = 3.5$

Figure 5.25 This diagram represents the idea of rolling two dice 1,000 times and recording the *mean* on each roll. The mean of the values on two dice is their sum divided by 2. This mean can range from $(1 + 1)/2 = 1$ to $(6 + 6)/2 = 6$.

six outcomes have roughly the same relative frequency, because the die is equally likely to land in each of the six possible ways. That is, the histogram shows a (nearly) *uniform distribution* (see Section 4.2). Using the methods described in Chapter 4, we can compute the mean and standard deviation for this distribution. It turns out that the distribution in Figure 5.24 has a mean of 3.41 and a standard deviation of 1.73.

Now suppose we roll *two* dice 1,000 times and record the *mean* of the two numbers that appear on each roll (Figure 5.25). To find the mean for a single roll, we add the two numbers and divide by 2. For example, if the two dice come up 3 and 5, the mean for the roll is $(3 + 5)/2 = 4$. Thus, the possible values of the mean on a roll of two dice are $1.0, 1.5, 2.0, \ldots, 5.0, 5.5, 6.0$.

Figure 5.26a shows a typical result of rolling two dice 1,000 times. The most common values in this distribution are central values such as 3.0, 3.5, and 4.0. These values are common because they can occur in several ways. For example, a mean of 3.5 can occur if the two dice land as 1 and 6, 2 and 5, 3 and 4, 4 and 3, 5 and 2, or 6 and 1. High and low values occur less frequently because they can occur in fewer ways. For example, a roll can have a mean of 1.0 only if both dice land showing a 1. Again, it is possible to compute the mean and standard deviation for this distribution, and they turn out to be 3.43 and 1.21, respectively.

What happens if we increase the number of dice we roll? Suppose we roll five dice 1,000 times and record the mean of the five numbers on each roll. A histogram for this experiment is

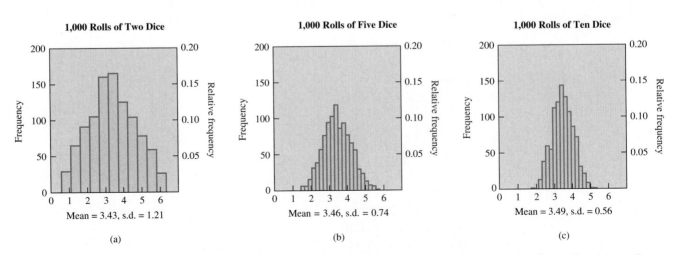

1,000 Rolls of Two Dice

Mean = 3.43, s.d. = 1.21

(a)

1,000 Rolls of Five Dice

Mean = 3.46, s.d. = 0.74

(b)

1,000 Rolls of Ten Dice

Mean = 3.49, s.d. = 0.56

(c)

Figure 5.26 Frequency and relative frequency distributions of sample means from rolling (a) two dice 1,000 times, (b) five dice 1,000 times, and (c) ten dice 1,000 times.

shown in Figure 5.26b. Once again we see that the central values around 3.5 occur most frequently, but the spread of the distribution is narrower than in the two previous cases. Computing the mean and standard deviation of this distribution, we get values of 3.46 and 0.74, respectively.

If we further increase the number of dice to ten on each of 1,000 rolls, we find the histogram in Figure 5.26c, which is even narrower. In this case, the mean is 3.49 and standard deviation is 0.56.

Table 5.2 summarizes the four experiments we've described. The table refers to a *distribution of means* because, in each of the dice rolling experiments, we recorded the mean of each of 1,000 rolls (that is, the mean of one die, of two dice on each roll, of five dice on each roll, or of ten dice on each roll). Thus, the mean for all 1,000 rolls in an experiment is *a mean of the distribution of means*. Similarly, the standard deviation for all 1,000 rolls in an experiment is a *standard deviation of the distribution of means*.

Table 5.2 Summary of Dice Rolling Experiments

Number of dice rolled each time	Mean of the distribution of means	Standard deviation of the distribution of means
1	3.41	1.73
2	3.43	1.21
5	3.46	0.74
10	3.49	0.56

A remarkable insight emerges from these four experiments. Rolling $n = 1$ die 1,000 times can be regarded as taking 1,000 samples of size $n = 1$ from the population of all possible dice rolls. Rolling $n = 2$ dice 1,000 times can be viewed as taking 1,000 samples of size $n = 2$. Likewise, rolling $n = 5$ and $n = 10$ dice 1,000 times is like taking 1,000 samples of size $n = 5$ and $n = 10$, respectively. Table 5.2 shows that as the sample size increases, the mean of the distribution of means approaches the value 3.5 and the standard deviation becomes smaller (making the distribution narrower). More important, the distribution looks more and more like a normal distribution as the sample size increases. This latter fact may seem surprising because we have taken samples from a *uniform* distribution (the outcomes of rolling a single die shown in Figure 5.24), *not* from a normal distribution. Nevertheless, the distribution of means clearly approaches a normal distribution for large sample sizes. This fact is a consequence of the **Central Limit Theorem**.

The Central Limit Theorem

Suppose we take many random samples of size n for a variable with any distribution (not necessarily a normal distribution) and record the distribution of the *means* of each sample. Then,

1. The distribution of means will be approximately a normal distribution for large sample sizes.

2. The mean of the distribution of means approaches the population mean, μ, for large sample sizes.

3. The standard deviation of the distribution of means approaches $\sigma/\sqrt{n}$ for large sample sizes, where σ is the standard deviation of the population.

TECHNICAL NOTE

(1) For practical purposes, the distribution of means will be nearly normal if the sample size is larger than 30. (2) If the original population is normally distributed, then the sample means will be normally distributed for *any* sample size n. (3) In the ideal case where the distribution of means is formed from *all* possible samples, the mean of the distribution of means *equals* μ and the standard deviation of the distribution of means equals $\sigma/\sqrt{n}$.

Be sure to note the very important adjustment, described by item 3 above, that must be made when working with samples or groups instead of individuals:

The standard deviation of the distribution of sample means is not the standard deviation of the population, σ, but rather σ/√n, where n is the size of the samples.

Time out to think

Confirm that the standard deviation of the distributions of means given in Table 5.2 for $n = 2$, 5, 10 agrees with the prediction of the Central Limit Theorem, given that $\sigma = 1.73$ (the population standard deviation found in Figure 5.24). For example, with $n = 2$, $\sigma/\sqrt{2} = 1.22 \approx 1.21$.

Let's summarize the ingredients of the Central Limit Theorem. We always start with a particular variable, such as the outcomes of rolling a die or weights of people, that varies randomly over a population. The variable has a certain mean, μ, and standard deviation, σ, which we may or may not know. This variable can have *any* sort of distribution, not necessarily normal. Now we take many samples of that variable, with n items in each sample, and find the mean of each sample (such as the mean value of n dice or the mean weight of a sample of n people). If we then make a histogram of the means from the many samples, we will see a distribution that is close to a normal distribution. The larger the sample size, n, the more closely the distribution of means approximates a normal distribution. Careful study of Figure 5.27 should help solidify these important ideas.

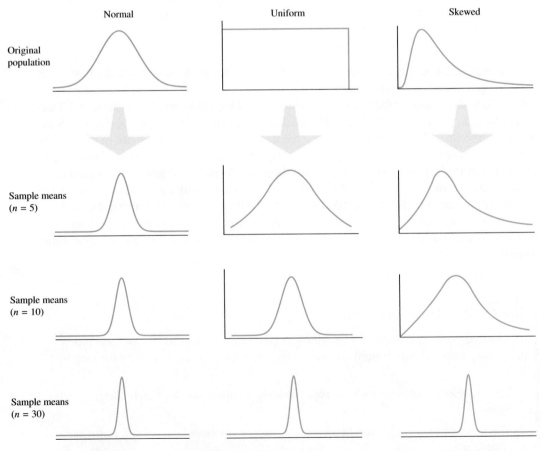

Figure 5.27 As the sample size increases ($n = 5$, 10, 30), the distribution of sample means approaches a normal distribution, regardless of the shape of the original distribution. The larger the sample size, the smaller is the standard deviation of the distribution of sample means.

EXAMPLE 1 Predicting Test Scores

You are a middle school principal and your 100 eighth-graders are about to take a national standardized test. The test is designed so that the mean score is $\mu = 400$ with a standard deviation of $\sigma = 70$. Assume the scores are normally distributed.

a. What is the likelihood that *one* of your eighth-graders, selected at random, will score below 380 on the exam?

b. Your performance as a principal depends on how well your entire *group* of eighth-graders scores on the exam. What is the likelihood that your group of 100 eighth-graders will have a *mean* score below 380?

SOLUTION

a. In dealing with an individual score, we use the method of standard scores discussed in Section 5.2. Given the mean of 400 and standard deviation of 70, a score of 380 has a standard score of

$$z = \frac{\text{data value} - \text{mean}}{\text{standard deviation}} = \frac{380 - 400}{70} = -0.29$$

According to Table 5.1, a standard score of -0.29 corresponds to the 38th percentile—that is, 38% of all students can be expected to score below 380. Thus, there is about a 38% chance that a randomly selected student will score below 380. Notice that we need to know that the scores have a normal distribution in order to make this calculation, because the table of standard scores applies only to normal distributions.

b. The question about the mean of a *group* of students must be handled with the Central Limit Theorem. According to this theorem, if we take random samples of size $n = 100$ students and compute the mean test score of each group, the distribution of means is approximately normal. Moreover, the mean of this distribution is $\mu = 400$ and its standard deviation is $\sigma/\sqrt{n} = 70/\sqrt{100} = 7$. With these values for the mean and standard deviation, the standard score for a mean test score of 380 is

> When you are listening to corn pop, are you hearing the Central Limit Theorem?
> —William A. Massey

$$z = \frac{\text{data value} - \text{mean}}{\text{standard deviation}} = \frac{380 - 400}{7} = -2.9$$

Table 5.1 shows that a standard score of -2.9 corresponds to the 0.19th percentile. In other words, only 0.19% of all random samples of 100 students will have a mean score of less than 380. Therefore, the chance that a randomly selected group of 100 students will have a mean score below 380 is 0.0019, or about 1 in 500. Notice that this calculation regarding the group mean did *not* depend on the individual scores' having a normal distribution.

This example has an important lesson. The likelihood of an *individual* scoring below 380 is almost 4 in 10 (38%), but the likelihood of a *group* of 100 students having a mean score below 380 is less than 1 in 500 (0.19%). In other words, there is more variation in the scores of individuals than in the means of groups of individuals. ∎

EXAMPLE 2 Salary Equity

The mean salary of the 9,000 employees at Holley.com is $\mu = \$26,400$ with a standard deviation of $\sigma = \$2,420$. A pollster samples 400 randomly selected employees and finds that the mean salary of the sample is $26,650. Is it likely that the pollster would get these results by chance, or does the discrepancy suggest that the pollster's results are suspect?

SOLUTION The question deals with the mean of a *group* of 400 individuals, which is a case for the Central Limit Theorem. The theorem tells us that if we select many groups of 400 individuals and compute the mean of each group, the distribution of means will be close to normal with a mean of $\mu = \$26,400$ and a standard deviation of $\sigma/\sqrt{n} = \$2,420/\sqrt{400} = \$121$. Within the distribution of means, a mean salary of $26,650 has a *standard score* of

$$z = \frac{\text{data value} - \text{mean}}{\text{standard deviation}} = \frac{\$26,650 - \$26,400}{\$121} = 2.07$$

In other words, if we assume that the sample is randomly selected, its mean salary is more than 2 standard deviations above the mean salary of the entire company. According to Table 5.1, a standard score of 2.07 lies near the 98th percentile. Thus, the mean salary of *this* sample is greater than the mean salary we would find in 98% of the possible samples of 400 workers. That is, the likelihood of selecting a group of 400 workers with a mean salary above $26,650 is about 2%, or 0.02. The mean salary of the sample is surprisingly high; perhaps the survey was flawed. ▪

Time out to think
Would a salary of $26,650 for an *individual* worker lie above or below the 98th percentile? Explain.

The Value of the Central Limit Theorem

The Central Limit Theorem allows us to say something about the mean of a group if we know the mean, μ, and the standard deviation, σ, of the entire population. This can be useful, but it turns out that the opposite problem is far more important.

Two major activities of statistics are making estimates of population means and testing claims about population means. Suppose we do *not* know the mean of a variable for the entire population. Is it possible to make a good estimate of the population mean (such as the mean income of all Internet users) knowing only the mean of a much smaller sample? As you can probably guess, being able to answer this type of question lies at the heart of statistical sampling, especially in polls and surveys. The Central Limit Theorem provides the key to answering such questions. We will return to this topic in Chapter 8.

Review Questions

1. Briefly explain the difference between a question asking about a data value for an individual in a population and one asking about a mean value for a group or sample drawn from a population. Give an example.

2. Explain how we find the *mean* for a roll of n dice. If we perform the rolls many times, how do we make a histogram for the distribution of means?

3. Briefly explain the Central Limit Theorem.

4. Give an example to show how the Central Limit Theorem implies that the distribution of means of a variable has a smaller standard deviation than the distribution of the variable itself. Explain why this fact should be expected.

Exercises

BASIC SKILLS AND CONCEPTS

1. **IQ SCORES AND THE CENTRAL LIMIT THEOREM.** IQ scores are normally distributed with a mean of 100 and a standard deviation of 16. Suppose that many samples of size n are taken from a large population of people and the mean IQ score is computed for each sample. Find the mean and standard deviation of the resulting distribution of sample means for $n = 100$ and for $n = 400$. Briefly explain why the standard deviation is different for the two values of n.

2. **SAT SCORES AND THE CENTRAL LIMIT THEOREM.** Assume that SAT scores are normally distributed with a mean of 500 and a standard deviation of 100. Suppose that many samples of size n are taken from a large population of students and the mean SAT score is computed for each sample. Find the mean and standard deviation of the resulting distribution of sample means for $n = 100$ and for $n = 400$. Briefly explain why the standard deviation is different for the two values of n.

3. **TWELVE-SIDED DICE AND THE CENTRAL LIMIT THEOREM.** Rolling a fair *twelve-sided* die produces a uniformly distributed set of numbers between 1 and 12 with a mean of 6.5 and a standard deviation of 3.452. Suppose that a sample of n twelve-sided dice are rolled many times and the mean of the outcomes is computed for each roll of n dice. Find the mean and standard deviation of the resulting distribution of sample means for $n = 36$ and for $n = 100$. State whether you would expect the distribution of means to be normal or nearly normal in each case. Briefly explain any differences in your answers for the two values of n.

4. **TEN-SIDED DICE AND THE CENTRAL LIMIT THEOREM.** Rolling a fair *ten-sided* die produces a uniformly distributed set of numbers between 1 and 10 with a mean of 5.5 and a standard deviation of 2.872. Suppose that a sample of n ten-sided dice are rolled many times and the mean of the outcomes is computed for each roll of n dice. Find the mean and standard deviation of the resulting distribution of sample means for $n = 64$ and for $n = 100$. State whether you would expect the distribution of means to be normal or nearly normal in each case. Briefly explain any differences in your answers for the two values of n.

FURTHER APPLICATIONS

5. **SCOTTISH CHEST DATA AND THE CENTRAL LIMIT THEOREM.** Consider the data on chest sizes of Scottish militiamen (see Exercise 27 of Section 5.2). The mean of this population is $\mu = 39.85$ inches and the standard deviation is $\sigma = 2.07$ inches.
   a. Suppose you select a random sample of $n = 25$ militiamen and compute their mean chest size. What is the likelihood that this mean is greater than 40 inches?
   b. Suppose you select a random sample of $n = 100$ militiamen and compute their mean chest size. What is the likelihood that this mean is less than 39.5 inches?
   c. Suppose you select a random sample of $n = 625$ militiamen and compute their mean chest size. What is the likelihood that this mean is greater than 40 inches?
   d. Suppose you select a random sample of $n = 1,600$ militiamen and compute their mean chest size. What is the likelihood that this mean is less than 39.7 inches?

6. **GENERIC VARIABLE.** Suppose a variable, such as height or blood pressure, is normally distributed across a large population with a mean of μ and a standard deviation of σ.
   a. Suppose we take many samples of size $n = 100$ and compute the mean of the variable for each sample. How does the standard deviation of the distribution of sample means compare to σ? Explain why.
   b. Suppose we take many samples of size $n = 1,000$ and compute the mean of the variable for each sample. How does the standard deviation of this distribution of sample means compare to the standard deviation of the distribution of sample means in part a? How does the standard deviation of this distribution of sample means compare to σ? Explain your answers.
   c. How does the standard deviation of the distribution of sample means change as we take larger and larger sample sizes?

AIRCRAFT AGES. In Exercises 7–10, assume that the ages of commercial aircraft are normally distributed with a mean of 13.0 years and a standard deviation of 7.9 years (data from the Aviation Data Services).

7. What percentage of individual aircraft have ages greater than 15 years? Assume that a random sample of 49 aircraft is selected and the mean age of the sample is computed. What percentage of sample means have ages greater than 15 years?

8. What percentage of individual aircraft have ages less than 10 years? Assume that a random sample of 84 aircraft is selected and the mean age of the sample is computed. What percentage of sample means have ages less than 10 years?

9. What percentage of individual aircraft have ages between 10 years and 16 years? Assume that a random sample of

81 aircraft is selected and the mean age of the sample is computed. What percentage of sample means are between 10 years and 16 years?

10. What percentage of individual aircraft have ages between 12.5 years and 13.5 years? Assume that a random sample of 400 aircraft is selected and the mean age of the sample is computed. What percentage of sample means are between 12.5 years and 13.5 years?

Running Times. According to *Runner's World* magazine, the finishing times in 10-kilometer races have a mean of 61 minutes and a standard deviation of 9 minutes. Assume that these times are normally distributed. Use this information to solve Exercises 11–14.

11. What percentage of individual finishing times are greater than 65 minutes? Assume that a random sample of 25 runners is selected and the mean finishing time of the sample is computed. What percentage of sample means are greater than 65 minutes?

12. What percentage of individual finishing times are less than 55 minutes? Assume that a random sample of 25 runners is selected and the mean finishing time of the sample is computed. What percentage of sample means are less than 55 minutes?

13. What percentage of individual finishing times are between 59 and 62 minutes? Assume that a random sample of 64 runners is selected and the mean finishing time of the sample is computed. What percentage of sample means are between 59 and 62 minutes?

14. What percentage of individual finishing times are between 60 and 62 minutes? Assume that a random sample of 36 runners is selected and the mean finishing time of the sample is computed. What percentage of sample means are between 60 and 62 minutes?

15. **Random Numbers.** Suppose a computer generates random numbers that are uniformly distributed between 0 and 1 with a mean of 0.500 and a standard deviation of 0.290. Suppose a sample of $n = 100$ random numbers is generated and its mean is computed. What is the probability that the sample will have each of the following means?
   a. A mean greater than 0.550
   b. A mean less than 0.480
   c. A mean between 0.450 and 0.550
   d. A mean between 0.480 and 0.530

16. **Blood Pressure in Women.** Systolic blood pressure for women between the ages of 18 and 24 is normally distributed with a mean of 114.8 (mm of mercury) and a standard deviation of 13.1.
   a. What is the likelihood that an individual woman has a blood pressure above 125?
   b. Suppose a random sample of $n = 300$ women is selected and the mean blood pressure for the sample is computed. What is the likelihood that the mean blood pressure for the sample will be above 125?
   c. Suppose a random sample of $n = 300$ women is selected and the mean blood pressure for the sample is computed. What is the likelihood that the mean blood pressure for the sample will be below 114?

Projects for the Web and Beyond

For useful links, select "Links for Web Projects" for Chapter 5 at www.aw.com/bbt.

17. **Central Limit Theorem on the Web.** Doing a Web search on "central limit theorem" will uncover many sites devoted to this subject. Find a site that has animated demonstrations of the Central Limit Theorem. Describe in your own words what you observed and how it illustrates the Central Limit Theorem.

18. **The Quincunx on the Web.** Do a Web search on "central limit theorem" or "quincunx" and find a site that has an animated demonstration of the quincunx (or Galton's board). Describe the quincunx and explain how it illustrates the Central Limit Theorem.

19. **Dice Rolling.** Demonstrate the Central Limit Theorem using dice, as discussed in this section. Give each person in your class as many dice as possible. Begin by rolling one die and making a histogram of the outcomes. Then let every person roll two dice and make a histogram of the mean of each roll. Increase the number of dice in each roll as long as dice and time allow. Comment on the appearance of the histogram at each stage.

Chapter Review Exercises

1. For each of the following situations, state whether the distribution of values is likely to be a normal distribution. Give a brief explanation justifying your choice.
   a. Numbers resulting from spins of a roulette wheel. (There are 38 equally likely slots with numbers 0, 00, 1, 2, 3, . . . , 36.)
   b. Incomes of Americans over the age of 18
   c. Scores on a test of manual dexterity

2. Assume that body temperatures of healthy adults are normally distributed with a mean of 98.20°F and a standard deviation of 0.62°F (based on data from University of Maryland researchers).
   a. If you have a body temperature of 99.00°F, what is your percentile score?
   b. Convert 99.00°F to a standard score (or z score).
   c. Is a body temperature of 99.00°F "unusual"? Why or why not?
   d. Fifty adults are randomly selected. What is the likelihood that the mean of their body temperatures is 97.98°F or lower?
   e. A person's body temperature is found to be 101.00°F. Is this result "unusual"? Why or why not? What should you conclude?
   f. What body temperature is the 95th percentile?
   g. What body temperature is the 5th percentile?
   h. Bellevue Hospital in New York City uses 100.6°F as the lowest temperature considered to indicate a fever. What percentage of normal and healthy adults would be considered to have a fever? Does this percentage suggest that a cutoff of 100.6°F is appropriate?
   i. If, instead of assuming that the mean body temperature is 98.20°F, we assume that the mean is 98.60°F (as many people believe), what is the chance of randomly selecting 106 people and getting a mean of 98.20°F or lower? (Continue to assume that the standard deviation is 0.62°F.) University of Maryland researchers did get such a result. What should we conclude?

FOCUS ON EDUCATION

What Do SAT Scores Mean?

The Scholastic Aptitude Test (SAT) has been taken by college-bound high school students since 1941, when 11,000 students took the first test. Today it is taken by approximately 1.2 million high school students every year. There are two parts to the SAT: the verbal test and the math test. The scores on each part of the SAT were originally scaled to have a mean of 500 and a standard deviation of 100. The maximum and minimum scores on each part are 800 and 200, respectively, which are 3 standard deviations above and below the mean.

Each year's test has common items with the previous year's test. The College Board (which designs the test) claims that valid comparisons can be made from year to year by analyzing the results of these common items. In other words, the Board claims that a score of 500 represents the same level of achievement on the test, no matter what year the test is taken.

Because scores from year to year are supposed to be comparable, SAT scores have been widely used to assess the general state of American education. Figure 5.28 shows the average scores on the verbal and math parts of the SAT between 1972 and 2000. If we ignore the period beginning with 1996 (when scores rose for reasons we'll discuss shortly), the total scores (combined male and female) for both parts declined from 1972 until about 1985, after which time they remained relatively flat.

If the scores from year to year are truly comparable, then students taking the SAT in recent years have not performed as well as students of a couple decades ago—particularly in verbal skills. But before we accept this conclusion, we must answer two important questions:

1. Are trends among the *sample* of high school students who take the SAT representative of trends for the *population* of all high school students?

2. Even with a short section of common questions that link one year to the next, the test changes every year. Can test scores from one year legitimately be compared to scores in other years?

Unfortunately, neither question can be answered with a clear yes. For example, many people argue that much of the long-term decline in SAT scores is the result of changes in the sample of students who take the test. Only about a third of high school graduates took the SAT in the 1970s, whereas over 40% of high school graduates took the SAT in the 1990s. If these samples represent the top tier of high school students, the decline in score averages might simply reflect the fact that a greater range of student abilities is represented in later samples. Because of the change in the sample, inference from sample trends to population trends is extremely difficult.

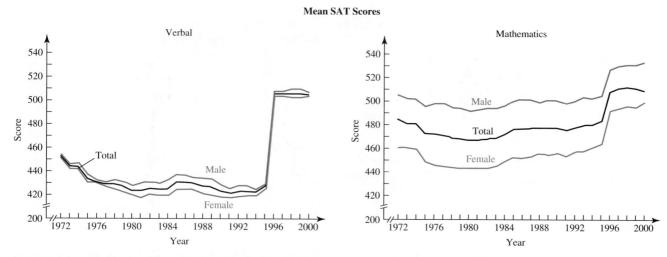

Figure 5.28   Verbal and mathematics SAT scores, 1972–2000. (Scores were "recentered" beginning in 1996.)

The second question is even more difficult. Even if some test questions stay the same from one year to the next, different material may be emphasized in high schools, thereby changing the likelihood that students will answer the same questions correctly from one year to the next.

Further complications arise when the test undergoes major changes. For example, in 1994 the use of calculators was allowed for the first time on the mathematics section, additional time was allotted for the entire test, and an entire section of prior verbal SATs was deleted (a section on antonyms). Perhaps it's not surprising that scores in mathematics increased moderately in both 1994 and 1995 relative to earlier years, and verbal scores rose moderately in 1995 (note the small upward blips in Figure 5.28, just before the much larger upward swing in 1996). Clearly, the change in the structure of the test makes it difficult to determine whether these rises in average scores reflect improvements in education.

The search for trends in SAT scores became even more difficult in 1996. The average scores on the SAT had fallen from around 500 in the 1960s to the mid-400s by the 1990s. In 1996, the College Board "recentered" SAT scores so that the average score on both the verbal and math sections once again became 500. The net effect of this change was to raise everyone's SAT scores significantly (explaining the large jumps in Figure 5.28). For example, a mathematics score of 500 in 1997 is the equivalent of only about 480 in 1995. Clearly, this change makes it far more difficult to do year-to-year comparisons of SAT scores. Further changes in the test are in the works. Spurred by colleges and universities that have questioned the value of the SAT, the College Board is currently considering a major overhaul of the test. A new test could be in use as early as 2006.

Average SAT scores are used not only to make comparisons over time, but also to compare results between different states or different school systems. This type of comparison can also be difficult. The first two columns of Table 5.3 show how several selected states ranked nationally in terms of average SAT scores. For example, Alabama's rank of 14 means it had the 14th highest average SAT score among the 50 states. However, as shown in the third column, different states had widely different percentages of high school students taking the SAT. The range in the table goes from 4% for Mississippi and Utah to 88% for Connecticut. If the students taking the SAT in any particular state represent top-tier students, then the meaning of the average SAT score is very different from state to state. In that case, for example, Utah's

Table 5.3 Rank of Selected States in SAT Scores and Percent of High School Students Taking the SAT

State	Rank	% taking SAT	Adjusted rank
Alabama	14	9%	37
Arkansas	18	6%	47
Colorado	23	28%	8
Connecticut	33	88%	14
Mississippi	16	4%	50
New Hampshire	27	78%	3
Utah	4	4%	42

SOURCE: *New York Times.*

average score really represents the average score for the top 4% of Utah high school students—which could not be expected to represent the average of *all* Utah high school students. In contrast, Connecticut's average score would be a fairly good measure of its average high school students, since the vast majority of students in Connecticut take the SAT. The final column of Table 5.3 shows an "adjusted ranking" based on this assumption about which students take the SAT in each state.

QUESTIONS FOR DISCUSSION

1. Do you think that comparisons of SAT scores over two years are meaningful? Over ten years? Overall, do you think it is true that students today have a lower level of achievement than did students a few decades ago? Why or why not?

2. Notice that scores for males have been consistently higher than scores for females. Why do you think this is the case? Do you think that changes in our education system could eliminate the gap? Defend your opinions.

3. Discuss personal experiences with the SAT among your classmates. Based on these personal experiences, what do you think the SAT is measuring? Do you think the test is a reasonable way to predict students' performance in college? Why or why not?

4. What were the statistical implications of recentering the SAT scores in 1996? (For example, how did it change the overall distribution of scores?) Do you think that recentering SAT scores was a good idea? Why or why not?

SUGGESTED READING

Crouse, James, and Trusheim, Dale. *The Case Against the SAT.* University of Chicago Press, 1988.

http://collegeboard.org

Lewin, Tamar. "College Board to Revise SAT." *New York Times,* March 23, 2002.

Tabor, Mary B.W. "S.A.T. Ranks for the States Disputed." *New York Times,* March 27, 1996.

FOCUS ON PSYCHOLOGY

Are We Smarter Than Our Parents?

Most kids tend to think that they're smarter than their parents, but is it possible that they're right? If you believe the results of IQ tests, not only are we smarter than our parents, on average, but our parents are smarter than our grandparents. In fact, almost all of us would have ranked as geniuses if we'd lived a hundred years ago. Of course, before any of us start touting our Einstein-like abilities, it would be good to investigate what lies behind this startling claim.

The idea of an IQ, which stands for *intelligence quotient*, was invented by French psychologist Alfred Binet (1857–1911). Binet created a test that he hoped would identify children in need of special help in school. He gave his test to many children, and then calculated each child's IQ by dividing the child's "mental age" by his or her physical age (and multiplying by 100). For example, a 5-year-old child who scored as well as an average 6-year-old was said to have a mental age of 6, and therefore an IQ of $6 \div 5 \times 100$, or 120. Note that, by this definition, IQ tests make sense only for children. However, later researchers, especially psychologists for the U.S. Army, extended the idea of IQ so that it could be applied to adults as well.

Today, IQ is defined by a normal distribution with a mean of 100 and a standard deviation of 16. According to the 68-95-99.7 rule, 68% of the people taking an IQ test get scores between 84 and 116, 95% get scores between 68 and 132, and 99.7% get scores between 52 and 148. Traditionally, psychologists classified people with an IQ below 70 (about 2 standard deviations below the mean of 100) as "intellectually deficient," and people who score above 130 (about 2 standard deviations above the mean) as "intellectually superior."

You're probably aware of the controversy that surrounds IQ tests, which boils down to two key issues:

- Do IQ tests measure intelligence or something else?

- If they do measure intelligence, is it something that is innate and determined by heredity or something that can be molded by environment and education?

A full discussion of these issues is too involved to cover here, but a recently discovered trend in IQ scores sheds light on these issues in a surprising way. As we'll see shortly, the trend is quite pronounced, but it was long hidden because of the way IQ tests are scored. There are several different, competing versions of IQ tests, and most of them are regularly changed and updated. But in all cases, the scores are adjusted to fit a normal distribution with a mean of 100 and standard deviation of 16. In other words, the scoring of an IQ test is essentially done in the same way that an instructor might grade an exam "on a curve." Because of this adjustment, the mean on IQ tests is *always* 100, which makes it impossible for measured IQ scores to rise and fall with time.

By the Way...

Binet himself assumed that intelligence could be molded and warned against taking his tests as a measure of any innate or inherited abilities. However, many later psychologists concluded that IQ tests could measure innate intelligence, which led to their being used for separating school children, military recruits, and many other groups of people according to supposed intellectual ability.

However, a few IQ tests have not been changed and updated over time, including some given by the military. In other cases, tests that have been updated sometimes still repeat old questions. In the early 1980s, a political science professor named Dr. James Flynn began to look at the raw, unadjusted scores on unchanged tests and questions. The results were astounding.

Dr. Flynn found that the raw scores have been steadily rising, although the precise amount of the rise varies somewhat with the type of IQ test. The highest rates of increase are found on tests that purport to measure abstract reasoning abilities (such as the "Raven's" tests). For these tests, Dr. Flynn found that the unadjusted IQ scores of people in industrialized countries have been rising at a rate of about 6 points per decade. In other words, a person who scored 100 on a test given in 2000 would have scored about 106 on a test in 1990, 112 on a test in 1980, and so on. Over a hundred years, this would imply a rise of some 60 points, suggesting that someone who scores an "intellectually deficient" IQ of 70 today would have rated an "intellectually superior" IQ of 130 a century ago.

This long-term trend toward rising scores on IQ tests is now called the *Flynn effect*. It is present for all types of IQ tests, though not always to the same degree as with abstract reasoning tests. For example, Figure 5.29 shows how results changed on one of the most widely used IQ tests (the Stanford-Binet test) between 1932 and 1997. Note that, in terms of unadjusted scores, the mean rose by 20 points during that time period. In other words, if people who scored an IQ of 100 on a 1997 test were instead scored on a 1932 test, they would rate an IQ of 120. As the figure shows, about one-fourth of the 1997 test-takers would have rated "intellectually superior" on the 1932 test.

Many other scientists have investigated the Flynn effect, and there seems to be little doubt that the effect is real. The implication is clear: Whatever IQ tests measure, people today really *do* have more of it than people just a few decades ago. If IQ tests measure intelligence, then it means we really are smarter than our parents (on average), who in turn are smarter than our grandparents (on average).

Of course, if IQ tests don't measure "intelligence" but only measure some type of skill, then the rise in scores may indicate only that today's children have more practice at that skill than past children. The fact that the greatest rise is seen on tests of abstract thinking lends some support to this idea. These tests often involve such problems as solving puzzles and

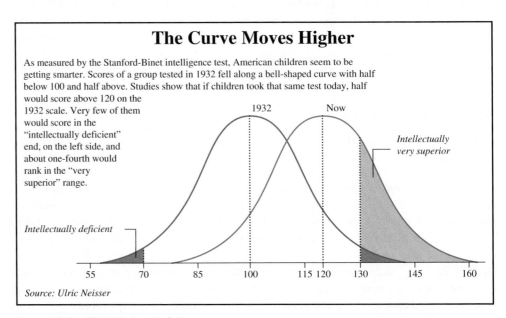

Figure 5.29   SOURCE: *New York Times.*

looking for patterns among sets of shapes, and these types of problems are now much more common in games than they were in the past.

While the Flynn effect does not answer the question of whether IQ tests measure intelligence, it may tell us one important thing: If IQs really have been rising as the Flynn effect suggests, then IQ must *not* be a primarily inherited trait, because inherited traits cannot change that much in just a few decades. Thus, if IQ tests are measuring intelligence, then intelligence can be molded by environmental factors. Conversely, if intelligence is hereditary, then IQ tests are not measuring it.

Dr. Flynn's discovery has already changed the way psychologists look at IQ tests, and it is sure to be an active topic of research in coming decades. Moreover, given the many uses to which modern society has put IQ tests, the Flynn effect is likely to have profound social and political consequences as well. So back to our starting question: Are we smarter than our parents? We really can't say, but we can certainly hope so, because it will take a lot of brainpower to solve the problems of the future.

QUESTIONS FOR DISCUSSION

1. Which explanation do you favor for the Flynn effect: that people are getting smarter or that people are merely getting more practice at the skills measured on IQ tests? Defend your opinion.

2. The rise in performance on IQ tests contrasts sharply with a steady decline in performance over the past few decades on tests that measure factual knowledge, such as the SAT. Think of several possible ways to explain these contrasting results, and form an opinion as to the most likely explanation.

3. Results on IQ tests tend to differ among different ethnic groups. Some people have used this fact to argue that some ethnic groups tend to be intellectually superior to others. Can such an argument still be supported in light of the Flynn effect? Defend your opinion.

4. Discuss some of the common uses of IQ tests. Do you think that IQ tests *should* be used for these purposes? Does the Flynn effect alter your thoughts about the uses of IQ tests? Explain.

SUGGESTED READING

Hall, Trish. "I.Q. Scores Are Up, and Psychologists Wonder Why." *New York Times*, February 24, 1998.

Neisser, Ulric (Editor). *The Rising Curve: Long-Term Gains in IQ and Related Measures*. American Psychological Association, 1998.

CHAPTER 6

Probability in Statistics

As discussed in Chapter 1, most statistical studies seek to learn something about a *population* from a much smaller *sample*. Thus, a key question in any statistical study is whether it is valid to generalize from a sample to the population. To answer this question, we must understand the likelihood, or probability, that what we've learned about the sample also applies to the population. In this chapter we will focus on a few basic ideas of probability that are commonly used in statistics. As you will see, these ideas of probability also have many applications in their own right.

LEARNING GOALS

1. Understand the concept of statistical significance and the essential role that probability plays in defining it.

2. Know how to find probabilities using theoretical and relative frequency methods and understand how to construct basic probability distributions.

3. Know how to find and apply expected values and understand the law of large numbers and the gambler's fallacy.

4. Distinguish between independent and dependent events and between overlapping and non-overlapping events, and be able to calculate *and* and either/or probabilities.

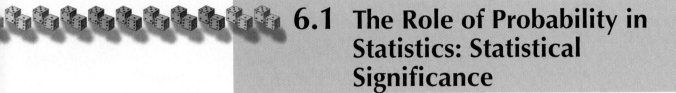

6.1 The Role of Probability in Statistics: Statistical Significance

To see why probability is so important in statistics, let's begin with a simple example of a coin toss. Suppose you are trying to test whether a coin is fair—that is, whether it is equally likely to land on heads or tails. If the coin is fair, there is a 50-50 chance of getting a head or a tail on any toss. Now, suppose you actually toss the coin 100 times and the results are 52 heads and 48 tails. Should you conclude that the coin is unfair? No, because we expect to see *roughly* 50 heads and 50 tails in every 100 tosses, with some variation in any particular sample of 100 tosses. In fact, if you were to toss the coin 100 more times, you might see 47 heads and 53 tails or something else that is reasonably close to 50-50. These small deviations from a perfect 50-50 split between heads and tails do not necessarily mean that the coin is unfair because they are what we *expect by chance*.

But suppose you toss the coin 100 times and the results are 20 heads and 80 tails. In this case, it's difficult to attribute to chance such a large deviation from 50-50. It's certainly possible that you witnessed a rare event. But it's more likely that there is another explanation, such as that the coin is not fair. When the difference between what is observed and what is expected is too unlikely to be explained by chance alone, we say the difference is *statistically significant*.

> ### Definition
>
> A set of measurements or observations in a statistical study is said to be **statistically significant** if it is unlikely to have occurred by chance.

EXAMPLE 1 Likely Stories?

a. *A detective in Detroit finds that 25 of the 62 guns used in crimes during the past week were sold by the same gun shop.* This finding is statistically significant. Because there are many gun shops in the Detroit area, having 25 out of 62 guns come from the same shop seems very unlikely to have occurred by chance.

b. *In terms of the global average temperature, five of the years between 1990 and 1999 were the five hottest years in the 20th century.* Having the five hottest years in 1990–1999 is statistically significant. By chance alone, any particular year in a century would have a 5 in 100, or 1 in 20, chance of being one of the five hottest years. Having five of those years come in the same decade is very unlikely to have occurred by chance alone. This statistical significance suggests that the world may be warming up.

c. *The team with the worst win-loss record in basketball wins one game against the defending league champions.* This one win is *not* statistically significant because although we expect a team with a poor win-loss record to lose most of its games, we also expect it to win occasionally, even against the defending league champions. ∎

From Sample to Population

Let's look at the idea of statistical significance in an opinion poll. Suppose that in a poll of 1,000 randomly selected people, 51% support the President. A week later, in another poll with a different randomly selected sample of 1,000 people, only 49% support the President. Should you conclude that the opinions of Americans changed during the one week between the polls?

You can probably guess that the answer is no. The poll results are *sample statistics* (see Section 1.1): 51% of the people in the first sample support the President. We can use this result to estimate the *population parameter,* which is the percentage of *all* Americans who support the President. At best, if the first poll were conducted well, it would say that the percentage of Americans who support the President is *close* to 51%. Similarly, the 49% result in the second poll means that the percentage of Americans supporting the President is *close* to 49%. Because the two sample statistics differed only slightly (51% versus 49%), it's quite possible that the real percentage of Americans supporting the President did not change at all. Instead, the two polls reflect expected and reasonable differences between the two samples.

In contrast, suppose the first poll found that 75% of the sample supported the President, and the second poll, taken a week later, found that only 30% supported the President. Assuming the polls were carefully conducted, it's highly unlikely that two groups of 1,000 randomly chosen people could differ so much by chance alone. In this case, we would look for another explanation. Perhaps Americans' opinions about the President really did change in the week between the polls.

In terms of statistical significance, the change from 51% to 49% in the first set of polls is *not* statistically significant, because we can reasonably attribute this change to the chance variations between the two samples. However, in the second set of polls, the change from 75% to 30% is statistically significant, because it is unlikely to have occurred by chance.

> **TECHNICAL NOTE**
>
> The difference between 49% and 51% is not statistically significant for typical polls, but it can be for a poll involving a very large sample size. In general, any difference can be significant if the sample size is large enough.

EXAMPLE 2   Statistical Significance in Experiments

A researcher conducts a double-blind experiment that tests whether a new herbal formula is effective in preventing colds. During a three-month period, the 100 randomly selected people in a treatment group take the herbal formula while the 100 randomly selected people in a control group take a placebo. The results show that 30 people in the treatment group get colds, compared to 32 people in the control group. Can we conclude that the herbal formula is effective in preventing colds?

SOLUTION   Whether a person gets a cold during any three-month period depends on many unpredictable factors. Therefore, we should not expect the number of people with colds in any two groups of 100 people to be exactly the same. In this case, the difference between 30 people getting colds in the treatment group and 32 people getting colds in the control group seems small enough to be explainable by chance. So the difference is not statistically significant, and we should not conclude that the treatment made any difference at all. ■

Quantifying Statistical Significance

In Example 2, we said that the difference between 30 colds in the treatment group and 32 colds in the control group was not statistically significant. This conclusion was fairly obvious because the difference was so small. But suppose that 24 people in the treatment group had colds instead of 30. Would the difference between 24 and 32 be large enough to be considered statistically significant? The definition of statistical significance that we've been using so far is too vague to answer this question. We need a way to quantify the idea of statistical significance.

In most cases, the issue of statistical significance can be addressed with one question:

> *Is the probability that the observed difference occurred by chance less than or equal to 0.05 (or 1 in 20)?*

For the moment, we will use the term *probability* in its everyday sense; in the next section, we will be more precise. If the probability that the observed difference occurred by chance is 0.05 or less, then the difference is *statistically significant at the 0.05 level.* If not, the observed difference is reasonably likely to have occurred by chance, so it is not statistically significant. The choice of 0.05 is somewhat arbitrary, but it's a figure that statisticians frequently use.

He that leaves nothing to chance will do few things ill, but will do very few things.
—George Savile Halifax

Sometimes the probability for claiming that an observed difference occurred by chance is even smaller. For example, we can claim statistical significance at the 0.01 level if the probability that an observed difference occurred by chance is 0.01 (or 1 in 100) or less.

> ### Quantifying Statistical Significance
> - If the probability of an observed difference occurring by chance is 0.05 (or 1 in 20) or less, the difference is statistically significant at the 0.05 level.
> - If the probability of an observed difference occurring by chance is 0.01 (or 1 in 100) or less, the difference is statistically significant at the 0.01 level.

You can probably see that caution is in order when working with statistical significance. We would expect roughly 1 in 20 trials to give results that are statistically significant at the 0.05 level even when the results actually occurred by chance. Thus, statistical significance at the 0.05 level—or at almost any level, for that matter—is *no guarantee* that an important effect or difference is present.

Time out to think

Suppose an experiment finds that people taking a new herbal remedy get fewer colds than people taking a placebo, and the results are statistically significant at the 0.01 level. Has the experiment proven that the herbal remedy works? Explain.

EXAMPLE 3   Polio Vaccine Significance

In the test of the Salk polio vaccine (see Section 1.1), 33 of the 200,000 children in the treatment group got paralytic polio, while 115 of the 200,000 in the control group got paralytic polio. The probability of this difference between the groups occurring by chance is less than 0.01. Describe the implications of this result.

SOLUTION   The results of the polio vaccine test are statistically significant at the 0.01 level, meaning that there is a 0.01 chance (or less) that the difference between the control and treatment groups occurred by chance. Therefore, we can be fairly confident that the vaccine really was responsible for the fewer cases of polio in the treatment group. (In fact, the probability of the Salk results occurring by chance is *much* less than 0.01, so researchers were quite convinced that the vaccine worked.) ∎

Review Questions

1. Describe in your own words the meaning of *statistical significance*. Give a few examples of results that are clearly statistically significant and a few examples of results that clearly are not statistically significant.

2. How does the idea of statistical significance apply to the question of whether results from a sample can be generalized to conclusions about a population? Explain.

3. Briefly describe how we quantify statistical significance. What does it mean for a result to be statistically significant at the 0.05 level? At the 0.01 level?

4. If a result is claimed to be statistically significant, does that automatically mean that it did not occur by chance? Why or why not? If a result is not statistically significant, does that automatically mean it occurred by chance? Explain.

5. Briefly summarize the role of probability in the ideas of statistical significance.

Exercises

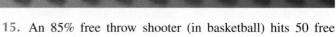

BASIC SKILLS AND CONCEPTS

SENSIBLE STATEMENTS? For Exercises 1–6, determine whether the given statement is sensible and explain why it is or is not.

1. The drunk driving fatality rate has statistical significance because it seriously affects so many people.

2. In a test of a technique of gender selection, 100 babies consist of at least 65 girls. Because there is only 1 chance in 500 of getting at least 65 girls among 100 babies, the result of 65 or more girls is statistically significant.

3. When 2,000 men and 2,000 women were tested for color blindness, the difference was statistically significant because 160 of the men were color blind whereas 5 of the women were color blind.

4. When 2,000 men and 2,000 women were surveyed for their opinions about the current president, the difference was statistically significant because 1,250 of the men felt that the President was doing a good job but only 1,245 of the women felt that way.

5. In a test of the effectiveness of a treatment for reducing headache pain, the difference between the treatment group and the control group (with no treatment) was found to be statistically significant. This means that the treatment will definitely ease headache pain for everyone.

6. In a test of differences in bacteria between a lake in Kenya and a lake in Belize, there can be no statistical significance because the comparison is not important.

SUBJECTIVE SIGNIFICANCE. For each event in Exercises 7–16, state whether the difference between what occurred and what you would have expected by chance is statistically significant. Discuss any implications of the statistical significance.

7. In 100 tosses of a coin, you observe 30 tails.

8. In 500 tosses of a coin, you observe 245 heads.

9. In 100 rolls of a six-sided die, a 3 appears 16 times.

10. In 100 rolls of a six-sided die, a 2 appears 28 times.

11. The last-place team in the conference wins eight hockey games in a row.

12. The first 20 cars you see during a trip are all convertibles.

13. A baseball team with a win/loss average of 0.650 wins 12 out of 20 games.

14. An 85% free throw shooter (in basketball) hits 24 out of 30 free throws.

15. An 85% free throw shooter (in basketball) hits 50 free throws in a row.

16. The first 15 people you encounter at a public meeting all appear to be over 70 years old.

FURTHER APPLICATIONS

17. **FUEL TESTS.** Thirty identical cars are selected for a fuel test. Half of the cars are filled with a regular gasoline, and the other half are filled with a new experimental fuel. The cars in the first group average 29.3 miles per gallon, while the cars in the second group average 35.5 miles per gallon. Discuss whether this difference seems statistically significant.

18. **STUDY SESSIONS.** The 40 students who attended regular study sessions throughout the semester had a mean score of 78.9 on the statistics final exam. The 30 students who did not attend the study sessions had a mean score of 77.8. Discuss whether this difference seems statistically significant.

19. **HUMAN BODY TEMPERATURE.** In a study by researchers at the University of Maryland, the body temperatures of 106 individuals were measured; the mean for the sample was 98.20°F. The accepted value for human body temperature is 98.60°F. The difference between the sample mean and the accepted value is significant at the 0.05 level.
    a. Discuss the meaning of the significance level in this case.
    b. If we assume that the mean body temperature is actually 98.6°F, the probability of getting a sample with a mean of 98.20°F or less is 0.000000001. Interpret this probability value.

20. **SEAT BELTS AND CHILDREN.** In a study of children injured in automobile crashes (*American Journal of Public Health,* Vol. 82, No. 3), those wearing seat belts had a mean stay of 0.83 day in an intensive care unit. Those not wearing seat belts had a mean stay of 1.39 days. The difference in means between the two groups is significant at the 0.0001 level. Interpret this result.

21. **SAT PREPARATION.** A study of 75 students who took an SAT preparation course (*American Education Research Journal,* Vol. 19, No. 3) concluded that the mean improvement on the SAT was 0.6 point. If we assume that the preparation course has no effect, the probability of getting a mean improvement of 0.6 point by chance is 0.08. Discuss whether this preparation course results in statistically significant improvement.

22. **WEIGHT BY AGE.** A National Health Survey determined that the mean weight of a sample of 804 men aged 25 to 34 was 176 pounds, while the mean weight of a sample of 1,657 men aged 65 to 74 was 164 pounds. The difference is significant at the 0.01 level. Interpret this result.

![globe icon] **PROJECTS FOR THE WEB AND BEYOND**

For useful links, select "Links for Web Projects" for Chapter 6 at www.aw.com/bbt.

23. **SIGNIFICANCE IN VITAL STATISTICS.** Visit a Web site that has vital statistics (for example, the U.S. Bureau of Census or the National Center for Health Statistics). Choose a question such as the following:

    • Are there significant differences in numbers of births among months?
    • Are there significant differences in numbers of natural deaths among days of the week?
    • Are there significant differences in infant mortality rates among selected states?
    • Are there significant differences in incidences of a particular disease among selected states?
    • Are there significant differences among the marriage rates in various states?

    Collect the relevant data and determine subjectively whether you think the observed differences are significant; that is, explain if they could occur by chance or provide some alternative explanations.

24. **LENGTHS OF RIVERS.** Using an almanac or the Internet, find the lengths of the principal rivers of the world. Construct a list of the leading digits only. Does any particular digit occur more often than the others? Does that digit occur significantly more often? Explain.

![globe icon] **IN THE NEWS**

1. **STATISTICAL SIGNIFICANCE.** Find a recent newspaper article on a statistical study in which the idea of statistical significance is used. Write a one-page summary of the study and the result that is considered to be statistically significant. Also include a brief discussion of whether you believe the result, given its statistical significance.

2. **SIGNIFICANT EXPERIMENT?** Find a recent news story about a statistical study that used an experiment to determine whether some new treatment was effective. Based on the available information, briefly discuss what you can conclude about the statistical significance of the results. Given this significance (or lack of), do you think the new treatment is useful? Explain.

3. **PERSONAL STATISTICAL SIGNIFICANCE.** Describe an incident in your own life that did not meet your expectation, defied the odds, or seemed unlikely to have occurred by chance. Would you call this incident statistically significant? To what did you attribute the event?

6.2 Basics of Probability

We have already seen that ideas of probability are fundamental to statistics, in part through the concept of statistical significance. We will return to the concept of statistical significance in Chapters 7 through 10. First, however, we need to develop some of the essential ideas of probability. Along the way, you will also see many other applications of probability in everyday life.

In probability, we work with processes that involve observations or measurements, such as rolling dice or drawing balls from a lottery barrel. An **outcome** is the most basic result of an observation or measurement. For example, in rolling two dice, rolling a 1 and a 4 is an outcome. In drawing six lottery balls, the sequence 8-16-23-25-30-38 is an outcome. However, there are times when a collection of outcomes all share some property of interest. Such a collection of outcomes is called an **event**. For example, rolling two dice and getting a

In this world, nothing is certain but death and taxes.

—Benjamin Franklin

sum of 5 is an event; this event occurs with the outcomes of rolling (1, 4), (2, 3), (3, 2), and (4, 1). In having a family with three children, having two boys is an event; it can occur in the following ways: BBG (Boy-Boy-Girl), BGB, or GBB.

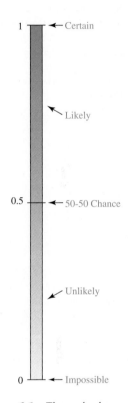

Figure 6.1   The scale shows various degrees of certainty as expressed by probabilities.

Definitions

Outcomes are the most basic possible results of observations or experiments.

An **event** is a collection of one or more outcomes that share a property of interest.

Mathematically, we express probabilities as numbers between 0 and 1. If an event is impossible, we assign it a probability of 0. For example, the probability of meeting a married bachelor is 0. At the other extreme, an event that is certain to occur is given a probability of 1. For example, according to the old saying from Benjamin Franklin, the probability of death and taxes is 1.

We write $P(\text{event})$ to mean the probability of an event. We often denote events by letters or symbols. For example, using H to represent the event of a head on a coin toss, we write $P(\text{H}) = 0.5$.

Expressing Probability

The probability of an event, expressed as $P(\text{event})$, is always between 0 and 1 inclusive. A probability of 0 means that the event is impossible, and a probability of 1 means that the event is certain.

Figure 6.1 shows the scale of probability values, along with common expressions of likelihood. It's helpful to develop the sense that a probability value such as 0.95 indicates that an event is very likely to occur, but is not certain. It should occur about 95 times out of 100. In contrast, a probability value of 0.01 describes an event that is very unlikely to occur. The event is possible, but will occur only about once in 100 times.

Three basic approaches to finding probabilities are the theoretical method, the relative frequency method, and estimating a subjective probability. We will look at each in turn.

Theoretical Probabilities

The **theoretical method** of finding probabilities involves basing the probabilities on a *theory*—or set of assumptions—about the process in question. For example, when we say that the probability of heads on a coin toss is 1/2, we are assuming that the coin is fair and is equally likely to land on heads or tails. Similarly, in rolling a fair die, we assume that all six outcomes are equally likely. As long as all outcomes are equally likely, we can use the following procedure to find theoretical probabilities.

Theoretical Method for Equally Likely Outcomes

1. Count the total number of possible outcomes.

2. Among all the possible outcomes, count the number of ways the event of interest, *A*, can occur.

3. Determine the probability, *P(A)*, from

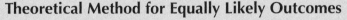

$$P(A) = \frac{\text{number of ways } A \text{ can occur}}{\text{total number of outcomes}}$$

By the Way...

Theoretical methods are also called *a priori* methods. The words *a priori* are Latin for "before the fact" or "before experience."

EXAMPLE 1 Guessing Birthdays

Suppose you select a person at random from a large group at a conference. What is the probability that the person selected has a birthday in July? Assume 365 days in a year.

SOLUTION If we assume that all birthdays are equally likely, we can use the three-step theoretical method.

Step 1: Each possible birthday represents an outcome, so the total number of possible outcomes is 365.

Step 2: July has 31 days, so 31 of the 365 possible outcomes represent the event of a July birthday.

Step 3: The probability that a randomly selected person has a birthday in July is

$$P(\text{July birthday}) = \frac{31}{365} \approx 0.0849$$

which is slightly more than 1 chance in 12.

COUNTING OUTCOMES

Suppose we toss two coins and want to count the total number of outcomes. The toss of the first coin has two possible outcomes: heads (H) or tails (T). The toss of the second coin also has two possible outcomes. The two outcomes for the first coin can occur with either of the two outcomes for the second coin. So the total number of outcomes for two tosses is $2 \times 2 = 4$; they are HH, HT, TH, and TT, as shown in the tree diagram of Figure 6.2a.

Time out to think

Explain why the outcomes for tossing one coin twice in a row are the same as those for tossing two coins at the same time.

We can now extend this thinking. If we toss three coins, we have a total of $2 \times 2 \times 2 = 8$ possible outcomes: HHH, HHT, HTH, HTT, THH, THT, TTH, and TTT, as shown in Figure 6.2b. This idea is the basis for the following counting rule.

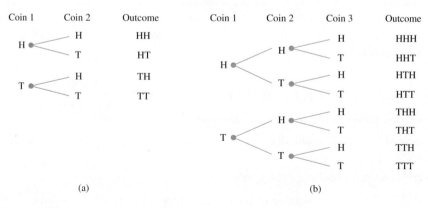

(a) (b)

Figure 6.2 Tree diagrams showing the outcomes of tossing (a) two and (b) three coins.

Counting Outcomes

Suppose process A has *a* possible outcomes and process B has *b* possible outcomes. Assuming the outcomes of the processes do not affect each other, the number of different outcomes for the two processes combined is $a \times b$. This idea extends to any number of processes. For example, if a third process C has *c* possible outcomes, the number of possible outcomes for the three processes combined is $a \times b \times c$.

EXAMPLE 2   Some Counting

a. How many outcomes are there if you roll a fair die and toss a fair coin?

b. What is the probability of rolling two 1's (snake eyes) when two fair dice are rolled?

SOLUTION

a. The first process, rolling a fair die, has six outcomes (1, 2, 3, 4, 5, 6). The second process, tossing a fair coin, has two outcomes (H, T). Therefore, there are $6 \times 2 = 12$ outcomes for the two processes (1H, 1T, 2H, 2T, . . . , 6H, 6T).

b. Rolling a single die has six equally likely outcomes. Therefore, when two fair dice are rolled, there are $6 \times 6 = 36$ different outcomes. Of these 36 outcomes, only one is the event of interest (two 1's). So the probability of rolling two 1's is

$$P(\text{two 1's}) = \frac{\text{number of ways two 1's can occur}}{\text{total number of outcomes}} = \frac{1}{36} = 0.0278 \qquad \blacksquare$$

EXAMPLE 3   Counting Children

What is the probability that, in a randomly selected family with three children, the oldest child is a boy, the second child is a girl, and the youngest child is a girl? Assume boys and girls are equally likely.

SOLUTION   There are two possible outcomes for each birth: boy or girl. For a family with three children, the total number of possible outcomes (birth orders) is $2 \times 2 \times 2 = 8$ (BBB, BBG, BGB, BGG, GBB, GBG, GGB, GGG). The question asks about one particular birth order (boy-girl-girl), so this is 1 of 8 possible outcomes. Therefore, this birth order has a probability of 1 in 8, or $1/8 = 0.125$. $\qquad \blacksquare$

Time out to think
How many different four-child families are possible if birth order is taken into account? What is the probability of a couple having a four-child family with four girls?

Relative Frequency Probabilities

The second way to determine probabilities is to *approximate* the probability of an event A by making many observations and counting the number of times event A occurs. This approach is called the **relative frequency** (or **empirical**) **method.** For example, if we observe that it rains an average of 100 days per year, we might say that the probability of rain on a randomly selected day is 100/365. We can use a general rule for this method:

> **Relative Frequency Method**
>
> 1. Repeat or observe a process many times and count the number of times the event of interest, *A*, occurs.
> 2. Estimate $P(A)$ by
>
> $$P(A) = \frac{\text{number of times } A \text{ occurred}}{\text{total number of observations}}$$

EXAMPLE 4   500-Year Flood

Geological records indicate that a river has crested above flood level four times in the past 2,000 years. What is the relative frequency probability that the river will crest above flood level next year?

SOLUTION   Based on the data, the probability of the river cresting above flood level in any single year is

$$\frac{\text{number of years with flood}}{\text{total number of years}} = \frac{4}{2,000} = \frac{1}{500}$$

Because a flood of this magnitude occurs on average once every 500 years, it is called a "500-year flood." The probability of having a flood of this magnitude in any given year is 1/500, or 0.002. ∎

Subjective Probabilities

The third method for determining probabilities is to estimate a **subjective probability** using experience, judgment, or intuition. For example, you could make a subjective estimate of the probability that a friend will be married in the next year. Weather forecasts are often subjective probabilities based on the experience of the forecaster, who must use current weather to make a prediction about weather in the coming days.

> **Three Approaches to Finding Probability**
>
> A **theoretical probability** is based on assuming that all outcomes are equally likely. It is determined by dividing the number of ways an event can occur by the total number of possible outcomes.
>
> A **relative frequency probability** is based on observations or experiments. It is the relative frequency of the event of interest.
>
> A **subjective probability** is an estimate based on experience or intuition.

EXAMPLE 5   Which Method?

Identify the method that resulted in the following statements.

a. The chance that you'll get married in the next year is zero.

b. Based on government data, the chance of dying in an automobile accident is 1 in 7,000 (per year).

c. The chance of rolling a 7 with a twelve-sided die is 1/12.

SOLUTION

a. This is a subjective probability because it is based on a feeling at the current moment.

b. This is a relative frequency probability because it is based on observed data on past automobile accidents.

c. This is a theoretical probability because it is based on assuming that a fair twelve-sided die is equally likely to land on any of its twelve sides. ∎

EXAMPLE 6   Hurricane Probabilities

Figure 6.3 shows a map released when Hurricane Floyd approached the southeast coast of the United States. The map shows "strike probabilities" in three different regions along the path of the hurricane. Interpret the term "strike probability" and discuss how these probabilities might have been determined.

SOLUTION   For a person in the 20–50% strike probability region, there is a 0.2 to 0.5 probability of being hit by the hurricane. Such a probability could not be determined by purely theoretical methods, because there is no simple model for hurricanes (as there is for coins and dice). The method used is partly empirical (the relative frequency method), based on records of past hurricanes with similar behavior. For example, forecasters may know that of ten past hurricanes with similar size, strength, and path, between two and five actually struck this particular region. Based on relative frequencies, this would lead to a 0.2 to 0.5 probability of a strike in this region. But certainly, in addition to using the evidence of past hurricanes, forecasters refined these probabilities with their own experience and intuition; thus, the probabilities are also partly subjective.

<aside>
By the Way...

Another approach to finding probabilities, called the *Monte Carlo method*, uses computer simulations. This technique essentially finds relative frequency probabilities; in this case, observations are made with the computer.
</aside>

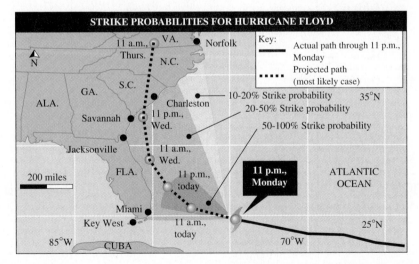

Figure 6.3   This map shows the actual path of Hurricane Floyd up to the time the map was made ("11 p.m., Monday"). It then shows the probabilities of various alternative paths, with the most likely path a dashed line. The map was used to help people along the southeast coast determine the likelihood of being struck by the hurricane. SOURCE: *Virginia Pilot.* ∎

Complementary Events

For any event *A*, the **complement** of *A* consists of all outcomes in which *A* does *not* occur. We denote the complement of *A* by $\overline{A}$ (read "A-bar"). For example, on a multiple-choice question with five possible answers, if event A is a correct answer, then event $\overline{A}$ is an incorrect answer. In this case, the probability of answering correctly with a random guess is 1/5, and the probability of not answering correctly is 4/5. Similarly, the probability of rolling a 6 with a single fair die is 1/6. Therefore, the probability of rolling anything *but* a 6 is 5/6. These examples show that the probability of *either A or $\overline{A}$* occurring is 1:

$$P(A) + P(\overline{A}) = 1$$

Therefore, we can give the following rule for complementary events.

Probability of the Complement of an Event

The **complement** of an event A, expressed as $\overline{A}$, consists of all outcomes in which A does not occur. The probability of $\overline{A}$ is given by

$$P(\overline{A}) = 1 - P(A)$$

EXAMPLE 7   Is Scanner Accuracy the Same for Specials?

In a study of checkout scanning systems, samples of purchases were used to compare the scanned prices to the posted prices. Table 6.1 summarizes results for a sample of 819 items. Based on these data, what is the probability that a regular-priced item has a scanning error? What is the probability that an advertised-special item has a scanning error?

Table 6.1 Scanner Accuracy

	Regular-priced items	Advertised-special items
Undercharge	20	7
Overcharge	15	29
Correct price	384	364

SOURCE: Ronald Goodstein, "UPC Scanner Pricing Systems: Are They Accurate?" *Journal of Marketing*, Vol. 58.

SOLUTION   We can let R represent a regular-priced item being scanned correctly. Because 384 of the 419 regular-priced items are correctly scanned,

$$P(R) = \frac{384}{419} = 0.916$$

The event of a scanning error is the complement of the event of a correct scan, so the probability of a regular-priced item being subject to a scanning error (either undercharged or overcharged) is

$$P(\overline{R}) = 1 - 0.916 = 0.084$$

Now let A represent an advertised-special item being scanned correctly. Of the 400 advertised-special items in the sample, 364 are scanned correctly. Therefore,

$$P(A) = \frac{364}{400} = 0.910$$

The probability of an advertised-special item being scanned incorrectly is

$$P(\overline{A}) = 1 - 0.910 = 0.090$$

The error rates are nearly equal. However, notice that most of the errors made with advertised-special items are not to the customer's advantage. With regular-priced items, over half of the errors are to the customer's advantage.

Probability Distributions

In Chapters 3 through 5, we worked with frequency and relative frequency distributions—for example, distributions of age or income. One of the most fundamental ideas in probability and statistics is that of a *probability distribution*. As the name suggests, a probability distribution is a distribution in which the variable of interest is associated with a probability.

Suppose you toss two coins simultaneously. The outcomes are the various combinations of a head and a tail on the two coins. Because *each* coin can land in two possible ways (heads or tails), the *two* coins can land in $2 \times 2 = 4$ different ways. Table 6.2 has a row for each of these four outcomes.

Table 6.2 Outcomes of Tossing Two Fair Coins			
Coin 1	Coin 2	Outcome	Probability
H	H	HH	1/4
H	T	HT	1/4
T	H	TH	1/4
T	T	TT	1/4

If we now count the number of heads and tails in each outcome, we see that the four outcomes represent only three different events: 2 heads (HH), 2 tails (TT), and 1 head and 1 tail (HT or TH). The probability of two heads is $P(\text{HH}) = 1/4 = 0.25$; the probability of two tails is $P(\text{TT}) = 1/4 = 0.25$; and the probability of one head and one tail is $P(\text{H and T}) = 2/4 = 0.50$. These probabilities result in a **probability distribution** that can be displayed as a table (Table 6.3) or a histogram (Figure 6.4). Note that the sum of all the probabilities must be 1 because one of the outcomes must occur.

Table 6.3 Tossing Two Coins	
Result	Probability
2 heads, 0 tails	0.25
1 head, 1 tail	0.50
0 heads, 2 tails	0.25
Total	**1**

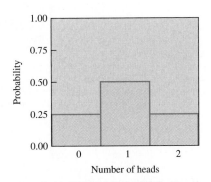

Figure 6.4 Histogram showing the probability distribution for the results of tossing two coins.

By the Way...

Another common way to express likelihood is to use odds. The odds *against* an event are the ratio of the probability that the event does not occur to the probability that it does occur. For example, the odds against rolling a 6 with a fair die are (5/6)/(1/6), or 5 to 1. The odds used in gambling are called *payoff odds*; they express your net gain on a winning bet. For example, suppose that the payoff odds on a particular horse at a horse race are 3 to 1. This means that for each $1 you bet on this horse, you will gain $3 if the horse wins.

> ## Making a Probability Distribution
>
> A **probability distribution** represents the probabilities of all possible events. To make a table of a probability distribution:
>
> **Step 1:** List all possible *outcomes*; use a table or figure if it is helpful.
>
> **Step 2:** Identify outcomes that represent the same *event*. Find the probability of each event.
>
> **Step 3:** Make a table in which one column lists each event and another column lists each probability. The sum of all the probabilities must be 1.

EXAMPLE 8   Tossing Three Coins

Make a probability distribution for the number of heads that occurs when three coins are tossed simultaneously.

SOLUTION   We apply the three-step process.

Step 1: The number of different outcomes when three coins are tossed is $2 \times 2 \times 2 = 8$. Figure 6.2b (page 232) shows how we find all eight possible outcomes, which are HHH, HHT, HTH, HTT, THH, THT, TTH, and TTT.

Step 2: There are four possible events: 0 heads, 1 head, 2 heads, and 3 heads. By looking at the eight possible outcomes, we find the following:

- Only one of the eight outcomes represents the event of 0 heads. Thus, its probability is 1/8.

- Three of the eight outcomes represent the event of 1 head (and 2 tails): HTT, THT, and TTH. This event therefore has a probability of 3/8.

- Three of the eight outcomes represent the event of 2 heads (and 1 tail): HHT, HTH, and THH. This event also has a probability of 3/8.

- Only one of the eight outcomes represents the event of 3 heads, so its probability is 1/8.

Step 3: We make a table with the four events listed in the left column and their probabilities in the right column. Table 6.4 shows the result.

Table 6.4 Tossing Three Coins	
Result	**Probability**
3 heads (0 tails)	1/8
2 heads (1 tail)	3/8
1 head (2 tails)	3/8
0 heads (3 tails)	1/8
Total	**1**

Time out to think

When you toss four coins, how many different outcomes are possible? If you record the number of heads, how many different events are possible?

EXAMPLE 9   American Households

The first two columns of Table 6.5 show data on household sizes in the United States. Make a table and graph to display the probability distribution for these data.

SOLUTION   The second column of Table 6.5 gives frequencies of various household sizes. We convert each frequency into a probability, using the relative frequency method. The probability of randomly selecting a household of a given size is the number of times that household size occurs divided by the total number of households. The results are shown in the column on the right side of Table 6.5.

Table 6.5   U.S. Households of Various Sizes

Household size	Number (thousands)	Probability
1	24,900	0.250
2	32,526	0.326
3	16,724	0.168
4	15,118	0.152
5	6,631	0.067
6	2,357	0.024
7 or more	1,372	0.014
Total	**99,628**	**1.001**

SOURCE: U.S. Bureau of the Census.

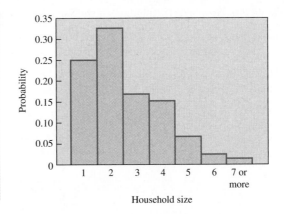

Figure 6.5   Probability distribution for household data.

It's always important to check that the probabilities in a distribution sum to 1. In this case, the total probability is not exactly 1 because of rounding errors. Figure 6.5 shows this probability distribution as a graph.   ∎

EXAMPLE 10   Two Dice Distribution

Make a probability distribution for the sum of the dice when two dice are rolled. Express the distribution as a table and as a histogram.

SOLUTION   Because there are six ways for each die to land, there are $6 \times 6 = 36$ outcomes of rolling two dice. We can enumerate all 36 outcomes by making a table in which we list one die along the rows and the other along the columns. In each cell, we show the sum of the numbers on the two dice.

Outcomes and Sums for the Roll of Two Dice						
	1	2	3	4	5	6
1	1 + 1 = 2	1 + 2 = 3	1 + 3 = 4	1 + 4 = 5	1 + 5 = 6	1 + 6 = 7
2	2 + 1 = 3	2 + 2 = 4	2 + 3 = 5	2 + 4 = 6	2 + 5 = 7	2 + 6 = 8
3	3 + 1 = 4	3 + 2 = 5	3 + 3 = 6	3 + 4 = 7	3 + 5 = 8	3 + 6 = 9
4	4 + 1 = 5	4 + 2 = 6	4 + 3 = 7	4 + 4 = 8	4 + 5 = 9	4 + 6 = 10
5	5 + 1 = 6	5 + 2 = 7	5 + 3 = 8	5 + 4 = 9	5 + 5 = 10	5 + 6 = 11
6	6 + 1 = 7	6 + 2 = 8	6 + 3 = 9	6 + 4 = 10	6 + 5 = 11	6 + 6 = 12

The possible sums range from 2 to 12. These are the *events* of interest in this problem. To find the probability of each sum, we divide the number of times the sum occurs by 36 (the total number of ways the dice can land). For example, there are five ways to get a sum of 8, so the probability of rolling a sum of 8 is 5/36. In a similar way, the probabilities of all the other results are summarized in Table 6.6. Note that the probabilities increase until we reach a sum of 7, which is the most likely sum. The probabilities decrease for sums greater than 7. As always, the probabilities add up to 1. The probability distribution is shown as a histogram in Figure 6.6.

Table 6.6 Probability Distribution for the Sum of Two Dice

Event (sum)	2	3	4	5	6	7	8	9	10	11	12	**Total**
Probability	$\frac{1}{36}$	$\frac{2}{36}$	$\frac{3}{36}$	$\frac{4}{36}$	$\frac{5}{36}$	$\frac{6}{36}$	$\frac{5}{36}$	$\frac{4}{36}$	$\frac{3}{36}$	$\frac{2}{36}$	$\frac{1}{36}$	**1**

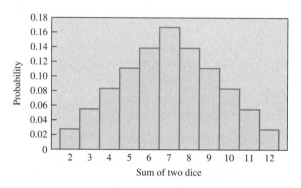

Figure 6.6 Histogram showing the probability distribution for the sum of two dice.

Review Questions

1. Describe and distinguish between an outcome and an event. Give an example in which the same event can occur through several different outcomes.

2. Explain the notation *P*(event). Why must the probability always be between 0 and 1? Interpret a probability of 0.02. Interpret a probability of 0.97.

3. Under what conditions can we use the theoretical method for finding probabilities? Give an example of the use of this method.

4. Suppose a process with *N* possible outcomes is repeated *R* times. How do we determine the total number of possible outcomes? Give an example in which you apply this rule.

5. Under what conditions can we use the relative frequency method for finding probabilities? Give an example of the use of this method.

6. What is a subjective probability? Why can't we make precise predictions with subjective probabilities?

7. Explain what we mean by a complementary event. What notation do we use for complementary events? How are the probabilities of an event and its complement related?

8. What is a probability distribution? How is it similar to and different from other types of distributions? How can a probability distribution be displayed? Give an example.

Exercises

BASIC SKILLS AND CONCEPTS

SENSIBLE STATEMENTS? For Exercises 1–6, determine whether the given statement is sensible and explain why it is or is not.

1. Because either there is life on Pluto or there is not, the probability of life on Pluto is 1/2.

2. The probability of guessing the correct answer to a multiple-choice test question is 0.2, and the probability of guessing a wrong answer is 0.75.

3. If the probability of an event A is 0.3, then the probability of the complement of A is 0.7.

4. If the probability of an event A is 1, then event A will definitely occur.

5. Jack estimates that the subjective probability of his being struck by lightning sometime next year is 1/2.

6. Jill estimates that the subjective probability of her being struck by lightning sometime next year is 1/1,000,000.

7. **OUTCOMES OR EVENTS.** For each observation described, state whether there is one way for the given observation to occur or more than one way for the given observation to occur (in which case it is an event).
   a. Tossing a head with a fair coin
   b. Tossing two heads with three fair coins
   c. Rolling a 6 with a fair die
   d. Rolling an even number with a fair die
   e. Rolling a sum of 7 with two fair dice
   f. Drawing a pair of kings from a regular deck of cards

8. **OUTCOMES OR EVENTS.** For each observation described, state whether there is one way for the given observation to occur or more than one way for the given observation to occur (in which case it is an event).
   a. Tossing a tail with a fair coin
   b. Tossing one tail with two fair coins
   c. Rolling at least one 5 with two fair dice
   d. Rolling an odd number with a fair die
   e. Drawing a jack from a regular deck of cards
   f. Drawing three aces from a regular deck of cards

THEORETICAL PROBABILITIES. In Exercises 9–18, use the theoretical method to determine the probability of the given outcomes and events. State any assumptions that you need to make.

9. Rolling an odd number with a fair die

10. Drawing an ace from a regular deck of cards

11. Meeting someone born on a Tuesday

12. Meeting someone born in March

13. Meeting someone with a phone number that ends in 0

14. Drawing a heart or a spade from a regular deck of cards

15. Meeting someone born between midnight and 1:00 a.m.

16. Randomly selecting a two-child family with two girl children

17. Randomly selecting a three-child family with exactly two girl children

18. Rolling two dice and getting a sum of 5

COMPLEMENTARY EVENTS. Determine the probability of the complementary events in Exercises 19–26. State any assumptions that you use.

19. What is the probability of not rolling a 4 with a fair die?

20. What is the probability of not drawing a spade from a regular deck of cards?

21. What is the probability of not tossing two heads with two fair coins?

22. What is the probability that a 76% free throw shooter will miss her next free throw?

23. What is the probability of meeting a person not born in January, February, or March?

24. What is the probability of a 0.320 hitter in baseball not getting a hit on his next at-bat?

25. What is the probability of not rolling a 4 or a 5 with a fair die?

26. What is the probability of not rolling an odd number with a fair die?

FURTHER APPLICATIONS

THEORETICAL PROBABILITIES. In Exercises 27–32, use the theoretical method to determine the probability of the given outcomes and events. State any assumptions that you need to make.

27. A bag contains 10 red M&Ms, 15 blue M&Ms, and 20 yellow M&Ms. What is the probability of drawing a red M&M? A blue M&M? A yellow M&M? Something besides a yellow M&M?

28. A drawer holds 4 pairs of black socks, 6 pairs of blue socks, and 12 pairs of white socks. If you reach in and pull out a matched pair of socks at random, what is the probability that you'll get a pair of blue socks? A pair of black socks? A pair of white socks? Anything but a pair of blue socks?

29. The New England College of Medicine uses an admissions test with multiple-choice questions, each with five possible answers, only one of which is correct. If you guess randomly on every question, what score might you expect to get (in percentage terms)?

30. The *Win a Fortune* game show uses a spinner that has six equal sectors labeled and colored as follows: 1—red, 2—blue, 3—white, 4—red, 5—blue, and 6—white. The spinning needle lands randomly in the six sectors. What is the probability that the outcome is red? An even number? Not white?

31. THREE-CHILD FAMILY. Suppose you randomly select a family with three children. Assume that births of boys and girls are equally likely. What is the probability that the family has each of the following?
a. Three girls
b. Two boys and a girl
c. A girl, a boy, and a boy, in that order
d. At least one girl
e. At least two boys

32. FOUR-CHILD FAMILY. Suppose you randomly select a family with four children. Assume that births of boys and girls are equally likely.
a. How many birth orders are possible? List all of them.
b. What is the probability that the family has four boys? Four girls?
c. What is the probability that the family has a boy, a girl, a boy, and a girl, in that order?
d. What is the probability that the family has two girls and two boys in any order?

RELATIVE FREQUENCY PROBABILITIES. Use the relative frequency method to estimate the probabilities in Exercises 33–38.

33. After recording the forecasts of your local weatherman for 30 days, you conclude that he gave a correct forecast 12 times. What is the probability that his next forecast will be correct?

34. What is the probability that a 0.340 (baseball) hitter will get a hit the next time he is at bat?

35. What is the probability of a 100-year flood this year?

36. Halfway through the season, a basketball player has hit 86% of her free throws. What is the probability that her next free throw will be successful?

37. Out of the U.S. population of 281 million (Census 2000), 306 people were struck by lightning in a single year. What is the probability of being struck by lightning during the course of a year?

38. A doctor diagnosed pneumonia in 100 patients and 34 of them actually had pneumonia. What is the probability that the next patient diagnosed with pneumonia will actually have pneumonia?

39. SENIOR CITIZEN PROBABILITIES. In the year 2000, there were 34.7 million people over 65 years of age out of a U.S. population of 281 million. In the year 2050, it is estimated that there will be 78.9 million people over 65 years of age out of a U.S. population of 394 million. Would your chances of meeting a person over 65 at random be greater in 2000 or 2050? Explain.

40. CITY COUNCIL PICKS. Suppose that a nine-member city council has four men and five women, seven Republicans and two Democrats. Suppose that a council member is chosen at random for an interview. Determine the probability of the following choices or state that the probability cannot be determined.
a. A Republican
b. A man
c. A woman Democrat
d. A man or a Republican

41. POLITICS AND GENDER. The following table gives the gender and political party of the 100 delegates at a political convention. Suppose you encounter a delegate at random. What is the probability that the delegate will *not* be a man? What is the probability that the delegate will *not* be a Republican?

	Women	Men
Republicans	21	28
Democrats	25	16
Independents	6	4

42. DECEPTIVE ODDS. The odds *for* an event A occurring are defined to be

$$\text{odds for } A = \frac{\text{probability that } A \text{ occurs}}{\text{probability that } A \text{ does not occur}}$$

Suppose event A has a 0.99 probability of occurring and event B has a 0.96 probability of occurring. Although both events are highly likely to occur, is it accurate to say that the odds for event A are four times greater than the odds for event B? Explain.

43. AGE AT FIRST MARRIAGE. The following table gives percentages of women and men married for the first time in several age categories (U.S. Census Bureau).

	Under 20	20–24	25–29	30–34	35–44	45–64	Over 65
Women	16.6	40.8	27.2	10.1	4.5	0.7	0.1
Men	6.6	36.0	34.3	14.8	7.1	1.1	0.1

a. What is the probability that a randomly encountered married woman was married, for the first time, between the ages of 35 and 44?

b. What is the probability that a randomly encountered married man was married, for the first time, before he was 20 years old?

c. Make a probability distribution, in the form of a histogram, for ages of first marriage for men and women.

44. FOUR-COIN PROBABILITY DISTRIBUTION.
a. Make a table similar to Table 6.2, showing all possible outcomes of tossing four coins at once.

b. Make a table similar to Table 6.4, showing the probability distribution for the events 4 heads, 3 heads, 2 heads, 1 head, and 0 heads.

c. What is the probability of getting two heads and two tails when you toss four coins at once?

d. What is the probability of tossing anything except four heads when you toss four coins at once?

e. Which event is most likely to occur?

45. COLORADO LOTTERY DISTRIBUTION. The histogram in Figure 6.7 shows the distribution of 5,964 Colorado lottery numbers (possible values range from 1 to 42).

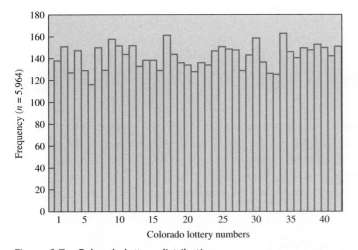

Figure 6.7 Colorado lottery distribution.

a. Assuming the lottery drawings are random, what would you expect the probability of any number to be?

b. Based on the histogram, what is the relative frequency probability of the most frequently appearing number?

c. Based on the histogram, what is the relative frequency probability of the least frequently appearing number?

d. Comment on the deviations of the empirical probabilities from the expected probabilities. Would you say these deviations are significant?

46. BIASED COINS. Suppose you have an extremely unfair coin: The probability of a head is 1/3, and the probability of a tail is 2/3.

a. If you toss the coin 36 times, how many heads and tails do you expect to see?

b. Suppose you toss two such weighted coins. Of the tosses in which the first coin shows heads, what proportion will occur with a head on the second coin?

c. Suppose you toss two such weighted coins. Of the tosses in which the first coin shows heads, what proportion will occur with a tail on the second coin?

d. Suppose you toss two such weighted coins. Of the tosses in which the first coin shows tails, what proportion will occur with a head on the second coin?

e. Suppose you toss two such weighted coins. Of the tosses in which the first coin shows tails, what proportion will occur with a tail on the second coin?

f. Make a probability distribution table that shows the probabilities of the outcomes HH, HT, TH, and TT.

PROJECTS FOR THE WEB AND BEYOND

For useful links, select "Links for Web Projects" for Chapter 6 at www.aw.com/bbt.

47. BLOOD GROUPS. The four major blood groups are designated A, B, AB, and O. Within each group there are two Rh types: positive and negative. Using library resources or the World Wide Web, find data on the relative frequency of blood groups, including the Rh types. Make a table showing the probability of meeting someone in each of the eight combinations of blood group and type.

48. AGE AND GENDER. The proportions of men and women in the population change with age. Using current data from a Web site, construct a table showing the probability of meeting a male or a female in each of these age categories: 0–5, 6–10, 11–20, 21–30, 31–40, 41–50, 51–60, 61–70, 71–80, over 80.

49. THUMB TACK PROBABILITIES. Find a standard thumb tack and practice tossing it onto a flat surface. Notice that there are two different outcomes: The tack can land point down or point up.

a. Toss the tack 50 times and record the outcomes.

b. Give the relative frequency probabilities of the two outcomes based on these results.

c. If possible, ask several other people to repeat the process. How well do your probabilities agree?

50. **THREE-COIN EXPERIMENT.** Toss three coins at once 50 times and record the outcomes in terms of the number of heads. Based on your observations, give the relative frequency probabilities of the outcomes. Do they agree with the theoretical probabilities? Explain and discuss your results.

51. **RANDOMIZING A SURVEY.** Suppose you want to conduct a survey involving a sensitive question that not all participants may choose to answer honestly (for example, a question involving cheating on taxes or drug use). Here is a way to conduct the survey and protect the identity of respondents. We will assume that the sensitive question requires a yes or no answer. First, ask all respondents to toss a fair coin. Then give the following instructions:

- If you toss a head, then answer the decoy question (YES/NO): Were you born on an even day of the month?
- If you toss a tail, then answer the real question of the survey (YES/NO).

The last step is to ask all participants to answer yes or no to the question they were assigned and then count the total numbers of yes and no responses.

a. Choose a particular question that may not produce totally honest responses. Conduct a survey in your class using this technique.

b. Given only the total numbers of yes and no responses to both questions, explain how you can estimate the number of people who answered yes and no to the real question.

c. Will the results computed in part b be exact? Explain.

d. Suppose the decoy question was replaced by these instructions: If you toss a head, then answer YES. Can you still determine the number of people who answered yes and no to the real question?

IN THE NEWS

1. **THEORETICAL PROBABILITIES.** Find a news article or research report that cites a theoretical probability. Provide a one-paragraph discussion.

2. **RELATIVE FREQUENCY PROBABILITIES.** Find a news article or research report that makes use of a relative frequency (or empirical) probability. Provide a one-paragraph discussion.

3. **SUBJECTIVE PROBABILITIES.** Find a news article or research report that refers to a subjective probability. Provide a one-paragraph discussion.

4. **PROBABILITY DISTRIBUTIONS.** Find a news article or research report that cites or makes use of a probability distribution. Provide a one-paragraph discussion.

6.3 Probabilities with Large Numbers

When we make observations or measurements that have random outcomes, it is impossible to predict any single outcome. We can state only the *probability* of a single outcome. However, when we make many observations or measurements, we expect the distribution of the outcomes to show some pattern or regularity. In this section, we use the ideas of probability to make useful statements about large samples. In the process, you will see one of the most important connections between probability and statistics.

The false ideas prevalent among all classes of the community respecting chance and luck illustrate the truth that common consent argues almost of necessity of error.

—Richard Proctor,
Chance and Luck (1887 textbook)

The Law of Large Numbers

If you toss a coin once, you cannot predict exactly how it will land; you can state only that the probability of a head is 0.5. If you toss the coin 100 times, you still cannot predict precisely how many heads will occur. However, you can reasonably expect to get heads *close to* 50% of the time. If you toss the coin 1,000 times, you can expect the proportion of heads to be even closer to 50%. In general, the more times you toss the coin, the closer the percentage of heads will be to exactly 50%. Thus, while individual events may be unpredictable, large collections of events should show some pattern or regularity. This principle is called the **law of large numbers** (sometimes called the *law of averages*).

The Law of Large Numbers

The **law of large numbers** (or law of averages) applies to a process for which the probability of an event A is $P(A)$ and the results of repeated trials are independent. It states:

If the process is repeated through many trials, the proportion of the trials in which event A occurs will be close to the probability $P(A)$. The larger the number of trials, the closer the proportion should be to $P(A)$.

We can illustrate the law of large numbers with a die-rolling experiment. The probability of a 1 on a single roll is $P(1) = 1/6 = 0.167$. To avoid the tedium of rolling the die many times, we can let a computer *simulate* random rolls of the die. Figure 6.8 shows a computer simulation of rolling a single die 5,000 times. The horizontal axis gives the number of rolls, and the height of the curve gives the proportion of 1's. Although the curve bounces around when the number of rolls is small, for larger numbers of rolls the proportion of 1's approaches the probability of 0.167—just as predicted by the law of large numbers.

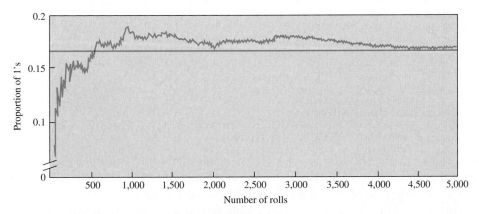

Figure 6.8   Results of computer simulation of rolling a die. Note that as the number of rolls grows large, the relative frequency of rolling a 1 gets close to the theoretical probability of 0.167.

EXAMPLE 1   Roulette

A roulette wheel has 38 numbers: 18 black numbers, 18 red numbers, and the numbers 0 and 00 in green.

a. What is the probability of getting a red number on any spin?

b. If patrons in a casino spin the wheel 100,000 times, how many times should you expect a red number?

SOLUTION

a. The theoretical probability of getting a red number on any spin is

$$P(A) = \frac{\text{number of ways red can occur}}{\text{total number of outcomes}} = \frac{18}{38} = 0.474$$

b. The law of large numbers tells us that as the game is played more and more times, the proportion of times that a red number appears should get closer to 0.474. Thus, in 100,000 tries, the wheel should come up red close to 47.4% of the time, or about 47,400 times.   ∎

Expected Value

The Providence Trust Company sells a special type of insurance in which it promises to pay you $100,000 in the event that you must quit your job because of serious illness. Based on data from past claims, the probability that a policyholder will be paid for loss of job is 1 in 500. Should the insurance company expect to earn a profit if it sells the policies for $250 each?

If the company sells only a few policies, the profit or loss is unpredictable. For example, selling 100 policies for $250 each would generate revenue of 100 × $250 = $25,000. If none of the 100 policyholders files a claim, the company will make a tidy profit. On the other hand, if the company must pay a $100,000 claim to even one policyholder, it will face a huge loss.

In contrast, if the company sells a large number of policies, the law of large numbers tells us that the proportion of policies for which claims will have to be paid should be very close to the 1 in 500 probability for a single policy. For example, if the company sells 1 million policies, it should expect that the number of policyholders collecting on a $100,000 claim will be close to

$$\underbrace{1{,}000{,}000}_{\substack{\text{number of} \\ \text{policies}}} \times \underbrace{\frac{1}{500}}_{\substack{\text{probability of} \\ \$100{,}000 \text{ claim}}} = 2{,}000$$

Paying these 2,000 claims will cost

$$2{,}000 \times \$100{,}000 = \$200 \text{ million}$$

This cost is an *average* of $200 for each of the 1 million policies. Thus, if the policies sell for $250 each, the company should expect to earn an average of $250 − $200 = $50 per policy. This amounts to a profit of $50 million on sales of 1 million policies.

We can find this same answer with a more formal procedure. Note that this insurance example involves two distinct events, each with a particular *probability* and *value* for the company:

1. In the event that a person buys a policy, the value to the company is the $250 price of the policy. The probability of this event is 1 because everyone who buys a policy pays $250.

2. In the event that a person is paid for a claim, the value to the company is −$100,000; it is negative because the company loses $100,000 in this case. The probability of this event is 1/500.

If we multiply the value of each event by its probability and add the results, we find the average or **expected value** of each insurance policy:

$$\text{expected value} = \underbrace{\$250}_{\substack{\text{value of} \\ \text{policy sale}}} \times \underbrace{1}_{\substack{\text{probability of} \\ \text{earning \$250 on sale}}} + \underbrace{(-\$100,000)}_{\substack{\text{value of} \\ \text{claim}}} \times \underbrace{\frac{1}{500}}_{\substack{\text{probability of} \\ \text{paying claim}}}$$

$$= \$250 - \$200 = \$50$$

This expected profit of $50 per policy is the same answer we found earlier. Keep in mind that this expected value is based on applying the law of large numbers. Thus, the company should expect to earn this amount per policy only if it sells a large number of policies. We can generalize this result to find the expected value in any situation that involves probability.

> The value of our expectation always signifies something in the middle between the best we can hope for and the worst we can fear.
>
> —Jacob Bernoulli,
> 17th-century mathematician

Expected Value

Consider two events, each with its own value and probability. The **expected value** is

$$\text{expected value} = \binom{\text{value of}}{\text{event 1}} \times \binom{\text{probability of}}{\text{event 1}} + \binom{\text{value of}}{\text{event 2}} \times \binom{\text{probability of}}{\text{event 2}}$$

The expected value formula can be extended to any number of events by including more terms in the sum.

Time out to think

Should the insurance company expect to see a profit of $50 on each individual policy? Should it expect a profit of $50,000 on 1,000 policies? Explain.

EXAMPLE 2   Lottery Expectations

Suppose that $1 lottery tickets have the following probabilities: 1 in 5 to win a free ticket (worth $1), 1 in 100 to win $5, 1 in 100,000 to win $1,000, and 1 in 10 million to win $1 million. What is the expected value of a lottery ticket? Discuss the implications.

SOLUTION   The easiest way to proceed is to make a table (below) of all the relevant events with their values and probabilities. We are calculating the expected value of a lottery ticket to *you;* thus, the ticket price has a negative value because it costs you money, while the values of the winnings are positive.

> **By the Way...**
>
> The mean of a probability distribution is really the same as the expected value. In families with five children, the expected value for the number of girls is 2.5, and the mean number of girls in such families is also 2.5. Expected value is used in a branch of mathematics called decision theory, employed in financial and business applications.

Lottery Outcomes			
Event	Value	Probability	Value × Probability
Ticket purchase	−$1	1	$(-\$1) \times 1 = -\$1.00$
Win free ticket	$1	$\frac{1}{5}$	$\$1 \times \frac{1}{5} = \$0.20$
Win $5	$5	$\frac{1}{100}$	$\$5 \times \frac{1}{100} = \$0.05$
Win $1,000	$1,000	$\frac{1}{100,000}$	$\$1,000 \times \frac{1}{100,000} = \$0.01$
Win $1 million	$1,000,000	$\frac{1}{10,000,000}$	$\$1,000,000 \times \frac{1}{10,000,000} = \$0.10$
			Sum of last column: −$0.64

The expected value is the sum of all the products *value* × *probability*, which the final column of the table shows to be −$0.64. Thus, averaged over many tickets, you should expect to lose 64¢ for each lottery ticket that you buy. If you buy, say, 1,000 tickets, you should expect to *lose* about 1,000 × $0.64 = $640.

Time out to think

Many states use lotteries to finance worthy causes such as parks, recreation, and education. Lotteries also tend to keep state taxes at lower levels. On the other hand, research shows that lotteries are played by people with low incomes. Do you think lotteries are good social policy? Do you think lotteries are good economic policy?

The Gambler's Fallacy

Consider a simple game involving a coin toss: You win $1 if the coin lands heads, and you lose $1 if it lands tails. Suppose you toss the coin 100 times and get 45 heads and 55 tails, putting you $10 in the hole. Are you "due" for a streak of better luck?

You probably recognize that the answer is *no*: Your past bad luck has no bearing on your future chances. However, many gamblers—especially compulsive gamblers—guess just the opposite. They believe that when their luck has been bad, it's due for a change. This mistaken belief is often called the **gambler's fallacy** (or the *gambler's ruin*).

Definition

The **gambler's fallacy** is the mistaken belief that a streak of bad luck makes a person "due" for a streak of good luck.

One reason people succumb to the gambler's fallacy is a misunderstanding of the law of large numbers. In the coin toss game, the law of large numbers tells us that the proportion of heads tends to be closer to 50% for larger numbers of tosses. But this does *not* mean that you are likely to recover early losses. To see why, study Table 6.7, which shows results from computer simulations of large numbers of coin tosses. Note that as the number of tosses increases, the percentage of heads gets closer to exactly 50%, just as the law of large numbers predicts. However, the difference between the number of heads and the number of tails continues to grow—meaning that in this particular set of simulations, the losses (the difference between the numbers of heads and tails) grow larger even as the proportion of heads approaches 50%.

Time is the dominant factor in gambling. Risk and time are opposite sides of the same coin, for if there were no tomorrow there would be no risk.

—Peter Bernstein,
Against the Gods

Table 6.7 Outcomes of Coin Tossing Trials			
Number of tosses	Number of heads	Percentage of heads	Difference between numbers of heads and tails
100	45	45%	10
1,000	470	47%	60
10,000	4,950	49.5%	100
100,000	49,900	49.9%	200

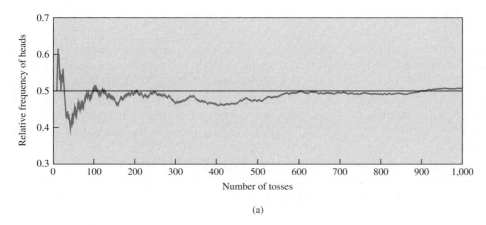

(a)

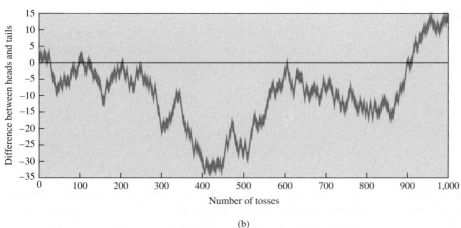

(b)

Figure 6.9   A computer simulation of 1,000 coin tosses. (a) This graph shows how the proportion of heads changes as the number of tosses increases. Note that the proportion approaches 0.5, as we expect from the law of large numbers. (b) This graph shows how the *difference* between the numbers of heads and tails changes as the number of tosses increases. Note that the difference can actually grow larger with more tosses, despite the fact that the proportion of heads approaches 0.5.

Figure 6.9 shows this result graphically for a computer simulation of a coin tossed 1,000 times. Although the percentage of heads approaches 50% over the course of 1,000 tosses, we see fairly large excursions away from equal numbers of heads and tails.

EXAMPLE 3   Continued Losses

You are playing the coin toss game in which you win $1 for heads and lose $1 for tails. After 100 tosses, you are $10 in the hole because you have 45 heads and 55 tails. You continue playing until you've tossed the coin 1,000 times, at which point you've gotten 480 heads and 520 tails. Is this result consistent with what we expect from the law of large numbers? Have you gained back any of your losses? Explain.

SOLUTION   The proportion of heads in your first 100 tosses was 45%. After 1,000 tosses, the proportion of heads has increased to 480 out of 1,000, or 48%. Because the proportion of heads moved closer to 50%, the results are consistent with what we expect from the law of large numbers. However, you've now won $480 (for the 480 heads) and lost $520 (for the 520 tails), for a net loss of $40. Thus, your losses *increased*, despite the fact that the proportion of heads grew closer to 50%. ∎

Everyone who bets any part of his fortune, however small, on a mathematically unfair game of chance acts irrationally. . . . The imprudence of a gambler will be the greater the larger part of his fortune which he exposes to a game of chance.

—Daniel Bernoulli,
18th-century mathematician

By the Way…

The same kind of thinking about streaks applies to selecting lottery numbers. There are no special combinations of lottery numbers that are more likely to be drawn than other combinations. However, people do pick special combinations (such as 1, 2, 3, 4, 5, 6) more often than ordinary combinations. Therefore, if you do win a lottery with an ordinary combination, you are less likely to split the prize with others, with the net effect being that your prize will be larger.

Streaks

Another common misunderstanding that contributes to the gambler's fallacy involves expectations about streaks. Suppose you toss a coin six times and see the outcome HHHHHH (all heads). Then you toss it six more times and see the outcome HTTHTH. Most people would say that the latter outcome is "natural" while the streak of all heads is surprising. But, in fact, both outcomes are equally likely. The total number of possible outcomes for six coins is $2 \times 2 \times 2 \times 2 \times 2 \times 2 = 64$, and every individual outcome has the same probability of 1/64.

Moreover, suppose you just tossed six heads and had to bet on the outcome of the next toss. You might think that, given the run of heads, a tail is "due" on the next toss. But the probability of a head or a tail on the next toss is still 0.50; the coin has no memory of previous tosses.

Time out to think

Is a family with six boys more or less likely to have a boy for the next child? Is a basketball player who has hit 20 consecutive free throws more or less likely to hit her next free throw? Is the weather on one day independent of the weather on the next (as assumed in the next example)? Explain.

EXAMPLE 4 Planning for Rain

A farmer knows that at this time of year in his part of the country, the probability of rain on a given day is 0.5. It hasn't rained in 10 days, and he needs to decide whether to start irrigating. Is he justified in postponing irrigation because he is due for a rainy day?

SOLUTION The 10-day dry spell is unexpected, and, like a gambler, the farmer is having a "losing streak." However, if we assume that weather events are independent from one day to the next, then it is a fallacy to expect that the probability of rain is any more or less than 0.5.

Review Questions

1. Explain the meaning of the law of large numbers. Does this law say anything about what will happen in a single observation or experiment? Why or why not?

2. In 10 tosses of a fair coin, would you be surprised to see six heads? In 1,000 tosses of a fair coin, would you be surprised to see 600 heads? Explain in terms of the law of large numbers.

3. What is an *expected value*? How do we compute an expected value?

4. Should we always expect to get the expected value? Why or why not?

5. What is the gambler's fallacy? Give an example.

Exercises

BASIC SKILLS AND CONCEPTS

SENSIBLE STATEMENTS? For Exercises 1–4, determine whether the given statement is sensible and explain why it is or is not.

1. Scott has just lost $500 playing roulette, but he reasons that, according to the law of large numbers, his bad luck will balance out and he will win more often, so he should start to place larger bets.

2. Kelly has access to a lottery machine, and she reasons that if she can buy a ticket for every possible combination of numbers in the state lottery she will definitely win the jackpot, so she plans to raise enough money to pay for all of those tickets.

3. In 10 tosses of a fair coin, the outcome was heads each time, so an outcome of tails is more likely on the next toss.

4. Lightning never strikes the same spot twice.

5. UNDERSTANDING THE LAW OF LARGE NUMBERS. Suppose you toss a fair coin 10,000 times. Should you expect to get exactly 5,000 heads? Why or why not? What does the law of large numbers tell you about the results you are likely to get?

6. SPEEDY DRIVER. A person who has a habit of driving fast has never had an accident or traffic citation. What does it mean to say that "the law of averages will catch up with him"? Is it true? Explain.

FURTHER APPLICATIONS

7. SHOULD YOU PLAY? Suppose someone gives you 5 to 1 odds that you cannot roll two even numbers with the roll of two fair dice. This means you win $5 if you succeed, and you lose $1 if you fail. What is the expected value of this game to you? Should you expect to win or lose the expected value in the first game? What can you expect if you play 100 times? Explain. (The table on page 239 will be helpful in finding the required probabilities.)

8. EXPECTED AIRLINE FARES. Find the expected value of an airline ticket to the airline company based on the way it prices seats. Suppose there are 150 seats on an airliner: 20 are in first class and are priced at $1,200; 45 are unrestricted coach and are priced at $750; 81 are restricted discount tickets and are priced at $320; and 4 are free for frequent flier awards. Assuming that the air-

craft is sold out, what is the expected value of a ticket to the airline?

9. EXTRA POINTS IN FOOTBALL. Football teams have the option of trying to score either 1 or 2 extra points after a touchdown. They can get 1 point by kicking the ball through the goal posts or 2 points by running or passing the ball across the goal line. For a recent year in the NFL, 1-point kicks were successful 94% of the time, while 2-point attempts were successful only 37% of the time. In either case, failure means zero points. Calculate the expected values of the 1-point and 2-point attempts. Based on these expected values, which option makes more sense in most cases? Can you think of any circumstances in which a team should make a decision different from what the expected values suggest? Explain.

10. INSURANCE CLAIMS. An actuary at an insurance company estimates from existing data that on a $1,000 policy, an average of 1 in 100 policyholders will file a $20,000 claim, an average of 1 in 200 policyholders will file a $50,000 claim, and an average of 1 in 500 policyholders will file a $100,000 claim.
    a. What is the expected value to the company for each policy sold?
    b. If the company sells 100,000 policies, can it expect a profit? Explain the assumptions of this calculation.

11. EXPECTED WAITING TIME. Suppose that you arrive at a bus stop randomly, so all arrival times are equally likely. The bus arrives regularly every 30 minutes without delay (say, on the hour and on the half hour). What is the expected value of your waiting time? Explain how you got your answer.

12. POWERBALL LOTTERY. The 21-state Powerball lottery advertises the following prizes and probabilities of winning for a single $1 ticket. Assume the jackpot has a value of $30 million one week. Note that there is more

Prize	Probability
Jackpot	1 in 80,089,128
$100,000	1 in 1,953,393
$5,000	1 in 364,042
$100	1 in 8,879
$100	1 in 8,466
$7	1 in 207
$7	1 in 605
$4	1 in 188
$3	1 in 74

than one way to win some of the monetary prizes (for example, two ways to win $100), so the table gives the probability for each way. What is the expected value of the winnings for a single lottery ticket? If you spend $365 per year on the lottery, how much can you expect to win or lose?

13. **Big Game.** The Multi-State Big Game lottery advertises the following prizes and probabilities of winning for a single $1 ticket. The jackpot is variable, but assume it has an average value of $3 million. Note that the same prize can be given to two outcomes with different probabilities. What is the expected value of a single lottery ticket to you? If you spend $365 per year on the lottery, how much can you expect to win or lose?

Prize	Probability
Jackpot	1 in 76,275,360
$150,000	1 in 2,179,296
$5,000	1 in 339,002
$150	1 in 9,686
$100	1 in 7,705
$5	1 in 220
$5	1 in 538
$2	1 in 102
$1	1 in 62

14. **Expected Number of Pizzas.** A survey of college students produced the following data on number of pizzas consumed per week. Find the expected value for the number of pizzas consumed by any student. Explain your work.

Number of pizzas	Number of students
0	75
1	60
2	40
3	20
4	10
5	5
6	1

15. **Average American Age.** The following table gives the distribution of Americans by age categories. Use the mid-

Age category	Percentage
0–13	20.0%
14–24	15.3%
25–34	13.6%
35–44	16.3%
45–64	22.2%
65 and over	12.6%

point of each age category (for example, use 6.5 for the 0–13 category and use 70 for 65 and over) to estimate the expected value of American ages.

16. **Mean Household Size.** It is estimated that 57% of Americans live in households with 1 or 2 people, 32% live in households with 3 or 4 people, and 11% live in households with 5 or more people. Explain how you would find the expected number of people in an American household. How is this related to the mean household size?

17. **Psychology of Expected Values.** In 1953, a French economist named Maurice Allais studied how people assess risk. Here are two survey questions that he used:

Decision 1

Option A: 100% chance of gaining $1,000,000

Option B: 10% chance of gaining $2,500,000; 89% chance of gaining $1,000,000; and 1% chance of gaining nothing

Decision 2

Option A: 11% chance of gaining $1,000,000 and 89% chance of gaining nothing

Option B: 10% chance of gaining $2,500,000 and 90% chance of gaining nothing

Allais discovered that for decision 1, most people chose option A, while for decision 2, most people chose option B.
   a. For each decision, find the expected value of each option.
   b. Are the responses given in the surveys consistent with the expected values?
   c. Give a possible explanation for the responses in Allais's surveys.

18. **House Edge in Roulette.** The probability of winning when you bet on a single number in roulette is 1 in 38. A $1 bet yields a net gain of $35 if it is a winner.
   a. Suppose that you bet $1 on the single number 23. What is your probability of winning? What is the expected value of this bet to you?
   b. Suppose that you bet $1 each on the numbers 8, 13, and 23. What is your probability of winning? What is the expected value of this bet to you? Remember that you lose your bet on the numbers that do not come up.
   c. Compare the results of parts a and b. Does the expected value change with the number of numbers on which you bet?

19. **GAMBLER'S FALLACY AND DICE.** Suppose you roll a die with a friend with the following rules: For every even number you roll, you win $1 from your friend; for every odd number you roll, you pay $1 to your friend.
    a. What are the chances of rolling an even number on one roll of a fair die? An odd number?
    b. Suppose that on the first 100 rolls, you roll 45 even numbers. How much money have you won or lost?
    c. Suppose that on the second 100 rolls, your luck improves and you roll 47 even numbers. How much money have you won or lost over 200 rolls?
    d. Suppose that over the next 300 rolls, your luck again improves and you roll 148 even numbers. How much money have you won or lost over 500 rolls?
    e. What was the percentage of even numbers after 100, 200, and 500 rolls? Explain why this game illustrates the gambler's fallacy.
    f. How many even numbers would you have to roll in the next 100 rolls to break even?

20. **BEHIND IN COIN TOSSING: CAN YOU CATCH UP?** Suppose that you toss a fair coin 100 times, getting 38 heads and 62 tails, which is 24 more tails than heads.
    a. Explain why, on your next toss, the *difference* in the numbers of heads and tails is as likely to grow to 25 as it is to shrink to 23.
    b. Extend your explanation from part a to explain why, if you toss the coin 1,000 more times, the final difference in the numbers of heads and tails is as likely to be larger than 24 as it is to be smaller than 24.
    c. Suppose that you continue tossing the coin. Explain why the following statement is true: If you stop at any random time, you always are more likely to have fewer heads than tails, in total.
    d. Suppose that you are betting on heads with each coin toss. After the first 100 tosses, you are well on the losing side (having lost the bet 62 times while winning only 38 times). Explain why, if you continue to bet, you will most likely remain on the losing side. How is this answer related to the gambler's fallacy?

PROJECTS FOR THE WEB AND BEYOND

For useful links, select "Links for Web Projects" for Chapter 6 at www.aw.com/bbt.

21. **ANALYZING LOTTERIES ON THE WEB.** Go to the Web site for all U.S. lotteries and study the summary of state and multi-state lottery odds and prizes. Pick five lotteries and determine the expected value for winnings in each case. Discuss your results.

22. **LAW OF LARGE NUMBERS.** Use a coin to simulate 100 births: Flip the coin 100 times, recording the results, and then convert the outcomes to genders of babies (tail = boy and head = girl). Use the results to fill in the following table. What happens to the proportion of girls as the sample size increases? How does this illustrate the law of large numbers?

Number of births	10	20	30	40	50	60	70	80	90	100
Proportion of girls										

IN THE NEWS

1. **LAW OF LARGE NUMBERS IN LIFE.** Describe a situation (or a news report) in which the law of large numbers is mentioned. Is the term used in an accurate way?

2. **PERSONAL LAW OF LARGE NUMBERS.** Describe a situation in which you personally have made use of the law of large numbers, either correctly or incorrectly. Why did you use the law of large numbers in this situation? Was it helpful?

3. **THE GAMBLER'S FALLACY IN LIFE.** Describe a situation in which you or someone you know has fallen victim to the gambler's fallacy. How could the situation have been dealt with correctly?

6.4 Combining Probabilities

The basic ideas of probability that we have discussed so far have already given us the power to understand some important statistical applications and solve problems that involve chance in daily life. In this section, we explore a few more important ideas of probability that extend the range of problems we can solve.

And Probabilities

> Chance favors only the prepared mind.
> —Louis Pasteur,
> 19th-century scientist

Suppose you toss two fair dice and want to know the probability that *both* will come up 4. One way to find the probability is to consider the two tossed dice as a *single* toss of two dice. Then we can find the probability using the *theoretical* method (see Section 6.2). Because a double-4 is 1 of 36 possible outcomes, its probability is 1/36.

Alternatively, we can consider the two dice individually. For each die, the probability of a 4 is 1/6. We find the probability that both dice show a 4 by multiplying the individual probabilities:

$$P(\text{double-4}) = P(4) \times P(4) = \frac{1}{6} \times \frac{1}{6} = \frac{1}{36}$$

By either method, the probability of rolling a double-4 is 1/36. In general, we call the probability of event A *and* event B occurring an ***and* probability** (or *joint probability*).

The advantage of the multiplication technique is that it can easily be extended to situations involving more than two events. For example, we might want to find the probability of getting 10 heads on 10 coin tosses or of having a baby *and* getting a pay raise in the same year. However, there is an important distinction that must be made when working with *and* probabilities. We must distinguish between events that are *independent* and events that are *dependent* upon each other. Let's investigate each case.

INDEPENDENT EVENTS

The repeated roll of a single die produces **independent events** because the outcome of one roll does not affect the probabilities of the other rolls. Similarly, coin tosses are independent. Deciding whether events are independent is important, but the answer is not always obvious. For example, in analyzing a succession of free throws of a basketball player, should we assume that one free throw is independent of the others? Whenever events *are* independent, we can calculate the *and* probability of two or more events by multiplying.

And Probability for Independent Events

Two events are **independent** if the outcome of one event does not affect the probability of the other event. Consider two independent events A and B with probabilities $P(A)$ and $P(B)$. The probability that A *and* B occur together is

$$P(A \text{ and } B) = P(A) \times P(B)$$

This principle can be extended to any number of independent events. For example, the probability of A, B, and a third independent event C is

$$P(A \text{ and } B \text{ and } C) = P(A) \times P(B) \times P(C)$$

Time out to think

Suppose a basketball player is shooting free throws. Are the different free throws independent events? Defend your opinion.

EXAMPLE 1   Three Coins

Suppose you toss three fair coins. What is the probability of getting three tails?

SOLUTION   Because coin tosses are independent, we multiply the probability of tails on each individual coin:

$$P(3 \text{ tails}) = \underbrace{P(\text{tails})}_{\text{coin 1}} \times \underbrace{P(\text{tails})}_{\text{coin 2}} \times \underbrace{P(\text{tails})}_{\text{coin 3}} = \frac{1}{2} \times \frac{1}{2} \times \frac{1}{2} = \frac{1}{8}$$

The probability that three tossed coins all land on tails is 1/8 (which we determined in Example 8 of Section 6.2 with much more work). ∎

EXAMPLE 2   Jury Selection

A nine-person jury is selected at random from a very large pool of people that has equal numbers of men and women. What is the probability of selecting an all-male jury?

SOLUTION   When a juror is selected, that juror is removed from the pool. However, because we are selecting a small number of jurors from a large pool, we can treat the selection of jurors as independent events. Assuming that equal numbers of men and women are available, the probability of selecting a single male juror is $P(\text{male}) = 0.5$. Thus, the probability of selecting 9 male jurors is

$$P(9 \text{ males}) = \underbrace{0.5 \times 0.5 \times \cdot \; \cdot \; \cdot \times 0.5}_{9 \text{ times}} = 0.5^9 = 0.00195$$

The probability of selecting an all-male jury by random selection is 0.00195, or roughly 2 in 1,000. (Note: Expressions of the form a^b, such as 0.5^9, can be evaluated on many calculators using the key marked x^y or y^x or ^.) ∎

By the Way...

In the late 1960s, the famed baby doctor Benjamin Spock was convicted by an all-male jury of encouraging draft resistance during the Vietnam War. His defense argued that a jury with women would have been more sympathetic.

DEPENDENT EVENTS

A batch of 15 phone cards contains 5 defective cards. If you select a card at random from the batch, the probability of getting a defect is 5/15. Now, suppose that you select a defect on the first selection and put it in your pocket. What is the probability of getting a defect on the second selection?

Because you've removed one defective card from the batch, the batch now contains only 14 cards, of which 4 are defective. Thus, the probability of getting a defective card on the second draw is 4/14. This probability is less than the 5/15 probability on the first selection because the first selection changed the contents of the batch. Because the outcome of the first event affects the probability of the second event, these are **dependent events.**

Calculating the probability for dependent events still involves multiplying the individual probabilities, but we must take into account how prior events affect subsequent events. In the case of the batch of phone cards, we find the probability of getting two defective cards in a row by multiplying the 5/15 probability for the first selection by the 4/14 probability for the second selection:

$$P(2 \text{ defectives}) = \underbrace{P(\text{defective})}_{\text{first selection}} \times \underbrace{P(\text{defective})}_{\text{second selection}} = \frac{5}{15} \times \frac{4}{14} = 0.0952$$

The probability of drawing two defective cards in a row is 0.0952, which is slightly less than (5/15) × (5/15) = 0.111, the probability we get if we replace the first card before the second selection.

TECHNICAL NOTE

P(B given A) is called a *conditional probability*. In some books, it is denoted P(B|A).

And Probability for Dependent Events

Two events are **dependent** if the outcome of one event affects the probability of the other event. The probability that dependent events A and B occur together is

$$P(A \text{ and } B) = P(A) \times P(B \text{ given } A)$$

where P(B given A) means the probability of event B given the occurrence of event A.

This principle can be extended to any number of individual events. For example, the probability of dependent events A, B, and C is

$$P(A \text{ and } B \text{ and } C) = P(A) \times P(B \text{ given } A) \times P(C \text{ given } A \text{ and } B)$$

EXAMPLE 3   Playing BINGO

The game of BINGO involves drawing labeled buttons from a bin at random, without replacement. There are 75 buttons, 15 for each of the letters B, I, N, G, and O.

a. What is the probability of drawing two B buttons in the first two selections?

b. What is the probability of drawing three O buttons in the first three selections?

SOLUTION

a. BINGO involves dependent events because removing a button changes the contents of the bin. The probability of drawing a B on the first draw is 15/75. If this occurs, 74 buttons remain in the bin, of which 14 are Bs. Therefore, the probability of drawing a B button on the second draw is 14/74. The probability of drawing two B buttons in the first two selections is

$$P(\text{B and B}) = \underbrace{P(\text{B})}_{\text{first draw}} \times \underbrace{P(\text{B})}_{\text{second draw}} = \frac{15}{75} \times \frac{14}{74} = 0.0378$$

b. By the reasoning in part a, the probability of drawing an O button on the first draw is 15/75; the probability of drawing an O button on the second draw, given that an O button has been removed, is 14/74; and the probability of drawing an O button on the third draw, given that two O buttons have been removed, is 13/73. Therefore, the probability of drawing three O buttons in the first three selections is

$$\frac{15}{75} \times \frac{14}{74} \times \frac{13}{73} = 0.00674$$

Time out to think

Without doing calculations, compare the probability of drawing three hearts in a row when the cards in the deck are replaced and shuffled after each draw to the probability of drawing three hearts in a row without replacement. Which probability is larger and why?

EXAMPLE 4   When Can We Treat Dependent Events as Independent Events?

A polling organization has a list of 1,000 people for a telephone survey. The pollsters know that 433 people out of the 1,000 are members of the Democratic Party. Assuming that a person cannot be called more than once, what is the probability that the first two people called will be members of the Democratic Party?

SOLUTION   This problem involves an *and* probability for dependent events: Once a person is called, that person cannot be called again. The probability of calling a member of the Democratic Party on the first call is 433/1,000. With that person removed from the calling pool, the probability of calling a member of the Democratic Party on the second call is 432/999. Therefore, the probability of calling two members of the Democratic Party on the first two calls is

$$\frac{433}{1,000} \times \frac{432}{999} = 0.1872$$

If we treat the two calls as independent events, the probability of calling a member of the Democratic Party is 433/1,000 on both calls. Then the probability of calling two members of the Democratic Party is

$$\frac{433}{1,000} \times \frac{433}{1,000} = 0.1875$$

which is nearly identical to the result assuming dependent events. In general, if relatively few items or people are selected from a large pool (in this case, 2 people out of 1,000), then dependent events can be treated as independent events with very little error. A common guideline is that independence can be assumed when the sample size is less than 5% of the population size. This practice is commonly used by polling organizations.   ■

> The importance of probability can only be derived from the judgment that it is rational to be guided by it in action.
> —John Maynard Keynes

EXAMPLE 5   Alarm Failures

A store has three alarm systems that operate independently of each other. Each system has a probability of 0.9 of working properly.

a. What event is the probability that, during a theft, all three alarm systems work properly?

b. What event is the complement of the event that all three alarm systems work properly?

c. What is the probability of the complement of the event that all three alarm systems work properly?

SOLUTION

a. The event that all three alarm systems work properly requires an *and* probability:
$$P(\text{three alarms work}) = 0.9 \times 0.9 \times 0.9 = 0.729$$

b. The complement of all three alarms working properly is that *at least one* alarm fails; note that the complement is *not* that all three alarms fail.

c. Using the rule for probabilities of complementary events, we have
$$P(\text{at least one alarm fails}) = 1 - P(\text{three alarms work})$$
$$= 1 - 0.729 = 0.271$$

There is roughly a 1 in 4 chance that at least one alarm fails.   ■

Either/Or Probabilities

So far we have considered the probability that one event occurs in a first trial *and* another event occurs in a second trial. Suppose we want to know the probability that, when one trial is conducted, *either* of two events occurs. In that case, we are looking for an **either/or probability**, such as the probability of having either a blue-eyed *or* a green-eyed baby or the probability of losing your home either to a fire *or* to a hurricane.

NON-OVERLAPPING EVENTS

A coin can land on either heads *or* tails, but it can't land on both heads *and* tails at the same time. When two events cannot possibly occur at the same time, they are said to be **non-overlapping** (or *mutually exclusive*). We can represent non-overlapping events with a Venn diagram in which each circle represents an event. If the circles do not overlap, it means that the corresponding events cannot occur together. For example, we show the possibilities of heads and tails in a coin toss as two non-overlapping circles because a coin cannot land on both heads and tails at the same time (Figure 6.10).

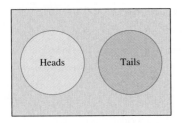

Figure 6.10 Venn diagram for non-overlapping events.

Suppose we roll a die once and want to find the probability of rolling a 1 *or* a 2. Because there are six equally likely outcomes and because two of those outcomes correspond to the events in question, we can use the theoretical method to conclude that the probability of rolling a 1 *or* a 2 is $P(1 \text{ or } 2) = 2/6 = 1/3$. Alternatively, we can find this probability by adding the individual probabilities $P(1) = 1/6$ and $P(2) = 1/6$. In general, we find the either/or probability of two non-overlapping events by adding the individual probabilities.

> ### Either/Or Probability for Non-overlapping Events
>
> Two events are **non-overlapping** if they cannot occur at the same time. If A and B are non-overlapping events, the probability that either A or B occurs is
>
> $$P(A \text{ or } B) = P(A) + P(B)$$
>
> This principle can be extended to any number of non-overlapping events. For example, the probability that either event A, event B, or event C occurs is
>
> $$P(A \text{ or } B \text{ or } C) = P(A) + P(B) + P(C)$$
>
> provided that A, B, and C are all non-overlapping events.

EXAMPLE 6   Either/Or Dice Probability

Suppose you roll a single die. What is the probability of getting an even number?

SOLUTION   The even outcomes of 2, 4, and 6 are non-overlapping because a single die can yield only one result. We know that $P(2) = 1/6$, $P(4) = 1/6$, and $P(6) = 1/6$. Therefore, the combined probability is

$$P(2 \text{ or } 4 \text{ or } 6) = P(2) + P(4) + P(6) = \frac{1}{6} + \frac{1}{6} + \frac{1}{6} = \frac{1}{2}$$

The probability of rolling an even number is 1/2. ∎

EXAMPLE 7   Either/Or Diseases

The two leading causes of death in the United States are heart disease and cancer. Based on the current population of the country and the number of deaths due to each disease, the probability of dying of heart disease in a given year is 0.00272 and the probability of dying of cancer in a given year is 0.00200. What is the probability of dying of either heart disease or cancer?

SOLUTION   If we assume that a person can die of either heart disease or cancer but not both, then we have two non-overlapping events. The probability of dying of either disease is

$$P(\text{death by heart disease or by cancer}) = 0.00272 + 0.00200$$
$$= 0.00472$$

The probability of dying of either heart disease or cancer in a given year is roughly 0.005, or 5 chances in 1,000. Of course, this assumes that a single probability applies to the entire population. In reality, the elderly are much more likely to die of these diseases than are younger people.   ∎

OVERLAPPING EVENTS

To improve tourism between France and the United States, the two governments form a committee consisting of 20 people: 2 American men, 4 French men, 6 American women, and 8 French women (as shown in Table 6.8). If you meet one of these people at random, what is the probability that the person will be *either* a woman *or* a French person?

Table 6.8 Tourism Committee		
	Men	Women
American	2	6
French	4	8

Twelve of the 20 people are French, so the probability of meeting a French person is 12/20. Similarly, 14 of the 20 people are women, so the probability of meeting a woman is 14/20. The sum of these two probabilities is

$$\frac{12}{20} + \frac{14}{20} = \frac{26}{20}$$

This cannot be the correct probability of meeting either a woman or a French person, because probabilities cannot be greater than 1. The Venn diagram in Figure 6.11 shows why simple addition was wrong in this situation. The left circle contains the 12 French people, the right circle contains the 14 women, and the American men are in neither circle. We see that there are 18 people who are either French or women (or both). Because the total number of people in the room is 20, the probability of meeting a person who is *either* French *or* a woman is 18/20 = 9/10.

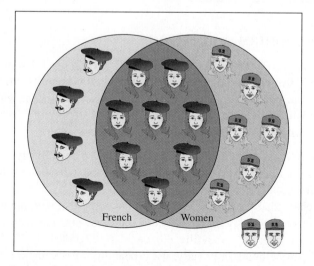

Figure 6.11 Venn diagram for overlapping events.

As the Venn diagram shows, simple addition was incorrect because the region in which the circles overlap contains 8 people who are *both* French *and* women. If we add the two individual probabilities, these 8 people get counted twice: once as women and once as French people. The probability of meeting one of these French women is 8/20. We can correct the double counting error by subtracting out this probability. Thus, the probability of meeting a person who is either French or a woman is

$$P(\text{woman or French}) = \underbrace{\frac{14}{20}}_{\substack{\text{probability} \\ \text{of a woman}}} + \underbrace{\frac{12}{20}}_{\substack{\text{probability of} \\ \text{a French person}}} - \underbrace{\frac{8}{20}}_{\substack{\text{probability of} \\ \text{a French woman}}} = \frac{18}{20} = \frac{9}{10}$$

which agrees with the result found by counting.

We say that meeting a woman and meeting a French person are **overlapping** (or *non–mutually exclusive*) events because both can occur at the same time. Generalizing the procedure we used in this example, we find the following rule.

Either/Or Probability for Overlapping Events

Two events A and B are **overlapping** if they can occur together. The probability that either A or B occurs is

$$P(A \text{ or } B) = P(A) + P(B) - P(A \text{ and } B)$$

The last term, $P(A \text{ and } B)$, corrects for the double counting of events in which A and B both occur together. Note that it is not necessary to use this formula. The correct probability can always be found by counting carefully and avoiding double counting.

Time out to think

Are the events of being born on a Wednesday or being born in Las Vegas overlapping? Are the events of being born on a Wednesday or being born in March overlapping? Are the events of being born on a Wednesday or being born on a Friday overlapping? Explain.

EXAMPLE 8 Minorities and Poverty

Pine Creek is an "average" American town: Of its 2,350 citizens, 1,950 are white, of whom 11%, or 215 people, live below the poverty level. Of the 400 minority citizens, 28%, or 112 people, live below the poverty level. (These percentages are consistent with national demographics.) If you visit Pine Creek, what is the probability of meeting (at random) a person who is *either* a minority *or* living below the poverty level?

SOLUTION Meeting a minority citizen and meeting a person living in poverty are overlapping events. It's useful to make a small table such as Table 6.9, showing how many citizens are in each of the four categories.

You should check that the figures in the table are consistent with the given data and that the total in all four categories is 2,350. Because there are 400 minority citizens, the probability of (randomly) meeting a minority citizen is 400/2,350 = 0.170. Because there are 215 + 112 = 327 people living in poverty, the probability of meeting a citizen in poverty is 327/2,350 = 0.139. The probability of meeting a person who is *both* a minority citizen and a

Table 6.9 Citizens in Pine Creek		
	In poverty	Above poverty
White	215	1,735
Minority	112	288

person living in poverty is 112/2,350 = 0.0477. According to the rule for overlapping events, the probability of meeting *either* a minority citizen *or* a person living in poverty is

$$P(\text{minority or poverty}) = 0.170 + 0.139 - 0.0477 = 0.261$$

The probability of meeting a citizen who is *either* a minority or a person living below the poverty level is about 1 in 4. Notice the importance of subtracting out the term that corresponds to meeting a person who is *both* a minority citizen and a person living in poverty.   ∎

Summary

Table 6.10 provides a summary of the formulas we've used in combining probabilities.

Table 6.10 Summary of Combining Probabilities			
And probability: independent events	*And* probability: dependent events	Either/or probability: non-overlapping events	Either/or probability: overlapping events
$P(A \text{ and } B) =$ $P(A) \times P(B)$	$P(A \text{ and } B) =$ $P(A) \times P(B \text{ given } A)$	$P(A \text{ or } B) =$ $P(A) + P(B)$	$P(A \text{ or } B) =$ $P(A) + P(B) - P(A \text{ and } B)$

Review Questions

1. What is an *and* probability? Give an example in which we might want to compute an *and* probability.

2. Distinguish between independent and dependent events, and describe how we compute *and* probabilities in each case. Give brief examples.

3. What is an either/or probability? Give an example in which we might want to find such a probability.

4. Distinguish between overlapping and non-overlapping events, and describe how we compute either/or probabilities in each case. Give brief examples.

Exercises

BASIC SKILLS AND CONCEPTS

SENSIBLE STATEMENTS? For Exercises 1–4, determine whether the given statement is sensible and explain why it is or is not.

1. The numbers 5, 17, 18, 27, 36, and 41 were drawn in the last lottery, so they should not be bet because they are now less likely to occur again.

2. The probability of flipping a coin and getting heads is 0.5. The probability of selecting a red card when one card is drawn from a shuffled deck is also 0.5. When flipping a coin and drawing a card, the probability of getting heads or a red card is 1.

3. The probability of event A is 0.4 and the probability of event A or B is 0.3.

4. When lottery numbers are drawn, the combination of 1, 2, 3, 4, 5, and 6 is less likely to be drawn than other combinations.

And **PROBABILITIES.** Determine whether the events in Exercises 5–12 are independent or dependent. Then find the probability of the event.

5. Drawing two successive aces from a regular deck, replacing the first card

6. Rolling four successive 6's with four rolls of a fair die

7. Discovering that your five best friends all have telephone numbers ending in 1

8. Selecting two defective computer chips from a large batch in which the defect rate is 1.5%

9. Having five girls in a row in the next five births

10. Drawing four clubs in a row from a regular deck, replacing each card after it has been drawn (13 of the 52 cards are clubs)

11. Having four boys in a row in the next four births

12. Rolling a 1, then a 2, then a 3, then a 4 with a fair die

FURTHER APPLICATIONS

13. **RADIO TUNES.** An MP3 player is loaded with 60 musical selections: 30 rock selections, 15 jazz selections, and 15 blues selections. The player is set on "random play," so selections are played randomly and can be repeated. What is the probability of each of the following events?

    a. The first four selections are all jazz.
    b. The first five selections are all blues.
    c. The first selection is jazz and the second is rock.
    d. Among the first four selections, none is rock.
    e. The second selection is the same song as the first.

14. **POLLING CALLS.** A telephone pollster has a list of names and telephone numbers for 45 voters, 20 of whom are listed as registered Democrats and 25 of whom are listed as registered Republicans. Calls are made in random order. Suppose you want to find the probability that the first two calls are to Republicans.

    a. Are these independent or dependent events? Explain.
    b. If you treat them as dependent events, what is the probability that the first two calls are to Republicans?
    c. If you treat them as independent events, what is the probability that the first two calls are to Republicans?
    d. Compare the results of parts b and c.

PROBABILITY AND COURT DECISIONS. The data in the following table show the outcomes of guilty and not-guilty pleas in 1,028 criminal court cases. Use the data to answer Exercises 15 and 16.

	Guilty plea	Not-guilty plea
Sent to prison	392	58
Not sent to prison	564	14

SOURCE: Brereton and Casper, "Does It Pay to Plead Guilty? Differential Sentencing and the Functioning of the Criminal Courts," *Law and Society Review*, Vol. 16, No. 1.

15. What is the probability that a randomly selected defendant either pled guilty or was sent to prison?

16. What is the probability that a randomly selected defendant either pled not guilty or was not sent to prison?

17. **BINGO.** The game of BINGO involves drawing numbered and lettered buttons at random from a barrel. The B numbers are 1–15, the I numbers are 16–30, the N numbers are 31–45, the G numbers are 46–60, and the O numbers are 61–75. Buttons are not replaced after they have been selected. What is the probability of each of the following events on the initial selections?

    a. Drawing a B button
    b. Drawing two B buttons in a row
    c. Drawing a B or an O
    d. Drawing a B, then a G, then an N, in that order
    e. Drawing anything but a B on five draws

18. **MEETING STRANGERS.** Suppose you walk into a room crowded with 100 people. Half are men, of whom 30 are American and 20 are English, and half are women, of whom 10 are American and 40 are English. Assume that you meet people randomly and don't meet the same person twice. What is the probability of each of the following events?
    a. Meeting an English man on your first encounter
    b. Meeting an American woman on your first encounter
    c. Meeting an English person or a woman on your first encounter
    d. Meeting an American on your first encounter
    e. Meeting an English person and then an American
    f. Meeting an American and then a woman

19. **DRUG TESTS.** An allergy drug is tested by giving 120 people the drug and 100 people a placebo. A control group consists of 80 people who were given no treatment. The number of people in each group who showed improvement appears in the table below.

	Allergy drug	Placebo	Control	Total
Improvement	65	42	31	**138**
No improvement	55	58	49	**162**
Total	**120**	**100**	**80**	**300**

   a. What is the probability that a randomly selected person in the study was given either the drug or the placebo?
   b. What is the probability that a randomly selected person either improved or did not improve?
   c. What is the probability that a randomly selected person either was given the drug or improved?
   d. What is the probability that a randomly selected person was given the drug and improved?

20. **POLITICAL PARTY AFFILIATIONS.** In a typical neighborhood (consistent with national demographics), the adults have the political affiliations listed in the table below. All figures are percentages.

	Republican	Democrat	Independent	Total
Men	17	15	18	**50**
Women	14	20	16	**50**
Total	**31**	**35**	**34**	**100**

   a. What is the probability that a randomly selected person in the town is a Republican or a Democrat?
   b. What is the probability that a randomly selected person in the town is a Republican or a woman?
   c. What is the probability that a randomly selected person in the town is an Independent or a man?
   d. What is the probability that a randomly selected person in the town is a Republican and a woman?

21. **PROBABILITY DISTRIBUTIONS AND GENETICS.** Many traits are controlled by a dominant gene, **A**, and a recessive gene, **a**. Suppose that two parents carry these genes in the proportion 3:1; that is, the probability of either parent giving the **A** gene is 0.75, and the probability of either parent giving the **a** gene is 0.25. Assume that the genes are selected from each parent randomly. To answer the following questions, you can imagine 100 trial "births."
    a. What is the probability that a child receives an **A** gene from both parents?
    b. What is the probability that a child receives an **A** gene from one parent and an **a** gene from the other parent? Note that this can occur in two ways.
    c. What is the probability that a child receives an **a** gene from both parents?
    d. Make a table showing the probability distribution for all events.
    e. If the combinations **AA** and **Aa** both result in the same dominant trait (say, brown hair) and **aa** results in the recessive trait (say, blond hair), what is the probability that a child will have the dominant trait?

AT LEAST ONCE PROBLEMS. A common problem asks for the probability that an event occurs at least once in a given number of trials. Suppose the probability of a particular event is p (for example, the probability of drawing a heart from a deck of cards is 0.25). Then the probability that the event occurs at least once in N trials is

$$1 - (1 - p)^N$$

For example, the probability of drawing at least one heart in 10 draws (with replacement) is

$$1 - (1 - 0.25)^{10} = 0.944$$

Use this rule to solve Exercises 22 and 23.

22. **THE BETS OF THE CHEVALIER DE MÈRE.** It is said that probability theory was invented in the 17th century to explain the gambling of a nobleman named the Chevalier de Mère.
    a. In his first game, the Chevalier bet on rolling at least one 6 with four rolls of a fair die. If played repeatedly, is this a game he should expect to win?
    b. In his second game, the Chevalier bet on rolling at least one double-6 with 24 rolls of two fair dice. If played repeatedly, is this a game he should expect to win?

23. **HIV AMONG COLLEGE STUDENTS.** Suppose that 3% of the students at a particular college are known to carry HIV.

   a. If a student has six sexual partners during the course of a year, what is the probability that at least one of them carries HIV?

   b. If a student has 12 sexual partners during the course of a year, what is the probability that at least one of them carries HIV?

   c. How many partners would a student need to have before the probability of an HIV encounter exceeded 50%?

PROJECTS FOR THE WEB AND BEYOND

For useful links, select "Links for Web Projects" for Chapter 6 at www.aw.com/bbt.

24. A classic probability problem involves a king who wants to increase the proportion of women in his kingdom. He decrees that after a mother gives birth to a son, she is prohibited from having any more children. The king reasons that some families will have just one boy whereas other families will have a few girls and one boy, so the proportion of girls will be increased. Use coin tossing to develop a simulation that emulates a kingdom that abides by this decree: "After a mother gives birth to a son, she will not have any other children." If this decree is followed, does the proportion of girls increase?

IN THE NEWS

1. Stephen Allensworth is a columnist for the *New York Daily News*, and he also publishes *Weekly Lottery News*. In a recent *New York Daily News* column, he provided tips for selecting numbers in New York State's Daily Numbers game. His system is based on the use of "cold digits," which are digits that hit once or not at all in a seven-day period. He made this statement: "That [system] produces the combos: 5–8–9, 7–8–9, 6–8–9, 0–8–9 and 3–8–9. These five combos have an excellent chance of being drawn this week. Good luck to all." Can this system work? Why or why not?

Chapter Review Exercises

In Exercises 1–6, use the data obtained from a test of Nicorette, a chewing gum designed to help people stop smoking. The following table is based on data from Merrell Dow Pharmaceuticals, Inc.

	Nicorette	Placebo
Mouth or throat soreness	43	35
No mouth or throat soreness	109	118

1. If 1 of the 305 subjects is randomly selected, find the probability of getting someone who used Nicorette.

2. If 1 of the 305 subjects is randomly selected, find the probability of getting someone who had no mouth or throat soreness.

3. If 1 of the 305 study subjects is randomly selected, find the probability of getting someone who used Nicorette or had mouth or throat soreness.

4. If 1 of the 305 study subjects is randomly selected, find the probability of getting someone who used the placebo or had no mouth or throat soreness.

5. If two *different* subjects are randomly selected, with replacement, find the probability that they both experienced mouth or throat soreness.

6. If two *different* subjects are randomly selected, without replacement, find the probability that they are both from the placebo group.

7. The Binary Computer Company manufactures computer chips used in DVD players. Those chips are made with a 27% yield, meaning that 27% of them are good and the others are defective.

   a. If one chip is randomly selected, find the probability that it is *not* good.

   b. If two chips are randomly selected, find the probability that they are both good.

   c. If five chips are randomly selected, what is the *expected number* of good chips?

   d. If five chips are randomly selected, find the probability that they are all good. If you did get five good chips among the five selected, would you continue to believe that the yield was 27%? Why or why not?

In Exercises 8–10, consider an event to be "unusual" if its probability is less than or equal to 0.05.

8. a. An instructor obsessed with the metric system insists that all multiple-choice questions have 10 different possible answers, one of which is correct. What is the probability of answering a question correctly if a random guess is made?

   b. Is it "unusual" to answer a question correctly by guessing?

9. a. A roulette wheel has 38 slots: One slot is 0, another slot is 00, and the other slots are numbered 1 through 36. If you bet all of your textbook money on the number 13 for one spin, what is the probability that you will win?

   b. Is it "unusual" to win when you bet on a single number in roulette?

10. a. A study of 400 randomly selected American Airlines flights showed that 344 arrived on time. What is the estimated probability of an American Airlines flight arriving late?

    b. Is it "unusual" for an American Airlines flight to arrive late?

FOCUS ON SOCIAL SCIENCE

Are Lotteries Fair?

Lotteries have become part of the American way of life. Most states now have legal lotteries, including multi-state lotteries such as Powerball and the Big Game. National statistics show that per capita (average per person) lottery spending is approaching $200 per year. Since many people do not play lotteries at all, this means that active players tend to spend much more than $200 per year.

The mathematics of lottery odds involves counting the various combinations of numbers that are winners. While these calculations can become complex, the essential conclusion is always the same: The probability of winning a big prize is infinitesimally small. Advertisements may make lotteries sound like a good deal, but the expected value associated with a lottery is always negative. On average, those who play regularly can expect to lose about half of what they spend.

Lottery proponents point to several positive aspects. For example, lotteries produce billions of dollars of revenue that states use for education, recreation, and environmental initiatives. This revenue allows states to keep tax rates lower than they would be otherwise. Proponents also point out that lottery participation is voluntary and enjoyed by a representative cross-section of society. Indeed, a recent Gallup poll shows that three-fourths of Americans approve of state lotteries (two-thirds approve of legal gambling in general).

This favorable picture is part of the marketing and public relations of state lotteries. For example, Colorado state lottery officials offer statistics on the age, income, and education of lottery players compared to the general population (Figure 6.12). Within a few percentage points, the age of lottery players parallels that of the population as a whole. Similarly, the histogram for the income of lottery players gives the impression that lottery players as a whole are typical citizens—with the exception of the bars for incomes of $15,000–$25,000 and $25,000–$35,000, which show that the poor tend to play more than we would expect for their proportion of the population.

Despite the apparent benefits of lotteries, critics have long argued that lotteries are merely an unfair form of taxation. Some support for this view comes from a recent report by the National Gambling Impact Study Commission and a *New York Times* study of lotteries in New Jersey. Both of these studies focus on the *amount* of money spent on lotteries by individuals.

The *New York Times* study was based on data from 48,875 people who had won at least $600 in New Jersey lottery games. (In an ingenious bit of sampling, these winners were taken to be a random sample of all lottery players; after all, lottery winners are determined randomly. However, the sample is not really representative of all lottery players because winners tend to buy more than an average number of tickets.) By identifying the home zip codes of the lottery players, researchers were able to determine whether players came from areas with high or low income, high or low average education, and various demographic characteristics. The overwhelming conclusion of the *New York Times* study is that lottery spending has a much

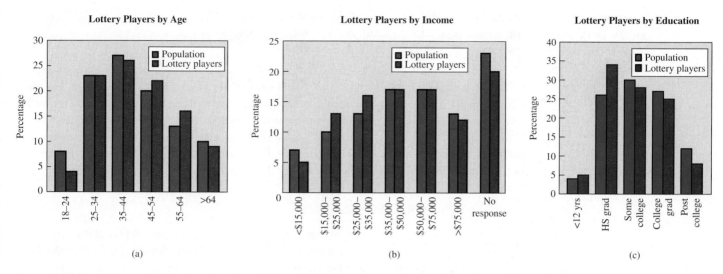

Figure 6.12 Three figures showing (a) age, (b) income, and (c) education of Colorado lottery players compared to population.

greater impact *in relative terms* on those players with lower incomes and lower educational background. For example, the following were among the specific findings:

- People in the state's lowest income areas spend five times as much of their income on lotteries as those in the state's highest income areas. Spending in the lowest income areas on one particular lottery game was $29 per $10,000 of annual income, compared to less than $5 per $10,000 of annual income in the highest income areas.

- The number of lottery sales outlets (where lottery tickets can be purchased) is nearly twice as high per 10,000 people in low-income areas as in high-income areas.

- People in areas with the lowest percentage of college education spent over five times as much per $10,000 of annual income as those in areas with the highest percentage of college education.

- Advertising and promotion of lotteries is focused in low-income areas.

Some of the results of the *New York Times* study are summarized in Figure 6.13. It suggests that while New Jersey has a progressive tax system (higher-income people pay a greater

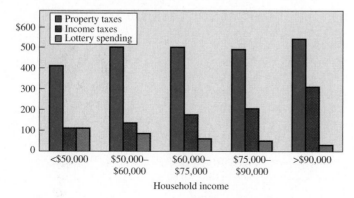

Figure 6.13 Taxes and lottery spending for New Jersey lottery winners (all figures are per $10,000 of income).

percentage of their income in taxes), the "lottery tax" is regressive. Moreover, the study also found that the areas that generate the largest percentage of lottery revenues do not receive a proportional share of state funding.

Similar studies reveal the same patterns in other states. The overall conclusions are inescapable: While lotteries provide many benefits to state governments, the revenue they produce comes disproportionately from poorer and less educated individuals. Indeed, a report by the National Gambling Impact Study Commission concluded that lotteries are "the most widespread form of gambling in the United States" and that state governments have "irresponsibly intruded gambling into society on a massive scale . . . through such measures as incessant advertising and the ubiquitous placement of lottery machines in neighborhood stores."

QUESTIONS FOR DISCUSSION

1. Study Figure 6.12. Do lottery players appear to be a typical cross-section of American society based on age? Based on income? Based on level of education? Explain. How does the "no response" category affect these conclusions?

2. Some lottery players use "systems" for choosing numbers. For example, they consult "experts" who tell them (sometimes for a fee) which numbers are popular or due to appear. Can such systems really improve your odds of winning the lottery? Why or why not? Do you think the use of such systems is related to educational background?

3. Considering all factors presented in this section and other facts that you can find, do you think lotteries are fair to poor or uneducated people? Should they remain legal? Should they be restricted in any way?

4. Find and study a particular lottery advertisement, and determine whether it is misleading in any way.

5. An anonymous quote circulated on the Internet read "Lotteries are a tax on people who are bad at math." Comment on the meaning and accuracy of this quote.

SUGGESTED READING

Pulley, Brett. "Living Off the Daily Dream of Winning a Lottery Prize." *New York Times,* May 22, 1999.

Safire, William. "Lotteries Are Losers." *New York Times,* June 21, 1999.

Walsh, James. *True Odds.* Meritt Publishing, 1996.

FOCUS ON LAW

Is DNA Fingerprinting Reliable?

DNA fingerprinting (also called DNA profiling or DNA identification) has become a major tool of law enforcement. It is used in criminal cases, in paternity cases, and even in the identification of human remains. (DNA fingerprinting was the primary way by which remains of victims were identified after the terrorist attacks on the World Trade Center in 2001.)

The scientific foundation for DNA identification has been in place for several decades. However, these ideas were first used in the courtroom only in 1986. The case involved a 17-year-old boy accused of the rape and murder of two schoolgirls in Narborough, in the Midlands of England. During his interrogation, the suspect asked for a blood test, which was sent to the laboratory of a noted geneticist, Alec Jeffreys, at the nearby University of Leicester. Using methods that he had already developed for paternity testing, Jeffreys compared the suspect's DNA to that found in samples from the victims. The tests showed that the rapes were committed by the same person, but not by the suspect in custody. The following year, after over 4,500 blood samples were collected, researchers made a positive identification of the murderer using Jeffreys' methods.

Word of the British case and Jeffreys' methods spread rapidly. The techniques were swiftly tested, commercialized, and promoted. Not surprisingly, the use of DNA identification also immediately met opposition and controversy.

To explore the essential roles that probability and statistics play in DNA identification, consider a simple eyewitness analogy. Suppose you are looking for a person who helped you out during a moment of need and you remember only three things about this person:

- The person was female.

- She had green eyes.

- She had long red hair.

If you find someone who matches this profile, can you conclude that this person is the woman who helped you? To answer, you need data telling you the probabilities that randomly selected indivivduals in the population have these characteristics. The probability that a person is female is about 1/2. Let's say that the probability of green eyes is about 0.06 (6% of the population has green eyes) and the probability of long red hair is 0.0075. If we assume that these characteristics occur independently of one another, then the probability that a randomly selected person has all three characteristics is

$$0.5 \times 0.06 \times 0.0075 = 0.000225$$

or about 2 in 10,000. This may seem relatively low, but it probably is not low enough to draw a definitive conclusion. For example, a profile matched by only 2 in 10,000 people will still be matched by some 200 people in a city with a population of 1 million. Clearly, if you want to be sure that you've found the right person, you'll need additional information for the profile.

DNA identification is based on a similar idea, but it is designed so that the probability of a profile match is much lower. The DNA of every individual is unique and is the same throughout the individual. A single physical trait is determined by a small piece of DNA called a *gene* at a specific *locus* (location) on a *chromosome*; humans have 23 chromosomes and over 30,000 genes (Figure 6.14). A gene can take two or more (often hundreds) of different forms, called *alleles* (pronounced a-leels). Different alleles give rise to variations of a trait (for example, different hair colors or different blood types). Not only can different alleles appear at a locus, but the corresponding piece of DNA can have different lengths in different people (called *variable number of tandem repeats* or VNTRs). The genetic evidence that is collected and analyzed in the lab consists of the allele lengths or allele types at five to eight different loci.

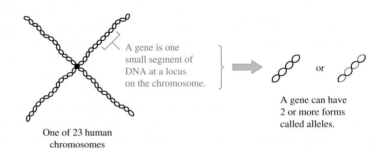

A gene is one small segment of DNA at a locus on the chromosome.

A gene can have 2 or more forms called alleles.

One of 23 human chromosomes

Figure 6.14 Diagram showing the relationship among chromosomes, alleles, genes, and loci.

Collecting genetic evidence (from samples of blood, tissues, hair, semen, or even saliva on a postage stamp) and analyzing it are straightforward, at least in theory. Nevertheless, the process is subject to both controversy and sources of error. Suppose, in our analogy, a person is found with "reddish brown" hair instead of "red" hair. Because many characteristics are continuous (not discrete) variables, should you rule this person out or assume that reddish brown is close enough? For this reason, the issue of *binning* becomes extremely important (see Section 3.1). You might choose to include all people with hair that is any shade of red, or you could choose a narrower bin—say, bright red hair only—which would give a more discriminating test.

The same issue arises in genetic tests. When allele types or lengths are measured in the lab, there is enough variability or error in the measurements that these variables are continuous. Bin widths need to be chosen, and the choice is the source of debate. Bins with a small width give a more refined test, exclude more suspects, and ultimately provide stronger evidence against a defendant.

Another source of scientific controversy is the assumption that the characteristics in a genetic profile are independent. Expert witnesses (geneticists and molecular biologists) have testified at great lengths about this point without agreeing. There does seem to be agreement that the assumption of independence does not introduce significant errors in light of other sources of error. But it is important to see how a difficult scientific question affects the mathematics involved: If characteristics are independent, then the multiplication rule for independent events is justified; if not, a different sort of probability needs to be used.

A third point of controversy concerns the choice of a reference population. Some argue that if genetic data from an entire population are used, they might not fairly represent ethnic subpopulations. For example, a particular allele may have a very different frequency in ethnic

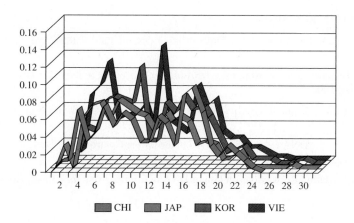

Figure 6.15 Binned frequency data for one allele, showing variability among four ethnic groups.
SOURCE: Kathyrn Roeder, "DNA Fingerprinting: A Review of the Controversy,"
Statistical Science, Vol. 9, No. 2, pp. 222–247.

Italians than in the entire U.S. population. Such a discrepancy would change the outcome of the probability calculation and could either strengthen or damage the case of a defendant.

Figure 6.15 shows an example of the reference population data used in genetic testing. The horizontal axis shows 30 bins for the different allele measurements, and the vertical axis shows the frequency for each bin. Of equal interest are the frequency curves for four different Asian subpopulations. The significant variation among the curves is the evidence for using specialized databases for subpopulations. However, until such databases are complete and there is supporting evidence within a case for using a specialized database, the choice is still to use the full population as a reference.

DNA technology has created new entwined pathways for scientists and legal scholars that are changing our society. Not surprisingly, statistical methods and thinking are an essential part of the change.

QUESTIONS FOR DISCUSSION

1. The result of a DNA test is considered physical (as opposed to circumstantial) evidence. Yet it is much more sophisticated and difficult to understand than a typical piece of physical evidence, such as a weapon or a piece of clothing. Do you believe that DNA evidence should be used in a criminal trial in which jury members may not fully understand how the evidence is collected and analyzed? Defend your opinion.

2. Suppose that an allele has a *greater* frequency in a subpopulation than in the full population. Explain how this would change the evidence for or against a defendant. Suppose that an allele has a *smaller* frequency in a subpopulation than in the full population. Explain how this would change the evidence for or against a defendant.

3. Evidence from blood tests can identify a suspect with a probability of about 1 in 200. Evidence from DNA tests often provides probabilities claimed to be on the order of 1 in 10 million. If you were a juror, would you accept such a probability as positive identification of a suspect?

4. Discuss a few other ways DNA tests can be useful, such as in settling issues of paternity. Overall, how much do you think DNA evidence is likely to affect our society in the future?

SUGGESTED READING

Budowle, B., and Lander, E. "DNA Fingerprinting Dispute Laid to Rest." *Nature,* October 27, 1994.

Coleman, Howard, and Swenson, Eric. *DNA in the Courtroom: A Trial Watcher's Guide.* GeneLex Corporation, 1994.

National Research Council (NRC). *DNA Technology in Forensic Science.* National Academy Press, 1992.

Roeder, Kathryn. "DNA Fingerprinting: A Review of the Controversy." *Statistical Science,* Vol. 9, No. 2, pp. 222–247.

"U.S. Panel Seeking Restrictions on Use of DNA in Courts." *New York Times,* April 14, 1992.

CHAPTER 7

The person who knows "how" will always have a job.
The person who knows "why" will always be his boss.

−Diane Ravitch

Correlation and Causality

Does smoking cause lung cancer? Does low unemployment lead to inflation? Does human use of fossil fuels cause global warming? A major goal of many statistical studies is to search for relationships among different variables so that researchers can then determine whether one factor *causes* another. Once a relationship is discovered, we can then try to determine whether there is an underlying cause. In this chapter, we will study relationships known as correlations and explore how they are important to the more difficult task of searching for causality.

LEARNING GOALS

1. Understand how to identify correlations with scatter diagrams and how to assess their type, strength, and statistical significance.

2. Be aware of important cautions concerning the interpretation of correlations, especially the effects of outliers, effects of grouping data, and the crucial point that correlation does not necessarily imply causality.

3. Understand how to find a line that best fits a data set and understand the limitations of using such a line to make predictions.

4. Understand the difficulty of establishing causality and a few guidelines that can be used in the search for causality.

7.1 Seeking Correlation

What does it mean when we say that smoking *causes* lung cancer? It certainly does *not* mean that you'll get lung cancer if you smoke a single cigarette. It does not even mean that you'll definitely get lung cancer if you smoke regularly and heavily for many years, since some heavy smokers do not get lung cancer. Rather, it is a *statistical* statement meaning that you are *much more likely* to get lung cancer if you smoke than if you don't smoke.

If you think a bit more deeply about this issue, you'll begin to understand how researchers learned that smoking causes lung cancer. The process began with informal observations, as doctors noticed that a surprisingly high proportion of their patients with lung cancer were smokers. This suggestion of a linkage led to carefully conducted studies in which researchers compared lung cancer rates among smokers and nonsmokers. (Such studies are observational, *case-control studies* in which the smokers represent the case group and the nonsmokers represent the control group; see Section 1.3.) These studies showed clearly that heavier smokers were more likely to get lung cancer. In more formal terms, we say that there is a correlation between the variables *amount of smoking* and *likelihood of lung cancer*. A **correlation** is a special type of relationship between variables, in which a rise or fall in one goes along with a corresponding rise or fall in the other.

> *Smoking is one of the leading causes of statistics.*
>
> —Fletcher Knebel

Definition

A **correlation** exists between two variables when higher values of one variable consistently go with higher values of another variable or when higher values of one variable consistently go with lower values of another variable.

Here are a few other examples of correlations:

- There is a correlation between the variables *height* and *weight* for people; that is, taller people tend to weigh more than shorter people.

- There is a correlation between the variables *demand for apples* and *price of apples;* that is, demand tends to decrease as price increases.

- There is a correlation between *practice time* and *skill* among piano players; that is, those who practice more tend to be more skilled.

It's important to realize that establishing a correlation between two variables does *not* mean that a change in one variable *causes* a change in the other. Thus, finding the correlation between smoking and lung cancer did not by itself prove that smoking causes lung cancer. We could imagine, for example, that some gene predisposes a person both to smoking and to lung cancer. Nevertheless, identifying the correlation was the crucial first step in learning that smoking causes lung cancer. We will discuss the difficult task of establishing causality later in this chapter. For now, we will concentrate on how we look for, identify, and interpret correlations.

By the Way...

Smoking is linked to many serious diseases besides lung cancer, including heart disease and emphysema. Smoking is also linked with less lethal health conditions such as premature skin wrinkling and sexual impotence.

Time out to think

Suppose there really were a gene that made people prone to both smoking and lung cancer. Explain why we would still find a strong correlation between smoking and lung cancer in that case, but would not be able to say that smoking causes lung cancer.

Table 7.1 Prices and Characteristics of a Sample of 23 Diamonds from Gem Dealers

Diamond	Price	Weight (carats)	Depth	Table	Color	Clarity
1	$6,958	1.00	60.5	65	3	4
2	$5,885	1.00	59.2	65	5	4
3	$6,333	1.01	62.3	55	4	4
4	$4,299	1.01	64.4	62	5	5
5	$9,589	1.02	63.9	58	2	3
6	$6,921	1.04	60.0	61	4	4
7	$4,426	1.04	62.0	62	5	5
8	$6,885	1.07	63.6	61	4	3
9	$5,826	1.07	61.6	62	5	5
10	$3,670	1.11	60.4	60	9	4
11	$7,176	1.12	60.2	65	2	3
12	$7,497	1.16	59.5	60	5	3
13	$5,170	1.20	62.6	61	6	4
14	$5,547	1.23	59.2	65	7	4
15	$7,521	1.29	59.6	59	6	2
16	$7,260	1.50	61.1	65	6	4
17	$8,139	1.51	63.0	60	6	4
18	$12,196	1.67	58.7	64	3	5
19	$14,998	1.72	58.5	61	4	3
20	$9,736	1.76	57.9	62	8	2
21	$9,859	1.80	59.6	63	5	5
22	$12,398	1.88	62.9	62	6	2
23	$11,008	2.03	62.0	63	8	3

NOTES: Weight is measured in carats (1 carat = 0.2 gram). Depth is defined as 100 times the ratio of height to diameter. Table is the size of the upper flat surface. (Depth and table determine "cut.") Color and clarity are each measured on standard scales, where 1 is best. For color, 1 = colorless and increasing numbers indicate more yellow. For clarity, 1 = flawless and 6 indicates that defects can be seen by eye.

Scatter Diagrams

Table 7.1 lists data for a sample of gem-store diamonds—their prices and several common measures that help determine their value. Because advertisements for diamonds often quote only their weights (in carats), we might suspect a correlation between the weights and the prices. We can look for such a correlation by making a **scatter diagram** (or *scatterplot*) showing the relationship between the variables *weight* and *price*.

Definition

A **scatter diagram** (or *scatterplot*) is a graph in which each point corresponds to the values of two variables.

The following procedure describes how to make the scatter diagram shown in Figure 7.1.

1. We assign one variable to each axis and label the axis with values that comfortably fit all the data. Here, we assign *weight* to the horizontal axis and *price* to the vertical axis. We choose a range of 0 to 2.5 carats for the *weight* axis and $0 to $16,000 for the *price* axis.

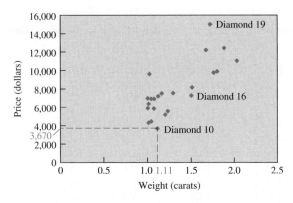

Figure 7.1   Scatter diagram showing the relationship between the variables *price* and *weight* for the diamonds in Table 7.1. The dashed lines show how we find the position of the point for Diamond 10.

By the Way...

When making a scatter diagram, we often have some reason to think that one variable depends at least in part on the other. For example, we might think that the price of a diamond depends in part on its weight. We then call *weight* the *explanatory variable* because it might help explain the price. We call *price* the *response variable* because it responds to changes in the explanatory variable. The explanatory variable is usually plotted on the horizontal axis and the response variable on the vertical axis.

2. For each diamond in Table 7.1, we plot a *single point* at the horizontal position corresponding to its weight and the vertical position corresponding to its price. For example, the point for Diamond 10 goes at a position of 1.11 carats on the horizontal axis and $3,670 on the vertical axis. The dashed lines on Figure 7.1 show how we locate this point.

3. (Optional) We can label the data points. The points representing Diamonds 16 and 19 are labeled in Figure 7.1. In this case, we've chosen not to label every point to avoid cluttering the diagram.

Scatter diagrams get their name because the way in which the points are scattered may reveal a relationship between the variables. In Figure 7.1, we see a general upward trend indicating that diamonds with greater weight tend to be more expensive. The correlation is not perfect. For example, the heaviest diamond is not the most expensive. But the overall trend seems fairly clear.

Time out to think
For practice interpreting scatter diagrams, identify the points in Figure 7.1 that represent Diamonds 3, 7, and 23.

EXAMPLE 1   Color and Price

Using the data in Table 7.1, create a scatter diagram to look for a correlation between a diamond's *color* and *price*. Comment on the correlation.

SOLUTION   Figure 7.2 shows the scatter diagram with color on the horizontal axis and price on the vertical axis. (You should check a few of the points against the data in Table 7.1.) The points appear much more scattered than in Figure 7.1. Nevertheless, you may notice a weak trend diagonally downward from the upper left toward the lower right. This trend represents a weak correlation in which diamonds with more yellow color (higher numbers for color) are less expensive. This trend is consistent with what we would expect, because colorless diamonds appear to sparkle more and are generally considered more desirable. ∎

Time out to think
Thanks to a large bonus at work, you have a budget of $6,000 for a diamond ring. A dealer offers you the following two choices for that price. One diamond weighs 1.20 carats and has color = 4. The other weighs 1.18 carats and has color = 3. Assuming all other characteristics of the diamonds are equal, which would you choose? Why?

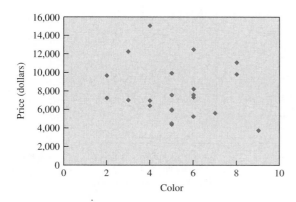

Figure 7.2 Scatter diagram for the color and price data in Table 7.1.

Types of Correlation

We have seen two examples of correlation. Figure 7.1 shows a fairly strong correlation between weight and price, while Figure 7.2 shows a weak correlation between color and price. We are now ready to generalize about types of correlation. Figure 7.3 shows eight scatter diagrams for variables called x and y. Note the following key features of these diagrams:

- Parts a to c of Figure 7.3 show **positive correlations**, in which the values of y tend to increase with increasing values of x. The correlation becomes stronger as we proceed from a to c. In fact, c shows a perfect positive correlation, in which all the points fall along a straight line.

- Parts d to f of Figure 7.3 show **negative correlations**, in which the values of y tend to decrease with increasing values of x. The correlation becomes stronger as we proceed from d to f. In fact, f shows a perfect negative correlation, in which all the points fall along a straight line.

- Part g of Figure 7.3 shows **no correlation** between x and y. In other words, values of x do not appear to be linked to values of y in any way.

- Part h of Figure 7.3 shows a **nonlinear relationship**, in which x and y appear to be related but the relationship does not correspond to a straight line. (*Linear* means along a straight line, and *nonlinear* means *not* along a straight line.)

TECHNICAL NOTE

In this text we use the term *correlation* only for *linear* relationships. Some statisticians refer to nonlinear relationships as "nonlinear correlations." There are techniques for working with nonlinear relationships that are similar to those described in this book for linear relationships.

Types of Correlation

Positive correlation: Both variables tend to increase (or decrease) together.

Negative correlation: The two variables tend to change in opposite directions, with one increasing while the other decreases.

No correlation: There is no apparent relationship between the two variables.

Nonlinear relationship: The two variables are related, but the relationship results in a scatter diagram that does not follow a straight-line pattern.

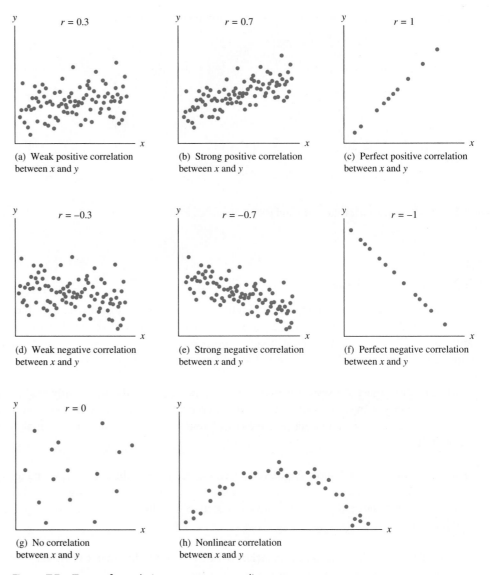

Figure 7.3   Types of correlation seen on scatter diagrams.

EXAMPLE 2   Life Expectancy and Infant Mortality

Figure 7.4 shows a scatter diagram for the variables *life expectancy* and *infant mortality* in 16 countries. What type of correlation does it show? Does this correlation make sense? Does it imply causality? Explain.

SOLUTION   The diagram shows a moderate negative correlation in which countries with *lower* infant mortality tend to have *higher* life expectancy. It is a *negative* correlation because the two variables vary in opposite directions. The correlation makes sense because we would expect that countries with better health care would have both lower infant mortality and higher life expectancy. However, it does *not* imply causality between infant mortality and life expectancy: We would not expect that a concerted effort to reduce infant mortality would increase life expectancy significantly unless it was part of an overall effort to improve health care. (Reducing infant mortality will *slightly* increase life expectancy because fewer people will die at very young ages, thereby raising the mean age of death for the population.)

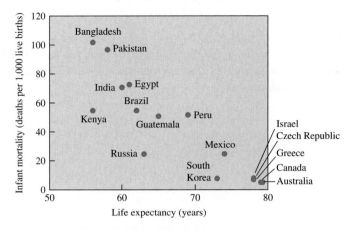

Figure 7.4   SOURCE: United Nations.

Measuring the Strength of a Correlation

For most purposes, it is enough to state whether a correlation is strong, weak, or nonexistent. However, sometimes it is useful to describe the strength of a correlation in more precise terms. Statisticians measure the strength of a correlation with a number called the **correlation coefficient**, represented by the letter r.

Correlation coefficients are easy to interpret, although they are tedious to calculate unless you use a calculator or computer. Look again at Figure 7.3, and notice that it shows the value of the correlation coefficient r for each scatter diagram. The correlation coefficient is always between -1 and 1. When points in a scatter diagram lie close to an ascending straight line, the correlation coefficient is positive and is close to 1. Similarly, points lying close to a descending straight line have a negative correlation coefficient with a value close to -1. Points that do not fit any type of straight-line pattern or that lie close to a *horizontal* straight line (indicating that the y values have no dependence on the x values) result in a correlation coefficient close to 0.

Properties of the Correlation Coefficient, *r*

- The value of the correlation coefficient is always between -1 and 1. That is,

$$-1 \leq r \leq 1$$

- A positive correlation has a positive correlation coefficient ($0 < r \leq 1$). The closer a scatter diagram's points lie to a *rising* straight line, the closer their correlation coefficient is to 1. A perfect positive correlation (in which all the points lie on an ascending straight line) has a correlation coefficient $r = 1$.

- A negative correlation has a negative correlation coefficient ($-1 \leq r < 0$). The closer a scatter diagram's points lie to a *descending* straight line, the closer their correlation coefficient is to -1. A perfect negative correlation (in which all the points lie on a descending straight line) has a correlation coefficient $r = -1$.

- If there is no correlation, the points do not follow any ascending or descending straight line pattern, and the value of *r* is close to 0.

TECHNICAL NOTE

For the methods of this section, there is a requirement that the two variables result in data having a "bivariate normal distribution." This basically means that for any fixed value of one variable, the corresponding values of the other variable have a normal distribution. This requirement is usually very difficult to check, so the check is often reduced to verifying that both variables result in data that are normally distributed.

Technical Note

Working with means tends to suppress individual variation, so a correlation coefficient computed with means may be considerably larger than one computed with individual values.

EXAMPLE 3 U.S. Farm Size

Figure 7.5 shows a scatter diagram for the variables *number of farms* and *mean farm size* in the United States. Each dot represents data from a single year between 1950 and 2000; on this diagram, the earlier years generally are on the right and the later years on the left. Estimate the correlation coefficient by comparing this diagram to those in Figure 7.3 and discuss the underlying reasons for the correlation.

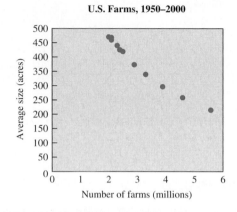

Figure 7.5 Farm size data. SOURCE: U.S. Department of Agriculture.

By the Way...

In 1900, over 40% of the U.S. population worked on farms; by 2000, less than 2% of the population worked on farms.

SOLUTION The scatter diagram shows a strong negative correlation that most closely resembles the scatter diagram in Figure 7.3f, suggesting a correlation coefficient around $r = -0.9$. The correlation shows that as the number of farms decreases, the size of the remaining farms increases. This trend reflects a basic change in the nature of farming: Prior to 1950, most farms were small family farms. Over time, these small farms have been replaced by large farms owned by agribusiness corporations. ∎

EXAMPLE 4 Accuracy of Weather Forecasts

The scatter diagrams in Figure 7.6 show two weeks of data comparing the actual high temperature for the day with the same-day forecast (part a) and the three-day forecast (part b). Estimate the correlation coefficient for each data set and discuss what these coefficients imply about weather forecasts.

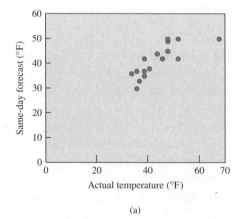

(a)

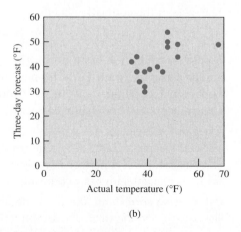

(b)

Figure 7.6 Comparison of actual high temperatures with (a) same-day and (b) three-day forecasts.

SOLUTION   If every forecast were perfect, each actual temperature would equal the corresponding forecasted temperature. This would result in all points lying on a straight line and a correlation coefficient of $r = 1$. In Figure 7.6a, in which the forecasts were made at the beginning of the same day, the points lie fairly close to a straight line, meaning that same-day forecasts are closely related to actual temperatures. By comparing this scatter diagram to the diagrams in Figure 7.3, we can reasonably estimate this correlation coefficient to be about $r = 0.8$. The correlation is weaker in Figure 7.6b, indicating that forecasts made three days in advance aren't as closely related to actual temperatures as same-day forecasts are. This correlation coefficient is about $r = 0.6$. Because the same-day forecast temperatures are more closely related to the actual temperatures, it appears that they are better predictors. (This assumes no systematic errors in the predictions.) ∎

EXAMPLE 5   Movie Success

Table 7.2 shows the production budget, gross receipts, and average viewer rating for 15 recent movies; the viewer rating comes from surveys of viewers who ranked the movie on a 10-point scale, where 1 is the worst and 10 is the best. Make a scatter diagram for the relationship between production budget and gross receipts. Estimate the correlation coefficient and discuss its meaning in this case. ("Gross receipts" is the total amount of money collected in movie theater ticket sales for the movie.)

Table 7.2 Movie Data (receipt data as of 2002)

Movie	Production budget (million of dollars)	Gross receipts (million of dollars)	Viewer rating
Lord of the Rings: The Fellowship of the Ring	109	289	9
The Mummy Returns	100	202	6.3
A.I.: Artificial Intelligence	100	79	7.1
Monsters, Inc.	100	252	8.2
Cast Away	85	234	7.4
Legally Blonde	18	96	6.8
Enemy at the Gates	85	51	7.3
Final Fantasy: The Spirits Within	137	32	6.8
The Last Castle	60	18	6.5
The Blair Witch Project	0.035	141	6.3
Shrek	70	268	8.2
Battlefield Earth	80	21	2.3
Ocean's Eleven	110	182	7.6
Good Will Hunting	10	138	8.5
Speed II	110	48	4.3

By the Way...

In percentage terms, the most profitable film of all time was the *Blair Witch Project*. Produced for $35,000 by two friends who came out of the University of Central Florida's film program, the movie grossed over $140 million—a gross of $4,000 for every $1 it cost to produce.

SOLUTION   Figure 7.7 shows the scatter diagram with production budget on the horizontal axis and gross receipts on the vertical axis. (You should check a few of the points against the data in Table 7.2.) The scatter diagram most closely resembles Figure 7.3g, which exhibits no correlation. Thus, the correlation coefficient is near $r = 0$. In other words, at least for the movies listed here, there is no correlation between the amount of money spent producing the movie and the amount of money it earned in gross receipts. ∎

Time out to think

For further practice, visually estimate the correlation coefficients for the data for diamond weight and price (Figure 7.1) and diamond color and price (Figure 7.2).

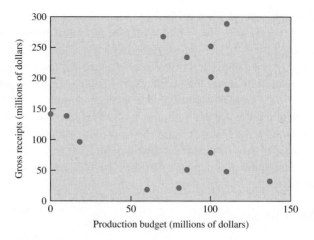

Figure 7.7 Scatter diagram for the data given in Table 7.2.

Statistical Significance of Correlations

Suppose that, for a class project, you collect personal data from a random sample of your fellow students. Two of the items you record for each person are height and the value of any coins in the person's possession. In analyzing these data, you discover a surprising correlation between height and coin value. Figure 7.8 shows the scatter diagram for your data.

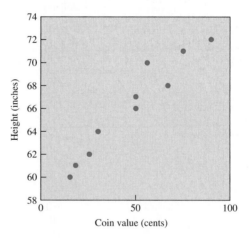

Figure 7.8 Data for the variables *height* and *coin value* among a sample of your fellow students.

Despite the strong correlation, it seems unreasonable to believe that height should generally be associated with the value of coins in a person's possession. Instead, this correlation is probably just a coincidence that occurred with this particular *sample;* we should suspect that it does not reflect a true correlation in the *population* of all students. This example points out that a strong correlation can appear by chance in any sample and does not necessarily mean that two variables are truly related in a population. A key question is the *statistical significance* of a correlation. Recall that a result is statistically significant if it is unlikely to have occurred by chance (see Section 6.1).

For most purposes we can qualitatively assess the statistical significance of a correlation by examining the strength of the correlation and the number of data points. For example:

- A strong correlation (that is, a correlation coefficient near 1 or −1) with many data points is likely to have a high level of statistical significance.

- A strong correlation with very few data points may not be statistically significant.

- A moderate or weak correlation with many data points may be statistically significant.

- A moderate or weak correlation with few data points generally is not statistically significant.

Time out to think

Consider two correlations with the same correlation coefficient of 0.9. The first correlation involves 10 data points and the second involves 100 data points. Explain why the second correlation has a higher level of statistical significance.

You can use Table 7.3 to assess statistical significance more quantitatively. For example, this table shows that for a correlation involving 10 data points, the correlation coefficient must be greater than 0.632 or less than −0.632 for statistical significance at the 0.05 level (Figure 7.9). As discussed in Section 6.1, this means that the probability of the observed correlation occurring by chance is 0.05 or less. Similarly, the correlation coefficient must be greater than 0.765 or less than −0.765 for statistical significance at the 0.01 level.

Table 7.3 Table for Assessing the Statistical Significance of Correlations

Number of data points	Correlation coefficient required for significance at the 0.05 level	Correlation coefficient required for significance at the 0.01 level
4	.950	.999
5	.878	.959
6	.811	.917
7	.754	.875
8	.707	.834
9	.666	.798
10	.632	.765
15	.514	.641
20	.444	.561
30	.361	.463
40	.312	.402
50	.279	.361
100	.196	.256

NOTE: If the computed value of the correlation coefficient is negative, ignore the negative sign. The correlation is then significant at the indicated level if the computed correlation coefficient is *greater* than the value in the table.

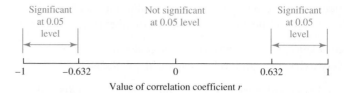

Figure 7.9 This diagram shows how we decide whether a correlation involving 10 data points is significant at the 0.05 level.

As always, remember that statistical significance is not the entire story. Even a very high level of statistical significance might have occurred by chance. For example, the correlation in Figure 7.8 turns out to be statistically significant at the 0.01 level, but it is still just a coincidence. Conversely, a real relationship may exist even when there appears to be no statistical significance; perhaps we did not have enough data or, just by chance, our sample did not reveal the relationship present in the population.

EXAMPLE 6 Inflation and Unemployment

Until the 1990s, most economists assumed that unemployment rates were negatively correlated with inflation rates. That is, when unemployment goes down, inflation goes up, and vice versa. Table 7.4 gives data on inflation and unemployment since 1990, and Figure 7.10 shows a scatter diagram for these data. (The unemployment data are based on randomly selected subjects and the inflation data are based on randomly selected consumer products and services; see Chapter 1 for a discussion of unemployment data and Chapter 2 for a discussion of the CPI.) Does the recent trend agree with the historical claim?

Table 7.4	Inflation and Unemployment	
Year	Unemployment rate	Inflation rate
1990	5.6	5.4
1991	6.8	4.2
1992	7.5	3.0
1993	6.9	3.0
1994	6.1	2.6
1995	5.6	2.8
1996	5.4	3.0
1997	4.9	2.3
1998	4.6	1.6
1999	4.2	2.2
2000	4.0	3.4
2001	4.2	1.8

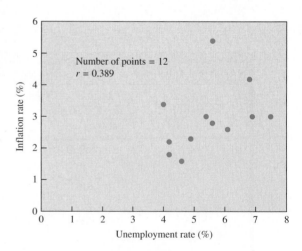

Figure 7.10 Unemployment and inflation rate data, 1990–2001. SOURCE: U.S. Bureau of Labor Statistics.

SOLUTION The graph shows a weak *positive* correlation between unemployment and inflation, with a correlation coefficient $r = 0.389$. However, for a graph with twelve data points, this correlation coefficient is well below the value required for significance at the 0.05 level (see Table 7.3; note that the required value for 10 points is $r = 0.632$ and for 15 points is $r = 0.514$, so the required value for 12 points must be between these two r values). Thus, the data show no significant correlation between inflation and unemployment. The data shown here do not support the longstanding belief among economists that inflation and unemployment are negatively correlated.

EXAMPLE 7   Gas Mileage

Table 7.5 shows weight and gas mileage (city and highway) for 20 different car models. Figure 7.11 shows a scatter diagram for the relationship between *weight* and *city gas mileage*. The correlation coefficient is $r = -0.871$. Evaluate the statistical significance of the correlation.

Table 7.5   Car Data

Car	Weight (pounds)	City mileage (mi/gal)	Highway mileage (mi/gal)
Chevrolet Camaro	3546	19	30
Chevrolet Cavalier	2795	23	31
Dodge Neon	2600	23	32
Ford Taurus	3515	19	27
Honda Accord	3245	23	30
Lincoln Continental	3930	17	24
Mercury Mystique	3115	20	29
Mitsubishi Eclipse	3235	22	33
Olds Aurora	3995	17	26
Pontiac Grand Am	3115	22	30
Toyota Camry	3240	23	32
Cadillac DeVille	4020	17	26
Chevrolet Corvette	3220	18	28
Chrysler Sebring	3175	19	27
Ford Mustang	3450	20	29
BMW 3-Series	3225	19	27
Ford Crown Victoria	3985	17	24
Honda Civic	2440	32	37
Mazda Protégé	2500	29	34
Hyundai Accent	2290	28	37

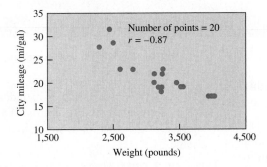

Figure 7.11   Scatter diagram for car data in Table 7.5.

SOLUTION   According to the instructions for Table 7.3, we ignore the negative sign on the correlation coefficient. We then see that for 20 data points, a correlation coefficient of 0.871 is well above the 0.561 required for significance at the 0.01 level. Thus, there is a significant negative correlation between weight and city gas mileage for these cars, suggesting that heavier cars get lower gas mileage. ∎

Calculating the Correlation Coefficient (Optional Section)

Here is a formula for calculating a (linear) correlation coefficient between data values for two variables x and y, where n is the number of pairs of data:

$$r = \frac{n \times \Sigma(x \times y) - (\Sigma x) \times (\Sigma y)}{\sqrt{n \times (\Sigma x^2) - (\Sigma x)^2} \times \sqrt{n \times (\Sigma y^2) - (\Sigma y)^2}}$$

The complexity of this formula is a hint that it is tedious to use; it is usually evaluated using a calculator or computer. (Because calculation of r is so common, this formula is included with almost every statistics software package and it is also included with many calculators.) Nevertheless, the formula is straightforward to use: Calculate each of the required sums and then substitute the values into the formula. Be sure to note that (Σx^2) and $(\Sigma x)^2$ are *not* equal: (Σx^2) tells you to first square all the values of the variable x and then add them; $(\Sigma x)^2$ tells you to add the x values first and then square this sum. In other words, perform the operation within the parentheses first. (Similarly, (Σy^2) and $(\Sigma y)^2$ are not the same.)

Review Questions

1. What is a correlation? Give three examples of pairs of variables that are positively correlated. Give three examples of pairs of variables that are negatively correlated.

2. What is a scatter diagram, and how do you make one? How can we use a scatter diagram to look for a correlation?

3. Define and distinguish among *positive correlation, negative correlation, no correlation,* and *nonlinear relationship*.

4. Describe the correlation coefficient and its meaning. How does it relate to the strength of a correlation?

5. Explain how Table 7.3 can be used to assess the statistical significance of a correlation.

Exercises

BASIC SKILLS AND CONCEPTS

SENSIBLE STATEMENTS? For Exercises 1–4, determine whether the given statement is sensible and explain why it is or is not.

1. There is a significant positive correlation between teachers' incomes and the amount of beer they consume, so it follows that paying teachers more money causes them to drink more beer.

2. A study in one city showed that, for different regions, as the number of police was increased, the crime rate decreased. Because this is a positive effect, we say that there is a positive correlation between police and crime.

3. If a study of two variables includes 20 pairs of data and the correlation coefficient is very close to -1, there appears to be a significant correlation between those variables.

4. If a study of two variables includes 20 pairs of data and the correlation coefficient is 0.5, there is a "50-50 chance" that there is a correlation between the two variables.

TYPES OF CORRELATION. Exercises 5–12 list pairs of variables. For each pair, state whether you believe the two variables are correlated. If you believe they are correlated, state whether the correlation is positive or negative and strong or weak. Explain your reasoning.

5. Height and shoe size

6. Rate of pedaling and speed of the bicycle

7. Number of cars on the road and air quality

8. Altitude on a mountain hike and surrounding air temperature

9. Blood alcohol level and reaction time (for an individual)

10. Weight of car and gas mileage

11. Weight of car and price of car

12. Retail price of an item and the number of items that can be sold in a day

SIGNIFICANCE OF CORRELATIONS. Exercises 13–20 each give a number of data points, n, and a correlation coefficient, r. For each case, state whether the correlation is significant at the 0.05 or 0.01 level and explain the meaning of your statement.

13. $n = 100$ points and $r = -0.22$.

14. $n = 50$ points and $r = 0.35$.

15. $n = 50$ points and $r = 0.35$.

16. $n = 20$ points and $r = 0.66$.

17. $n = 20$ points and $r = 0.58$.

18. $n = 30$ points and $r = -0.35$.

19. $n = 9$ points and $r = -0.81$.

20. $n = 8$ points and $r = -0.76$.

FURTHER APPLICATIONS

21. **WORLD MEAT AND GRAIN PRODUCTION.** The scatter diagram in Figure 7.12 shows the relationship between world production of meat and world production of grain, both measured in kilograms per person. Each data point represents one year between 1950 and 2000. Estimate the correlation coefficient and discuss the underlying reasons for the correlation.

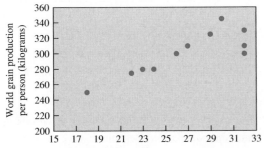

Figure 7.12 Scatter diagram for meat and grain production, 1950–2000. SOURCE: U.S. Department of Agriculture.

22. **TWO-DAY FORECAST.** Figure 7.13 shows a scatter diagram in which the actual high temperature for the day is compared with a forecast made two days in advance. The data are from the same two weeks as those in Figure 7.6. Estimate the correlation coefficient and discuss what these data imply about weather forecasts. Do you think you would get similar results if you made similar diagrams for other two-week periods? Why or why not?

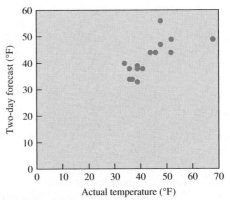

Figure 7.13

23. **SAFE SPEEDS?** Consider the following table showing speed limits and death rates for automobile accidents in selected countries in the 1980s.

Country	Death rate (per 100 million vehicle miles)	Speed limit (miles per hour)
Norway	3.0	55
United States	3.3	55
Finland	3.4	55
Britain	3.5	70
Denmark	4.1	55
Canada	4.3	60
Japan	4.7	55
Australia	4.9	65
Netherlands	5.1	60
Italy	6.1	75

SOURCE: D. J. Rivkin, *New York Times*.

a. Make a scatter diagram of the data.

b. Briefly characterize the correlation in words (for example, strong positive correlation, weak negative correlation) and estimate the correlation coefficient of the data. (Or calculate the correlation coefficient exactly with the aid of a calculator or software.)

c. In the newspaper, these data were presented in an article titled "Fifty-five mph speed limit is no safety guarantee." Based on the data, do you agree with this claim? Explain.

24. **POPULATION GROWTH.** Consider the following table showing percent change in population and birth rate (per 1,000 of population) for ten states between 1980 and 1990.

State	Percent change in population	Birth rate
Nevada	50.1%	16.3
California	25.7%	16.9
New Hampshire	20.5%	12.5
Utah	17.9%	21.0
Colorado	14.0%	14.6
Minnesota	7.3%	13.7
Montana	1.6%	12.3
Illinois	0%	15.5
Iowa	−4.7%	13.0
West Virginia	−8.0%	11.4

SOURCE: U.S. Bureau of Census and Department of Health and Human Services.

a. Make a scatter diagram for the data.
b. Briefly characterize the correlation in words and estimate the correlation coefficient.
c. Overall, does birth rate appear to be a good predictor of a state's population growth rate? If not, what other factor(s) must be affecting the growth rate?

25. **MOST VALUABLE PLAYERS.** Consider the following table showing the number of home runs and batting average for baseball's Most Valuable Players (NL = National League and AL = American League).

Player	Home runs	Batting average
Barry Larkin (1995 NL)	15	.319
Mo Vaughn (1995 AL)	39	.300
Ken Caminiti (1996 NL)	40	.326
Juan Gonzalez (1996 AL)	47	.314
Larry Walker (1997 NL)	49	.366
Ken Griffey Jr. (1997 AL)	56	.304
Sammy Sosa (1998 NL)	66	.308
Juan Gonzalez (1998 AL)	45	.318
Chipper Jones (1999 NL)	45	.319
Ivan Rodriguez (1999 AL)	35	.332
Jeff Kent (2000 NL)	33	.334
Jason Giambi (2000 AL)	43	.333
Barry Bonds (2001 NL)	73	.328
Ichiro Suzuki (2001 AL)	8	.350

a. Make a scatter diagram for the data.
b. Briefly characterize the correlation in words and estimate the correlation coefficient.
c. Do these data suggest that a high batting average is a good predictor of home runs? Explain.

26. **MOVIE DATA.** Consider the following table showing total box office receipts and total attendance for all American films, 1990–2000.

Year	Total receipts (billions of dollars)	Total attendance (billions)
1990	5.0	1.18
1991	4.8	1.14
1992	4.9	1.17
1993	5.2	1.24
1994	5.4	1.29
1995	5.5	1.26
1996	5.9	1.34
1997	6.4	1.39
1998	7.0	1.48
1999	7.5	1.47
2000	7.7	1.42

SOURCE: Motion Picture Association of America.

a. Make a scatter diagram for the data.
b. Briefly characterize the correlation in words and estimate the correlation coefficient.
c. How does the fact that the price of movies has increased since 1990 affect these data? If you were a movie executive, what general conclusions from these data would be important to you?

27. **TV TIME.** Consider the following table showing the average hours of television watched in households in five categories of annual income.

Household income	Weekly TV hours
Less than $30,000	56.3
$30,000–$40,000	51.0
$40,000–$50,000	50.5
$50,000–$60,000	49.7
More than $60,000	48.7

SOURCE: Nielsen Media Research.

a. Make a scatter diagram for the data. To locate the dots, use the midpoint of each income category. Use a value of $25,000 for the category "less than $30,000," and use $70,000 for "more than $60,000."
b. Briefly characterize the correlation in words and estimate the correlation coefficient.
c. Suggest a reason why families with higher incomes watch less TV. Do you think these data imply that you can increase your income simply by watching less TV? Explain.

28. JANUARY WEATHER. Consider the following table showing January mean monthly precipitation and mean daily high temperature for 10 Northern Hemisphere cities (National Oceanic and Atmospheric Administration).

City	Mean daily high temperature for January (°F)	Mean January precipitation (inches)
Athens	54	2.2
Bombay	88	0.1
Copenhagen	36	1.6
Jerusalem	55	5.1
London	44	2.0
Montreal	21	3.8
Oslo	30	1.7
Rome	54	3.3
Tokyo	47	1.9
Vienna	34	1.5

SOURCE: *The New York Times Almanac.*

a. Make a scatter diagram for the data.
b. Briefly characterize the correlation in words and estimate the correlation coefficient.
c. Can you draw any general conclusions about January temperatures and precipitation from these data? Explain.

29. RETAIL SALES. Consider the following table showing total sales (revenue) and profits for the 10 largest retailers in the United States.

Company	Total sales (billions of dollars)	Profits (billions of dollars)*
Wal-Mart	193.3	6.30
Kroger	49.0	0.88
Home Depot	45.7	2.58
Sears	40.9	1.34
Kmart	37.0	−0.24
Target	36.9	1.26
Albertson's	36.8	0.77
JCPenney	33.0	−0.71
Costco	32.2	0.63
Safeway	32.0	0.09

*Negative values reflect losses rather than profits.
SOURCE: Fortune.com.

a. Make a scatter diagram for the data.
b. Briefly characterize the correlation in words and estimate the correlation coefficient.
c. Discuss your observations. Does higher sales volume necessarily translate into greater earnings? Why or why not?

30. CALORIES AND INFANT MORTALITY. Consider the following table showing mean daily caloric intake (all residents) and infant mortality rate (per 1,000 births) for 10 countries.

Country	Mean daily calories	Infant mortality rate (per 1,000 births)
Afghanistan	1,523	154
Austria	3,495	6
Burundi	1,941	114
Colombia	2,678	24
Ethiopia	1,610	107
Germany	3,443	6
Liberia	1,640	153
New Zealand	3,362	7
Turkey	3,429	44
United States	3,671	7

a. Make a scatter diagram for the data.
b. Briefly characterize the correlation in words and estimate the correlation coefficient.
c. Discuss any patterns you observe and any general conclusions that you can reach.

31. OTHER DIAMOND CORRELATIONS. Using the data in Table 7.1, make scatter diagrams for the correlations between depth and price, table and price, and clarity and price. How do these correlations compare to that between weight and price? Do you think any of these correlations help to explain why the correlation between weight and price is not perfect? Explain.

32. DIAMOND SIGNIFICANCE. The correlation coefficients for the diamond data in Table 7.1 are $r = 0.77$ for the correlation between weight and price, and $r = -0.163$ for the correlation between color and price. Evaluate the statistical significance of these correlations. (Hint: Use the row in Table 7.3 for 20 data points, even though Table 7.1 actually has 23 data points.)

33. HIGHWAY MILEAGE. Using the data in Table 7.5, make scatter diagrams for highway mileage versus weight and highway mileage versus city mileage. Does highway mileage appear to correlate as well with weight as did city mileage in Example 7? Do highway and city mileage correlate? Explain.

34. MOVIE SUCCESS. Using the data in Table 7.2, make a scatter diagram for the relationship between production budget and viewer rating of movies. Estimate the correlation coefficient. Based on these data, do you think a large production budget is likely to result in a movie with a high viewer rating? Explain.

35. **PROPERTIES OF THE CORRELATION COEFFICIENT.** In addition to those listed in the text, the correlation coefficient has several other useful mathematical properties. By studying the formula for the correlation coefficient, explain why the following properties are true.

    a. The correlation coefficient remains unchanged if we interchange the variables x and y.

    b. The correlation coefficient remains unchanged if we change the units used to measure x, y, or both.

PROJECTS FOR THE WEB AND BEYOND

For useful links, select "Links for Web Projects" for Chapter 7 at www.aw.com/bbt.

36. **UNEMPLOYMENT AND INFLATION.** Use the Bureau of Labor Statistics Web page to find monthly unemployment rates and inflation rates over the past year. Make a scatter diagram for the data. Do you see any trends? Contrast your results with the results in Example 6.

37. **SUCCESS IN THE NFL.** Find last season's NFL team statistics. Make a table showing the following for each team: number of wins, average yards gained on offense per game, and average yards allowed on defense per game. Make scatter diagrams to explore the correlations between offense and wins and between defense and wins. Discuss your findings. Do you think that there are other team statistics that would yield stronger correlations with the number of wins?

38. **STATISTICAL ABSTRACT.** Explore the "frequently requested tables" at the Web site for the *Statistical Abstract of the United States*. Choose data that are of interest to you and explore at least two correlations. Briefly discuss what you learn from the correlations.

39. **HEIGHT AND ARM SPAN.** Select a sample of at least eight people and measure each person's height and arm span. For arm span, the person should stand with arms extended like the wings on an airplane. Using the paired sample data, construct a scatter diagram and estimate or calculate the value of the correlation coefficient. What do you conclude?

40. **HEIGHT AND PULSE RATE.** Select a sample of at least eight people and record each person's pulse rate by counting the number of heartbeats in 1 minute. Also record each person's height. Using the paired sample data, construct a scatter diagram and estimate or calculate the value of the correlation coefficient. What do you conclude?

IN THE NEWS

1. **CORRELATIONS IN THE NEWS.** Find a recent news report that discusses some type of correlation. Describe the correlation. Does the article give any sense of the strength of the correlation? Does it suggest that the correlation reflects any underlying causality? Briefly discuss whether you believe the implications the article makes with respect to the correlation.

2. **YOUR OWN POSITIVE CORRELATIONS.** Give examples of two variables that you expect to be positively correlated. Explain why the variables are correlated and why the correlation is (or is not) important.

3. **YOUR OWN NEGATIVE CORRELATIONS.** Give examples of two variables that you expect to be negatively correlated. Explain why the variables are correlated and why the correlation is (or is not) important.

7.2 Interpreting Correlations

Researchers sifting through statistical data are constantly looking for meaningful correlations, and the discovery of a new and surprising correlation often leads to a flood of news reports. You may recall hearing about some of these discovered correlations: oat bran consumption correlated with reduced risk of heart disease; cell phone use correlated with increased risk of brain cancer; eating less correlated with increased longevity; and the addition of air bags to cars correlated with increased deaths among infants in low-speed collisions. Unfortunately, the task of *interpreting* such correlations is far more difficult than discovering them in the first place. Long after the news reports have faded, we may still be unsure of whether the correlations are significant and, if so, whether they tell us anything of practical importance. In this section, we discuss some of the common difficulties associated with interpreting correlations.

Statistics show that of those who contract the habit of eating, very few survive.

—Wallace Irwin

Beware of Outliers

Take a look at the scatter diagram in Figure 7.14. Your eye probably tells you that there is a positive correlation in which larger values of *x* tend to mean larger values of *y*. Indeed, if you calculate the correlation coefficient for these data, you'll find that it is a relatively high $r = 0.880$. Moreover, because the diagram shows 10 data points, Table 7.3 tells us that this correlation has statistical significance at the 0.01 level. All in all, Figure 7.14 appears to show a very strong correlation.

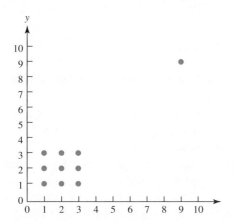

Figure 7.14 How does the outlier affect the correlation?

However, if you place your thumb over the data point in the upper right corner of Figure 7.14, the apparent correlation disappears. In fact, without this data point, the correlation coefficient is zero! In other words, removing this one data point changes the correlation coefficient from $r = 0.880$ to $r = 0$.

This example shows that correlations can be very sensitive to outliers. Recall that an *outlier* is a data value that is extreme compared to most other values in a data set (see Section 4.1). We must therefore examine outliers and their effects carefully before interpreting a correlation. On the one hand, if the outliers are mistakes in the data set, they can produce apparent correlations that are not real or mask the presence of real correlations. On the other hand, if the outliers represent real and correct data points, they may be telling us about relationships that would otherwise be difficult to see.

Note that while we should examine outliers carefully, we should *not* remove them unless we have strong reason to believe that they do not belong in the data set. Even in that case, good research principles demand that we report the outliers along with an explanation of why we thought it legitimate to remove them.

EXAMPLE 1 Masked Correlation

You've conducted a study to determine how the number of calories a person consumes in a day correlates with time spent in vigorous bicycling. Your sample consisted of ten professional women cyclists, all of approximately the same height and weight. Over a period of two weeks, you asked each woman to record the amount of time she spent cycling each day and what she ate on each of those days. You used the eating records to calculate the calories consumed each day. Figure 7.15 shows a scatter diagram with each woman's mean time spent cycling on the horizontal axis and mean caloric intake on the vertical axis. Do higher cycling times correspond to higher intake of calories?

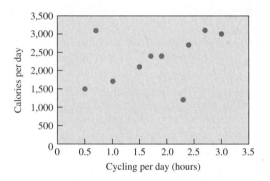

Figure 7.15 Data from the cycling study.

SOLUTION   If you look at the data as a whole, your eye will probably tell you that there is a positive correlation in which greater cycling time tends to go with higher caloric intake. But the correlation is very weak, with a correlation coefficient of $r = 0.374$. According to Table 7.3, this correlation is not statistically significant. However, notice that two points are outliers: one representing a cyclist who cycled about a half-hour per day and consumed more than 3,000 calories, and the other representing a cyclist who cycled more than 2 hours per day on only 1,200 calories. It's difficult to explain the two outliers, given that all the women in the sample are similar in height and weight. Thus, we might suspect that these two women either recorded their data incorrectly or were not following their usual habits during the two-week study. There may be other plausible explanations, but in any case it is reasonable to question whether the outliers are valid data points. Figure 7.16 shows the scatter diagram without the two outliers.

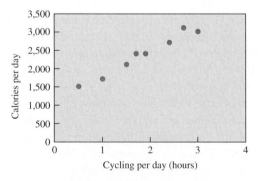

Figure 7.16 The data from Figure 7.15 without the two outliers.

Now the correlation appears very strong and suggests that the number of calories consumed rises by a little more than 500 calories for each hour of cycling. Of course, we should *not* remove the outliers unless we confirm our suspicion that they were invalid data points.   ∎

Beware of Inappropriate Grouping

Correlations can also be misinterpreted when data are grouped inappropriately. In some cases, grouping data hides correlations. Consider a study in which researchers seek a correlation between hours of TV watched per week and high school grade point average (GPA). They collect the 21 data pairs in Table 7.6.

The scatter diagram (Figure 7.17) shows virtually no correlation; the correlation coefficient for the data is about $r = -0.063$. The apparent conclusion is that TV viewing habits are unrelated to academic achievement. However, one astute researcher realizes that some of the students watched mostly educational programs, while others tended to watch comedies, dramas, and movies. She therefore divides the data set into two groups, one for the students who watched mostly educational television and one for the other students. Table 7.7 shows her results with the students divided into these two groups.

Table 7.6 Hours of TV and High School GPA	
Hours per week of TV	GPA
2	3.2
4	3.0
4	3.1
5	2.5
5	2.9
5	3.0
6	2.5
7	2.7
7	2.8
8	2.7
9	2.5
9	2.9
10	3.4
12	3.6
12	2.5
14	3.5
14	2.3
15	3.7
16	2.0
20	3.6
20	1.9

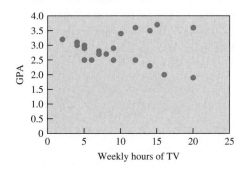

Figure 7.17   The full set of data concerning TV and GPA shows virtually no correlation.

Table 7.7 Hours of TV and High School GPA—Grouped Data			
Group 1: watched educational programs		Group 2: watched regular TV	
Hours per week of TV	GPA	Hours per week of TV	GPA
5	2.5	2	3.2
7	2.8	4	3.0
8	2.7	4	3.1
9	2.9	5	2.9
10	3.4	5	3.0
12	3.6	6	2.5
14	3.5	7	2.7
15	3.7	9	2.5
20	3.6	12	2.5
		14	2.3
		16	2.0
		20	1.9

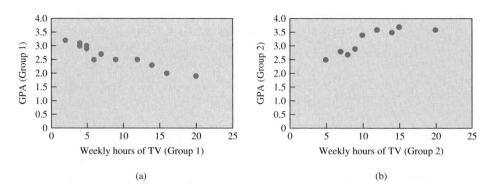

Figure 7.18   These scatter diagrams show the same data as Figure 7.17, separated into two groups identified in Table 7.7.

Now we find two very strong correlations (Figure 7.18): a strong positive correlation for the students who watched educational programs ($r = 0.855$) and a strong negative correlation for the other students ($r = -0.951$). The moral of this story is that the original data set hid an important (hypothetical) correlation between TV and GPA: Watching educational TV correlated positively with GPA and watching non-educational TV correlated negatively with GPA. Only when the data were grouped appropriately could this discovery be made.

In other cases, a data set may show a stronger correlation than actually exists among subgroups. Consider the data in Table 7.8, collected by a consumer group studying the relationship between the weights and prices of cars. Figure 7.19 shows the scatter diagram.

Figure 7.19   Car weights and prices.

Table 7.8 Car Weights and Prices (hypothetical data)	
Weight (lb)	Price ($)
1,500	9,500
1,600	8,000
1,700	8,200
1,750	9,500
1,800	9,200
1,800	8,700
3,000	29,000
3,500	25,000
3,700	27,000
4,000	31,000
3,600	25,000
3,200	30,000

The data set as a whole shows a strong correlation; the correlation coefficient is $r = 0.949$. However, on closer examination, we see that the data fall into two rather distinct categories corresponding to light and heavy cars. If we analyze these subgroups separately, neither shows any correlation: The light cars alone (top six) have a correlation coefficient $r = 0.019$ and the heavy cars alone (bottom six) have a correlation coefficient $r = -0.022$. You can see the problem by looking closely at Figure 7.19. The apparent correlation of the full data set occurs because of the separation between the two clusters of points; there's no correlation within either cluster.

Time out to think

Suppose you were shopping for a compact car. If you looked at only the overall data and correlation coefficient from Figure 7.19, would it be reasonable to consider weight as an important factor in price? What if you looked at the data for light and heavy cars separately? Explain.

CASE STUDY   Fishing for Correlations

Oxford physician Richard Peto submitted a paper to the British medical journal *Lancet* showing that heart attack victims had a better chance of survival if they were given aspirin within a few hours after their heart attacks. The editors of *Lancet* asked Peto to break down the data into subsets, to see whether the benefits of the aspirin were different for different groups of patients. For example, was aspirin more effective for patients of a certain age or for patients with certain dietary habits?

Breaking the data into subsets can reveal important facts, such as whether men and women respond to the treatment differently. However, Peto felt that the editors were asking him to divide his sample into too many subgroups. He therefore objected to the request, arguing that it would result in purely coincidental correlations. Writing about this story in the *Washington Post,* journalist Rick Weiss said, "When the editors insisted, Peto capitulated, but among other things he divided his patients by zodiac birth signs and demanded that his findings be included in the published paper. Today, like a warning sign to the statistically uninitiated, the wacky numbers are there for all to see: Aspirin is useless for Gemini and Libra heart-attack victims but is a lifesaver for people born under any other sign."

The moral of this story is that a "fishing expedition" for correlations can often produce them. That doesn't make the correlations meaningful, even though they may appear significant by standard statistical measures.   ■

Correlation Does *Not* Imply Causality

Perhaps the most important caution about interpreting correlations is one we've already mentioned: ***Correlation does not necessarily imply causality.*** In fact, correlations can appear to be significant for many reasons, but most can be grouped into one of three general categories.

Possible Explanations for a Correlation

1. The correlation *may* be a *coincidence.*
2. Both correlation variables might be directly influenced by some *common underlying cause.*
3. One of the correlated variables may actually be a *cause* of the other. But note that, even in this case, it may be just one of several causes.

We've already discussed examples of all three explanations. The correlation between height and coin value in Figure 7.8 is certainly a coincidence. The correlation between infant mortality and life expectancy in Figure 7.4 is an example of a common underlying cause: Both variables respond to the underlying variable *quality of health care.* The correlation between smoking and lung cancer reflects the fact that smoking causes lung cancer (see the discussion in Section 7.4).

By the Way...

Another famous (or infamous) indicator of the stock market comes from the world of fashion: When women's skirt hemlines rise, the stock market supposedly is also about to rise.

Caution about causality is particularly important in light of the fact that many statistical studies are designed to look for causes. Because these studies generally begin with the search for correlations, it's tempting to think that the work is over as soon as a correlation is found. However, as we will discuss in Section 7.4, establishing causality is very difficult and often impossible. So beware of false claims of causality.

EXAMPLE 2   How to Get Rich in the Stock Market (Maybe)

Every financial advisor has a strategy for predicting the direction of the stock market. Most focus on fundamental economic data, such as interest rates and corporate profits. But an alternative strategy relies on a remarkable and well-known correlation between the Super Bowl winner in January and the direction of the stock market for the rest of the year: The stock market tends to rise when a team from the old, pre-1970 NFL wins the Super Bowl, and tends to fall when the winner is not from the old NFL. This correlation successfully matched 28 of the first 32 Super Bowls to the stock market. Suppose that the Super Bowl just ended and the winner was the Detroit Lions, an old NFL team. Should you invest all your spare cash (and maybe even some that you borrow) in the stock market?

SOLUTION   Based on the reported correlation, you might be tempted to invest since the old-NFL winner suggests a rising stock market over the rest of the year. However, this investment would make sense only if you believed that the Super Bowl result actually *causes* the stock market to move in a particular direction. This is clearly preposterous, and the correlation is undoubtedly a coincidence. If you are going to invest, don't base your investment on this correlation.   ∎

CASE STUDY   Oat Bran and Heart Disease

If you buy a product that contains oat bran, there's a good chance that the label will tout the healthful effects of eating oats. Indeed, several studies have found correlations in which people who eat more oat bran tend to have lower rates of heart disease. But does this mean that everyone should eat more oats?

Not necessarily. Just because oat bran consumption is correlated with reduced risk of heart disease does not mean that it *causes* reduced risk of heart disease. In fact, the question of causality is quite controversial in this case. Other studies suggest that people who eat a lot of oat bran tend to have generally healthful diets. Thus, the correlation between oat bran consumption and reduced risk of heart disease may be a case of a common underlying cause: Having a healthy diet leads people both to consume more oat bran and to have a lower risk of heart disease. In that case, for some people, adding oat bran to their diets might be a *bad* idea because it could cause them to gain weight, and weight gain is associated with *increased* risk of heart disease.

This example shows the importance of using caution when considering issues of correlation and causality. It may be a long time before medical researchers know for sure whether adding oat bran to your diet actually causes a reduced risk of heart disease.   ∎

Useful Interpretations of Correlation

In discussing uses of correlation that might lead to wrong interpretations, we have described the effects of outliers, inappropriate groupings, fishing for correlation, and incorrectly concluding that correlation implies causality. But there are many correct and useful interpretations of correlation. For example, in Section 7.1 we showed how correlation is used to determine the prices of diamonds. In other applications, correlation has been used to establish a relationship between population size and the weight of plastic discarded as garbage. Correlation has been used to establish a relationship between the durations of eruptions of Old Faithful geyser and the intervals between eruptions. In general, correlation plays a prominent and important role in a variety of fields, including meteorology, medical research, business, economics, market research, advertising, psychology, and computer science. We will see in Section 7.3 that once a correlation between two variables has been established, we can then proceed to identify the nature of the relationship so that useful predictions can be made.

Review Questions

1. Briefly discuss the effects of outliers in correlations. How can outliers produce apparent correlations that are not real? How can they mask the presence of real correlations? Under what circumstances should we remove outliers from consideration in a correlation?

2. Briefly discuss how grouped data can affect correlations. How can grouped data hide real correlations? How can grouped data show a stronger correlation than actually exists among individuals?

3. Explain why a correlation does not necessarily imply causality.

4. Describe the three general categories of explanation for a correlation. Give an example of each.

Exercises

BASIC SKILLS AND CONCEPTS

SENSIBLE STATEMENTS? For Exercises 1–4, determine whether the given statement is sensible and explain why it is or is not.

1. Based on a study showing a significant correlation between the weight of a car (in pounds) and its fuel consumption rate (in miles per gallon), we can conclude that increasing a car's weight causes the fuel consumption rate to decrease.

2. Studies have shown that there is a correlation between exercise and health. But correlation does not imply causality, so we should not exercise because doing so would not result in better health.

3. Because studies show a correlation between drinking alcohol and car crashes, we can conclude that drinking alcohol causes car crashes.

4. Given 20 pairs of data, suppose we add an additional pair that constitutes an outlier. The effect of the outlier on the correlation coefficient will be very small because the outlier represents only 1 of 21 pairs of data.

CORRELATION AND CAUSALITY. Exercises 5–14 make statements about a correlation. In each case, state the correlation clearly. (For example, there is a positive correlation between variable A and variable B.) Then state whether the correlation is most likely due to coincidence, a common underlying cause, or a direct cause. Explain your answer.

5. In one U.S. city, the crime rate increased at the same time that the number of people in prison increased.

6. The less Jack exercises, the more he weighs.

7. Over the past three decades, the number of miles of freeways in Los Angeles has grown, and traffic congestion has worsened.

8. Astronomers have discovered that, with the exception of a few nearby galaxies, all galaxies in the universe are moving away from us. Moreover, the farther the galaxy, the faster it is moving away. That is, the more distant a galaxy, the greater the speed at which it is moving away from us.

9. In one high school class, final grades were found to be correlated with the age of the student's car. That is, the older the student's car, the higher the student's grades.

10. In some studies, the incidence of melanoma (the most dangerous form of skin cancer) increases with latitude.

11. When gasoline prices rise, the number of airline passengers increases.

12. When deer hunting permit costs were gradually increased in one region, the incidence of mountain lion sightings increased.

13. Over a period of 20 years, the number of ministers and priests increased, as did attendance at movies.

14. A grocer notices that if she raises the price of peaches, fewer peaches are sold in a single day.

FURTHER APPLICATIONS

15. OUTLIER EFFECTS. Consider the scatter diagram in Figure 7.20.

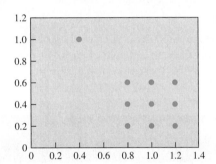

Figure 7.20

a. Which point is an outlier? Ignoring the outlier, estimate or compute the correlation coefficient for the remaining points.
b. Now include the outlier. How does the outlier affect the correlation coefficient? Estimate or compute the correlation coefficient for the complete data set.

16. OUTLIER EFFECTS. Consider the scatter diagram in Figure 7.21.

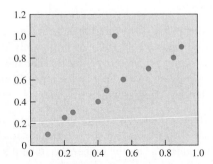

Figure 7.21

a. Which point is an outlier? Ignoring the outlier, estimate or compute the correlation coefficient for the remaining points.
b. Now include the outlier. How does the outlier affect the correlation coefficient? Estimate or compute the correlation coefficient for the complete data set.

17. INFLATION OUTLIERS. Make a scatter diagram for the data in Table 7.4, but without the data points for 1990 and 1991. Compare your diagram to the diagram for the complete data set in Figure 7.10. Would you say that the 1990 and 1991 data points are outliers in the data set? Why or why not? How does removing these two points affect the correlation in this case? Briefly discuss any conclusions you reach about inflation and unemployment in the 1990s.

18. TITANIC OUTLIER. The movie *Titanic* had a production budget of $200 million and gross receipts of $601 million. Add this data point to the scatter diagram in Figure 7.7 (which is based on data in Table 7.2). How does the addition of *Titanic* change the correlation? Should you now conclude that movies with higher production budgets tend to make more money? Why or why not?

19. GROUPED SHOE DATA. The following table gives measurements of weight and shoe size for 10 people (including both men and women).

Weight (lb)	Shoe size
105	6
112	4.5
115	6
123	5
135	6
155	10
165	11
170	9
180	10
190	12

a. Make a scatter diagram for the data. Estimate or compute the correlation coefficient. Based on this correlation coefficient, would you conclude that shoe size and weight are correlated? Explain.
b. You later learn that the first five data values in the table are for women and the next five are for men. How does this change your view of the correlation? Is it still reasonable to conclude that shoe size and weight are correlated?

20. GROUPED TEMPERATURE DATA. The following table shows the average January high temperature and the average July high temperature for 10 major cities around the world.

City	January high	July high
Berlin	35	74
Geneva	39	77
Kabul	36	92
Montreal	21	78
Prague	34	74
Auckland	73	56
Buenos Aires	85	57
Sydney	78	60
Santiago	85	59
Melbourne	78	56

a. Make a scatter diagram for the data. Estimate or compute the correlation coefficient. Based on this correlation coefficient, would you conclude that January and July temperatures are correlated for these cities? Explain.

b. Notice that the first five cities in the table are in the Northern Hemisphere and the next five are in the Southern Hemisphere. How does this change your view of the correlation? Would you now conclude that January and July temperatures are correlated for these cities? Explain.

21. **BIRTH AND DEATH RATES.** Figure 7.22 shows the birth rates and death rates for different countries, measured in births per 1,000 population and deaths per 1,000 population.

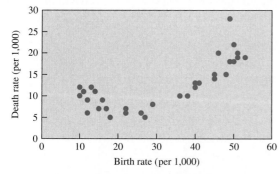

Figure 7.22   Birth and death rates for different countries. SOURCE: United Nations.

a. Estimate the correlation coefficient and discuss whether there is a strong correlation between the variables.

b. Notice that there appear to be two groups of data points within the full data set. Make a reasonable guess as to the makeup of these groups. In which group might you find a relatively wealthy country like Sweden? In which group might you find a relatively poor country like Uganda?

c. Assuming that your guess about groups in part b is correct, do there appear to be correlations within the groups? Explain. How could you confirm your guess about the groups?

22. **READING AND TEST SCORES.** The following (hypothetical) data set gives the number of hours 10 sixth-graders read per week and their performance on a standardized verbal test (maximum 100).

Reading time per week	Verbal test score
1	50
1	65
2	56
3	62
3	65
4	60
5	75
6	50
10	88
12	38

a. Make a scatter diagram for these data. Estimate or compute the correlation coefficient. Based on this correlation coefficient, would you conclude that reading time and test scores are correlated? Explain.

b. Suppose you learn that five of the children read only comic books while the other five read regular books. Make a guess as to which data points fall in which group. How could you confirm your guess about the groups?

c. Assuming that your guess in part b is correct, how would it change your view of the correlation between reading time and test scores? Explain.

 PROJECTS FOR THE WEB AND BEYOND

For useful links, select "Links for Web Projects" for Chapter 7 at www.aw.com/bbt.

23. **FOOTBALL-STOCK UPDATE.** Find data for recent years concerning the Super Bowl winner and the end-of-year change in the stock market (positive or negative). Do recent results still agree with the correlation described in Example 2? Explain.

24. a. Describe a real situation in which there is a positive correlation that is the result of coincidence.

b. Describe a real situation in which there is a positive correlation that is the result of a common underlying cause.

c. Describe a real situation in which there is a positive correlation that is the result of a direct cause.

d. Describe a real situation in which there is a negative correlation that is the result of coincidence.

e. Describe a real situation in which there is a negative correlation that is the result of a common underlying cause.

f. Describe a real situation in which there is a negative correlation that is the result of a direct cause.

IN THE NEWS

1. **MISINTERPRETED CORRELATIONS.** Find a recent news report in which you believe that a correlation may have been misinterpreted. Describe the correlation, the reported interpretation, and the problems you see in the interpretation.

2. **WELL-INTERPRETED CORRELATIONS.** Find a recent news report in which you believe that a correlation has been presented with a reasonable interpretation. Describe the correlation and the reported interpretation, and explain why you think the interpretation is valid.

7.3 Best-Fit Lines and Prediction

Suppose you are lucky enough to win a 1.5-carat diamond in a contest. Based on the correlation between weight and price in Figure 7.1, it should be possible to predict the approximate value of the diamond. We need only study the graph carefully and decide where a point corresponding to 1.5 carats is most likely to fall. To do this, it is helpful to draw a **best-fit line** (also called a *regression line*) through the data, as shown in Figure 7.23. This line is a "best fit" in the sense that, according to a standard statistical measure (which we will discuss later in this section), the data points lie closer to this line than to any other straight line that we could draw through the data.

By the Way...

The term *regression* comes from an 1877 study by Sir Francis Galton. He found that the heights of boys with short or tall fathers were closer to the mean than were the heights of their fathers. He therefore said that the heights of the children *regress* toward the mean, from which we get the term *regression*. The term is now used even for data that have nothing to do with a tendency to regress toward a mean.

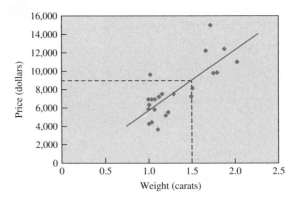

Figure 7.23 Best-fit line for the data from Figure 7.1.

Definition

The **best-fit line** (or *regression line*) on a scatter diagram is a line that lies closer to the data points than any other possible line (according to a standard statistical measure of closeness).

Of all the possible straight lines that can be drawn on a diagram, how do you know which one is the best-fit line? In many cases, you can make a good estimate of the best-fit line simply by looking at the data and drawing the line that visually appears to pass closest to all the data points. This method involves drawing the best-fit line "by eye." As you might guess, there are methods for calculating the precise equation of a best-fit line (see the optional section later in this chapter), and many computer programs and calculators can do these calculations automatically. For our purposes in this book, a fit by eye will generally be sufficient.

Predictions with Best-Fit Lines

We can now use the best-fit line in Figure 7.23 to predict the price of a diamond that weighs 1.5 carats. As indicated by the dashed lines in the figure, the best-fit line predicts that the diamond will cost about $9,000. Notice, however, that two actual data points in the figure correspond to 1.5-carat diamonds, and both of these diamonds cost less than $9,000. Thus, although the predicted price of $9,000 sounds reasonable, it is certainly not guaranteed. In fact, the degree of scatter among the data points in this case tells us that we should *not* trust the best-fit line to predict accurately the price for any individual diamond. Instead, the prediction is meaningful only in a statistical sense: Assuming the correlation is significant, it tells us that if we examined many 1.5-carat diamonds, their mean price would be about $9,000.

This is only the first of several important cautions about interpreting predictions with best-fit lines. A second caution is to beware of using best-fit lines to make predictions that go beyond the bounds of the available data. Figure 7.24 shows a best-fit line for the correlation between infant mortality and longevity from Figure 7.4. According to this line, a country with a life expectancy of more than about 80 years would have a *negative* infant mortality rate, which is impossible.

> It is a capital mistake to theorize before one has data.
>
> —Arthur Conan Doyle

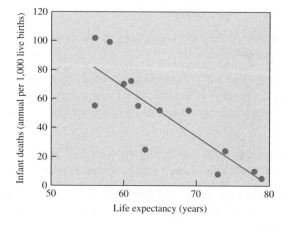

Life Expectancy and Infant Deaths

Figure 7.24   A best-fit line for the correlation between infant mortality and longevity from Figure 7.4.
SOURCE: United Nations.

A third caution is to avoid using best-fit lines from old data sets to make predictions about current or future results. For example, economists studying historical data found a strong negative correlation between unemployment and the rate of inflation. According to this correlation, inflation should have risen dramatically in the mid-1990s when the unemployment rate fell below 6% and then below 5%. But inflation remained low, showing that the correlation from old data did not continue to hold.

> It's tough to make predictions, especially about the future.
>
> —Yogi Berra

Fourth, a correlation discovered with a sample drawn from a particular population cannot generally be used to make predictions about other populations. For example, we can't expect that the correlation between aspirin consumption and heart attacks in an experiment involving only men will also apply to women.

Fifth, remember that we can draw a best-fit line through any data set, but that line is meaningless when the correlation is not significant or when the relationship is nonlinear. For example, there is no significant linear correlation between shoe size and IQ. Thus, we should not use shoe size to predict IQ.

Cautions in Making Predictions from Best-Fit Lines

1. Don't expect a best-fit line to give a good prediction unless the correlation is significant. If the sample points lie very close to the best-fit line, the correlation is very strong and the prediction is more likely to be accurate. If the sample points lie away from the best-fit line by substantial amounts, the correlation is weak and predictions tend to be much less accurate.

2. Don't use a best-fit line to make predictions beyond the bounds of the data points to which the line was fit.

3. A best-fit line based on past data is not necessarily valid now and might not result in valid predictions of the future.

4. Don't make predictions about a population that is different from the population from which the sample data were drawn.

5. Remember that a best-fit line should not be used for predictions when there is no significant correlation or when the relationship is nonlinear.

EXAMPLE 1   Valid Predictions?

State whether the prediction (or implied prediction) should be trusted in each of the following cases, and explain why or why not.

a. You've found a best-fit line for a correlation between the number of hours per day that people exercise and the number of calories they consume each day. You've used this correlation to predict that a person who exercises 18 hours per day would consume 15,000 calories per day.

b. There is a well-known but weak correlation between SAT scores and college grades. You use this correlation to predict the college grades of your best friend from her SAT scores.

c. Historical data have shown a strong negative correlation between national birth rates and affluence. That is, countries with greater affluence tend to have lower birth rates. These data predict a high birth rate in Russia.

d. A study in China has discovered correlations that are useful in designing museum exhibits that Chinese children enjoy. A curator suggests using this information to design a new museum exhibit for Atlanta-area school children.

e. Scientific studies have shown a very strong correlation between children ingesting lead and mental retardation. Based on this correlation, paints containing lead were banned.

f. Based on a large data set, you've made a scatter diagram for salsa consumption (per person) versus years of education. The diagram shows no significant correlation, but you've drawn a best-fit line anyway. The line predicts that someone who consumes a pint of salsa per week has at least 13 years of education.

SOLUTION

a. No one exercises 18 hours per day on an ongoing basis, so this much exercise must be beyond the bounds of any data collected. Therefore, a prediction about someone who exercises 18 hours per day should not be trusted.

b. The fact that the correlation between SAT scores and college grades is weak means there is much scatter in the data. While a best-fit line may allow predictions that are valid in a statistical sense, it cannot be used to make specific predictions about individuals, at least not with much accuracy.

c. We should not assume that the historical data still apply today. In fact, Russia had a very low birth rate throughout the 1990s, despite having a very low level of affluence.

d. The suggestion to use information from the Chinese study for an Atlanta exhibit assumes that predictions made from correlations in China also apply to Atlanta. However, given the cultural differences between China and Atlanta, the curator's suggestion should not be considered without more information to back it up.

e. Given the strength of the correlation and the severity of the consequences, this prediction and the ban that followed seem quite reasonable. In fact, later studies established lead as an actual *cause* of mental retardation, making the rationale behind the ban even stronger.

f. Because there is no significant correlation, the best-fit line and any predictions made from it are meaningless.

> ## By the Way...
>
> In the United States, lead was banned from house paint in 1978 and from food cans in 1991, and a 25-year phaseout of lead in gasoline was completed in 1995. Nevertheless, a 1997 report from the Centers for Disease Control still estimated that one million children under age 6 have enough lead in their blood to damage their health. Major sources of ongoing lead hazards include paint in older housing and soil near major roads, which has high lead content from past use of leaded gasoline.

EXAMPLE 2   Will Women Be Faster Than Men?

Figure 7.25 shows data and best-fit lines for both men's and women's world record times in the 1-mile race. Based on these data, predict when the women's world record will be faster than the men's world record. Comment on the prediction.

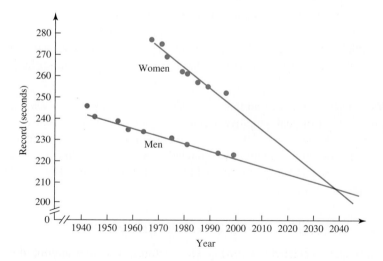

Figure 7.25 World record times in the mile (men and women).

SOLUTION   If we accept the best-fit lines as drawn, the women's world record will equal the men's world record by about 2040. However, this is *not* a valid prediction because it is based on extending the best-fit lines beyond the range of the actual data. It's certainly not out of the question that the women's record will be faster than the men's record in 2040. But by the same reasoning, both world records will eventually be zero, implying that someone will finish the race when the starting gun is fired! ∎

The Correlation Coefficient and Best-Fit Lines

Earlier, we discussed the correlation coefficient as one way of measuring the strength of a correlation. We can also use the correlation coefficient to say something about the validity of predictions with best-fit lines.

For mathematical reasons that we will not discuss here, the *square* of the correlation coefficient, or r^2, is the proportion of the variation in a variable that is accounted for by the best-fit line (or, more technically, by the linear relationship that the best-fit line expresses). For example, the correlation coefficient for the diamond weight and price data (see Figure 7.23) turns out to be $r = 0.777$. If we square that value, we get $r^2 = 0.604$, which we can interpret as follows: About 60% of the variation in the diamond prices is accounted for by the best-fit line relating weight and price. That leaves 40% of the variation in price that must be due to other factors, presumably such things as depth, table, color, and clarity—which is why predictions made with the best-fit line in Figure 7.23 are not very precise. A best-fit line can give precise predictions only in the case of a perfect correlation ($r = 1$ or $r = -1$); in this case, we find $r^2 = 1$, which means that 100% of the variation in a variable can be accounted for by the best-fit line. In this special case of $r^2 = 1$, predictions should be exactly correct, except for the fact that the sample data might not be a true representation of the population data.

TECHNICAL NOTE

Statisticians often call r^2 the coefficient of determination.

> ### Best-Fit Lines and r^2
>
> The *square* of the correlation coefficient, or r^2, is the proportion of the variation in a variable that is accounted for by the best-fit line.

EXAMPLE 3   Retail Hiring

You are the manager of a large department store. Over the years, you've found a reasonably strong correlation between your September sales and the number of employees you'll need to hire for peak efficiency during the holiday season. The correlation coefficient is 0.950. This year your September sales are fairly strong. Should you start advertising for help based on the best-fit line?

SOLUTION   In this case we find that $r^2 = 0.950^2 = 0.903$, which means that 90% of the variation in the number of peak employees can be accounted for by a linear relationship with September sales. That leaves only 10% of the variation in the number of peak employees unaccounted for. Because 90% is so high, we conclude that the best-fit line accounts for the data quite well, so it is a good idea to use it to predict the number of employees you'll need for this year's holiday season. ∎

EXAMPLE 4   Voter Turnout and Unemployment

Political scientists are interested in knowing what factors affect voter turnout in elections. One such factor is the unemployment rate. Data collected in presidential election years since 1964

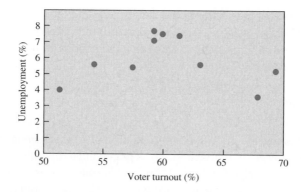

Figure 7.26   Data on voter turnout and unemployment, 1964–2000.
SOURCE: U.S. Bureau of Labor Statistics.

show a very weak negative correlation between voter turnout and the unemployment rate, with a correlation coefficient $r = -0.09$ (Figure 7.26). Based on this correlation, should we use the unemployment rate to predict voter turnout in the next presidential election?

SOLUTION   First, notice from the figure that there are ten data points in the scatter diagram. According to Table 7.3, the correlation coefficient of -0.09 is not sufficiently negative for significance at the 0.05 level. Moreover, the square of the correlation coefficient is $r^2 = (-0.09)^2 = 0.0081$, which means that less than 1% of the variation in the data is accounted for by the best-fit line. Thus, nearly all of the variation in the data must be explained by other factors. We conclude that unemployment is *not* a reliable predictor of voter turnout. ∎

Multiple Regression

If you've ever purchased a diamond, you might have been surprised that we found such a weak correlation between color and price in Figure 7.2. Surely a diamond cannot be very valuable if it has poor color quality. Perhaps color helps to explain why the correlation between weight and price is not perfect. For example, maybe differences in color explain why two diamonds with the same weight can have different prices. To check this idea, it would be nice to look for a correlation between the price and some combination of *weight and color together*.

Time out to think
Let's check this idea in Table 7.1. Notice, for example, that Diamonds 4 and 5 have nearly identical weights, but Diamond 4 costs only $4,299 while Diamond 5 costs $9,589. Can differences in their color explain the different prices? Study other examples in Table 7.1 in which two diamonds have similar weights but different prices. Overall, do you think that the correlation with price would be stronger if we used weight and color together instead of either one alone? Explain.

All who drink his remedy recover in a short time, except those whom it does not help, who all die. Therefore, it is obvious that it fails only in incurable cases.
—Galen, Roman "doctor"

There is a method for investigating a correlation between one variable (such as price) and a *combination* of two or more other variables (such as weight and color). The technique is called **multiple regression**, and it essentially allows us to find a *best-fit equation* that relates three or more variables (instead of just two). Because it involves more than two variables, we cannot make simple diagrams to show best-fit equations for multiple regression. However, it is still possible to calculate a measure of how well the data fit a linear equation. The most common measure in multiple regression is the *coefficient of determination*, denoted

TECHNICAL NOTE

Because the value of the coefficient of determination, R^2, gets larger as more variables are included, an "adjusted R^2" is often used instead. The adjusted R^2 changes the value of R^2 by taking into account the sample size and number of variables.

R^2. It tells us how much of the scatter in the data is accounted for by the best-fit equation. If R^2 is close to 1, the best-fit equation should be very useful for making predictions within the range of the data values. If R^2 is close to zero, then predictions with the best-fit equation are essentially useless.

Definition

The use of **multiple regression** allows the calculation of a best-fit equation that represents the best fit between one variable (such as price) and a *combination* of two or more other variables (such as weight and color). The coefficient of determination, R^2, tells us the proportion of the scatter in the data accounted for by the best-fit equation.

In this book, we will not describe methods for finding best-fit equations by multiple regression. However, you can use the value of R^2 to interpret results from multiple regression. For example, the correlation between price and *weight and color together* results in a value of $R^2 = 0.79$. This is somewhat higher than the $r^2 = 0.61$ that we found for the correlation between price and weight alone. Statisticians who study diamond pricing know that they can get stronger correlations by including additional variables in the multiple regression (such as depth, table, and clarity). Given the billions of dollars spent annually on diamonds, you can be sure that statisticians play prominent roles in helping diamond dealers realize the largest possible profits.

EXAMPLE 5 Alumni Contributions

You've been hired by your college's Alumni Association to help estimate how much the association can reasonably expect to gain in a new fund-raising drive. The director suggests that you use data concerning past contributions and alumni income level, which have a correlation coefficient of 0.6. Is this a good strategy? Can you suggest a better one?

SOLUTION   The correlation coefficient $r = 0.6$ means that $r^2 = 0.36$, so a linear relationship between donations and alumni income accounts for only 36% of the variation. Thus, using the best-fit line to predict the amounts of money that will be donated by individual alumni won't yield very accurate predictions. A better strategy would be to use a multiple regression equation that includes other factors that might influence donations, such as alumni majors, distance of current home from the college, years since graduation, and membership in a fraternity or sorority. It's possible that a well-chosen multiple regression equation will produce a much stronger correlation with donations than any correlation based on only two variables. ∎

Finding Equations for Best-Fit Lines (Optional Section)

The mathematical technique for finding the equation of a best-fit line is based on the following basic ideas. If we draw *any* line on a scatter diagram, we can measure the *vertical* distance of every data point from that line. One measure of how well the line fits the data is the *sum of the squares* of these vertical distances. A large sum means that the vertical distances of data points from the line are fairly large and hence the line is not a very good fit. A small sum means the data points lie close to the line and the fit is good. Of all possible lines, the best-fit line is the line that minimizes the sum of the squares of the vertical distances. Because of this property, the best-fit line is sometimes called the *least squares line*.

By the Way...

One study of alumni donations found that, in developing a multiple regression equation, one should include these variables: income, age, marital status, whether the donor belonged to a fraternity or sorority, whether the donor is active in alumni affairs, donor's distance from the college, and the nation's unemployment rate, used as a measure of the economy (Bruggink and Siddiqui, "An Econometric Model of Alumni Giving: A Case Study for a Liberal Arts College," *The American Economist*, Vol. 39, No. 2).

You may recall that the equation of any straight line can be written in the general form

$$y = mx + b$$

where m is the *slope* of the line and b is the *y-intercept* of the line. The formulas for the slope and y-intercept of the best-fit line are as follows:

$$\text{slope} = m = \frac{n \times \Sigma(x \times y) - (\Sigma x) \times (\Sigma y)}{n \times (\Sigma x^2) - (\Sigma x)^2}$$

$$y\text{-intercept} = b = \frac{\Sigma y}{n} - \left(m \times \frac{\Sigma x}{n} \right) = \bar{y} - (m \times \bar{x})$$

As usual, $\bar{x}$ represents the mean of all the values of the variable x, $\bar{y}$ represents the mean of all the values of the variable y, and n is the number of pairs of data. Because these formulas are tedious to use by hand, we usually use calculators or computers to find the slope and y-intercept of best-fit lines.

Review Questions

1. What is a best-fit line? How can you make a prediction from a best-fit line?

2. Briefly describe five cautions about making predictions from best-fit lines, with examples of each.

3. What does the square of the correlation coefficient (r^2) tell us about a best-fit line? Give examples of its meaning.

4. What is multiple regression, and when is it useful?

Exercises

BASIC SKILLS AND CONCEPTS

SENSIBLE STATEMENTS? For Exercises 1–4, determine whether the given statement is sensible and explain why it is or is not.

1. It is found that there is a correlation between the value of a car and its age. Using the equation of the best-fit line, it is projected that a 25-year-old car is worthless.

2. Using the equation of the best-fit line for data revealing a correlation between the height and the weight of women, it is estimated that a 2,000-pound woman would be 112.5 inches tall.

3. It is found that, for a sample of men, there is a correlation between the time spent watching television and the number of times the remote control is used. Using the equation of the best-fit line, it is estimated that a woman watching television for 1 hour would use the remote control 18 times.

4. Although it is found that there is no correlation between IQ and hat size, the equation of the best-fit line is found for a sample of 50 adults. Using that equation, it is estimated that someone with a hat size of 6 will have an IQ of 104.

FURTHER APPLICATIONS

BEST-FIT LINES ON SCATTER DIAGRAMS. Do the following for Exercises 5–10.

a. Add a best-fit line to the given scatter diagram.
b. Estimate or compute r and r^2. Based on your value for r^2, tell how much of the variation in the variable can be accounted for by the best-fit line.
c. Briefly discuss whether you could make valid predictions from this best-fit line.

5. Use the scatter diagram for color and price in Figure 7.2.

6. Use the scatter diagram for life expectancy and infant mortality in Figure 7.4.

7. Use the scatter diagram for number of farms and size of farms in Figure 7.5.

8. Use both scatter diagrams for actual and predicted temperature in Figure 7.6.

9. Use the scatter diagram for production budget and gross receipts of movies in Figure 7.7.

10. Use the scatter diagram for weight and city mileage in Figure 7.11.

BEST-FIT LINES. Exercises 11–18 refer to the tables in the Exercises for Section 7.1. In each case, do the following.

a. Construct a scatter diagram and, based on visual inspection, draw the best-fit line by eye.

b. Briefly discuss the strength of the correlation. Estimate or compute r and r^2. Based on your value for r^2, tell how much of the variation in the variable can be accounted for by the best-fit line.

c. Identify any outliers on the diagram and discuss their effects on the strength of the correlation and on the best-fit line.

d. For this case, do you believe that the best-fit line gives reliable predictions outside of the range of the data on the scatter diagram? Explain.

11. Use the data in Exercise 23 of Section 7.1.

12. Use the data in Exercise 24 of Section 7.1.

13. Use the data in Exercise 25 of Section 7.1.

14. Use the data in Exercise 26 of Section 7.1.

15. Use the data in Exercise 27 of Section 7.1. To locate the dots, use the midpoint of each income category; use a value of $25,000 for the category "less than $30,000," and use a value of $70,000 for the category "over $60,000."

16. Use the data in Exercise 28 of Section 7.1.

17. Use the data in Exercise 29 of Section 7.1.

18. Use the data in Exercise 30 of Section 7.1.

PROJECTS FOR THE WEB AND BEYOND

For useful links, select "Links for Web Projects" for Chapter 7 at www.aw.com/bbt.

19. **LEAD POISONING.** Research lead poisoning, its sources, and its effects. Discuss the correlations that have helped researchers understand lead poisoning. Discuss efforts to prevent it.

20. **WORLDWIDE POPULATION INDICATORS.** The following table gives five population indicators for eleven selected countries. Study these data and try to identify possible correlations. Doing additional research if necessary, discuss the possible correlations you have found, speculate on the reasons for the correlations, and discuss whether they suggest a causal relationship. Birth and death rates are per 1,000 population; fertility rate is per woman.

Country	Birth rate	Death rate	Life expectancy	Percent urban	Fertility rate
Afghanistan	50	22	43	20	6.9
Argentina	21	8	72	88	2.6
Australia	15	7	78	85	1.9
Canada	14	7	78	77	1.6
Egypt	29	8	64	45	3.4
El Salvador	30	6	68	45	3.1
France	13	9	78	73	1.6
Israel	21	7	77	91	2.8
Japan	10	7	79	78	1.5
Laos	45	15	51	22	6.7
United States	16	9	76	76	2.0

SOURCE: *The New York Times Almanac.*

IN THE NEWS

1. **PREDICTIONS IN THE NEWS.** Find a recent news report in which a correlation is used to make a prediction. Evaluate the validity of the prediction, considering all of the cautions described in this section. Overall, do you think the prediction is valid? Why or why not?

2. **BEST-FIT LINE IN THE NEWS.** Although scatter diagrams are rare in the news, they are not unheard of. Find a scatter diagram of any kind in a news article (recent or not). Draw a best-fit line by eye. Discuss what predictions, if any, can be made from your best-fit line.

3. **YOUR OWN MULTIPLE REGRESSION.** Come up with an example from your own life or work in which a multiple regression analysis might reveal important trends. Without actually doing any analysis, describe in words what you would look for through the multiple regression and how the answers you find might be useful.

7.4 The Search for Causality

A correlation may suggest causality, but by itself a correlation *never* establishes causality. Much more evidence is required to establish that one factor *causes* another. Earlier, we found that a correlation between two variables may be the result of either (1) coincidence, (2) a common underlying cause, or (3) one variable actually having a direct influence on the other. Thus, the process of establishing causality is essentially a process of ruling out the first two explanations.

In principle, we can rule out the first two explanations by conducting experiments:

- We can rule out coincidence by repeating the experiment many times or using a large number of subjects in the experiment. Because coincidences occur randomly, they should not occur consistently in many subjects or experiments. Thus, while a coincidence may confuse the situation in a few trials of an experiment, it is unlikely to confuse the experiment after many trials.

- We can rule out a common underlying cause by controlling the experiment to eliminate the effects of confounding variables (see Section 1.3). In this way, *only* the variables of interest vary between the treatment and control groups. If the controls rule out confounding variables, then any remaining effects must be caused by the variables being studied.

Unfortunately, these ideas are often difficult to put into practice. In the case of ruling out coincidence, it may be too time-consuming or expensive to repeat an experiment a sufficient number of times. To rule out a common underlying cause, the experiment must control for *everything* except the variables of interest, and this is often impossible. Moreover, there are many cases in which experiments are impractical or unethical, so we can gather only observational data. Because observational studies cannot definitively establish causality, we must find other ways of trying to establish causality.

Establishing Causality

Suppose you have discovered a correlation and suspect causality. How can you test your suspicion? Let's return to the issue of smoking and lung cancer. The strong correlation between smoking and lung cancer did not by itself prove that smoking causes lung cancer. In principle, we could have looked for proof with a controlled experiment. But such an experiment would be unethical, since it would require forcing a group of randomly selected people to smoke cigarettes. So how was smoking established as a cause of lung cancer?

The answer involves several lines of evidence. First, researchers found correlations between smoking and lung cancer among many groups of people: women, men, and people of different races and cultures. Second, among groups of people that seemed otherwise identical, lung cancer was found to be rarer in nonsmokers. Third, people who smoked more and for longer periods of time were found to have higher rates of lung cancer. Fourth, when researchers accounted for other potential causes of lung cancer (such as exposure to radon gas or asbestos), they found that almost all the remaining lung cancer cases occurred among smokers.

These four lines of evidence made a strong case, but still did not rule out the possibility that some other factor, such as genetics, predisposes people both to smoking and to lung cancer. But two additional lines of evidence made this possibility highly unlikely. First, animal experiments with randomly chosen treatment and control groups also showed a correlation between inhalation of cigarette smoke and lung cancer, which seems to rule out a genetic factor,

The truth is rarely pure and never simple.
—Oscar Wilde

at least in the animals. Second, biologists studying cell cultures (that is, small samples of human lung tissue) discovered the basic process by which ingredients in cigarette smoke can create cancer-causing mutations. This process does not appear to depend in any way on specific genetic factors, making it all but certain that lung cancer is caused by smoking and not by any preexisting genetic factor.

The following box summarizes these ideas about establishing causality. Generally speaking, the case for causality is stronger when more of these guidelines are met.

By the Way...

The first four methods to the right are called *Mill's methods* after John Stuart Mill (1806–1873). Mill was a leading scholar of his time and an early advocate of women's right to vote. In philosophy, the four methods are called, respectively, the methods of agreement, difference, concomitant variation, and residues.

Guidelines for Establishing Causality

If you suspect that a particular variable (the suspected cause) is causing some effect:

1. Look for situations in which the effect is correlated with the suspected cause even while other factors vary.

2. Among groups that differ only in the presence or absence of the suspected cause, check that the effect is similarly present or absent.

3. Look for evidence that larger amounts of the suspected cause produce larger amounts of the effect.

4. If the effect might be produced by other potential causes (besides your suspected cause), make sure that the effect still remains after accounting for these other potential causes.

5. If possible, test the suspected cause with an experiment. If the experiment cannot be performed with humans for ethical reasons, consider doing the experiment with animals, cell cultures, or computer models.

6. Try to determine the physical mechanism by which the suspected cause produces the effect.

Time out to think

There's a great deal of controversy concerning whether animal experiments are ethical. What is your opinion of animal experiments? Defend your opinion.

CASE STUDY Air Bags and Children

By the mid-1990s, passenger-side air bags had become commonplace in cars. Statistical studies showed that the air bags saved many lives in moderate- to high-speed collisions. But a disturbing pattern also appeared. In at least some cases, young children, especially infants and toddlers in child car seats, were killed by air bags in low-speed collisions.

At first, many safety advocates found it difficult to believe that air bags could be the cause of the deaths. But the observational evidence became stronger, meeting the first four guidelines for establishing causality. For example, the greater risk to infants in child car seats fit Guideline 3, because it indicated that being closer to the air bags increased the risk of death. (A child car seat sits on top of the built-in seat, thereby putting a child closer to the air bags than the child would be otherwise.)

To seal the case, safety experts undertook experiments using dummies. They found that children, because of their small size, often sit where they could be easily hurt by the explosive opening of an air bag. The experiments also showed that an air bag could impact a child car seat hard enough to cause death, thereby revealing the physical mechanism by which the deaths occurred.

With the physical mechanism understood, it was possible to develop strategies to prevent deaths. First, it was immediately obvious that the best prevention would be to keep children away from the air bags. This led to new guidelines telling parents that child car seats should *never* be used on the front seat, and that children under age 12 should sit in the back seat if possible. Second, the findings led to additional experiments suggesting that air bags could provide the same safety with less explosive openings. Air bags that open less forcefully are now required in all new cars. (There is still controversy over whether such air bags compromise safety for adults.) Third, the fact that the risk for children turned out to be related to their small size suggested that small adults would also be at risk. This led to recommendations that small adult drivers (such as short women) sit as far back as possible from the air bags and that they use only cars with the newer air bags that open less forcefully. ■

By the Way…

Many states and countries have laws concerning where children sit in cars. In Belgium, for example, it is illegal for a child under age 12 to sit in the front seat if the back seat is open.

CASE STUDY   ## What Causes Global Warming?

Statistical measurements suggest that the average surface temperature of the entire Earth, or the *global average temperature,* has risen slightly (by about 1°F) in the past 50 years. But what is causing this so-called *global warming?*

Most scientists attribute the rise to the increasing concentration of carbon dioxide and other greenhouse gases in our atmosphere (see "Focus on Environment" in Chapter 3). Comparative studies of the Earth and other planets, particularly Venus and Mars, show that greenhouse gases meet the first three guidelines for establishing causality. Indeed, there is virtually no doubt that the average temperature of each planet is strongly influenced by its atmospheric concentration of greenhouse gases. The case for greenhouse gases also meets Guideline 6, as scientists have clearly established the physical mechanism by which these gases can cause global warming. (Basically, greenhouse gases slow the escape of heat from a planet's surface, thereby causing the surface to warm up.) Thus, as the suspected cause of global warming, greenhouse gases meet at least four of the six guidelines for establishing causality.

Nevertheless, we cannot be certain that the observed global warming of the past 50 years has been caused by greenhouse gases. The trouble comes mainly from Guideline 4, which tells us to account for other potential causes. One such potential cause is an increase in heating of the Earth by the Sun. Some evidence suggests that the Sun has become slightly warmer during the past century, thereby increasing the amount of heat the Earth receives by a few hundredths of 1%. If true, this might account for some or all of the observed global warming.

The best way to sort out the causes would be with an experiment (Guideline 5), but it is impossible to perform experiments with entire planets. Instead, researchers run experiments with *computer models* of planets that account for the way weather and climate work. In this way, researchers can compare results among models with varying amounts of heat from the Sun and varying amounts of greenhouse gases. Unfortunately, the models made by different research groups tend to give different results. Unless everyone can agree on what constitutes a valid computer model, at least some controversy about the cause of global warming is likely to persist. ■

By the Way…

Some greenhouse gases are present naturally in the Earth's atmosphere. Without them, the global average temperature would be a frigid −10°F; with them, the global average temperature is about 59°F. Thus, the greenhouse effect is a good thing for life on Earth. On Venus, where the temperature would be pleasantly warm with an Earth-like atmosphere, a very strong greenhouse effect produces a global average temperature of about 870°F—proving that it's possible to have too much of a good thing.

CASE STUDY   Ozone and CFCs

While some people still dispute the human connection to global warming, evidence is much more definitive for another environmental issue: the depletion of the Earth's atmospheric ozone. Ozone is a gas that exists naturally at high altitudes, where it absorbs harmful ultraviolet radiation from the Sun. Ozone depletion (a reduction in the total amount of ozone) could pose a serious problem by allowing more of this ultraviolet radiation to reach the ground, where it can cause mutations, cancers, and deaths of both plants and animals.

By the early 1970s, experiments had shown that human-made chemicals called *chlorofluorocarbons* (CFCs) could destroy ozone. CFCs were invented in the 1930s and have been widely used in air conditioners and refrigerators, in the manufacture of packaging foam, as propellants in spray cans, in industrial solvents used in the computer industry, and for many other purposes. Because of their many everyday uses, it seemed unwise to ban CFCs on the basis of a few experiments, especially since there was no evidence at the time that any ozone depletion was actually occurring.

The situation changed in the mid-1980s with the discovery of the so-called *ozone hole* over Antarctica (Figure 7.27). Researchers found that the formation of this hole (generally lasting from August through November) correlated strongly with the presence of chlorine and fluorine—both components of CFCs—in the atmosphere. They soon identified the physical mechanism that led to the fairly extreme ozone depletion over Antarctica compared to other places. (The mechanism involves chemical reactions that occur on the surfaces of ice crystals that form in the Antarctic clouds.)

By the Way…

Paul Crutzen, Mario Molina, and F. Sherwood Rowland received the 1995 Nobel Prize in chemistry for their work in discovering the linkage between CFCs and ozone destruction.

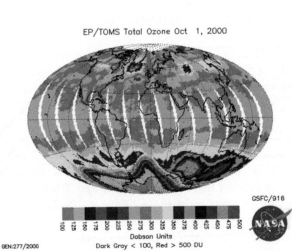

Figure 7.27   This map shows ozone distribution on October 1, 2000, as measured from a satellite. The key explains how colors represent ozone concentrations. The regions of gray, pink, and purple around Antarctica represent the ozone hole, where ozone concentrations are far below normal. The white strips on the map are regions where the satellite did not collect data. SOURCE: NASA/NOAA Earth Probe TOMS data.

With CFCs confirmed as a cause of damage to the Earth's ozone, negotiations soon began on international treaties that now ban the manufacture of CFCs. Unfortunately, these treaties have not by themselves put the problem of ozone depletion behind us. Most of the CFCs made during the 20th century are either still in use or in discarded appliances and materials in landfills, and probably will leak into the atmosphere over the next few decades. In addition, because substitute chemicals tend to be more expensive than CFCs, there is now a large international black market for CFCs manufactured in violation of the international treaties. ∎

Time out to think

Suppose your car air conditioner is broken and will cost $200 to fix. You hear about a shop that can fix it for only $100—undoubtedly by using black market CFCs. Given that your single air conditioner won't make much difference to the ozone layer, is it worth the $100 difference to avoid the black market CFCs? Defend your opinion.

Hidden Causality

So far we have discussed how to establish causality after first discovering a correlation. Sometimes, however, correlations—or the lack of a correlation—can hide an underlying causality. As the next case study shows, such hidden causality often occurs because of confounding variables.

CASE STUDY   Cardiac Bypass Surgery

Cardiac bypass surgery is performed on people who have severe blockage of arteries that supply the heart with blood (the coronary arteries). If blood flow stops in these arteries, a patient may suffer a heart attack and die. Bypass surgery essentially involves grafting new blood vessels onto the blocked arteries so that blood can flow around the blocked areas. By the mid-1980s, many doctors were convinced that the surgery was prolonging the lives of their patients.

However, a few early case-control studies turned up a disconcerting result: Statistically, the surgery appeared to be making little difference. In other words, patients who had the surgery seemed to be faring no better on average than similar patients who did not have it. If this were true, it meant that the surgery was not worth the pain, risk, and expense involved.

Because these results flew in the face of what many doctors thought they had observed in their own patients, researchers began to dig more deeply. Soon, they found confounding variables that had not been accounted for in the early studies. For example, they found that patients getting the surgery tended to have more severe blockage of their arteries, apparently because doctors recommended the surgery more strongly to these patients. Because these patients were in worse shape to begin with, a comparison of longevity between them and other patients was not really valid.

More important, the research soon turned up substantial differences in the results among patients who had the surgery in different hospitals. In particular, a few hospitals were achieving remarkable success with bypass surgery and their patients fared far better than patients who did not have the surgery or had it at other hospitals. Clearly, the surgical techniques used by doctors at the successful hospitals were somehow different and superior. Doctors studied the differences to ensure that all doctors could be trained in the superior techniques.

In summary, the confounding variables of *amount of blockage* and *surgical technique* had prevented the early studies from finding a real correlation between cardiac bypass surgery and prolonged life. Today, cardiac bypass surgery is accepted as a *cause* of prolonged life in patients with blocked coronary arteries. It is now among the most common types of surgery, and it typically adds *decades* to the lives of the patients who undergo it. ∎

Confidence in Causality

Consider the case study concerning the cause of global warming. Substantial global warming (say, a few degrees Fahrenheit) would almost certainly cause consequences such as a rise in sea level, an increase in the frequency of severe storms, changes in weather patterns that

would reduce agricultural production on many existing farms, and extinctions of plant and animal species that cannot adapt to the changing climate. Clearly, if human activity is *causing* global warming, we'd be wise to change our activities so as to stop it. However, the causality has not been clearly established. Moreover, the required changes would probably cost trillions of dollars, since they would involve reducing our dependence on fossil fuels (oil, coal, and natural gas). Given the huge cost and the uncertainty as to whether the changes would even make a difference, how can we set sensible policies?

In an ideal world, we would continue to study the issue until we could establish whether human activity is the cause of global warming. However, we have seen that it is difficult to establish causality and often impossible to *prove* causality beyond all doubt. We are therefore forced to make decisions about global warming, and many other important issues, despite great uncertainty about cause and effect.

In other areas of statistics, accepted techniques help us deal with uncertainty by allowing us to calculate a numerical level of confidence or significance. But there are no accepted ways to assign such numbers to the uncertainty that comes with questions of causality. Fortunately, another area of study has dealt with practical problems of causality for hundreds of years: our legal system. You may be familiar with the following three broad ways of expressing a legal level of confidence.

Broad Levels of Confidence in Causality

Possible cause: We have discovered a correlation, but cannot yet determine whether the correlation implies causality. In the legal system, possible cause (such as thinking that a particular suspect possibly caused a particular crime) is often the reason for starting an investigation.

Probable cause: We have good reason to suspect that the correlation involves cause, perhaps because some of the guidelines for establishing causality are satisfied. In the legal system, probable cause is the general standard for getting a judge to grant a warrant for a search or wiretap.

Cause beyond reasonable doubt: We have found a physical model that is so successful in explaining how one thing causes another that it seems unreasonable to doubt the causality. In the legal system, cause beyond reasonable doubt is the usual standard for convictions and generally demands that the prosecution have shown how and why (essentially the physical model) the suspect committed the crime. Note that beyond *reasonable* doubt does *not* mean beyond *all* doubt.

While these broad levels remain fairly vague, they give us at least some common language for discussing confidence in causality. If you study law, you will learn much more about the subtleties of interpreting these terms. However, because statistics has little to say about them, we will not discuss them much further in this book.

Time out to think

Given what you know about global warming, do you think that human activity is a possible cause, probable cause, or cause beyond reasonable doubt? Defend your opinion. Based on your level of confidence in the causality, how would you recommend setting policies with regard to global warming?

Review Questions

1. How can an experiment prove causality? Why can't an observational study prove causality?

2. Briefly describe each of the six guidelines presented in this section for establishing causality. Give an example of the application of each guideline.

3. Briefly explain how confounding variables can create situations in which an underlying causality is hidden.

4. Briefly describe three levels of confidence in causality and how they can be useful when we do not have absolute proof of causality.

Exercises

BASIC SKILLS AND CONCEPTS

SENSIBLE STATEMENTS? For Exercises 1–4, determine whether the given statement is sensible and explain why it is or is not.

1. Although a correlation by itself generally does not establish causality, if a sample of paired data from two variables yields a correlation coefficient of 1, then we can conclude that one of the variables has a direct causal link with the other variable.

2. If there is no correlation between head circumference and IQ, then head circumference cannot be a single direct cause of IQ.

3. After finding a significant correlation between exposure to tobacco smoke and the cotinine level in the body, we can establish that the exposure causes the cotinine if we can rule out coincidence as a possible explanation.

4. After finding a significant correlation between exposure to tobacco smoke and the cotinine level in the body, we can establish that the exposure causes the cotinine if we can rule out coincidence and common underlying cause as possible explanations.

PHYSICAL MODELS. Exercises 5–8 describe generally accepted causal connections. Suggest a physical model that explains each connection.

5. Running out of gas causes a car to stop.

6. Being unable to breathe for half an hour causes death.

7. Dropping a book causes it to fall.

8. Inflating a balloon and then letting it go (without tying the end) causes it to fly about the room.

FURTHER APPLICATIONS

9. IDENTIFYING CAUSES: HEADACHES. You are trying to identify the cause of late-afternoon headaches that plague you several days each week. For each of the following tests and observations, explain which of the six guidelines for establishing causality you used and what you concluded. Then summarize your overall conclusion based on all the observations.

   • The headaches occur only on days that you go to work.
   • If you stop drinking Coke at lunch, the headaches persist.
   • In the summer, the headaches occur less frequently if you open the windows of your office slightly. They occur even less often if you open the windows of your office fully.

10. SMOKING AND LUNG CANCER. There is a strong correlation between tobacco smoking and incidence of lung cancer, and most physicians believe that tobacco smoking causes lung cancer. Yet, not everyone who smokes gets lung cancer. Briefly describe how smoking could cause cancer when not all smokers get cancer.

11. OTHER LUNG CANCER CAUSES. Several things besides smoking have been shown to be probabilistic causal factors in lung cancer. For example, exposure to asbestos and exposure to radon gas, both of which are found in many homes, can cause lung cancer. Suppose that you meet a person who lives in a home that has a high radon level and insulation that contains asbestos. The person tells you, "I smoke, too, because I figure I'm doomed to lung cancer anyway." What would you say in response? Explain.

12. **LONGEVITY OF ORCHESTRA CONDUCTORS.** A famous study in *Forum on Medicine* concluded that the mean lifetime of conductors of major orchestras was 73.4 years, about 5 years longer than that of all American males at the time. The author claimed that a life of music *causes* a longer life. Evaluate the claim of causality and propose other explanations for the longer life expectancy of conductors.

13. **OLDER MOMS.** A study reported in *Nature* claims that women who give birth later in life tend to live longer. Of the 78 women who were at least 100 years old at the time of the study, 19% had given birth after their 40th birthday. Of the 54 women who were 73 years old at the time of the study, only 5.5% had given birth after their 40th birthday. A researcher stated that "if your reproductive system is aging slowly enough that you can have a child in your 40s, it probably bodes well for the fact that the rest of you is aging slowly too." Was this an observational study or an experiment? Does the study suggest that later child bearing *causes* longer lifetimes or does later child bearing reflect an underlying cause? Comment on how persuasive you find the conclusions of the report.

14. **HIGH-VOLTAGE POWER LINES.** Suppose that people living near a particular high-voltage power line have a higher incidence of cancer than people living farther from the power line. Can you conclude that the high-voltage power line is the cause of the elevated cancer rate? If not, what other explanations might there be for it? What other types of research would you like to see before you conclude that high-voltage power lines cause cancer?

15. **GUN CONTROL.** Those who favor gun control often point to a correlation between the availability of handguns and murder rates to support their position that gun control would save lives. Does this correlation, by itself, indicate that handgun availability causes a higher murder rate? Suggest some other factors that might support or weaken this conclusion.

16. **VASECTOMIES AND PROSTATE CANCER.** An article entitled "Does Vasectomy Cause Prostate Cancer?" (*Chance*, Vol. 10, No. 1) reports on several large studies that found an increased risk of prostate cancer among men with vasectomies. In the absence of a direct cause, several researchers attribute the correlation to *detection bias*, in which men with vasectomies are more likely to visit the doctor and thereby more likely to have any prostate cancer found by the doctor. Briefly explain how this detection bias could affect the claim that vasectomies cause prostate cancer.

 # PROJECTS FOR THE WEB AND BEYOND

For useful links, select "Links for Web Projects" for Chapter 7 at www.aw.com/bbt.

17. **AIR BAGS AND CHILDREN.** Starting from the Web site of the National Highway Traffic Safety Administration, research the latest studies on the safety of air bags, especially with regard to children. Write a short report summarizing your findings and offering recommendations for improving child safety in cars.

18. **GLOBAL WARMING.** Find recent information about global warming and its potential consequences. Discuss the evidence linking human activity to global warming. In light of your findings, suggest how we should deal with the issue of global warming.

19. **OZONE DEPLETION.** Research the latest information about ozone depletion and efforts to combat it. Summarize your findings and suggest what should be done in the future to help prevent ozone depletion.

20. **DIETARY FIBER AND CORONARY HEART DISEASE.** In the largest study of how dietary fiber prevents coronary heart disease (CHD) in women (*Journal of the American Medical Association*, Vol. 281, No. 21), researchers detected a reduced risk of CHD among women who have a high-fiber diet. Find the research paper, summarize its findings, and discuss whether a cause for the correlation is proposed.

21. **THERAPEUTIC PRAYER?** The question of whether prayer has therapeutic power has been of interest at least since 1872, when the biologist and statistician Francis Galton declared it a "subject of legitimate scientific inquiry." The 1988 study by R. C. Byrd (reprinted in *Alternative Therapies in Health and Medicine*) is one of the more recent to claim that there is a positive therapeutic effect of prayer on the recovery of patients. Not surprisingly, other studies reached the opposite conclusion. An article by I. Tessman and J. Tessman (*Skeptical Inquirer*, March/April 2000) cites statistical flaws in the Byrd study. Find both of these articles and any other relevant sources. Summarize the positions of both sides of the debate and give a conclusion based on your reading.

22. **COFFEE AND GALLSTONES.** Writing in the *Journal of the American Medical Association* (Vol. 281, No. 22), researchers reported finding a negative correlation between incidence of gallstone disease and coffee consumption in men. Find the research paper, summarize its findings, and discuss whether a cause for the correlation is proposed.

23. ALCOHOL AND STROKE. Researchers reported in the *Journal of the American Medical Association* (Vol. 281, No. 1) that moderate alcohol consumption is correlated with a decreased risk of stroke in people 40 years of age and older. (Heavy consumption of alcohol was correlated with deleterious effects.) Find the research paper, summarize its findings, and discuss whether a cause for the correlation is proposed.

24. TOBACCO LAWSUITS. Tobacco companies have been the subject of many lawsuits related to the dangers of smoking. Research one recent lawsuit. What were the plaintiffs trying to prove? What statistical evidence did they use? How well do you think they established causality? Did they win? Summarize your findings in one to two pages.

IN THE NEWS

1. CAUSATION IN THE NEWS. Find a recent news report in which a statistical study has led to a conclusion of causation. Describe the study and the claimed causation. Do you think the claim of causation is legitimate? Explain.

2. LEGAL CAUSATION. Find a news report concerning an ongoing legal case, either civil or criminal, in which establishing causality is important to the outcome. Briefly describe the issue of causation in the case and how the ability to establish or refute causality will influence the outcome of the case.

Chapter Review Exercises

In Exercises 1–4, refer to the data in the table. The data, provided by geologist Rick Hutchinson and the National Park Service, were obtained from the Old Faithful geyser in Yellowstone National Park. The duration values are times (in seconds) of an eruption, the interval values are times (in minutes) to the next eruption, and the height values are heights (in feet) of the eruptions.

Duration	Interval	Height
240	86	140
237	86	154
122	62	140
267	104	140
113	62	160
258	95	140
232	79	150
105	62	150
276	94	160
248	79	155
243	86	125
241	85	136
214	86	140
114	58	155
272	89	130

1. After each eruption, rangers at Yellowstone National Park provide an estimate of the "interval" value, which is the time before the next eruption of Old Faithful. Construct a scatter diagram that uses the given values of interval and duration. Based on the result, does there appear to be a correlation between duration and interval?

2. Construct a scatter diagram that uses the given values of interval and height. Based on the result, does there appear to be a correlation between interval and height?

3. The computed value of the correlation coefficient is 0.933 for the interval/duration data. Refer to Table 7.3 and interpret that result. What do you conclude about the signifi-

cance of the correlation between interval and duration? What percent of the variation in interval can be explained by the variation in duration?

4. The computed value of the correlation coefficient is -0.382 for the interval/height data. Refer to Table 7.3 and interpret that result. What do you conclude about the significance of the correlation between interval and height? What percent of the variation in interval can be explained by the variation in height?

5. Based on a study in Sweden, several newspapers reported that "living near power lines causes leukemia in children." What data are likely to be the basis for such a claim? What is fundamentally wrong with the claim that proximity to power lines causes leukemia?

6. For 10 pairs of sample data, the correlation coefficient is computed to be $r = -1$. What do you know about the scatter diagram?

7. In a study of randomly selected subjects, it is found that there is a significant correlation between household income and number of visits to dentists. Is it valid to conclude that higher incomes cause people to visit dentists more often? Is it valid to conclude that more visits to dentists cause people to have higher incomes? How might the significant correlation be explained?

8. You are considering the most expensive purchase that you are likely to make: the purchase of a home. Identify at least five different variables that are likely to affect the actual value of a home. Among the variables that you have identified, which single variable is likely to have the greatest influence on the value of the home? Identify a variable that is likely to have little or no effect on the value of a home.

9. A researcher collects paired sample data and computes the value of the linear correlation coefficient, with a result of 0. Based on that value, he concludes that there is no relationship between the two variables. What is wrong with this conclusion?

FOCUS ON EDUCATION

What Helps Children Read?

Everyone has an idea about how best to teach reading to children. Some advocate a phonetic approach, teaching students to "sound out" words. Some advocate a "whole language" approach, teaching students to recognize words from their context. Others advocate a combination of these approaches—or something else entirely. These differing ideas would be unimportant if they were merely opinions. But in a nation that spends roughly a *trillion dollars* per year on education, differing approaches to teaching reading involve major political confrontations among groups with different special interests. A change in politics can cause a sudden change in school policies. For example, in 1998, the California legislature passed laws making public school funding contingent upon the school's moving away from a whole language approach to reading.

The huge stakes involved in teaching reading demand statistics to measure the effectiveness of various approaches. Politically, at least, the most important educational statistics are those that come from the National Assessment of Educational Progress (NAEP), often known more simply as "the Nation's Report Card." The NAEP is an ongoing survey of student achievement conducted by a government agency, the National Center for Education Statistics, with authorization and funding from the U.S. Congress.

The NAEP uses stratified random sampling (see Chapter 1) to choose representative samples of 4th-, 8th-, and 12th-grade students of varying ethnicity, family income, type of school attended, and so on. Students chosen for the samples are given tests designed to measure their academic achievement in a particular subject area, such as reading, mathematics, or history. Samples are chosen on both state and national levels. Overall, a few thousand students are chosen for each test. Results from NAEP tests inevitably make the newspaper, with articles touting improvements or decrying drops in test scores. They also have political impact. For example, California's move away from whole language occurred after it ranked 45th among the 50 states in reading on the NAEP tests.

But what really causes improvement in reading performance? Researchers begin by searching for correlations between reading performance and other factors. Sometimes the correlations are clear, but offer no direction for improving reading. For example, parental education is clearly correlated with reading achievement: Children with more highly educated parents tend to read more proficiently than those with uneducated parents. But this correlation doesn't offer much guidance for the schools, since children cannot replace their parents. Other times the correlations may suggest ways to improve reading. For example, students who report reading more pages daily in school and for homework tend to score higher than students who read fewer pages. This suggests that schools should assign more reading.

Of course, the high stakes involved in education make education statistics particularly prone to misinterpretation or misuse. Consider just a few of the problems that make the NAEP reading tests difficult to interpret:

- They are "standardized tests" that are the same for all students tested and tend to be mostly multiple choice. Some people believe that such tests are inevitably biased and cannot truly measure reading ability.

- Because the tests generally don't affect students' grades, some students may not take the tests seriously, in which case test results may not reflect actual reading ability. For example, some test administrators have reported students making designs with the multiple-choice "bubbles" on their tests, rather than trying to answer the questions to the best of their ability.

- State-by-state comparisons may not be valid if the makeup of the student population varies significantly among states. For example, California has a relatively large percentage of students for whom English is a second language. Some people believe that this explained the state's low test scores, rather than anything to do with teaching techniques.

- There is some evidence of cheating on the part of the *adults* involved in the NAEP tests by, for example, choosing samples that are not truly representative but instead skewed toward students who read better. This cheating may be motivated by the fact that individual schools, school districts, and states are ranked according to NAEP results. High scores can lead to rewards for teachers and administrators in the form of increased funding or higher salaries, while low scores may lead to various punitive actions.

You can probably think of a dozen other problems that make it difficult to interpret NAEP results. Thus, it should not be surprising that reading continues to be a huge political battleground. So what can you do, as an individual, to help a child to read? Fortunately, the NAEP studies also reveal a few correlations that are fairly uncontroversial and agree with common sense. For example, higher reading performance correlates with each of the following factors:

- more total reading, both for school and for pleasure

- more choice in reading—that is, allowing children to pick their own books to read

- more writing, particularly of extended pieces such as essays or long letters

- more discussion of reading material with friends and family

- watching less television

These correlations give at least some guidance on how to help a child learn to read and should be good starting points for discussions of how to increase literacy. Of course, politicians and special interest groups will probably find ways to make these results fit whatever preconceived agenda they might have. So, if you have strong opinions about teaching techniques, you can join the political battles that will probably continue for decades to come.

QUESTIONS **FOR DISCUSSION**

1. One clear result of the NAEP reading tests is that students in private schools tend to score significantly higher than students in public schools. Does this imply that private schools are "better" than public schools? Defend your opinion.

2. Do you think that standardized tests like those in the NAEP are valid ways to measure academic achievement? Why or why not?

3. Currently, the NAEP tests are given to only a few thousand of the millions of school children in the United States. Some people advocate giving similar tests to all students, on

either a voluntary or a mandatory basis. Do you think such "standardized national testing" is a good idea? Why or why not?

4. One correlation that has not yet been studied carefully is the correlation between computer use and reading. Do you think that using a computer and the Internet helps or hurts children in terms of learning to read? Why?

SUGGESTED READING

Lemann, Nicholas. "The Reading Wars." *The Atlantic Monthly,* November, 1997.

The NAEP 1998 Reading Report Card for the Nations and States. U.S. Department of Education, March, 1999.

FOCUS ON PUBLIC HEALTH

What Do Disease Clusters Mean?

In the movie *A Civil Action* (based on the book of the same title by Jonathan Harr), John Travolta plays a lawyer suing on behalf of families of children with leukemia in Woburn, Massachusetts. These children were contracting leukemia at a rate well above the national average. The lawyer played by Travolta helped to discover that local companies had illegally dumped chemicals that reached the town's water supply. Some of these chemicals are known to cause cancer in laboratory animals. Putting these facts together seems to create quite a strong case for a link between the chemical dumping and the Woburn leukemia cases.

But consider an opposing view. The high rate of leukemia in Woburn represents what statisticians call a *disease cluster:* a place where a disease occurs at a rate significantly above the national average. Just as we don't expect all people to be of average height, we shouldn't expect all towns and groups of people to have the average rate of any disease. By random chance, some places should have much higher rates of any particular disease than other places. Thus, the mere existence of a disease cluster in Woburn does not prove that anything unusual caused the disease.

Now consider the chemicals in the water supply. Because the cases of leukemia that led to the lawsuits occurred years before the water was carefully sampled, we have no way of knowing the chemical concentrations that were present at the time. However, we know that many toxic chemicals are found in at least trace amounts nearly everywhere in the modern world. Thus, possible chemical causes of leukemia could be found anywhere. Perhaps the only reason researchers focused on these chemicals in Woburn was because of the disease cluster, which we've already said might be nothing more than a random coincidence. Furthermore, the chemicals are proven only to cause cancer in animals, not necessarily in humans. These new considerations clearly weaken the causal link between chemical exposure and leukemia.

This case points out the statistical difficulties posed by disease clusters. The discovery of a disease cluster leads to an almost immediate search for significant correlations between possible causes and the disease. If you look hard enough, you are almost certain to find *some* correlation, but it is much more difficult to establish cause. In fact, cases in which a variable correlated with a disease cluster has been established as a cause are rare. Some of the most famous examples include establishing radiation exposure as the cause of clusters of cancer and birth defects among survivors of the atomic bombs in Japan, establishing inhalation of coal dust as the cause of clusters of black lung disease among coal miners, and establishing asbestos inhalation as the cause of clusters of lung cancer among workers in the asbestos industry.

Moreover, in many cases of disease clusters, correlations that initially appear strong unravel upon further examination. In a famous case in Sweden, researchers found a correlation between child leukemia and how close the children lived to high-voltage power lines. This spurred several other studies that found similar correlations between high-voltage power lines and clusters of cancer. However, a recent meta-analysis seems to show that, on a larger scale, no such correlation exists. If this result is correct, then the multimillion-dollar search for a linkage between high-voltage power lines and cancer has been nothing more than a wild goose chase.

Ultimately, the key to thinking about disease clusters is to remember that we need much more than correlation alone before we can confidently claim causality. Returning to the case of the Woburn leukemia, there is at least some additional evidence of causality. For example, the children who developed leukemia were found to have mothers who drank water from particular wells in the city, and these wells were more prone to chemical contamination than others. Is this additional evidence enough to establish causality? The lawyers on both sides thought so, and the companies accused of the dumping agreed to pay millions of dollars to the affected families and to finance an expensive cleanup of the water supply.

QUESTIONS FOR DISCUSSION

1. Critics of lawsuits related to disease clusters complain that the legal standard for causality is much lower than the scientific standard. As a result, they say, companies often settle lawsuits rather than fight them in court, even when science says the companies did nothing wrong. Do you agree? Defend your opinion.

2. Although current evidence suggests that there is no linkage between living near high-voltage power lines and cancer, real estate values for houses near these power lines remain significantly lower than the prices of similar homes just a few blocks away. Would you consider such homes a bargain? Why or why not?

3. Overall, do you think it makes sense for researchers to focus their attention on disease clusters? If so, how much effort should be expended, both in time and in dollars, before giving up a particular hunt for causality? If not, why not?

SUGGESTED READING

Braus, Patricia. "Why Does Cancer Cluster?" *American Demographics,* March 1996.

Johnson, Kevin V. "Cancer Clusters Are Difficult to Nail Down." *USA Today,* April 13, 1999.

Kolata, Gina. "Probing Disease Clusters: Easier to Spot than Prove." *New York Times,* January 31, 1999.

Seligman, Dan. "Defeated by the Odds." *Forbes Magazine,* November 30, 1998.

CHAPTER 8

Some people hate the very name statistics, but I find them full of beauty and interest. Whenever they are not brutalized, but delicately handled by the higher methods, and are warily interpreted, their power of dealing with complicated phenomena is extraordinary.

—Sir Francis Galton (1822–1912)

From Samples to Populations

Did you ever wonder how the outcome of a national election can be accurately predicted hours before the polls close? Or how a large retailer can make critical marketing decisions based on a survey of only a few hundred people? These examples illustrate perhaps the most powerful aspect of statistics: the capability to use information gathered from a small sample to make predictions about a much larger population. This process, called *making inferences,* is the subject of this chapter.

LEARNING GOALS

1. Understand the fundamental ideas of a sampling distribution and know how distributions of sample means and proportions are formed.

2. Estimate population means and compute the associated margins of error and confidence intervals.

3. Estimate population proportions and compute the associated margins of error and confidence intervals.

325

8.1 Sampling Distributions

Consider the following statements taken from recent news articles or research reports.

- The mean daily protein consumption by Americans is 67 grams.

- Nationwide, the mean hospital stay after delivery of a baby decreased from 3.2 days in 1980 to the current mean of 2.0 days.

- Thirty percent of high school girls in this country believe they would be happier being married than not being married.

- About 5% of all American children live with a grandparent.

We hear or read statements like these every day. They make a claim about a large population of individuals, and yet it is clear that not everyone in the population could possibly have been surveyed or measured. For example, the third statement is not based on a survey of every high school girl in the country. It is based on a random sample of high school girls. The responses of the girls in the sample were then used to make a claim about the entire population of high school girls. How is it possible to *infer* from a random sample a very general conclusion about the population? These questions go to the heart of the branch of statistics called **inferential statistics**.

Notice that two different types of claims are made in the previous statements. The first two statements give estimates of a *mean* of a quantity—mean protein consumption of 67 grams and mean hospital stays of 3.2 days and 2.0 days. The last two statements say something about a *proportion* of the population—30% of high school girls and 5% of American children. These means and proportions pertain to the entire population, so they are *population parameters* (see Section 1.1). In this chapter, we will discuss methods for estimating population parameters.

Public opinion in this country is everything.

— Abraham Lincoln

Sample Means: The Basic Idea

Much of the work in this and the next chapter involves selecting a sample from a population, analyzing the sample, and then drawing conclusions about the population based on what was learned from the sample. In order to gain some experience with this important process, we will begin with an example.

Table 8.1 lists the weights of the five starting players (Bryant, Fox, Green, Harper, and O'Neal) on a recent Los Angeles Lakers basketball team; we will refer to each player by the first letter of his last name. For the purposes of this example, we regard these five players as *the entire population* (with a mean of 242.4 pounds); in other words, it is the complete population of starters for the Lakers. This is a very small population by statistical standards, but its small size will enable us to look carefully at the sampling process. Samples drawn from this population of 5 players can range in size from $n = 1$ (one player out of the five) to $n = 5$ (all five players).

With a sample size of $n = 1$, there are five different samples that could be selected: Each player is a sample. The mean of each sample of size $n = 1$ is simply the weight of the player in the sample. Figure 8.1 shows a histogram of the means of the five samples; it is called a **distribution of sample means**, because it shows the means of all five samples of size $n = 1$. The distribution of sample means created by this process is an example of a **sampling distribution**. This term simply refers to a distribution of a sample statistic, such as a mean, taken from *all* possible samples of a particular size. Notice that the mean of the five sample means is the mean of the entire population:

$$\frac{215 + 242 + 225 + 215 + 315}{5} = 242.4 \text{ pounds}$$

This demonstrates a general rule: *The mean of a distribution of sample means is the population mean.*

Table 8.1 Weights for the Population of Five Starters on the Los Angeles Lakers	
Player	Weight (pounds)
B (Bryant)	215
F (Fox)	242
G (Green)	225
H (Harper)	215
O (O'Neal)	315

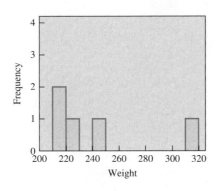

Figure 8.1 Sampling distribution for sample size $n = 1$.

Let's move on to samples of size $n = 2$, in which each sample consists of 2 different players. With 5 players, there are 10 different samples of size $n = 2$. Each sample has its own mean. For example, the sample of players B and F has a mean of $(215 + 242)/2 = 228.5$ pounds. Table 8.2 lists the 10 samples with their means, and Figure 8.2 shows the distribution of these 10 sample means. Again, notice that the mean of the distribution of sample means is equal to the population mean, 242.4 pounds.

Table 8.2 Sample Means for Basketball Example; Sample Size $n = 2$	
Sample	Mean
BF	228.5
BG	220.0
BH	215.0
BO	265.0
FG	233.5
FH	228.5
FO	278.5
GH	220.0
GO	270.0
HO	265.0
Mean	**242.4**

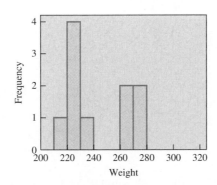

Figure 8.2 Sampling distribution for sample size $n = 2$.

TECHNICAL NOTE

Tables 8.2, 8.3, and 8.4 involve sampling *without replacement*, which has the advantage of avoiding duplication whenever the same item is selected more than once. However, in statistics we are particularly interested in selection *with replacement* for these reasons: In large populations, it makes no significant difference whether we sample with replacement or without replacement, but sampling with replacement results in independent events that are unaffected by previous outcomes; independent events are easier to analyze and they result in simpler formulas.

Ten different samples of size $n = 3$ are also possible in a population of 5 players. Table 8.3 shows these samples and their means, and Figure 8.3 shows the distribution of these sample means. Again, the mean of the distribution of sample means is equal to the population mean, 242.4 pounds.

Table 8.3 Sample Means for Basketball Example; Sample Size $n = 3$	
Sample	Mean
BFG	227.3
BFH	224.0
BFO	257.3
BGH	218.3
BGO	251.7
BHO	248.3
FGH	227.3
FGO	260.7
FHO	257.3
GHO	251.7
Mean	**242.4**

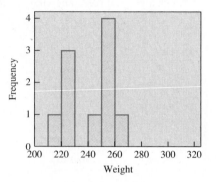

Figure 8.3   Sampling distribution for sample size $n = 3$.

With a sample size of $n = 4$, only 5 different samples are possible. Table 8.4 shows these samples and their means, and Figure 8.4 shows the distribution of these sample means.

Table 8.4 Sample Means for Basketball Example; Sample Size $n = 4$	
Sample	Mean
BFGH	224.25
BFGO	249.25
BFHO	246.75
BGHO	242.50
FGHO	249.25
Mean	**242.4**

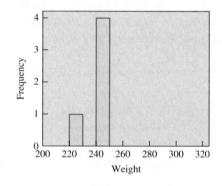

Figure 8.4 Sampling distribution for sample size $n = 4$.

Finally, for a population of 5 players, there is only one possible sample of size $n = 5$: the entire population. In this case, the distribution of sample means is just a single bar (Figure 8.5). Again the mean of the distribution of means is the population mean, 242.4 pounds.

To summarize, when we work with *all* possible samples of a population of a given size, the mean of the distribution of sample means is always the population mean. A closer look at the distributions in Figures 8.1 to 8.5 also reveals that as the sample size increases, the distribution narrows and clusters around the mean. In fact, if we looked at a large population and larger sample sizes, we would find that the distribution of sample means looks more and more

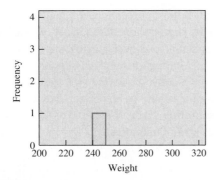

Figure 8.5 Sampling distribution for sample size $n = 5$.

like a normal distribution as the sample size increases. All of these observations are consequences of the Central Limit Theorem; see Section 5.3.

Sample Means with Larger Populations

In reality, we rarely work with populations as small as five individuals. To explore the process of selecting samples and forming distributions of sample means in a more realistic setting, let's consider a somewhat larger population.

Imagine that you work for the computer services department of a small college. In order to develop networking strategies, you survey all 400 students at the college to determine how many hours per week they spend on the Internet. The responses (hours per week) are shown at the bottom of the page.

You could, of course, calculate the mean of all 400 responses. You would find that it is 3.88 hours per week. This mean is the true **population mean**, because it is the mean for the entire population of 400 students; we denote the population mean by the Greek letter μ (pronounced "mew"). Similarly, a calculation shows that the population standard deviation is $\sigma = 2.40$.

3.4	6.8	6.7	3.4	0.0	5.0	5.4	1.8	0.7	1.6	2.1	3.5	3.4	6.4	7.2	1.8	7.4	3.0	4.0	5.2
1.2	7.8	7.0	0.4	7.2	4.8	3.6	8.0	5.4	6.4	3.5	5.3	4.7	5.4	5.6	3.8	0.1	2.4	0.5	4.0
4.5	8.0	4.2	1.0	6.2	7.1	3.8	0.7	5.5	1.7	2.6	1.6	0.7	1.3	6.5	2.4	3.0	0.3	2.2	0.4
1.9	5.0	2.0	5.3	7.5	5.0	0.3	7.4	6.0	4.3	1.3	0.8	7.2	6.6	0.2	3.4	1.6	2.2	3.0	4.5
5.5	5.3	6.5	0.1	0.3	4.2	2.2	6.2	7.3	3.1	5.4	1.3	6.3	4.5	7.1	5.8	6.1	0.5	0.4	4.1
7.0	6.0	1.1	0.8	1.4	2.9	7.3	0.8	2.7	0.6	3.0	0.7	2.8	6.5	1.9	3.6	1.6	2.6	2.6	6.6
6.8	6.1	3.6	1.4	7.7	5.2	3.8	6.0	2.2	7.5	6.7	4.4	4.1	7.3	5.2	5.7	6.7	2.4	0.6	6.7
1.0	2.3	0.7	1.2	4.5	3.3	4.2	2.1	5.9	3.0	7.2	7.9	2.5	7.1	8.0	6.7	4.1	4.9	0.0	3.1
6.0	0.5	4.2	2.7	0.1	1.4	2.1	2.5	3.9	5.8	5.9	2.7	2.8	3.7	7.3	0.7	6.9	4.4	0.7	1.6
3.1	2.1	7.4	3.6	6.5	2.9	5.4	3.9	3.0	0.8	0.3	0.8	3.3	0.8	8.0	5.6	7.1	1.3	0.2	5.2
7.8	4.7	7.2	0.9	5.1	0.9	1.7	1.2	0.4	6.9	0.6	3.0	3.6	6.1	1.6	6.0	3.8	0.4	1.1	4.0
3.8	4.0	1.8	0.9	1.1	3.9	1.7	1.7	2.6	0.1	4.0	1.4	1.9	0.9	0.2	4.2	4.7	0.2	5.3	2.2
5.8	7.5	5.8	5.2	3.9	3.4	7.3	4.1	0.5	7.9	7.7	7.7	5.0	2.3	7.8	2.3	5.6	6.5	7.9	5.0
2.0	5.5	5.4	6.6	6.7	4.4	7.2	2.5	4.9	7.0	2.1	7.2	4.1	1.2	6.2	3.3	6.3	2.3	4.9	2.2
6.4	7.2	0.1	5.3	3.0	0.7	1.5	1.2	1.1	7.4	5.1	7.2	7.2	3.0	7.1	4.5	6.7	7.2	7.2	0.9
2.9	4.3	2.5	0.7	7.6	3.9	0.7	5.8	6.6	3.4	0.3	6.5	7.5	0.7	6.1	6.1	4.8	1.9	1.9	5.0
1.1	7.8	6.8	4.9	3.0	6.5	5.2	2.2	5.1	3.4	4.7	7.0	3.8	5.7	6.8	1.2	1.7	6.5	0.1	4.3
6.3	1.2	0.8	0.7	0.6	7.0	4.0	6.6	6.9	0.5	4.3	1.0	0.5	3.1	0.9	2.3	5.7	6.7	7.3	0.5
0.3	0.9	2.4	2.5	7.8	5.6	3.2	0.7	5.4	0.0	5.7	0.3	7.2	5.1	2.5	3.2	3.1	2.8	5.0	5.6
3.1	0.7	0.5	3.9	2.6	7.3	1.4	1.2	7.1	5.5	3.1	5.0	6.8	6.5	1.7	2.1	7.3	4.0	2.2	5.6

In typical statistical applications, populations are huge and it is impractical or expensive to survey every individual in the population; consequently, we rarely know the true population mean, μ. Therefore, it makes sense to consider using the mean of a *sample* to estimate the mean of the entire population. Although a sample is easier to work with, it cannot possibly represent the entire population exactly. Therefore, we should not expect an estimate of the population mean obtained from a sample to be perfect. The error that we introduce by working with a sample is called **sampling error**. We can explore this idea by considering samples drawn from the 400 responses about the Internet.

Sampling Error

The error that is introduced when a random sample is used to estimate a population parameter is called a **sampling error**. It does not include other sources of error, such as those due to biased sampling, bad survey questions, or recording mistakes.

Time out to think

Would you expect the sampling error to increase or decrease if the sample size were increased? Explain.

Suppose you select a random sample of $n = 32$ responses from the data set on page 329 and calculate their mean. For example, the random sample might be

1.1	7.8	6.8	4.9	3.0	6.5	5.2	2.2	5.1	3.4	4.7
7.0	3.8	5.7	6.5	2.7	2.6	1.4	7.1	5.5	3.1	5.0
6.8	6.5	1.7	2.1	1.2	0.3	0.9	2.4	2.5	7.8	

The mean of this sample is $\bar{x} = 4.17$; we use the standard notation $\bar{x}$ to denote this mean. We say that $\bar{x}$ is a *sample statistic* because it comes from a sample of the entire population. Thus, $\bar{x}$ is called a **sample mean**.

Notation for Population and Sample Means

n = sample size

μ = population mean

$\bar{x}$ = sample mean

Our goal is to use a single sample mean to estimate the population mean, but first it's useful to explore further samples. Suppose you collect additional samples for the purpose of studying the behavior of sample means. Here is one such sample:

1.8	0.4	4.0	2.4	0.8	6.2	0.8	6.6	5.7	7.9	2.5
3.6	5.2	5.7	6.5	1.2	5.4	5.7	7.2	5.1	3.2	3.1
5.0	3.1	0.5	3.9	3.1	5.8	2.9	7.2	0.9	4.0	

For this sample, the sample mean is $\bar{x} = 3.98$.

Now you have two sample means that don't agree with each other, and neither one agrees with the true population mean. Suppose you decide to select many more samples of 32 responses (an option that usually doesn't exist in practice). For each sample, you calculate the sample mean, $\bar{x}$. To get a good picture of all these sample means, you could make a histogram showing the number of samples that have various sample means. Figure 8.6 shows a histogram that results from 100 different samples, each with 32 students. As we expect from the Central Limit Theorem, this histogram is very close to a *normal distribution* and its mean is very close to the population mean, $\mu = 3.88$.

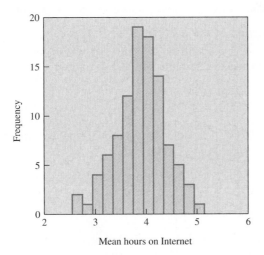

Figure 8.6 A distribution of 100 sample means, with a sample size of $n = 32$, appears close to a normal distribution with a mean of 3.88.

Time out to think

Suppose you choose only one sample of size $n = 32$. According to Figure 8.6, are you more likely to choose a sample with a mean less than 2.5 or a sample with a mean less than 3.5? Explain.

In our earlier example with the population of 5 basketball players, we were able to examine *all* possible samples of each size from $n = 1$ to $n = 5$. We cannot accomplish the same task for a population of 400; it turns out, for example, that the number of possible samples of size $n = 32$ is nearly a trillion trillion trillion trillion (more precisely, it is 2×10^{47}). However, from Figure 8.6, we can already see that the distribution of sample means for $n = 32$ would be nearly normal. Once again, we are seeing an important property of the distribution of sample means: The distribution of sample means approaches a normal distribution for large sample sizes and the mean of the distribution of sample means equals the population mean, μ.

In practice, we often have only *one* sample available. In that case, its sample mean, $\bar{x}$, is our best estimate of the population mean, μ. Fortunately, as we will soon see, it is still possible to say something about how well $\bar{x}$ approximates μ.

The Distribution of Sample Means

The **distribution of sample means** is the distribution that results when we find the means of *all* possible samples of a given size. The larger the sample size, the more closely this distribution approximates a normal distribution. In all cases, the mean of the distribution of sample means equals the population mean. If only one sample is available, its sample mean, $\bar{x}$, is the best estimate for the population mean, μ.

TECHNICAL NOTE

A common guideline is to assume that the distribution of sample means is close to normal if the sample size is greater than 30.

We can extend this idea to see the important role that the normal distribution plays. Consider again the distribution of sample means in Figure 8.6. If we were to include *all* possible samples of size $n = 32$, this distribution would have these characteristics (all found from the Central Limit Theorem in Section 5.3):

- The distribution of sample means is approximately a normal distribution.

- The mean of the distribution of sample means is 3.88 (the mean of the population).

- The standard deviation of the distribution of sample means depends on the population standard deviation and the sample size. The population standard deviation is $\sigma = 2.40$ and the sample size is $n = 32$, so the standard deviation of sample means is

$$\frac{\sigma}{\sqrt{n}} = \frac{2.40}{\sqrt{32}} = 0.42$$

Suppose we select the following random sample of 32 responses from the 400 responses given earlier:

5.8	7.5	5.8	5.2	3.9	3.4	7.3	4.1	0.5	7.9	7.7
7.7	5.0	2.3	7.8	2.3	5.0	6.8	6.5	1.7	2.1	7.3
4.0	2.2	5.6	4.7	5.3	3.5	6.5	3.4	6.6	5.0	

The mean of this sample is $\bar{x} = 5.01$. Given that the mean of the distribution of sample means is 3.88 and the standard deviation is 0.42, the sample mean of $\bar{x} = 5.01$ has a *standard score* of

$$z = \frac{\text{sample mean} - \text{pop. mean}}{\text{standard deviation}} = \frac{5.01 - 3.88}{0.42} = 2.7$$

(Recall that we use z to denote a standard score; see Section 5.2 for review.) Thus, the sample we have selected has a standard score of $z = 2.7$, which means that it is 2.7 standard deviations above the mean of the sampling distribution. From Table 5.1, this standard score corresponds to the 99.65th percentile, so the probability of selecting another sample with a mean *less* than 5.01 is about 0.9965. It follows that the probability of selecting another sample with a mean *greater* than 5.01 is about $1 - 0.9965 = 0.0035$. Apparently, the sample we selected is rather extreme within this distribution. This example shows that when we have a normal sampling distribution, we can determine whether a particular sample is somewhat ordinary or very rare.

By the Way...

The total number of U.S. farms decreased from 2,440,000 in 1980 to 2,172,000 in 2000. The mean acreage per farm increased from 426 in 1980 to 434 in 2000.

Time out to think

Suppose a sample mean is in the 95th percentile. Explain why the probability of randomly selecting another sample with a mean greater than the first mean is 0.05.

EXAMPLE 1   Sampling Farms

Texas has roughly 225,000 farms, more than any other state in the United States. The actual mean farm size is $\mu = 582$ acres and the standard deviation is $\sigma = 150$ acres. Thus, for random samples of $n = 100$ farms, the distribution of sample means has a mean of 582 acres and a standard deviation of 15 acres. (The standard deviation of 15 is computed from $\sigma/\sqrt{n} = 150/\sqrt{100} = 15$.) What is the probability of selecting a random sample of 100 farms with a mean greater than 600 acres?

SOLUTION A sample mean of $\bar{x} = 600$ acres has a standard score of

$$z = \frac{\text{sample mean } - \text{ pop. mean}}{\text{standard deviation}} = \frac{600 - 582}{15} = 1.2$$

According to Table 5.1, this standard score is in the 88th percentile, so the probability of selecting a sample with a mean less than 600 acres is about 0.88. Thus, the probability of selecting a sample with a mean greater than 600 acres is about 0.12. ∎

Sample Proportions

Much of what we have learned about distributions of sample means carries over to distributions of *sample proportions*. We can see the parallels by returning to the survey of 400 students described earlier. Suppose your goal is to determine the proportion (or percentage) of all 400 students who own a laptop computer. Each Y (for yes) or N (for no) below is one person's answer to the question "Do you own a laptop computer?"

```
Y N Y Y N Y N N Y N Y Y N N N Y Y Y Y N Y N Y N Y N Y Y Y N Y Y Y N Y Y N N N Y Y Y N Y N Y N Y N Y Y Y Y
Y Y N Y Y Y N Y Y N N N Y Y Y N Y N Y N Y N Y N Y Y Y Y Y N Y Y Y N Y N Y N N Y N Y Y N N N Y Y Y Y N Y N Y N Y
N Y Y Y Y N Y N Y N Y N Y Y Y Y N Y Y Y N Y Y N N N Y Y Y N Y N Y N Y N Y N Y Y Y Y Y N Y Y N Y N N Y N Y Y N N
N N N Y Y Y N Y N Y N Y N Y N Y Y Y Y Y Y N Y Y Y N Y N N Y N Y Y N N N Y Y Y Y N Y N Y N Y N Y Y Y N Y Y N Y Y
N Y Y N Y N N Y N Y Y Y N N N Y Y Y N N N Y Y Y N Y N Y N Y N Y N Y Y Y Y Y Y Y N Y N Y N Y N Y Y Y N Y Y N Y Y
N Y Y Y N Y Y N Y Y Y Y N N N Y Y Y N N N Y Y Y Y N Y N Y N Y N Y Y N Y N Y N Y N N Y N Y N Y N Y Y Y Y Y Y Y N
Y N Y N Y Y Y Y Y Y N Y N Y N Y N Y Y Y N Y N N N Y N Y Y N N N Y Y Y N N N N Y Y Y N Y N Y N N Y Y Y N Y Y N Y Y
Y Y Y N Y N Y N Y Y Y Y N Y Y N Y N Y N N Y N Y Y N N N Y Y Y N N N Y Y Y N Y N Y N Y N Y Y Y N Y Y N Y Y
```

If you counted carefully, you would find that 240 of the 400 responses are Y's, so the *exact* proportion of laptop owners in the 400-student population is

$$p = \frac{240}{400} = 0.6$$

This *population proportion, $p = 0.6$*, is another example of a *population parameter*. Of the 400 students in the population, it is the true proportion of laptop computer owners.

Time out to think

Give another survey question that would result in a population proportion rather than a population mean.

Once again, in typical statistical problems, it is impractical or expensive to survey every individual in the population. Therefore, it's reasonable to consider the idea of using a random sample of, say, $n = 32$ people to estimate the population proportion. Suppose you randomly draw 32 responses from the list of Y's and N's to generate the following sample:

```
Y N Y Y N Y Y N Y N Y Y Y N Y Y Y Y N N Y Y Y N N Y Y N Y Y Y N Y Y
```

The proportion of Y responses in this list is

$$\hat{p} = \frac{21}{32} = 0.656$$

This proportion is another example of a *sample statistic*. In this case, it is a **sample proportion** because it is the proportion of laptop computer owners within a *sample;* we use the symbol $\hat{p}$ (read "*p*-hat") to distinguish this sample proportion from the population proportion, *p*.

Our goal is to determine how well a *single* sample proportion, $\hat{p}$, approximates the population proportion, p. For the moment, imagine that you have the luxury of selecting another sample of $n = 32$ responses; this sample produces the list

Y Y N N N Y Y N Y Y Y N N Y N Y Y N Y N Y N Y N Y Y N Y Y N N Y N Y

The proportion of Y responses in this list is

$$\hat{p} = \frac{18}{32} = 0.563$$

There are three sorts of opinion: informed, uninformed and inconsequential. Uninformed opinions encompass most of American opinion.

— Paul Talmey, pollster

Suppose you decide to select *many* more samples of 32 responses (again, an option that usually doesn't occur in practice). For each sample, you calculate the sample proportion, $\hat{p}$. If you draw a histogram of the many sample proportions, it will show the number of samples that have particular values of $\hat{p}$ from 0 through 1. Figure 8.7 shows such a histogram, in this case resulting from 100 samples of size $n = 32$. As we found for sample means, this distribution of sample proportions is very close to a normal distribution. Furthermore, the mean of this distribution is very close to the population proportion of 0.6.

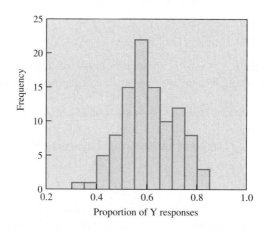

Figure 8.7 The distribution of 100 sample proportions, with a sample size of 32, appears to be close to a normal distribution.

Again, we are seeing the Central Limit Theorem at work. Suppose it were possible to select *all* possible samples of size $n = 32$. The resulting distribution would be called a **distribution of sample proportions**. The mean of this distribution equals the population proportion exactly, and this distribution approaches a normal distribution as the sample size increases.

In practice, we often have only one sample to work with. In that case, the best estimate for the population proportion, p, is the sample proportion, $\hat{p}$.

> ### The Distribution of Sample Proportions
>
> The **distribution of sample proportions** is the distribution that results when we find the proportions ($\hat{p}$) in *all* possible samples of a given size. The larger the sample size, the more closely this distribution approximates a normal distribution. In all cases, the mean of the distribution of sample proportions equals the population proportion. If only one sample is available, its sample proportion, $\hat{p}$, is the best estimate for the population proportion, p.

EXAMPLE 2   Analyzing a Sample Proportion

Consider the distribution of sample proportions shown in Figure 8.7. Assume that its mean is $p = 0.6$ and its standard deviation is 0.1. Suppose you randomly select the following sample of 32 responses:

Y Y N Y Y Y Y N Y Y Y Y Y Y Y N Y Y N Y Y Y N Y Y N Y Y N Y N Y Y

Compute the sample proportion, $\hat{p}$, for this sample. How far does it lie from the mean of the distribution? What is the probability of selecting another sample with a proportion greater than the one you selected?

SOLUTION   The proportion of Y responses in this sample is

$$\hat{p} = \frac{24}{32} = 0.75$$

Using a mean of 0.6 and a standard deviation of 0.1, we find that the sample statistic, $\hat{p} = 0.75$, has a standard score of

$$z = \frac{\text{sample proportion} - \text{pop. proportion}}{\text{standard deviation}} = \frac{0.75 - 0.6}{0.1} = 1.5$$

Thus, the sample proportion is 1.5 standard deviations above the mean of the distribution. Using Table 5.1, we see that a standard score of 1.5 corresponds to the 93rd percentile. The probability of selecting another sample with a proportion less than the one we selected is about 0.93. Thus, the probability of selecting another sample with a proportion greater than the one we selected is about $1 - 0.93 = 0.07$. In other words, if we were to select 100 random samples of 32 responses, we should expect to see only 7 samples with a higher proportion than the one we selected. ∎

> **TECHNICAL NOTE**
>
> The standard deviation of sample proportions is
>
> $$\sqrt{\frac{p(1 - p)}{n}}$$
>
> If we don't know the value of the population proportion, p, we can estimate that standard deviation by using the sample proportion, $\hat{p}$. For example, if $\hat{p} = 0.75$ and $n = 32$, we get
>
> $$\sqrt{\frac{0.75(1 - 0.75)}{32}} = 0.0765$$
>
> or 0.1 rounded.

Review Questions

1. Describe the process that leads to a distribution of sample means.

2. Explain what happens to the distribution of sample means as the sample size increases. What is the mean of the distribution of sample means?

3. Given only one sample mean, what is the best estimate of the population mean?

4. Describe the process that leads to a distribution of sample proportions.

5. Explain what happens to the distribution of sample proportions as the size of the sample increases. What is the mean of the distribution of sample proportions?

6. Given only one sample proportion, what is the best estimate of the population proportion?

Exercises

BASIC SKILLS AND CONCEPTS

SENSIBLE STATEMENTS? For Exercises 1–4, determine whether the given statement is sensible and explain why it is or is not.

1. A sampling error occurs when a pollster incorrectly records a survey response.

2. When a random sample is used to estimate a population mean, the sample mean tends to become a better estimate of the population mean as the sample size increases.

3. In a Gallup poll of 491 randomly selected adults, 319 indicated that they were in favor of the death penalty for a person convicted of murder. Based on the sample proportion of 319/491, the best estimate of the population proportion is also 319/491 (or 0.650).

4. In a Gallup poll of 491 randomly selected adults, 319 indicated that they were in favor of the death penalty for a person convicted of murder. If a larger sample were obtained, the sample proportion would be likely to be larger than 319/491 (or 0.650).

5. SAMPLE AND POPULATION MEANS. The mean birth rate in Massachusetts is 13.2 births per 1,000. Suppose you select a sample of five towns in Massachusetts that results in the following birth rates:

    12.3 13.1 14.3 14.2 13.8

What is the sample mean of birth rates? Does your sample appear to be representative of the population? Explain.

6. SAMPLE AND POPULATION MEANS. The mean household income in Alaska is $55,000. Suppose you select a sample of five households in Alaska that results in the following data:

    $49,500 $57,000 $54,000 $48,000 $52,000

What is the sample mean of household incomes? Does your sample appear to be representative of the population? Discuss.

7. ESTIMATING POPULATION MEANS. You select a random sample of 150 students at a small college attended by 3,200 students. Within your sample, you find that the mean study time per week is 17.8 hours. Based on this sample statistic, what is the best estimate for the mean study time for all students at the college? Would you be more confident of your estimate if you sampled 300 students? Explain.

8. ESTIMATING POPULATION MEANS. Suppose you select a random sample of 1,200 people attending a Chicago White Sox baseball game. Within the sample, the mean distance traveled to get to the game is 15.8 miles. Based on this sample statistic, what is the best estimate for the mean travel time for all people at the game? Would you be more confident of your estimate if you sampled 2,000 people? Explain.

FURTHER APPLICATIONS

9. DISTRIBUTION OF SAMPLE MEANS. Suppose you know that Durable tires have a mean lifetime of 45,000 miles. You also know that the distribution of sample means for samples of size 400 is normal with a mean of 45,000 miles and a standard deviation of 1,100 miles. Suppose you take a sample of 400 tires and find that the mean is 46,500 miles.
   a. How many standard deviations is the sample mean from the mean of the distribution of sample means?
   b. In general, what is the probability that a random sample has a mean greater than 46,500?

10. DISTRIBUTION OF SAMPLE MEANS. Suppose you know that the mean number of people per household in the United States is 2.64. You also know that the distribution of sample means for samples of 500 households is normal with a mean of 2.64 and a standard deviation of 0.06. Suppose you select a random sample of $n = 500$ households and determine that the mean number of people per household for this sample is 2.55.
   a. How many standard deviations is the sample mean from the mean of the distribution of sample means?
   b. What is the probability that a second sample would be selected with a mean less than 2.55?

11. SAMPLE AND POPULATION PROPORTIONS. Suppose that, in a suburb of 12,345 people, 6,523 people moved there within the last five years. You survey 500 people and find that 245 of the people in your sample moved to the suburb in the last five years.
   a. What is the population proportion of people who moved to the suburb in the last five years?
   b. What is the sample proportion of people who moved to the suburb in the last five years?
   c. Does your sample appear to be representative of the population? Discuss.

12. SAMPLE AND POPULATION PROPORTIONS. Suppose that, in a school with 1,348 students, 137 students are left-handed. You survey 100 students and find that 11 of the students in your sample are left-handed.

a. What is the population proportion of left-handed students?
b. What is the sample proportion of left-handed students?
c. Does your sample appear to be representative of the population? Discuss.

13. **ESTIMATING POPULATION PROPORTIONS.** You select a random sample of 150 people at a medical convention attended by 1,608 people. Within your sample, you find that 73 people have traveled from abroad. Based on this sample statistic, estimate how many people at the convention traveled from abroad. Would you be more confident of your estimate if you sampled 300 people? Explain.

14. **ESTIMATING POPULATION PROPORTIONS.** A random sample of 1,320 people is selected from the 4,500 people attending a World Cup soccer game. Within the sample, 103 people are supporting the visiting team. Based on this sample statistic, estimate how many people at the game support the visiting team. Would you be more confident of your estimate if you sampled 2,000 people? Explain.

15. **DISTRIBUTION OF SAMPLE PROPORTIONS.** Suppose you know that 34% of the students at your university are nonresident students. You also know that the distribution of sample proportions of samples of 200 students is normal with a mean of 0.34 and a standard deviation of 0.03. Suppose you select a random sample of 200 students and find that the proportion of nonresident students in the sample is 0.32.
a. How many standard deviations is the sample proportion from the mean of the distribution of sample proportions?
b. What is the probability that a second sample would be selected with a proportion less than 0.32?

16. **DISTRIBUTION OF SAMPLE PROPORTIONS.** Suppose you know that 42% of the employees at a large computer manufacturing company are women. You also know that the distribution of sample proportions is normal with a mean of 0.42 and a standard deviation of 0.21. Suppose you select a random sample of employees and find that the proportion of women in the sample is 0.45.
a. How many standard deviations is the sample proportion from the mean of the distribution of sample proportions?
b. What is the probability that a second sample would be selected with a proportion greater than 0.45?

17. **FORMING SAMPLING DISTRIBUTIONS.** Five states and their areas (in thousands of square miles) are given in the following table. Consider these five states to be the entire population from which samples will be selected. Find the distribution of sample means for sample sizes $n = 1, 2, 3, 4$, and 5. Find the mean of the distribution of sample means in each case. Compare the means of the distributions of sample means to the population mean.

State	Area (thousands of sq. mi)
A (Alabama)	52
C (Connecticut)	5
G (Georgia)	60
M (Maine)	33
O (Oregon)	97

18. **FORMING SAMPLING DISTRIBUTIONS.** At the end of a practice session, the five starters on a hockey team have scored the following numbers of points.

Player	Points
A	101
B	87
C	75
D	66
E	62

Find the distribution of sample means for sample sizes $n = 1, 2, 3, 4$, and 5. Find the mean of the distribution of sample means in each case. Compare the means of the distributions of sample means to the population mean.

PROJECTS FOR THE WEB AND BEYOND

19. **DISTRIBUTIONS OF SAMPLE MEANS.** Consider the data set on page 329, showing weekly hours on the Internet for a population of 400 people. Discuss methods for selecting a random sample from this population. Let each person in the class select a random sample of $n = 10$ individuals from the population and find the mean of his/her sample. Find the mean of the individual sample means and compare it to the population mean of 3.88 hours. Repeat the process with samples of size $n = 20$. How does the mean of the sample means compare to the population mean?

20. **DISTRIBUTIONS OF SAMPLE PROPORTIONS.** Consider the data set on page 333, showing yes/no responses to a survey. Discuss methods for selecting a random sample from this population. Let each person in the class select a random sample of $n = 10$ responses from the population and find the proportion of yes responses for his/her sample. Find the mean of the individual sample proportions and compare it to the population proportion of 0.6. Repeat the process with samples of size $n = 20$. How does the mean of the sample proportions compare to the population proportion?

IN THE NEWS

1. **SAMPLE MEANS IN THE NEWS.** Find a news or research report in which a sample mean is cited. Discuss how it is used to estimate a population mean.

2. **SAMPLE PROPORTIONS IN THE NEWS.** Find a news or research report in which a sample proportion is cited. Discuss how it is used to estimate a population proportion.

8.2 Estimating Population Means

In Section 8.1, we saw how distributions of sample means and sample proportions arise. We also saw that, if we have a single random sample from a population, its sample mean is our best estimate of the population mean. But how good is this "best estimate"? In this section, we will investigate how we can make quantitative statements about uncertainty when inferring a population mean from a sample mean; we will return to the topic of population proportions in the next section.

Estimating a Population Mean: The Basics

Plants and animals constantly produce new cells, not only for growth, but also to replace aging cells. The production of new cells requires proteins, the building blocks of living organisms. For this reason, proteins are an essential part of our diet. How much protein should you eat per day? Many nutritional organizations and government agencies provide recommended daily allowances (RDAs) not only for protein, but also for vitamins, minerals, carbohydrates, and fat. These RDAs differ according to the source, and they change over time as new research is done. Most recommendations for protein consumption are clustered around a value of 55–60 grams per day for men and 45–50 grams per day for women who are not pregnant or nursing.

Given these recommendations for how much protein we *should* eat, it is interesting to ask how much protein Americans actually *do* eat. Figure 8.8 shows partial results of a large

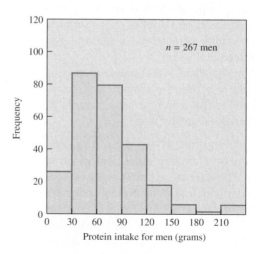

Figure 8.8 Histogram of daily protein intake for men taken from a sample of $n = 267$ men.
SOURCE: National Center for Health Statistics.

survey of the nutritional habits of Americans (Third National Health and Nutrition Examination Survey, or NHANES III, conducted by the National Center for Health Statistics). The original study involved roughly 30,000 participants who were surveyed on many different aspects of their health and diet. The histogram shows the average daily intake of protein, in grams, for a sample of $n = 267$ men taken from the study.

Our goal is to use the average protein intake for this sample of $n = 267$ men to make an inference about the average protein intake for the population of all American men. The sample data have a mean of $\bar{x} = 77.0$ grams and a standard deviation of $s = 58.6$ grams. As discussed in Section 8.1, when we have only a single sample, the sample mean is our best estimate of the population mean, μ. However, we do not expect the sample mean to be equal to the population mean, because there is likely to be some sampling error. Therefore, in order to make an inference about the population mean, we need some way to describe how well we expect it to be represented by the sample mean. The most common method for doing this is by way of *confidence intervals*. We discussed the use of confidence intervals in Chapter 1. Now, we are ready to look at how confidence intervals are computed.

The idea of a confidence interval comes directly from the work we did with sampling distributions. If the distribution of sample means is normal with a mean of μ, then 95% of all sample means lie within 1.96 standard deviations of the population mean (see Chapter 5); for our purposes in this book, we will approximate this as 2 standard deviations. A **confidence interval** is a range of values likely to contain the true value of the population mean. In this book, we will work only with a 95% confidence level, although other levels of confidence (such as 90% or 99%) are sometimes used in statistics. We find the 95% confidence interval from the sample mean by working with the margin of error, as defined in the following box.

95% Confidence Interval for a Population Mean

The **margin of error** for the 95% confidence interval is

$$\text{margin of error} = E \approx \frac{2s}{\sqrt{n}}$$

where s is the standard deviation of the sample. We find the **95% confidence interval** by adding and subtracting the margin of error from the sample mean. That is, the 95% confidence interval ranges

from $(\bar{x} - \text{margin of error})$ to $(\bar{x} + \text{margin of error})$

We can write this confidence interval more formally as

$$\bar{x} - E < \mu < \bar{x} + E$$

or more briefly as

$$\bar{x} \pm E$$

TECHNICAL NOTE

The precise formula for the margin of error uses 1.96 rather than 2. Also, one consequence of the Central Limit Theorem is that the distribution of sample means has a standard deviation of $\sigma/\sqrt{n}$, but s may be used in place of σ as long as the sample is large enough (typically $n > 30$, but the size requirement depends on the nature of the actual population distribution). The formula for margin of error can be extended to other confidence levels. For example, for a 90% confidence level, replace the 2 in the formula by 1.645. For a 99% confidence level, replace the 2 in the formula by 2.575.

The sample mean, $\bar{x}$, lies at the center of the confidence interval, which extends a distance equal to the margin of error in either direction (Figure 8.9). We say that we are 95% confident that the confidence interval contains the true value of the population mean. This statement should be carefully interpreted as follows: If we were to repeat the process of obtaining samples and constructing such confidence intervals many times, 95% of the confidence intervals would contain the value of the population mean.

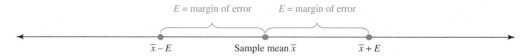

95% confidence interval

E = margin of error E = margin of error

$\bar{x} - E$ Sample mean $\bar{x}$ $\bar{x} + E$

Figure 8.9 The 95% confidence interval extends a distance equal to the margin of error on either side of the sample mean.

EXAMPLE 1   Computing the Margin of Error

Compute the margin of error and find the 95% confidence interval for the protein intake sample of $n = 267$ men, which has a sample mean of $\bar{x} = 77.0$ grams and a sample standard deviation of $s = 58.6$ grams.

SOLUTION   The sample size is $n = 267$ and the standard deviation for the sample is $s = 58.6$, so the margin of error is

$$E \approx \frac{2s}{\sqrt{n}} = \frac{2 \times 58.6}{\sqrt{267}} = 7.2$$

The sample mean is $\bar{x} = 77.0$ grams, so the 95% confidence interval extends approximately from $77.0 - 7.2 = 69.8$ grams to $77.0 + 7.2 = 84.2$ grams. We write this result more formally as

$$69.8 \text{ grams} < \mu < 84.2 \text{ grams}$$

or more simply as 77.0 ± 7.2 grams. We are 95% confident that the range of values from 69.8 grams to 84.2 grams contains the true value of the population mean. It is interesting to note that the entire confidence interval, from 69.8 grams to 84.2 grams, is significantly greater than the recommended daily protein allowance for men of 55 to 60 grams—suggesting that actual protein consumption is significantly greater than recommended.   ■

> **TECHNICAL NOTE**
>
> If we use the precise formula (with 1.96 instead of 2), we find that the confidence interval limits are 70.0 and 84.0.

Interpreting the Confidence Interval

Figure 8.10 shows a visual interpretation of the confidence interval. Imagine that we have, say, 20 different samples. Each sample has a different sample mean, $\bar{x}$, with a confidence interval about the sample mean. We can never know for sure which intervals contain the population mean. However, on average, 95% of the samples, or 19 out of the 20 samples, will have a confidence interval that captures the true population mean. An occasional confidence interval (about 5% of those generated) does not capture the true population mean. Thus, there is a 0.95 probability that the confidence interval computed from the original sample captures the true population mean.

Be careful—it is easy to interpret or state a confidence interval incorrectly. It is tempting to claim that a 95% confidence interval means there is a 0.95 probability that the population mean falls within the confidence interval. But this is not true: Even though we may not know the actual population mean, μ, it is a fixed number, not a variable. Thus, either it *does* fall within the confidence interval or it *does not* fall within the confidence interval. We cannot talk about the probability that μ falls within the confidence interval. To recap, the correct interpretation of a 95% confidence interval is this: If we repeat the process of obtaining samples and constructing confidence intervals, in the long run 95% of the confidence intervals will contain the true population mean.

EXAMPLE 2   Constructing a Confidence Interval

A study finds that the average time spent by eighth-graders watching television is 6.7 hours per week, with a margin of error of 0.4 hour (for 95% confidence). Construct and interpret the 95% confidence interval.

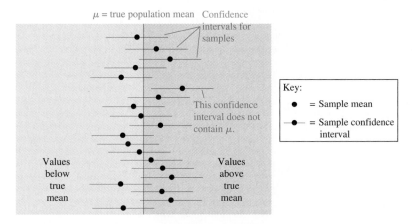

Figure 8.10 This figure illustrates the idea behind confidence intervals. The central vertical line represents the true population mean, μ. Each of the 20 horizontal lines represents the 95% confidence interval for a particular sample, with the sample mean marked by the dot in the center of the confidence interval. With a 95% confidence interval, we expect that 95% of all samples will give a confidence interval that contains the population mean, as is the case in this figure, for 19 of the 20 confidence intervals do indeed contain the population mean. We expect that the population mean will not be within the confidence interval in 5% of the cases; here, 1 of the 20 confidence intervals (the sixth from the top) does not contain the population mean.

SOLUTION   The best estimate of the population mean is the sample mean, $\bar{x} = 6.7$ hours. We find the confidence interval by adding and subtracting the margin of error from the sample mean, so the interval extends from $6.7 - 0.4 = 6.3$ hours to $6.7 + 0.4 = 7.1$ hours. Thus, we can claim with 95% confidence that the average time spent watching television for the entire population of eighth graders is between 6.3 and 7.1 hours, or

$$6.3 \text{ hours} < \mu < 7.1 \text{ hours}$$

If 100 random samples of the same size were taken, we would expect the confidence intervals of 95 of those samples to contain the population mean.   ∎

EXAMPLE 3   Protein Intake for Women

The NHANES III nutritional study also produced data on protein intake for women. Figure 8.11 shows a histogram for a random sample of $n = 264$ women (a small part of the entire sample). The mean of these data is $\bar{x} = 59.6$ grams and the standard deviation is $s = 30.5$ grams. Estimate the population mean and give a 95% confidence interval. Comment on how these values compare to the recommended daily allowance (RDA) for women of 45–50 grams.

SOLUTION   The sample mean, $\bar{x} = 59.6$ grams, is our best estimate of the population mean. To find the 95% confidence interval, we must first find the margin of error using the sample size ($n = 264$) and sample standard deviation ($s = 30.5$ grams):

$$E \approx \frac{2s}{\sqrt{n}} = \frac{2 \times 30.5 \text{ grams}}{\sqrt{264}} = 3.8 \text{ grams}$$

Thus, the 95% confidence interval ranges from $59.6 - 3.8 = 55.8$ grams to $59.6 + 3.8 = 63.4$ grams, or

$$55.8 \text{ grams} < \mu < 63.4 \text{ grams}$$

We can say with 95% confidence that the interval from 55.8 grams to 63.4 grams contains the population mean. Based on this result, we conclude that the population mean, which represents

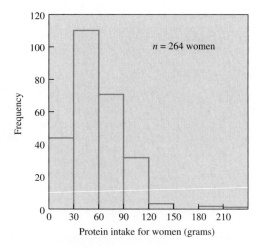

Figure 8.11 Histogram for daily protein intake for 264 women.
SOURCE: National Center for Health Statistics.

actual protein consumption by women, is greater than the recommended daily allowance of 45–50 grams for women. ∎

Time out to think

Recall that the standard deviation of the data on protein intake for the sample of men was $s = 58.6$ grams — almost double the standard deviation ($s = 30.5$ grams) for the sample of women. How does this difference affect the margins of error? Speculate on why the standard deviations are different.

EXAMPLE 4   Garbage Production

A study conducted by the Garbage Project at the University of Arizona analyzed the contents of garbage discarded by $n = 62$ households; the households ranged in size from 2 to 11 members. The histogram in Figure 8.12 shows the total weekly garbage production (in pounds) for

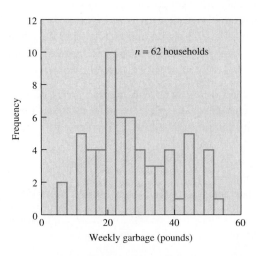

Figure 8.12 Histogram for total garbage production for $n = 62$ households.

the households in the sample. (The complete study gives the breakdown of garbage by various categories.) The mean for the sample is $\bar{x} = 27.4$ pounds and the standard deviation is $s = 12.5$ pounds. Estimate the population mean for weekly garbage production with a 95% confidence interval. Comment on the conclusion of the study.

SOLUTION   The sample mean, $\bar{x} = 27.4$ pounds, is our best estimate of the population mean. We use the margin of error formula to find that

$$E \approx \frac{2s}{\sqrt{n}} = \frac{2 \times 12.5 \text{ grams}}{\sqrt{62}} = 3.2 \text{ pounds}$$

Thus, the 95% confidence interval ranges from $27.4 - 3.2 = 24.2$ pounds to $27.4 + 3.2 = 30.6$ pounds, or

$$24.2 \text{ pounds} < \mu < 30.6 \text{ pounds}$$

With 95% confidence, we can claim that the interval from 24.2 pounds to 30.6 pounds contains the mean amount of garbage discarded by American households in a week. Two observations might be made about this conclusion. First, the wide range of household sizes produces significant variation in the data. This variation is reflected in a large standard deviation that produces a large margin of error. The conclusion might be more meaningful if it were given in terms of individual household sizes. Second, the relatively small sample size (62 households) also contributes to the large margin of error. A more reliable conclusion could be obtained from a larger sample. ∎

EXAMPLE 5   Mean Body Temperature

A study by University of Maryland researchers investigated the body temperatures of $n = 106$ subjects. The sample mean of the data set is $\bar{x} = 98.20°\text{F}$ and the standard deviation for the sample is $s = 0.62°\text{F}$. Estimate the population mean body temperature with a 95% confidence interval.

SOLUTION   The sample mean, $\bar{x} = 98.20°\text{F}$, is our best estimate of the population mean body temperature. The margin of error is

$$E \approx \frac{2s}{\sqrt{n}} = \frac{2 \times 0.62°\text{F}}{\sqrt{106}} = 0.12°\text{F}$$

Thus, the 95% confidence interval ranges from $98.20°\text{F} - 0.12°\text{F} = 98.08°\text{F}$ to $98.20°\text{F} + 0.12°\text{F} = 98.32°\text{F}$, or

$$98.08°\text{F} < \mu < 98.32°\text{F}$$

We interpret this result as follows: If we were to select many different samples of size $n = 106$ and compute confidence intervals for all of the samples, we expect that 95% of the confidence intervals would contain the true population mean. Notice that the commonly cited mean body temperature for humans (98.6°F) is *not* contained in the confidence interval. Thus, based on this sample, it is likely that the accepted mean body temperature is wrong. ∎

Choosing Sample Size

In planning statistical surveys and experiments, we often know in advance the margin of error we would like to achieve. For example, we might want to estimate the mean cost of a new car to within $200. We can estimate the sample size, n, needed to ensure this margin of error by solving the margin of error formula ($E \approx 2s/\sqrt{n}$) for n. With a little bit of algebra, we find

$$n \approx \left(\frac{2s}{E}\right)^2$$

By the Way...

According to the Garbage Project data, the leading component of most household garbage is paper (comprising a third to a half of the total garbage), followed by food and glass. Also, there is a significant correlation between the weight of the discarded plastic and the size of a household, so discarded plastic could be used to estimate the population of a region.

The exact sample size formula uses the population standard deviation, σ, in place of s. In practice, we rarely know the population standard deviation, because we study only samples. Therefore, to use the exact sample size formula, we usually estimate the population standard deviation based on previous studies, pilot studies, or educated guesses. The size of any actual sample must be a whole number, so we round the result of the sample size formula *up* to the nearest whole number. Any sample larger than this size will give us a margin of error as small as or smaller than the one we seek.

Choosing the Correct Sample Size

In order to estimate the population mean with a specified margin of error of at most E, the size of the sample should be at least

$$n = \left(\frac{2\sigma}{E}\right)^2$$

where σ is the population standard deviation (usually estimated).

EXAMPLE 6   Mean Housing Costs

You want to study housing costs in the country by sampling recent house sales in various (representative) regions. Your goal is to provide a 95% confidence interval estimate of the housing cost. Previous studies suggest that the population standard deviation is about $7,200. What sample size (at a minimum) should be used to ensure that the sample mean is within

a. $500 of the true population mean?

b. $100 of the true population mean?

SOLUTION

a. With $E = \$500$ and σ estimated as $7,200, the minimum sample size that meets the requirements is

$$n = \left(\frac{2\sigma}{E}\right)^2 = \left(\frac{2 \times 7,200}{500}\right)^2 = 28.8^2 = 829.4$$

Because the sample size must be a whole number, we conclude that the sample should include *at least* 830 prices.

b. With $E = \$100$ and $\sigma = \$7,200$, the minimum sample size that meets the requirements is

$$n = \left(\frac{2\sigma}{E}\right)^2 = \left(\frac{2 \times 7,200}{500}\right)^2 = 144^2 = 20,736$$

Notice that to decrease the margin of error by a factor of 5 (from $500 to $100), we must increase the sample size by a factor of 25. That is why achieving greater accuracy generally comes with a high cost. ∎

Time out to think
If you decide you want a smaller margin of error for a confidence interval, should you increase or decrease the sample size? Explain.

Review Questions

1. How is the margin of error for a sample proportion calculated? Once you have a margin of error, how is it used to form a confidence interval?

2. Explain the meaning of a 95% confidence interval for a population mean.

3. Explain how to find the sample size required to obtain a certain margin of error.

Exercises

BASIC SKILLS AND CONCEPTS

SENSIBLE STATEMENTS? For Exercises 1–4, determine whether the given statement is sensible and explain why it is or is not.

1. When sample data are used to estimate a population mean, a confidence interval is found to be 98.20.

2. When the mean height of all adult males in the United States is estimated, the required sample size is calculated from the size of the population.

3. When sample data are used to estimate the value of a population mean, the margin of error increases as the sample size increases.

4. When sample data are used to estimate the value of a population mean, the 95% confidence interval is found to be $5.0 < \mu < 10.0$. Based on that result, the sample mean is calculated to be 7.5.

MARGINS OF ERROR AND CONFIDENCE INTERVALS. In Exercises 5–8, assume that population means are to be estimated from the samples described. In each case, use the sample results to approximate the margin of error and 95% confidence interval.

5. Sample size = 400, sample mean = 120, sample standard deviation = 10

6. Sample size = 100, sample mean = 15, sample standard deviation = 5

7. Sample size = 2,500, sample mean = 160, sample standard deviation = 20

8. Sample size = 256, sample mean = 20, sample standard deviation = 16

SAMPLE SIZES. In Exercises 9–12, give an estimate of the sample size needed to obtain the specified margin of error for a 95% confidence interval. The sample standard deviation is given.

9. Margin of error = 0.02, standard deviation = 1

10. Margin of error = 2.0, standard deviation = 10

11. Margin of error = 0.01, standard deviation = 0.25

12. Margin of error = 5.0, standard deviation = 60

FURTHER APPLICATIONS

13. TEXTBOOK PRICES. You want to estimate the mean price of textbooks at your college. A sample of 81 textbooks has a mean price of $51.25 and a standard deviation of $7.75. Estimate the mean price of textbooks at your college. Give the 95% confidence interval.

14. SAT SCORES. A sample of 900 first-year students at a large university has a mean SAT mathematics score of 525 with a standard deviation of 88. Estimate the mean SAT mathematics score for the entire first-year class. Give the 95% confidence interval.

15. TIME TO GRADUATION. Data from the National Center for Education Statistics on 4,400 college graduates show that the mean time required to graduate with a

bachelor's degree is 5.15 years with a standard deviation of 1.68 years. Estimate the mean time required to graduate for all college graduates. Give the 95% confidence interval.

16. GARBAGE PRODUCTION. Based on a sample of 62 households, the following table gives the mean weight (in pounds) and standard deviation for paper, glass, and plastic discarded in one week (data from the Garbage Project at the University of Arizona). For each category, estimate the mean weight for the entire population of American households. Give the 95% confidence interval.

	Sample mean	Sample standard deviation
Paper	9.42	4.12
Glass	3.75	3.11
Plastic	1.91	1.07

17. STARTING SALARIES. According to data from the U.S. Census Bureau, the mean annual starting salary for graduates with a bachelor's degree is $38,112. For graduates with only a high school diploma, the mean annual starting salary is $22,154. Suppose you survey 900 recent college graduates and find that the mean starting salary for the sample is $35,000 with a standard deviation of $4,500. You also survey 900 recent high school graduates and find that the mean starting salary for the sample is $26,500 with a standard deviation of $3,900. Estimate the mean starting salary for all college graduates and for all high school graduates. Find the confidence intervals for both estimates. Does either of the confidence intervals include the population mean reported by the Census Bureau?

18. TV SURVEY. Nielsen Media Research wishes to estimate the mean number of hours that high school students spend watching TV on a weekday. A margin of error of 0.25 hour is desired. Past studies suggest that a population standard deviation of 1.7 hours is reasonable. Estimate the minimum sample size required to estimate the population mean with the stated accuracy.

19. HOUSING PRICES. A government survey conducted to estimate the mean price of houses in a large metropolitan area is designed to have a margin of error of $10,000. Pilot studies suggest that the population standard deviation is $65,500. Estimate the minimum sample size needed to estimate the population mean with the stated accuracy.

20. CONSUMER SPENDING. The Bureau of Labor Statistics conducts a Consumer Expenditure Survey to measure the amounts spent by households in many different categories

(food, housing, apparel, transportation, health care, entertainment). In 2000, the mean annual spending in all categories was $38,045. Suppose a margin of error of $1,500 is desired in this survey. Estimate the minimum sample size needed to estimate the population mean with the stated accuracy. Assume that a population standard deviation of $25,500 is expected.

21. FAMILY SIZE. You select a random sample of $n = 31$ families in your neighborhood and find the following family sizes (number of people in the family):

2 3 6 5 4 2 3 3 1 2 3
2 3 4 5 3 1 3 3 4 7 3
2 3 2 2 3 4 1 5 2

a. What is the mean family size for the sample?
b. What is the standard deviation for the sample?
c. What is the best estimate for the mean family size for the population of all American families?
d. What is the 95% confidence interval for the estimate?
e. Comment on the reliability of the estimate.

22. TV SETS. A random sample of $n = 31$ households is asked the number of TV sets in the household. The responses are as follows.

1 0 2 3 2 3 4 2 1 1 2
4 3 2 3 3 0 1 0 1 3 2
4 3 2 1 4 0 1 2 3

a. What is the mean number of TVs for the sample?
b. What is the standard deviation for the sample?
c. What is the best estimate for the mean number of TVs for the population of all American households?
d. What is the 95% confidence interval for the estimate?
e. Comment on the reliability of the estimate.

PROJECTS FOR THE WEB AND BEYOND

For useful links, select "Links for Web Projects" for Chapter 8 at www.aw.com/bbt.

23. CAR AGES. Assume that you want to estimate the mean age of cars driven by students at your college. A previous study shows that the standard deviation of those ages is approximately 3.7 years. How many ages must you randomly select in order to be 95% confident that your sample mean is within 1 year of the population mean? Using that sample size, collect your own sample data, consisting of the ages of cars driven by students at

your college. Then use the methods of this section to construct a 95% confidence interval. Write a statement summarizing your results.

24. **POLLING ORGANIZATIONS.** Three leading public polling organizations are the Gallup Organization, Harris Poll, and Yankelovich Partners. Visit their Web sites. Describe the history of each organization and the polling services it provides. Which organization has the best description on its Web site of its polling methods?

25. **NETWORK POLLS.** All of the major television networks conduct regular polls on a variety of issues. Visit the Web site of at least one major network and gather the results of a particular poll that involves estimation of a population mean. Be sure to include all information that is given about the sample size, margin of error, and confidence intervals.

26. **INTERNATIONAL CORRUPTION.** Transparency International uses surveys to determine a Bribe Payer's Index

and Corruption Perception Index that measure the degree of corruption in many countries worldwide. Visit the Transparency International Web site and review the extensive documentation describing the methods used by this organization. Discuss the results and their validity.

1. **ESTIMATING POPULATION MEANS.** Find a news article or report in which a population mean is estimated from a sample. The article should include a margin of error and/or a confidence interval. Discuss the methods used in the study and how the conclusions were reached.

8.3 Estimating Population Proportions

We now proceed in much the same way as we did in the previous section, except that the goal is to estimate a population *proportion* (rather than a mean). Many well-known polls and surveys rely on the techniques that we will discuss. For example, the Nielsen ratings estimate the proportion of the population tuned in to certain radio and television shows, the monthly unemployment figures released by the Bureau of Labor Statistics are estimates of the proportion of Americans who are unemployed, and the opinion polls that dominate American politics estimate the proportion of the population that supports a particular candidate or measure.

The Basics of Estimating a Population Proportion

The Bureau of Labor Statistics estimates the unemployment rate from a monthly survey of 60,000 households (see Example 2 in Section 1.1). The unemployment rate for this sample is the *sample proportion* (the proportion of people in the sample who are unemployed), denoted $\hat{p}$. The sample proportion is the best estimate for the *population proportion* (the proportion of people in the population who are unemployed), denoted p.

Just as with population means, an estimate of the population proportion can be better understood if we can say something about its accuracy. Again, we use margins of error and confidence intervals. The only change from estimating population means is in the definition of the margin of error, which is given in the following box. Figure 8.13 shows how we interpret the confidence interval.

TECHNICAL NOTE

The precise formula for the margin of error uses 1.96 rather than 2. The formula can be extended to other confidence levels. For a 90% confidence level, replace the 2 in the formula by 1.645. For a 99% confidence level, replace the 2 in the formula by 2.575. The margin of error formula given here requires $np \geq 5$ and $n(1 - p) \geq 5$, conditions that are usually met easily in practice. (Techniques covered in more advanced texts work for small sample sizes.)

95% Confidence Interval for a Population Proportion

For a population proportion, the margin of error for the 95% confidence interval is

$$E \approx 2\sqrt{\frac{\hat{p}(1 - \hat{p})}{n}}$$

where $\hat{p}$ is the sample proportion. The 95% confidence interval ranges

from $\hat{p}$ − margin of error   to $\hat{p}$ + margin of error

We can write this confidence interval more formally as

$$\hat{p} - E < p < \hat{p} + E$$

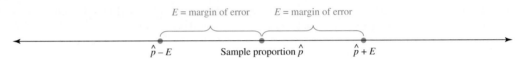

95% confidence interval

E = margin of error E = margin of error

$\hat{p} - E$ Sample proportion $\hat{p}$ $\hat{p} + E$

Figure 8.13 The confidence interval extends a distance equal to the margin of error on either side of the sample proportion, $\hat{p}$.

EXAMPLE 1   Unemployment Rate

The Bureau of Labor Statistics finds 2,160 unemployed people in a sample of $n = 60,000$ people. Estimate the population unemployment rate and give a 95% confidence interval.

SOLUTION   The sample proportion is the unemployment rate for the sample:

$$\hat{p} = \frac{2,160}{60,000} = 0.036$$

This is the best estimate for the population unemployment rate. The margin of error is

$$E \approx 2\sqrt{\frac{\hat{p}(1 - \hat{p})}{n}} = 2\sqrt{\frac{0.036(1 - 0.036)}{60,000}} = 0.0015$$

(The approximation is valid because of the large sample size.) The 95% confidence interval ranges from $0.0360 - 0.0015 = 0.0345$ to $0.0360 + 0.0015 = 0.0375$, or

$$0.0345 < p < 0.0375$$

We can have 95% confidence that the interval from 3.45% to 3.75% contains the true unemployment rate for the population. We interpret this result as follows: If we computed confidence intervals for many samples of size $n = 60,000$, we should expect 95% of the confidence intervals to contain the true population proportion. ∎

By the Way...

Here's how the Bureau of Labor Statistics describes uncertainty in its unemployment survey: A sample is not a total count and the survey may not produce the same results that would be obtained from interviewing the entire population. But the chances are 90 out of 100 that the monthly estimate of unemployment from the sample is within 230,000 of the figure obtainable from a total census. Since monthly unemployment totals have ranged between about 6 and 11 million in recent years, the possible error resulting from sampling is not large enough to distort the total unemployment picture.

EXAMPLE 2   TV Nielsen Ratings

The Nielsen ratings for television use a random sample of households. A Nielsen survey results in an estimate that a women's World Cup soccer game had 72.3% of the entire viewing audience. Assuming that the sample consists of $n = 5,000$ randomly selected households, find the margin of error and the 95% confidence interval for this estimate.

SOLUTION   The sample proportion, $\hat{p} = 72.3\% = 0.723$, is the best estimate of the population proportion. The margin of error is

$$E \approx 2\sqrt{\frac{\hat{p}(1 - \hat{p})}{n}} = 2\sqrt{\frac{0.723(1 - 0.723)}{5,000}} = 0.013$$

The 95% confidence interval is $0.723 - 0.013 < p < 0.723 + 0.013$, or

$$0.710 < p < 0.736$$

With 95% confidence, we conclude that between 71.0% and 73.6% of the entire viewing audience watched the women's World Cup soccer game.   ∎

EXAMPLE 3   Gallup Polls

Since 1935, the Gallup Organization has been a leader in the measurement and analysis of people's attitudes, opinions, and behavior. Although Gallup is best known for the Gallup Poll, the company also provides marketing and management research for large corporations. In a recent survey, 1,016 randomly selected adults were asked:

> *As you may know, former major league player Pete Rose is ineligible for baseball's Hall of Fame due to charges that he had gambled on baseball games. Do you think he should or should not be eligible for admission to the Hall of Fame?*

Among those surveyed, 59% believed that Pete Rose should be eligible. A smaller sample of 628 adults identified themselves as "baseball fans." Among the baseball fans, 62% believed that he should be eligible for admission to the Hall of Fame. Find the margin of error and confidence interval for each sample. The survey of baseball fans cited a margin of error of "no more than 5 percentage points." Is this claim consistent with the sample size?

SOLUTION   A sample with $n = 1,016$ respondents and a sample proportion of $\hat{p} = 0.59$ has a margin of error of

$$E \approx 2\sqrt{\frac{\hat{p}(1 - \hat{p})}{n}} = 2\sqrt{\frac{0.59(1 - 0.59)}{1,016}} = 0.031$$

or about 3 percentage points. Thus, the 95% confidence interval ranges from $59 - 3 = 56\%$ to $59 + 3 = 62\%$.

A sample with $n = 628$ respondents and a sample proportion of $\hat{p} = 0.62$ has a margin of error of

$$E \approx 2\sqrt{\frac{\hat{p}(1 - \hat{p})}{n}} = 2\sqrt{\frac{0.62(1 - 0.62)}{628}} = 0.039$$

or about 4 percentage points. The 95% confidence interval extends from $62 - 4 = 58\%$ to $62 + 4 = 66\%$. Note that the cited margin of error of "no more than 5 percentage points" is, in fact, an overestimate.   ∎

Choosing Sample Size

Designers of surveys and polls often specify a certain level of accuracy for the results. For example, it might be desirable to estimate a population proportion with a 95% confidence interval and a margin of error of no more than 1.5 percentage points. In such situations, it's

TECHNICAL NOTE

You can derive the $E \approx 1/\sqrt{n}$ formula from the more precise formula given earlier for the margin of error by replacing the product $\hat{p}(1 - \hat{p})$ by its maximum possible value of 0.25. This approximation overestimates the actual margin of error and is most accurate when p is near 0.5.

necessary to determine how large the sample must be to guarantee this accuracy. As long as we use a 95% level of confidence, we can work with this simplified, approximate formula for the margin of error:

$$E \approx \frac{1}{\sqrt{n}}$$

This formula gives a conservative (higher than necessary) estimate for the margin of error. Solving for n yields the sample size needed to achieve a margin of error E:

$$n \approx \frac{1}{E^2}$$

Any sample size equal to or larger than this value will suffice.

Choosing the Correct Sample Size

In order to estimate a population proportion with a 95% degree of confidence and a specified margin of error of E, the size of the sample should be at least

$$n = \frac{1}{E^2}$$

EXAMPLE 4　Minimum Sample Size for Survey

You plan a survey to estimate the proportion of students on your campus who carry a cell phone regularly. How many students should be in the sample if you want (with 95% confidence) a margin of error of no more than 4 percentage points?

SOLUTION　Note that 4 percentage points means a margin of error of 0.04. From the given formula, the minimum sample size is

$$n = \frac{1}{E^2} = \frac{1}{0.04^2} = 625$$

You should survey at least 625 students.　∎

EXAMPLE 5　Yankelovich Poll

Yankelovich Partners is an international public opinion and marketing research firm. The company does regular polls for *TIME Magazine* and CNN News. The results of its polls can be found in the monthly publication *Yankelovich Monitor*. A recent poll concluded that 51% of all households have a computer, with a margin of error of 3.5 percentage points. What sample size, at a minimum, must have been used in this poll?

SOLUTION　A margin of error of 3.5 percentage points (or 0.035) could be achieved with a sample size of

$$n = \frac{1}{E^2} = \frac{1}{0.035^2} = 816.3$$

Because we must round up to the next larger whole number, we conclude that at least 817 households were surveyed.　∎

Review Questions

1. How is the margin of error for a sample proportion calculated? How is the margin of error used to form a confidence interval?

2. Explain the meaning of a 95% confidence interval for a population proportion.

3. Explain how to find the sample size required to obtain a certain margin of error.

Exercises

BASIC SKILLS AND CONCEPTS

SENSIBLE STATEMENTS? For Exercises 1–4, determine whether the given statement is sensible and explain why it is or is not.

1. When sample data are used to estimate a population proportion, a confidence interval is found to be 0.237.

2. When the proportion of all golfers in the United States who are left-handed is estimated, the required sample size is calculated from the size of the population of golfers.

3. When sample data are used to estimate the value of a population proportion, the margin of error becomes smaller as the sample size increases.

4. A sample of prison inmates will likely provide a good estimate of the proportion of adults in the United States who watch the television show *60 Minutes*, if the sample is large enough.

MARGINS OF ERROR AND CONFIDENCE INTERVALS. In Exercises 5–8, assume that population proportions are to be estimated from the samples described. In each case, find the approximate margin of error and 95% confidence interval.

5. Sample size = 400, sample proportion = 0.5

6. Sample size = 100, sample proportion = 0.1

7. Sample size = 2,500, sample proportion = 0.9

8. Sample size = 256, sample proportion = 0.6

SAMPLE SIZE. In Exercises 9–12, estimate the minimum sample size needed to achieve the given margin of error.

9. $E = 0.02$

10. $E = 0.05$

11. $E = 0.04$

12. $E = 0.035$

FURTHER APPLICATIONS

13. **NIELSEN RATINGS.** Nielsen Media Research uses samples of 5,000 households to rank TV shows. Suppose Nielsen reports that *NFL Monday Night Football* had 35% of the TV audience. What is the 95% confidence interval for this result?

14. **WHAT'S IMPORTANT?** A study done by researchers at Columbia University used a telephone survey of 500 adults. It revealed that 99% of those surveyed ranked "loving family relationships" as important or very important, while 98% ranked financial security as important or very important. Third on the list was religious or spiritual fulfillment (86%), followed by a satisfying sex life (82%) and job satisfaction (79%). The margin of error for the study was reported as 4 percentage points.
    a. Is the reported margin of error consistent with the sample size?
    b. Make a sketch showing the estimates of the five population parameters cited in the study on a number line. Then draw the confidence interval around each estimate. Do any of the five areas appear to be more important than the others?

15. **HAZING OF ATHLETES.** A study done by researchers at Alfred University concluded that 80% of all student-athletes in this country have been subjected to some form of hazing. The study is based on responses from 1,400 athletes. What are the margin of error and 95% confidence interval for the study?

16. **STUDENT OPINIONS.** The annual survey of first-year college students, conducted by the Higher Education Research Institute at UCLA, asks approximately 276,000 students about their attitudes on a variety of subjects. According to a recent survey, 51% of first-year students believe that abortion should be legal (down from 65% in 1990) and 40% believe that casual sex is acceptable (down from 50% in 1975). What are the margins of error and 95% confidence intervals for these estimates?

17. **STATE HABITS.** The Centers for Disease Control and Prevention conducts surveys to compare the habits and attitudes of citizens in various states. For each of the following results, give the 95% confidence interval.
    a. 17.2% of those surveyed in Colorado are sedentary (the lowest in the country); sample size = 1,500.
    b. 13.2% of those surveyed in Utah smoke (the lowest in the country); sample size = 2,500.
    c. 22.9% of those surveyed in Wisconsin are binge drinkers; sample size = 3,500.

18. **IMPORTANT ISSUES.** In an ABC/*Washington Post* survey, 1,526 randomly selected Americans were asked to list the most important issues for the 2000 elections. Education (79%), the economy (74%), and managing the budget (74%) were listed as very important issues. The margin of error cited for the survey was 3 percentage points. Is the margin of error consistent with the sample size?

19. **DRUGS IN MOVIES.** A study by Stanford University researchers for the Office of National Drug Control Policy and the Department of Health and Human Services concluded that 98% of the top rental films involve drugs, drinking, or smoking. Assume that this study is based on the top 400 rental films.
    a. Use the results of this sample to estimate the proportion of all films that involve drugs, drinking, or smoking.
    b. What is the 95% confidence interval?
    c. Do you believe that the top 400 films represent a random sample? Explain.

20. **TEEN PRESSURE.** A study commissioned by the U.S. Department of Education concluded that 44% of teenagers cite grades as their greatest source of pressure. The study was based on responses from 1,015 teenagers. What is the 95% confidence interval?

21. **ELECTION PREDICTIONS.** In a random sample of 1,600 people from a large city, it is found that 900 support the current mayor in the upcoming election. Based on this sample, would you claim that the mayor will win a majority of the votes? Explain.

22. **ELECTION PREDICTIONS.** In a random sample of 2,500 people from a large city, it is found that 1,300 support the current mayor in the upcoming election. Based on this sample, would you claim that the mayor will win a majority of the votes? Explain. What conclusion would be reached from a sample of 250 people with 130 supporting the mayor?

23. **PRE-ELECTION POLLS.** Prior to a statewide election for the U.S. Senate, three polls are conducted. In the first poll, 780 of 1,500 voters favored candidate Martinez. In the second poll, 1,285 of 2,500 voters favored Martinez. In the third poll, 1,802 of 3,500 voters favored Martinez. Find the 95% confidence intervals for all three polls. Discuss Martinez's prospects for victory based on these polls.

24. **UNEMPLOYMENT SURVEY.** The Bureau of Labor Statistics estimates the unemployment rate in the United States monthly by surveying 60,000 individuals.
    a. In one month, 3.4% of the 60,000 individuals surveyed are found to be unemployed. Find the margin of error for this estimate. Is the precision (nearest tenth of a percent) reasonable? Explain.
    b. Suppose that the number of individuals surveyed were increased by a factor of four (to 240,000). By how much would the margin of error change?
    c. Suppose that the number of individuals surveyed were decreased by a factor of one-fourth (to 15,000). By how much would the margin of error change?

25. **OPINION POLL.** A poll finds that 54% of the population approves of the job that the President is doing; the poll has a margin of error of 4% (assuming a 95% degree of confidence).
    a. What is the 95% confidence interval for the true population percentage that approves of the President's performance?
    b. What was the size of the sample for this poll?

26. **CONCEALED WEAPONS.** Two-thirds (or 66.6%) of 626 Colorado residents polled by Talmey-Drake Research & Strategy Inc. said they backed a bill pending in the legislature that would standardize laws on granting concealed-weapon permits to gun owners. The bill would force local law enforcement to grant such permits to anyone who can legally carry a gun. The margin of error in the poll was reported as 4 percentage points.
    a. Is the reported margin of error consistent with the sample size for this estimate?
    b. What sample size would be needed to give a margin of error of 2 percentage points?

27. **IMPROVING THE MARGIN OF ERROR.** In general, if one wishes to decrease the margin of error by a factor of 2 in an estimation of a population proportion (say, from $E = 0.02$ to 0.01), by what factor should the sample size be increased? Explain.

PROJECTS FOR THE WEB AND BEYOND

For useful links, select "Links for Web Projects" for Chapter 8 at www.aw.com/bbt.

28. **WHO'S THE VICE PRESIDENT?** Assume that you want to estimate the proportion of students at your college who can correctly identify the Vice President of the United States. How many students must you randomly select in order to be 95% confident that your sample proportion is within 0.1 of the population proportion? Using that sample size, collect your own sample data by randomly selecting and surveying students at your college. Then use the methods of this section to construct a 95% confidence interval. Write a statement summarizing your results.

29. **NIELSEN METHODS.** Visit the Nielsen Media Research Web site and report on the actual methods used to estimate population proportions and confidence intervals in Nielsen ratings.

30. **NETWORK POLLS.** All of the major television networks conduct regular polls on a variety of issues. Visit the Web sites of the major networks and gather the results of a particular poll that involves estimation of a population proportion. Be sure to include all information that is given about the sample size, margin of error, and confidence intervals. Include any details about the actual polling procedure.

IN THE NEWS

1. **ESTIMATING POPULATION PROPORTIONS.** Find a news article or report in which a population proportion is estimated from a sample. The article should include a margin of error and/or a confidence interval. Discuss the methods used in the study and how the conclusions were reached.

Chapter Review Exercises

1. Studies of skulls from past cultures can reveal important information, such as whether there was interbreeding between cultures. Researchers obtained a sample of 35 skulls of Egyptian males who lived around 1850 B.C. The maximum breadth of each skull was measured, with the results $\bar{x} = 134.5$ millimeters and $s = 3.48$ millimeters (based on data from *Ancient Races of the Thebaid* by Thomson and Randall-Maciver).

   a. Use these sample results to construct a 95% confidence interval estimate of the population mean.

   b. Write a statement that correctly interprets the confidence interval found in part a.

   c. What is the margin of error?

   d. The sample mean of $\bar{x} = 134.5$ millimeters was obtained from one specific sample of 35 skulls. If many different samples of 35 such skulls are obtained and their means are recorded, what do we know about the distribution of the sample means?

   e. If we could somehow obtain more skulls so that the sample size was increased beyond 35, what would be the effect on the confidence interval limits? Would they get closer together or farther apart, or would they not change?

   f. Using a different set of sample data, we find a 95% confidence interval $15 < \mu < 25$. What is wrong with this interpretation: "There is a 95% chance that the population mean will fall between 15 and 25"?

2. We want to estimate the mean IQ score for the population of statistics professors. We know that for people randomly selected from the general population, the standard deviation of IQ scores is 16.

   a. Using a standard deviation of 16, how many statistics professors must we randomly select for IQ tests if we want to have 95% confidence that the sample mean is within 2 IQ points of the population mean?

   b. How is our estimate of the mean IQ of statistics professors affected if the sample size is larger than necessary? Smaller than necessary?

   c. Is the actual standard deviation of IQ scores for statistics professors likely to be equal to 16, more than 16, or less than 16? Explain. If we used the actual value instead of 16 in part a, how would the answer to part a be affected? Would it be the same, smaller, or larger?

3. Each year, billions of dollars are spent at theme parks owned by Disney, Universal Studios, Sea World, Busch Gardens, and others. A survey of 1,233 people who took trips revealed that 111 of them included a visit to a theme park (based on data from the Travel Industry Association of America).

   a. What single value is the best estimate of the proportion of people who visit a theme park when they take a trip?

   b. Find a 95% confidence interval estimate of the proportion of all people who visit a theme park when they take a trip.

   c. Write a statement that correctly interprets the result found in part b.

4. a. You have been hired by General Motors to determine the proportion of car owners who plan to purchase a new car within the next 365 days. Assuming that you want to be 95% confident that your sample proportion is within 0.04 of the true population proportion, how many people must you survey?

   b. Suppose that, in conducting the survey described in part a, you find that half of the people called refuse to answer survey questions because they believe that you are trying to sell them something. If you proceed by calling twice as many people so that your sample size is large enough, will your results be good? Explain.

FOCUS ON MEDIA

Did NBC Lose $66 Million to a Nielsen Sampling Error?

The major television networks pay more than $10 million per year to obtain audience data from Nielsen Media Research. But this does not mean the networks are happy with the service. In fact, the networks have been embroiled in a major dispute about the accuracy of Nielsen's statistical studies, claiming that Nielsen systematically underestimates true audience size. Lower audience size means the networks can't charge as much for advertising spots. Moreover, when Nielsen reports that the audience for a particular show was smaller than was promised by a network to an advertiser, the network must give the advertiser free advertising spots as compensation for the broken promise. The networks claim that Nielsen underestimates cost them tens of millions of dollars in lost advertising revenue. For one fall television season alone, NBC accused Nielsen of costing it at least $66 million in lost revenue.

To understand the nature of the dispute, we need to understand how Nielsen measures audience size. As discussed in Chapter 1, Nielsen Media Research uses a sample of 5,000 homes in its survey. Nielsen equips each television in these 5,000 homes with a "people meter" that continuously monitors when each television set is turned on and what channel it is tuned to. Thus, the data for what the *television* is doing should be quite accurate.

But what Nielsen really needs to know is how many people are watching a particular television at a particular time, and something about the personal characteristics of each person (such as age and gender). To gather these data, Nielsen assigns a personal viewing button on the people meter to each person living in each sample household. In order for the people meter to record that a particular individual is watching the television, the individual must push his or her personal viewing button. The people meter also has buttons for guests, who must enter their age and gender when they push a guest button.

The heart of the dispute between the networks and Nielsen comes down to button pushing. The networks claim that many people in Nielsen's sample are not pushing their buttons on the people meter. In that case, the sample data would be incorrect because some people who are watching the television do not get counted. Note that this is a *systematic error*, because failure to push a button always means that Nielsen's audience estimate will be low. The networks are particularly concerned about viewers in the 18–34 age category, because these viewers are especially valuable to advertisers. Nielsen's data show that the number of 18–34-year-olds watching prime-time network television at any given moment dropped significantly during the 1990s, with a particularly steep decline in 1998 when NBC made the $66 million accusation (Figure 8.14).

The networks say that Nielsen's reported drop-off in viewing is so large as to be implausible, and claim instead that it reflects increasing reluctance among the 18–34-year-olds to

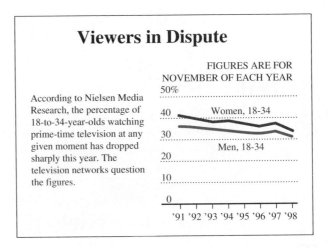

Figure 8.14. SOURCE: *New York Times.*

push people meter buttons. They support this claim with results from an experimental new audience-measuring service (known as "Smart"). Nielsen counters that it has conducted small independent surveys, using techniques other than the people meters, which confirm the reported drop-offs. Thus, Nielsen believes that its data are accurate and not subject to systematic errors. Given the huge amounts of money at stake, it's unlikely that the dispute between the networks and Nielsen will be resolved any time soon.

QUESTIONS FOR DISCUSSION

1. Based on the television viewing habits of 18–34-year-olds whom you know personally, do you think that Nielsen's reported drop-off in network television viewing is realistic? Explain.

2. Nielsen has said that some of its independent surveys suggest that the people meters are overestimating, rather than underestimating, the number of viewers. Can you think of any ways that systematic errors could cause the people meters to overestimate the audience size? Explain.

3. Nielsen's survey is also subject to participation bias (see Section 1.4) because not all of the households randomly selected by Nielsen agree to have the people meters in their homes. For example, in 1990 only 47% of the people contacted by Nielsen agreed to participate, down from 68% in 1980. (In other words, Nielsen must initially select considerably more than 5,000 homes in order to get 5,000 homes that agree to take the people meters.) How might this participation bias be affecting the accuracy of Nielsen's audience size estimates?

4. If you worked for Nielsen, would you suggest an alternative method for collecting data besides the use of the people meters? If so, what? If not, why not?

SUGGESTED READING

Carter, Bill. "Networks Battle Nielsen as Young Viewers Turn Up Missing." *New York Times,* December 22, 1998.

Carter, Bill. "TV Networks Are Scrambling to Deal with Era of New Media." *New York Times,* May 17, 1999.

Nielsen Media Research. *2000 Report on Television.* Nielsen Media Research Communications Department.

FOCUS ON LITERATURE

How Many Words Did Shakespeare Know?

Imagine that you go to an orientation party for international students. During the course of the evening, you meet 12 Swedish people, 9 Chinese people, 6 French people, 4 Israelis, 3 Koreans, and 1 Iranian. You know there are people of other nationalities whom you did *not* meet at the party. Based only on the people you met, is it possible to estimate the *total* number of nationalities represented at the party—including those of the people you did not meet? Thanks to ideas of sampling, the answer is yes.

The party problem may be a bit frivolous. However, essentially the same question arose when Oxford University marine biologist Charles Paxton wondered how many "sea monsters" (creatures more than 2 meters in length) remain to be discovered. In this case, the nationalities at the party correspond to species of sea monsters. Using statistical methods, Paxton was able to estimate that, in addition to the roughly 220 sea monsters already known, another 47 wait to be discovered.

Similar methods have been used to analyze the works of Shakespeare. Statisticians Bradley Efron and Ronald Thisted wondered about the number of words Shakespeare actually knew, which must have been larger than the number he used in his writings. Here, the nationalities at the party correspond to different words in Shakespeare's plays and poems. The data collected at the party can be regarded as the *first sample*. For the Shakespeare question, the first sample consists of the complete known works of Shakespeare—specifically, the numbers of words that are used in these works once, twice, three times, and so forth. Table 8.5 shows a (small) part of the first sample. For example, the table says that in the works of Shakespeare, 14,376 words were used exactly once, 4,343 words were used exactly twice, and so forth. (The full table is much larger and continues far beyond 10 occurrences.)

Given the full table for the first sample, we can now ask a hypothetical question. Suppose a second, new and different sample of Shakespeare's works of the same size as the first sample were discovered. How many words could we expect to find in the second sample that were *not* used in the first sample? We would expect there to be fewer new words in the second sample, because in the first sample *every* first occurrence of a word is new, even a common word like "the"; in the second sample, those common words are no longer new. Efron and Thisted estimated that 11,430 words would appear in the second sample that did not appear in the first sample.

They repeated this argument with a third sample, fourth sample, fifth sample, and so on. With each new sample, the number of new words decreases, but the total number of words used (among all samples) increases. Efron and Thisted eventually found that the number of new words approached about 35,000. This means that in addition to the 31,534 words that

Table 8.5 Numbers of Words in Complete Works of Shakespeare Used from One to Ten Times	
Occurrences	Number of words
1	14,376
2	4,343
3	2,292
4	1,463
5	1,043
6	837
7	638
8	519
9	430
10	364

Shakespeare knew and used, there were approximately 35,000 words that he knew but didn't use. Thus, they estimated that Shakespeare knew approximately 66,500 words.

The analysis of Efron and Thisted, which required advanced methods, was done in 1976. More than ten years later, the ideas were put to practical use when a new sonnet by an unknown author of Shakespeare's time period was discovered. As before, the complete volume of Shakespeare's works was considered the first sample. Now, the second sample was the new sonnet with 429 words. The same statistical method was used to predict that, if Shakespeare was the author, the new sonnet should have seven new words that did not appear in the complete works of Shakespeare. In fact, the new sonnet had nine new words that did not previously appear. Similarly, the method predicted that the new sonnet should have four words that were used exactly once in the complete works. In fact, there were seven words that were used exactly once in the complete works. And the number of words in the sonnet that were used exactly twice in the complete works was predicted to be three, when in fact there were five such words. The authors concluded that the agreement between the predictions and the actual word frequencies in the sonnet was good enough to attribute authorship to Shakespeare.

These statistical methods have been used to distinguish the works of Shakespeare from those of other Elizabethan writers such as Marlowe, Donne, and Jonson. They have been used to strengthen the belief that James Madison wrote certain of the Federalist Papers whose authorship was in doubt. They have even been used to establish the order of Plato's works.

This example illustrates the remarkable applicability of statistics to problems in a seemingly unrelated discipline. Perhaps more important, it demonstrates how two disciplines as distant from each other as literature and marine biology can be related by a common statistical thread.

QUESTIONS FOR DISCUSSION

1. Comment on whether you believe that literature is enhanced by statistical analysis. For example, is your appreciation of Shakespeare improved by knowing how many words Shakespeare knew?

2. Do you believe that statistical analysis is useful for identifying authors of "lost works"?

3. Suggest another discipline, besides biology and literature, in which the ideas described in this Focus could be used to estimate an unknown quantity.

SUGGESTED READING

Efron, B., and Thisted, R. "Estimating the Number of Unknown Species: How Many Words Did Shakespeare Know?" *Biometrika*, Vol. 63, No. 3, 1976, pp. 435–437.

Morin, Richard. "Unconventional Wisdom: Statistics, the King's Deer, and Monsters of the Sea." *Washington Post*, March 7, 1999.

Snell, Laurie. Chance News 8.03.

CHAPTER 9

Hypothesis Testing

Everyone makes claims. Advertisers make claims about their products. Universities claim their programs are superb. Governments claim their programs are effective. Lawyers make claims about a suspect's guilt or innocence. Medical diagnoses are claims about the presence or absence of disease. Pharmaceutical companies make claims about the effectiveness of their drugs. But how do we know whether any of these claims are true? Statistics offers a way to test many claims, through a powerful set of techniques that go by the name of *hypothesis testing*.

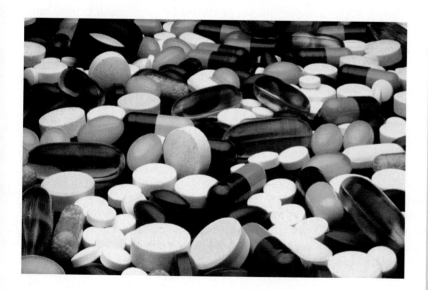

LEARNING GOALS

1. Know the fundamental ideas of hypothesis testing, including the purpose of a hypothesis test and the meaning of statistical significance.

2. Be able to formulate the null and alternative hypotheses for left-, right-, and two-tailed tests and know how to organize the information needed for a test.

3. Know how to conduct and interpret left- and right-tailed tests for claims made about population means, including how to determine significance levels and *P*-values.

4. Know how to conduct and interpret two-tailed tests for claims made about population means and understand type I and type II errors.

5. Know how to conduct and interpret hypothesis tests for claims made about population proportions, including how to determine significance levels and *P*-values.

9.1 Fundamentals of Hypothesis Testing

A company called ProCare Industries, Ltd. once claimed that its product, called Gender Choice, could greatly increase a couple's chances of having a baby girl (by "up to 80%"). Suppose that, as a researcher for the Food and Drug Administration, you decide to test the claim that Gender Choice makes the chances of having a girl much greater than 50%, or 0.50. Because it's not practical to survey all parents who have used Gender Choice, you select a random sample of 100 newborns whose parents used the product.

How would you decide whether the Gender Choice claim was true? Depending on the results of the 100 births, the decision might be easy. For example, suppose that the 100 births resulted in 52 girls and 48 boys. Although 52 girls is greater than half (50 girls), the difference is relatively small. It could easily occur by chance and is not significant. In contrast, if the 100 births resulted in 97 girls and 3 boys, we would probably conclude that Gender Choice works: It is so unlikely that 97 girls would occur by chance that we would appear to have evidence supporting the Gender Choice claim.

In these two instances (52 girls and 97 girls out of 100 births, respectively), a decision about significance is fairly obvious. But now suppose that there are 64 girls among the 100 births in the sample. The issue of significance is not obvious in this case, so we need a method for determining whether this sample offers evidence that Gender Choice works. More generally, our goal is to find a way to determine whether the results observed in any sample are significant or whether they could be expected by chance.

Statistics offers a procedure for evaluating whether sample results are significant. Here is how the procedure works for the case of a sample with 64 girls in 100 births:

- We begin by assuming that Gender Choice does *not* work; that is, it does *not* increase the percentage of girls. If this is true, then we should expect about 50% girls among the population of all births to parents using the product.

- We next ask how likely it is that a random *sample* of 100 births would result in 64% baby girls, assuming that the percentage of baby girls among the *population* of all parents using Gender Choice is 50%.

- If we find that a random sample of births is fairly likely to have 64% girls, then we do not have evidence that Gender Choice works. By contrast, if we find that a random sample of births is unlikely to have 64% girls, then we conclude that the result is due to something other than chance, suggesting that the product is effective. (But this would not *prove* that the product is effective, because there might be other explanations for the significant result.)

By the Way...

The Gender Choice product consisted of printed instructions and a thermometer. It was based on the principle that the gender of a baby can be determined through careful timing of conception. There was no evidence that the product actually worked, and it is no longer available.

Time out to think

Suppose Gender Choice is effective and you select a new random sample of 1,000 births. Based on the first sample (with 64 girls among 100 births), how many baby girls would you expect in the new sample? Next, suppose Gender Choice is *not* effective; how many baby girls would you expect in a random sample of 1,000 births?

This example illustrates the basic idea of *hypothesis testing.* We start with a claim or a statement about a population parameter. For the cases we discuss in this chapter, the population parameter may be either a *population proportion* (such as the proportion of baby girls) or a *population mean* (such as the mean weight of babies). In the Gender Choice example, the claim is about a population proportion—the proportion of baby girls. The starting assumption for the test—that the product does not increase the percentage of baby girls—is called the **null hypothesis** and is denoted H_0 (read "H-naught"). In this case, the null hypothesis is the claim that the proportion of baby girls born to the population of parents using Gender Choice is 0.50, or 50%. Using p to represent the population proportion, we can express this null hypothesis as

$$H_0 \text{ (null hypothesis): } p = 0.50$$

Because we are interested in the company's claim that Gender Choice *increases* the percentage of girls from its natural value of about 50%, it might seem that the null hypothesis should be that the population proportion is *less than or equal to* 0.5. However, with the methods discussed in this book, we can test only a *single value* of the population proportion. Therefore, the null hypothesis will always state that the population parameter is *equal to* a claimed value.

The **alternative hypothesis** (sometimes called a *research hypothesis*), denoted by H_a, is a claim that can be supported only if the null hypothesis is rejected. In our current example, the alternative hypothesis is the claim that the proportion of baby girls born to parents using Gender Choice is *greater than* 0.50. This alternative hypothesis can be expressed as

$$H_a \text{ (alternative hypothesis): } p > 0.50$$

A different alternative hypothesis might be the claim that the proportion of girls was *less than* 0.50, but this case is not considered here because we are testing the claim that the proportion of baby girls is *greater than* 0.50.

In general, there are two possible outcomes to any hypothesis test:

- rejecting the null hypothesis, H_0

- not rejecting the null hypothesis, H_0

Note that "accepting the null hypothesis" is not listed as a possible outcome. At best, we can only find statistical evidence that the null hypothesis is *false*, in which case we can safely reject it. In the absence of such evidence, we choose not to reject the null hypothesis. This allows the possibility that the null hypothesis is true, but does not absolutely prove it to be true.

> ## By the Way...
>
> The word *null* comes from the Latin *nullus*, meaning "nothing." A null hypothesis often states that there is no special effect or difference.

Time out to think

As an illustration of the difficulty of proving the truth of a null hypothesis, explain why it is easier for a biologist to prove that a rare species of animal does exist than to prove that it does not exist. Is it ever possible to prove that a species (say, Bigfoot) does not exist?

In the Gender Choice example, the two possible decisions are

- to reject the null hypothesis and conclude that the proportion of baby girls is greater than 0.50 when parents use Gender Choice

- not to reject the null hypothesis, in which case we don't have enough evidence to conclude that the proportion of baby girls is greater than 0.50.

Note that the null and alternative hypotheses must always be formulated based on prior beliefs or expectations *before* the sample is selected for testing. Although it may be tempting, it is inappropriate to examine sample data before formulating the hypotheses.

Definitions

A **hypothesis** is a claim about a population parameter, such as a population proportion, p, or population mean, μ.

The **null hypothesis**, or H_0, gives a specific value for a population parameter. When testing a claim about a population mean or population proportion, we write null hypotheses in the form

$$H_0: \text{population parameter} = \text{claimed value}$$

The **alternative hypothesis**, or H_a, is a statement that the population parameter has a value that somehow differs from the value claimed in the null hypothesis.

A **hypothesis test** is a standard procedure for testing a claim about the value of a population parameter. There are always two possible outcomes of a hypothesis test: (1) *reject* the null hypothesis, which lends support to the alternative hypothesis, or (2) *do not reject* the null hypothesis. Note that the latter possibility does not *prove* that the null hypothesis is true.

EXAMPLE 1 Formulating Hypotheses

The manufacturer of a new fuel-conserving car advertises that the car averages 38 miles per gallon on the highway. A consumer group claims that the true mean is less than 38 mi/gal. State the null and alternative hypotheses for a hypothesis test.

SOLUTION The null hypothesis must state that the population parameter, in this case a population mean, is *equal* to some specific value. The claimed population mean is the advertised mileage for all of the new cars, $\mu = 38$ miles per gallon. This claim is the null hypothesis. The alternative hypothesis is the consumer group's claim that the true gas mileage is *less* than advertised, or $\mu < 38$ miles per gallon. (A different alternative hypothesis might claim $\mu > 38$ miles per gallon, but it is not of interest here because we are testing the consumer group's claim.) To summarize:

$$H_0 \text{ (null hypothesis): } \mu = 38 \text{ miles per gallon}$$
$$H_a \text{ (alternative hypothesis): } \mu < 38 \text{ miles per gallon}$$

Statistical Significance

When it is not necessary to make a decision, it is necessary not to make a decision.

—Lord Falkland

Let's return to the Gender Choice example. The key question is this: Is the difference between the proportion of baby girls claimed by the null hypothesis (0.50) and the proportion observed in the sample (0.64) significant in the sense that it is not likely to occur by chance? If the difference is unlikely to occur by chance, it is *statistically significant* (see Section 6.1). In that case, we would conclude that the null hypothesis is false. By contrast, if the difference is too small to be statistically significant, we cannot reject the null hypothesis.

Let's work with a level of significance of 0.05. In the case of hypothesis testing, a 0.05 significance level means that a sample result is significant if its probability is 0.05 or less, assuming the null hypothesis is true. If its probability is greater than 0.05, we say that the sample result is not significant and could easily occur by chance.

Here are two examples. First, suppose there are 52 girls in a sample of 100 births to couples using Gender Choice. Under the assumption of the null hypothesis (that the true population proportion is 50% girls), we can use methods not yet discussed to find that the probability of getting at least 52 girls in a random sample of 100 is 0.382. Because 0.382 is greater than 0.05, this result is not significant at the 0.05 level and we cannot reject the null hypothesis. Thus, this sample does not lend support to the claim that Gender Choice is effective.

For the second example, suppose there are 64 girls in the sample of 100 births. Under the assumption that the null hypothesis is true, the probability of a sample result this high or higher turns out to be only 0.00332. Because 0.00332 is less than 0.05, the result *is* significant. Thus, the sample with 64 girls among 100 births gives us good reason to reject the null hypothesis, thereby providing support for the alternative hypothesis that, among couples who use Gender Choice, girls occur at a rate that is greater than 50%.

In these examples, we are using the following *rare event rule,* which is the basis for hypothesis testing.

Rare Event Rule

If, under a given assumption (such as the null hypothesis), the probability of a particular event at least as extreme as the observed event is very small (such as 0.05 or less), we conclude that the assumption is probably not correct.

EXAMPLE 2   Coin Tossing Significance

To test whether coins are biased against heads, each of three people tosses a coin 100 times. Which of the following results is significant at the 0.05 level? At the 0.01 level? The probability for each result is given.

a. The first person gets 30 heads in 100 tosses; the probability of getting 30 or fewer heads with a fair coin is 0.00004.

b. The second person gets 40 heads in 100 tosses; the probability of getting 40 or fewer heads with a fair coin is 0.028.

c. The third person gets 47 heads in 100 tosses; the probability of getting 47 or fewer heads with a fair coin is 0.31.

SOLUTION   The null hypothesis is that the coin is fair and the expected proportion of heads is 0.5.

a. Because the 0.00004 probability of tossing 30 or fewer heads in 100 tosses is less than both 0.05 and 0.01, this result is significant at both the 0.05 level and the 0.01 level. We have good reason to conclude that the coin is biased against heads.

b. Because the 0.028 probability of 40 or fewer heads lies between 0.01 and 0.05, this result is significant at the 0.05 level, but not at the 0.01 level. We have good reason to conclude that the coin is biased against heads.

c. The probability of tossing 47 or fewer heads in 100 tosses is 0.31, which is greater than 0.05. Thus, this result is not significant at either the 0.05 level or the 0.01 level. This event could easily have occurred by chance with a fair coin, and we have no reason to conclude that the coin is biased against heads.   ∎

P-Values

We need one more important idea before we cover hypothesis testing in more detail. Once again, the Gender Choice example is helpful. Suppose we assume that the null hypothesis is true (the proportion of baby girls for all parents using Gender Choice is 0.50). We then determine that the probability of getting at least 64 girls out of 100 is 0.00332. This tells us that, if the null hypothesis is true, the probability of selecting a sample as extreme as the one we observed is about 3 in 1,000—which is quite small. This probability of 0.00332 is called a **P-value** (short for probability value). Because this probability is less than 0.05, the difference between the sample proportion and the claimed proportion is significant at the 0.05 level.

> ## Definition
>
> The **P-value** for a hypothesis test of a claim about a population parameter is the probability of selecting a sample at least as extreme as the observed sample, assuming that the null hypothesis is true.

The usefulness of a P-value is that it not only determines whether an observed sample result is significant at a certain level, but also gives us an idea of *how* significant the sample result is. A small P-value indicates that the observed result is unlikely and provides evidence to reject the null hypothesis. A large P-value indicates that the sample result is not unusual or that it could easily occur by chance; in this case, we cannot reject the null hypothesis. We will discuss how to compute P-values later in the chapter.

EXAMPLE 3   Significance and Birth Weight

A county health official believes that the mean birth weight of male babies at a local hospital is *greater than* the national average of 3.39 kilograms. A random sample of 145 male babies born at that hospital has a mean birth weight of 3.61 kilograms. Assuming that the mean birth weight of all male babies born at the hospital is 3.39 kilograms, a calculation shows that the probability of selecting a sample with a mean birth weight of 3.61 kilograms or more is 0.032.

a. Formulate the null and alternative hypotheses.

b. What is the P-value for this sample?

c. Is the difference between the population mean (3.39 kilograms) and the observed mean (3.61 kilograms) significant at the 0.05 level?

SOLUTION

a. The null hypothesis is the claim that the mean birth weight of all male babies born at this hospital is the national average of 3.39 kilograms:

$$H_0 \text{ (null hypothesis): } \mu = 3.39 \text{ kilograms}$$

The alternative hypothesis (formulated *before* the sample is selected) is the claim of the health official:

$$H_a \text{ (alternative hypothesis): } \mu > 3.39 \text{ kilograms}$$

b. The *P*-value is 0.032; it is the probability of randomly selecting a sample with a mean of at least 3.61 kilograms (assuming that the population mean is really 3.39 kilograms).

c. Because the *P*-value is less than 0.05, the difference is significant at the 0.05 level. Based on the sample, there is sufficient evidence to reject the null hypothesis, thereby supporting the alternative hypothesis that the mean weight for all male babies born at the hospital is greater than 3.39 kilograms. ∎

Legal and Medical Analogies of Hypothesis Testing

Two analogies—one legal and one medical—might help clarify the idea of hypothesis testing. In American courts of law, the fundamental principle is that a defendant is presumed innocent until proven guilty. Imagine that you are a juror in a criminal case and must listen to the evidence and decide on the guilt or innocence of a defendant. Because innocence is the initial assumption, the null hypothesis is the statement that the defendant is innocent. Thus, we have

H_0: The defendant is innocent.

H_a: The defendant is guilty.

The job of the prosecutor is to present evidence so compelling that the jury is persuaded to reject the null hypothesis. In other words, if the evidence (which is analogous to the sample data) is inconsistent with the null hypothesis, then the jury may find grounds for conviction, which amounts to rejecting H_0. Note that finding a person innocent (accepting H_0) is not an option. In fact, in our legal system, defendants are found to be guilty or not guilty, but they are never found to be innocent. A verdict of not guilty means the evidence is not sufficient to establish guilt, but it does not prove innocence.

Lawyers are just like physicians: what one says, the other contradicts.

—Sholem Aleichem

Time out to think

Consider two situations. In one, you are a juror in a case in which the defendant could be fined a maximum of $200. In the other, you are a juror in a case in which the defendant could receive the death penalty. Compare the significance levels you would use in the two situations. In each situation, what are the consequences of wrongly rejecting the null hypothesis?

In the case of making a medical diagnosis of some disease, a physician generally starts with the assumption of normal health (no disease) and then looks for evidence that a disease is actually present. Thus, we have

H_0: The disease is absent.

H_a: The disease is present.

The aim of a physician is to collect enough evidence (the sample data) to reject the null hypothesis and conclude that disease is present.

Time out to think

You are a physician diagnosing one patient with flu symptoms and another patient with terminal cancer symptoms. Compare the significance levels you would use in the two situations. In each situation, what are the consequences of wrongly rejecting the null hypothesis?

Review Questions

1. In testing a population parameter, what does the null hypothesis claim? How do we look for evidence for the alternative hypothesis?

2. What are the two possible conclusions of a hypothesis test?

3. Explain the meaning of statistical significance in the context of hypothesis testing.

4. Explain in general terms the meaning of a *P-value*.

5. In a court of law, what is the null hypothesis? What is the null hypothesis for a medical diagnosis?

Exercises

BASIC SKILLS AND CONCEPTS

SENISBLE STATEMENTS? For Exercises 1–4, determine whether the given statement is sensible and explain why it is or is not.

1. We can never support a claim that a gender-selection technique favors girls if sample results include fewer girls than boys.

2. A claim that a gender-selection technique favors girls will be supported whenever sample results include more girls than boys.

3. In testing a claim about the mean amount of aspirin in tablets, a researcher states the null hypothesis as $\mu > 300$ mg.

4. Testing a claim that the mean amount of aspirin in tablets is equal to 300 milligrams results in sufficient data to prove that the null hypothesis of $\mu = 300$ mg is true.

5. WHAT IS SIGNIFICANT? Imagine that your average body weight has been 130 pounds for a long time.
   a. One day you get on your scale and find that you weigh 132 pounds (an increase of 1.5%). Would you consider this increase statistically significant or would you attribute it to random fluctuations?
   b. One day you get on your scale and find that you weigh 140 pounds (an increase of 7.7%). Would you consider this increase statistically significant or would you attribute it to random fluctuations?
   c. How much of a gain would you need to see before you considered it statistically significant and looked for an explanation? Explain.

6. WHAT IS SIGNIFICANT? Imagine that you record the gas mileage of your car every time you buy gasoline and you know that the average mileage is 29 miles per gallon.

   a. One day you see that your gas mileage is 28 miles per gallon. Would you consider this change statistically significant and look for an explanation, or is it the sort of variation that you would attribute to chance?
   b. One day you see that your gas mileage is 22 miles per gallon. Would you consider this change statistically significant and look for an explanation, or is it the sort of variation that you would attribute to chance?
   c. About how much of a decrease would you need to see before you considered it statistically significant and looked for an explanation? Explain.

FURTHER APPLICATIONS

FORMULATING HYPOTHESES. In Exercises 7–10, state the null and alternative hypotheses for a test of significance.

7. A principal claims that the percentage of left-handed people in her school is less than 11%.

8. The coach of a soccer team claims that the mean height of the players on his team is less than 70 inches (which is the mean height of American males).

9. The owners of a retirement community claim that the mean lifetime of the women in their community is greater than 78 years (which is the mean lifetime of all American women).

10. The Food and Drug Administration claims that the amount of vitamin C in tablets produced by a company is less than the advertised 500 milligrams.

P-VALUES AND BIRTHS. Assume that among all births in the population, 50% result in boys and 50% result in girls. The following table shows the probability of various numbers of male babies in a random sample of 80 births. Use this information for Exercises 11–14, and assume that we are testing for a bias against males.

Number of births	Probability
30 or fewer	0.007
34 or fewer	0.036
36 or fewer	0.060
38 or fewer	0.080

11. A random sample of 80 births has 30 male babies. Is this result significant at the 0.05 level? What is the P-value for this result?

12. A random sample of 80 births has 34 male babies. Is this result significant at the 0.01 level? What is the P-value for this result?

13. A random sample of 80 births has 36 male babies. Is this result significant at the 0.05 level? What is the P-value for this result?

14. A random sample of 80 births has 38 male babies. Is this result significant at the 0.05 level? What is the P-value for this result?

P-VALUES AND DICE. The table below shows the probability of various numbers of 6's in 100 rolls of a fair die. Use this information for Exercises 15–17, and assume that we are testing for a bias against 6's.

Number of sixes	Probability
8 or fewer	0.006
10 or fewer	0.020
12 or fewer	0.052

15. In 100 rolls of a fair die, 8 rolls are 6's. Is this result significant at the 0.05 level? What is the P-value for this result?

16. In 100 rolls of a fair die, 10 rolls are 6's. Is this result significant at the 0.05 level? What is the P-value for this result?

17. In 100 rolls of a fair die, 12 rolls are 6's. Is this result significant at the 0.05 level? What is the P-value for this result?

PROJECTS FOR THE WEB AND BEYOND

18. Many professional journals, such as *Journal of the American Medical Association*, contain articles that include information about formal tests of hypotheses. Find such an article and identify the null hypothesis and alternative hypotheses. In simple terms, state the objective of the hypothesis test and the conclusion that was reached.

19. Select a particular quarter and test the claim that it favors heads when flipped. State the null and alternative hypotheses; then flip the quarter 100 times. Applying only common sense, what do you conclude about the claim that the quarter favors heads?

1. **HYPOTHESIS TESTING IN THE NEWS.** Find a news article or research report that describes (perhaps not explicitly) a hypothesis test for a population mean or proportion. Attach the article and summarize the method used.

9.2 Setting Up Hypothesis Tests

In Section 9.1, we sketched the basic outline of a hypothesis test. In this and the following sections, we will fill in the details and define the essential terminology until we arrive at a complete procedure.

Formulating the Hypotheses

Columbia College advertises that the mean starting salary of its graduates is $29,000 (which is above the national average). The Committee for Truth in Advertising, an independent organization, suspects that this claim may be exaggerated and decides to conduct a hypothesis test. As we have seen, a hypothesis test works with two opposing claims: the null hypothesis and the alternative hypothesis. In this example, the null hypothesis is the college's claim that the

TECHNICAL NOTE

As discussed earlier, the hypothesis test requires one specific value of μ. For this reason, we will always state H_0 with an equality and will conduct the test under the assumption that the mean *equals* the value given in the null hypothesis. However, we will reject the null hypothesis only if there is sufficient evidence to support the alternative hypothesis.

mean starting salary for all graduates is $29,000. Denoting a population mean by μ, we can express this claim as

$$H_0: \mu = \$29{,}000$$

The alternative hypothesis expresses the suspicion of the independent organization that the mean starting salary is *less than* $29,000. Thus, the alternative hypothesis for this example can be expressed as

$$H_a: \mu < \$29{,}000$$

Having formed its hypotheses, the Committee for Truth in Advertising selects a random sample of 100 recent graduates from the college. The mean salary of the graduates in the sample turns out to be $27,000. Based on the results of this survey, should we doubt the claim of Columbia College? In other words, if we assume that the college's claim is true (the mean salary of all graduates is $29,000), how likely is it that we would select a sample with a mean salary of $27,000?

Before answering this question, let's consider variations on this example. In other situations, it might be appropriate to claim that the mean salary is *greater than* $29,000 or *different from* $29,000. Thus, depending on the specific problem, the alternative hypothesis can take the forms

$$H_a \text{ (alternative hypothesis): } \mu < \$29{,}000$$
$$H_a \text{ (alternative hypothesis): } \mu > \$29{,}000$$
$$H_a \text{ (alternative hypothesis): } \mu \neq \$29{,}000$$

The first form ("less than") leads to a **left-tailed test** because, as we will see, it requires testing whether the population mean lies to the *left* (lower values) of the claimed value. Similarly, the second form ("greater than") leads to a **right-tailed test.** The third form ("different from") results in a **two-tailed test** because it tests whether the population mean lies significantly far to *either* side of the claimed value. (Left-tailed tests and right-tailed tests are discussed in Section 9.3; two-tailed tests are discussed in Section 9.4.) Much of the art of hypothesis testing lies in choosing the correct form of the alternative hypothesis. We will discuss this issue later in the chapter.

Time out to think

Suppose you were a recruiter for Columbia College and wanted to investigate whether the mean starting salary of graduates is actually more than the claimed $29,000. Which form of the alternative hypothesis would you use?

Forms of the Null and Alternative Hypotheses

The **null hypothesis**, or H_0, gives a specific value for a population parameter. Thus, it has the form that includes equality:

$$H_0: \text{population parameter} = \text{claimed value}$$

The **alternative hypothesis**, or H_a, has one of the following forms:

(left-tailed) H_a: population parameter $<$ claimed value
(right-tailed) H_a: population parameter $>$ claimed value
(two-tailed) H_a: population parameter $\neq$ claimed value

The hypotheses should be formulated before sample data are analyzed, and one should never test a hypothesis using the same data that suggested the hypothesis.

EXAMPLE 1   Identifying Hypotheses

In each case, give the population mean or population proportion and state the null and alternative hypotheses for a hypothesis test.

a. The Ohio Department of Health claims that the average stay in Ohio hospitals after childbirth is greater than 2.0 days, which is the national average.

b. A wildlife biologist working in the African savanna claims that the actual proportion of female zebras in the region is different from the accepted proportion of 50%.

SOLUTION

a. The population is all women in Ohio who have recently given birth. The null hypothesis claims that the mean hospital stay for the population equals 2.0 days. The Health Department suspects that the mean stay in Ohio is *greater than* 2.0 days. Thus, the hypotheses are

$$H_0 \text{ (null hypothesis): } \mu = 2.0 \text{ days}$$
$$H_a \text{ (alternative hypothesis): } \mu > 2.0 \text{ days}$$

This choice of an alternative hypothesis leads to a right-tailed test.

b. In this case, the claim involves a population proportion. The claimed population proportion is $p = 0.5$, which becomes the null hypothesis. The wildlife biologist claims that the actual population proportion is different from (either greater or less than) 0.5. The hypotheses are

$$H_0 \text{ (null hypothesis): } p = 0.5$$
$$H_a \text{ (alternative hypothesis): } p \neq 0.5$$

This choice of an alternative hypothesis leads to a two-tailed test.

The Idea Behind the Hypothesis Test

As discussed in Section 9.1, all hypothesis tests are based on the assumption that the null hypothesis is true. With the Columbia College example, the null hypothesis is the college's claim that the mean salary for all graduates (the population mean) is $29,000; the alternative hypothesis is the suspicion that the mean starting salary is less than $29,000.

Recall that the hypothesis test has two possible outcomes: Either we reject the null hypothesis and conclude that the mean starting salary is less than $29,000, or we do not reject the null hypothesis, in which case we cannot conclude that the mean is less than $29,000.

The evidence we will use to decide between these two possibilities comes from the single sample with $n = 100$ recent graduates and a mean of $27,000. Using the notation introduced in Chapter 8, we write

$$\text{sample mean} = \bar{x} = \$27,000$$

In addition to the sample mean, $\bar{x}$, and the sample size, n, the hypothesis test also requires the population standard deviation, σ, which in the case of large samples may be approximated by the standard deviation of the sample, s. These three quantities ($\bar{x}$, n, and s) are the *sample statistics* for the test.

The key step is finding the probability of observing this particular sample mean given that the null hypothesis is true. In other words, how likely is it that we would select a sample with a mean of $27,000 or less when the mean for the whole population is $29,000? Here is the crucial observation, based on our work with sampling distributions in Chapter 8: The observed sample mean ($\bar{x} = \$27,000$) is just one point in a distribution of sample means. Furthermore, assuming the null hypothesis is true, this distribution of sample means is approximately normal with a mean of $\mu = \$29,000$.

As shown in Figure 9.1, if the sample mean lies near the assumed population mean, then the probability of observing it is not small, and there is no reason to reject the null hypothesis. If the sample mean lies far from the assumed population mean, then the probability of observing it is small, which gives reason to reject the null hypothesis. We will return to the details of the process and be precise about "near" and "far" in the next section.

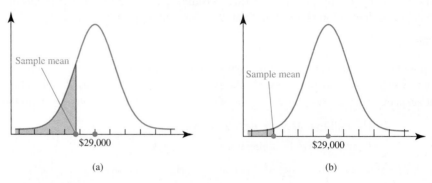

(a)   (b)

Figure 9.1   (a) A sample mean near the assumed population mean gives no reason to reject the null hypothesis. (b) If the sample mean lies far from the assumed population mean, we have reason to reject the null hypothesis.

> ## Hypothesis Test Requirements
>
> We need the following information to conduct a hypothesis test:
>
> 1. the claimed value of the population parameter. This value may be either a population mean, μ, or a population proportion, p.
>
> 2. the sample mean, $\bar{x}$, or the sample proportion, $\hat{p}$
>
> 3. the sample size, n
>
> In the case of a population mean, we also need
>
> 4. the population standard deviation, σ, but for large samples we can approximate it by the sample standard deviation, s.
>
> We use this information to determine the probability of observing the sample statistic found in the study, *assuming* the null hypothesis is true. Based on this probability, we decide whether to reject the null hypothesis.

When you have eliminated the impossible, whatever remains, however improbable, must be the truth.

—Sir Arthur Conan Doyle,
The Sign of Four

EXAMPLE 2   Mean Rental Car Mileage

Sid Symington is the owner of Statewide Car Rental Company, which has a large fleet of cars. He wants to test the claim that the mean annual mileage for the population of all cars in his fleet is more than 11,725 miles (which is the mean annual mileage for all cars in the United States). He selects a random sample of $n = 225$ cars from his fleet. The mean annual mileage for this sample is $\bar{x} = 12,100$ miles. A good friend makes the following observation to Sid: If you assume that the mean annual mileage for all cars in the fleet is, in fact, $\mu = 11,725$ miles, then the probability of selecting a 225-car sample with a mean annual mileage of 12,100 miles or more is 0.01.

a. Formulate the null and alternative hypotheses for this situation.

b. Is the difference between the claimed mean mileage, $\mu = 11{,}725$ miles, and the observed sample mean, $\bar{x} = 12{,}100$ miles, significant at the 0.05 level?

c. Do these observations give you sufficient evidence to reject the null hypothesis?

SOLUTION

a. The null hypothesis states that the mean annual mileage for the population of all cars in the fleet is 11,725 miles:

$$H_0: \mu = 11{,}725 \text{ miles}$$

Because Sid suspects that the mean mileage for his fleet is more than 11,725 miles, his alternative hypothesis is

$$H_a: \mu > 11{,}725 \text{ miles}$$

b. The probability (of 0.01) is less than 0.05. Therefore, the difference between the claimed mean and the observed sample mean is significant at the 0.05 level.

c. There is sufficient evidence to reject the null hypothesis at the 0.05 significance level, which means that there is sufficient evidence to support the claim that the mean annual mileage of his fleet is greater than 11,725 miles.   ∎

EXAMPLE 3   Pre-Election Polls

A Senate candidate hires a pollster to determine her prospects in an upcoming election. The goal is to test the claim that, among the population of all voters, the candidate has more than $p = 0.5$ (50%) of the vote. The pollster selects a random sample of 400 prospective voters. Within the sample, the proportion of voters who support the candidate is $\hat{p} = 0.51$ (51% of the voters). The pollster gives the candidate the following fact: If you assume that the proportion of people in the population who support you is $p = 0.5$, the probability of selecting a sample in which the proportion of your supporters is at least $\hat{p} = 0.51$ is 0.345.

a. Formulate the null and alternative hypotheses for this situation.

b. Is the difference between $p = 0.5$ and the observed sample proportion, $\hat{p} = 0.51$, significant at the 0.05 level?

c. Do these observations give the candidate evidence to reject the null hypothesis?

d. How would the interpretation of the test change if the proportion of voters in the sample supporting the candidate was $\hat{p} = 0.55$? The probability of selecting a sample with $\hat{p}$ at least 0.55 is 0.0228.

SOLUTION

a. We are testing a population *proportion* (rather than a population mean). The null hypothesis must be the claim that the population proportion is equal to some claimed value—in this case, $p = 0.5$:

$$H_0: p = 0.5$$

The candidate would like to conclude that she has the support of more than 50% of the voters (a majority). Therefore, her alternative hypothesis is

$$H_a: p > 0.5$$

b. The probability of 0.345 is *not* less than 0.05, so the difference between the sample proportion and the claimed population proportion is not significant at the 0.05 level. Let's interpret this result carefully. It says that the sample we have selected is not particularly unexpected. If we were to select 100 samples of 400 voters, roughly 35% of the samples would have proportions of $\hat{p} = 0.51$ or more. This says that the sample is quite consistent with the null hypothesis that the population proportion is $p = 0.5$.

c. The lack of significance in the results implies that we do *not* have grounds to reject the claim that $p = 0.5$. Thus, based on this sample, the candidate cannot reasonably expect a majority of voters to support her.

d. Because the probability of selecting a sample with $\hat{p} \geq 0.55$ is 0.0228, which is less than 0.05, the result is significant at the 0.05 level. Thus, the candidate has good reason to reject the null hypothesis of receiving only 50% of the vote. This evidence supports the alternative hypothesis, which means she is likely to have more than 50% of the vote. ∎

Review Questions

1. Give an example of a situation that requires testing a population *mean* for significance. Give an example of a situation that requires testing a population *proportion* for significance.

2. What are the three forms of the alternative hypothesis?

3. For testing a population mean, what information is needed from the sample?

4. For testing a population proportion, what information is needed from the sample?

Exercises

BASIC SKILLS AND CONCEPTS

SENSIBLE STATEMENTS? For Exercises 1–4, determine whether the given statement is sensible and explain why it is or is not.

1. In a test of the claim that statistics professors have a mean IQ greater than 100, the alternative hypothesis is $\mu > 100$.

2. In a test of a claim about a population mean, the most relevant sample statistic is the sample mean, $\bar{x}$.

3. In a test of the claim that, among newborn babies, the proportion of boys is greater than 0.5, the null hypothesis is $p > 0.5$.

4. In a test of the claim that the proportion of left-handed adults is greater than 0.08, the alternative hypothesis is $\mu > 0.08$.

FURTHER APPLICATIONS

FORMULATING HYPOTHESES. In Exercises 5–14, formulate the null and alternative hypotheses for a hypothesis test. Iden-

tify the population and describe how to select a random sample for a hypothesis test. State in clear terms the conclusion that should be given if the null hypothesis is rejected.

5. A consumer group claims that the mean amount of preservative added to a brand of potato chips exceeds the amount listed on the packages.

6. A high school principal claims that the mean mathematics SAT score of female seniors at his school is less than the national average of 495.

7. The director of public works for a small town claims that the mean water usage among households exceeds 1,675 gallons per month, the level required for the water supply to last through the year.

8. A manufacturer of rubber bands claims that the mean lifetime of its rubber bands far exceeds 9 months, the mean lifetime of all other rubber bands.

9. The owner of a chain of hardware stores opens new franchises only in prosperous towns. He claims that Moon

Valley is a good spot for a new franchise because the mean per capita income is above the national average of $25,598.

10. The Food and Drug Administration claims that the amount of active ingredient in a new antibiotic is less than 500 milligrams.

11. A college football coach claims that the average weight of his linemen is greater than 275 pounds, the average weight of all college linemen.

12. An agricultural research company claims that the yield for its new breed of corn is greater than 204 bushels per acre, the yield of the nearest competitor.

13. A chemical company wishes to defend its claim that the levels of pollution downstream from its plant are less than the critical level specified by the Environmental Protection Agency.

14. A middle school principal claims that his students read for more than 14 hours per week, the average for the school district.

PROJECTS FOR THE WEB AND BEYOND

For useful links, select "Links for Web Projects" for Chapter 9 at www.aw.com/bbt.

15. Assume you want to test the claim that, among newborn babies, the proportion of boys is greater than 0.5. State the null and alternative hypotheses. Use the Web to find relevent data that might be used in such a hypothesis test. Do the sample data appear to support the given claim? Explain.

IN THE NEWS

1. **SETTING UP A HYPOTHESIS TEST.** Find a news article or research report that makes a specific claim that could, at least in principle, be tested with a hypothesis test. Describe how you would set up the hypothesis test and list all the information you would need to conduct the test.

9.3 Hypothesis Tests for Population Means

The previous sections laid the groundwork for the complete process of hypothesis testing. In this section, we develop a systematic procedure for *one-tailed tests* (left-tailed or right-tailed) for population means.

Finding the Standard Score

Recall that we formulated the null and alternative hypotheses for the Columbia College example as follows:

H_0: μ = $29,000, or the mean starting salary for *all* graduates at the college is $29,000.

H_a: μ < $29,000, or the mean starting salary for *all* graduates at the college is less than $29,000.

Having formulated the hypotheses, we now look at the sample statistics. The mean starting salary of the sample of graduates is $\bar{x}$ = $27,000 and the sample size is n = 100. Let's assume that the standard deviation of the salaries in the sample is s = $6,150. (Given the sample data, all three sample statistics can be computed fairly easily.) To summarize, here are the three required sample statistics for the hypothesis test:

$$\text{sample size} = n = 100$$
$$\text{sample mean} = \bar{x} = \$27,000$$
$$\text{sample standard deviation} = s = \$6,150$$

As discussed in the previous section, the key step is finding the probability of observing a sample with a mean of $\bar{x} = \$27{,}000$ or less given that the null hypothesis is true. Again, imagine that we select *many* samples of $n = 100$ graduates at the college and compute the mean starting salary for each sample. If the mean starting salary for the entire population really is $29{,}000$, then the mean of this distribution of sample means has a nearly normal distribution centered about $29{,}000$.

By the Central Limit Theorem (Section 5.3), the standard deviation of this distribution of sample means is

$$\frac{\sigma}{\sqrt{n}} \approx \frac{s}{\sqrt{n}} = \frac{\$6{,}150}{\sqrt{100}} = \$615$$

Figure 9.2 shows this distribution of sample means. The actual sample (with $\bar{x} = \$27{,}000$) is just one of the many samples from this distribution. The question now becomes, What is the *probability* of selecting a single sample with a mean of $27{,}000$ or less from this distribution?

We can answer this question by using standard scores (see Chapter 5). The mean of the distribution of sample means is $29{,}000$ and the standard deviation is 615, so the standard score for a value of $27{,}000$ is

$$z = \frac{\bar{x} - \mu}{\sigma/\sqrt{n}} = \frac{\$27{,}000 - \$29{,}000}{\$615} = -3.25$$

From Table 5.1, a standard score of -3.25 is below the 0.13 *percentile*, which corresponds to a probability below 0.0013. In fact, the probability of selecting a sample with a mean less than or equal to $27{,}000$ is about 0.0006, which is extremely low. In other words, this sample provides strong evidence for rejecting the null hypothesis.

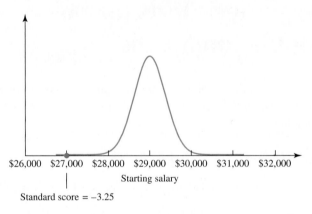

Figure 9.2 The distribution of sample means for starting salaries has a mean of $29{,}000$ and a standard deviation of 615. For a particular sample with a mean of $27{,}000$, the standard score is -3.25.

Calculations for a Hypothesis Test

Given the sample mean, $\bar{x}$, the sample standard deviation, s, the sample size, n, and the claimed population mean, μ, the following quantities are needed for the hypothesis test:

$$\text{standard deviation for the distribution of sample means} = \frac{\sigma}{\sqrt{n}} \approx \frac{s}{\sqrt{n}}$$

$$\text{standard score for the sample mean, } z = \frac{\bar{x} - \mu}{\sigma/\sqrt{n}}$$

Stating the Results: Level of Significance

We can now be more precise about deciding to reject or not reject the null hypothesis. We will carry out the hypothesis test at the 0.05 significance level. In Table 5.1, we see that the 5th percentile has a standard score between $z = -1.6$ and $z = -1.7$; using more precise tables, we would find that the 5th percentile has a standard score of $z = -1.645$. Therefore, a left-tailed hypothesis test is significant at the 0.05 level if the standard score of the sample mean is less than or equal to $z = -1.645$, which is the *critical value* for this hypothesis test.

For the Columbia College example, the standard score of $z = -3.25$ for the sample mean is less than the critical value of $z = -1.645$ (Figure 9.3). Thus, we have reason to reject the null hypothesis at the 0.05 significance level.

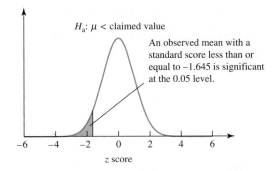

Figure 9.3 The observed sample mean of $\bar{x} = \$27,000$ has a standard score of -3.25, which is less than the critical value of $z = -1.645$.

The Columbia College example involved a *left-tailed* test because the alternative hypothesis had the form of H_a: $\mu <$ claimed value. A similar argument applies to right-tailed tests with alternative hypotheses of the form H_a: $\mu >$ claimed value. In such cases, significance at the 0.05 level requires that the sample mean lie at or *above* the 95th percentile—which requires a standard score *greater than* or equal to $z = 1.645$ (Figure 9.4). Table 9.1 summarizes our findings.

Table 9.1   Testing for Significance at the 0.05 Level

Type of test	Form of H_a	Reject H_0 if standard score is
Left-tailed test	H_a: $\mu <$ claimed value	$z \leq -1.645$
Right-tailed test	H_a: $\mu >$ claimed value	$z \geq 1.645$

TECHNICAL NOTE

To test at a 0.01 significance level in a right-tailed test or a left-tailed test, replace $z = 1.645$ by $z = 2.33$ in Table 9.1. The values of z for which H_0 is rejected form the *rejection region* or *critical region*.

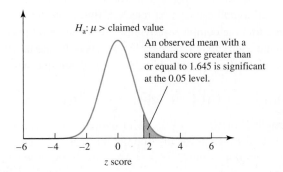

Figure 9.4 For a right-tailed test with H_a: $\mu >$ claimed value, the sample mean must have a standard score $z \geq 1.645$ for significance at the 0.05 level.

Time out to think

Suppose that, in right-tailed tests, one study finds a sample mean with $z = 2$ and another study finds a sample mean with $z = 10$. Note that both are significant at the 0.05 level. But which result is more significant? Explain.

EXAMPLE 1   Testing Aspirin

The Medassist Pharmaceutical Company bottles aspirin tablets in batches of 500 tablets. It is imperative that each bottle have at least 500 tablets. As a regulator for the Food and Drug Administration (FDA), you want to test the claim that consumers are getting *less than* the advertised 500 tablets per bottle.

a. Give the null and alternative hypotheses for a hypothesis test.

b. You select a random sample of $n = 200$ bottles and find that they have a mean of $\bar{x} = 499$ tablets per bottle and a standard deviation of $s = 7$ tablets. Find the standard score for the observed sample mean. State the conclusion of the hypothesis test.

SOLUTION

a. The null hypothesis is the claim of the pharmaceutical company that all bottles contain a mean of 500 tablets:

$$H_0: \mu = 500 \text{ tablets per bottle}$$

Your suspicion that there are fewer than 500 tablets per bottle forms the alternative hypothesis:

$$H_a: \mu < 500 \text{ tablets per bottle}$$

Note that this is a left-tailed test.

b. We use $s = 7$ as an approximation to the population standard deviation, σ. (This approximation is valid because the sample size is relatively large, $n > 30$.) The standard deviation of the distribution of sample means is

$$\frac{\sigma}{\sqrt{n}} \approx \frac{s}{\sqrt{n}} = \frac{7}{\sqrt{200}} = 0.49$$

The standard score of the sample mean with respect to the claimed mean is

$$z = \frac{\bar{x} - \mu}{\sigma/\sqrt{n}} = \frac{499 - 500}{0.49} = -2.0$$

The sample mean is 2 standard deviations below the claimed mean. Because this standard score is less than the critical value of -1.645 for a left-tailed test, the result is statistically significant and the null hypothesis can be rejected. Based on the sample, there is sufficient evidence to conclude that the company provides fewer than 500 tablets per bottle. Note that this result is significant even though the sample mean is only one tablet less than the claimed mean, thanks to the relatively large sample size and small sample standard deviation. ∎

> As a general rule for such large samples, if the difference between the expected value and the observed number is greater than two or three standard deviations, then the [null] hypothesis would be suspect to a social scientist.
>
> —Supreme Court,
> Castaneda v. Partida, 1977

EXAMPLE 2   Mean Rental Car Mileage (Revisited)

Remember Sid (Example 2 of Section 9.2), who owns a car rental company and tests the claim that the mean annual mileage for the population of all cars in his fleet is greater than 11,725 miles. He uses a random sample of $n = 225$ cars from his fleet. The sample mean is $\bar{x} = 12,100$ miles and the sample standard deviation is $s = 2,415$ miles. Is the difference between the claimed mean and the sample mean significant at the 0.05 level?

SOLUTION   We use the standard deviation of the sample, $s = 2{,}415$ miles, to approximate the population standard deviation, σ. The standard deviation of the distribution of sample means is

$$\frac{\sigma}{\sqrt{n}} \approx \frac{s}{\sqrt{n}} = \frac{2{,}415}{\sqrt{225}} = 161$$

We can now find the standard score for the observed mean:

$$z = \frac{\bar{x} - \mu}{\sigma/\sqrt{n}} = \frac{12{,}100 - 11{,}725}{161} = 2.33$$

This standard score is greater than the critical value of $z = 1.645$ (Figure 9.5). Based on this sample, there is sufficient evidence to reject the null hypothesis that $\mu = 11{,}725$ miles at the 0.05 significance level. Thus, Sid has reason to believe that the mean mileage for his fleet is greater than the national average. Note that this is a right-tailed test.

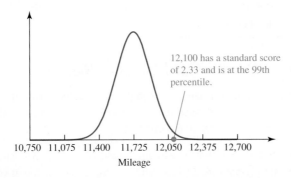

12,100 has a standard score of 2.33 and is at the 99th percentile.

Figure 9.5 A distribution of sample means with a mean of $\mu = 11{,}725$ miles and a standard deviation of $s = 161$ miles. A sample mean of $\bar{x} = 12{,}100$ miles has a standard score of 2.33 and is at the 99th percentile.

Stating the Results: *P*-Values

Significance levels and standard scores allow us to make a yes/no decision about rejecting the null hypothesis. However, it is common practice (particularly in many research journals) to be more specific about the strength of the evidence for rejecting the null hypothesis. We introduced the idea of a *P*-value in Section 9.1. Now, we need to be more precise about it.

In the Columbia College example, we saw that the sample mean, $\bar{x} = \$27{,}000$, has a standard score of $z = -3.25$. This standard score lies below the 0.1 percentile and, in fact, corresponds to a probability of about 0.0006. This probability is the *P-value* for the test. Recall that if the *P*-value is less than or equal to the significance level of the test, then we are justified in rejecting the null hypothesis. In this case, the *P*-value is less than the significance level of 0.05, but we also see *how much less* it is. This is the extra power of a *P*-value. Table 9.2 gives some commonly used interpretations of various *P*-values.

Because *P*-values are probabilities, we can visualize them with a graph of a normal curve. Figure 9.6 shows typical normal distributions of sample means. For a left-tailed test (Figure 9.6a), the *P*-value is the shaded area under the curve, which corresponds to the probability of selecting a sample with a mean less than or equal to the sample mean. Similarly, for a right-tailed test (Figure 9.6b), the *P*-value is the shaded area under the curve, which now corresponds to the probability of selecting a sample with a mean greater than or equal to the observed mean.

The P-value . . . is often used as a gauge of the degree of contempt in which the null hypothesis deserves to be held.

—Robert Abelson, *Statistics as a Principled Argument*

Table 9.2 Interpretation of P-values	
P-value	**Interpretation**
Less than 0.01	Test is highly statistically significant and offers strong evidence against H_0.
0.01 to 0.05	Test is statistically significant and offers moderate evidence against H_0.
Greater than 0.05	Test is not statistically significant and does not offer sufficient evidence against H_0.

Time out to think

Suppose you calculate a P-value and reject the null hypothesis at the 0.01 significance level. Explain why this means that you also reject the null hypothesis at the 0.05 significance level. If you do not reject the null hypothesis at the 0.01 significance level, does this mean that you would not reject it at the 0.05 significance level either? Explain.

EXAMPLE 3   Mean Rental Car Mileage (Revisited)

Sid, the owner of the car rental company in Example 2, selected a sample with a mean annual mileage of $\bar{x} = 12,100$ miles. He determined that the standard score of this sample mean is $z = 2.33$. Find and interpret the P-value for this test.

SOLUTION   Table 5.1 shows that a standard score of $z = 2.33$ corresponds to the 99th percentile. Therefore, the probability of selecting a sample with a mean of $\bar{x} = 12,100$ miles or more is 0.01, which is the P-value for the test (see Figure 9.5). The low P-value offers strong evidence against the null hypothesis, suggesting that Sid is correct in believing that his cars are driven more than the national average of 11,725 miles. ∎

Putting It All Together

Having looked at the pieces of left- and right-tailed hypothesis tests of a population mean, we can now put it all together into a four-step procedure.

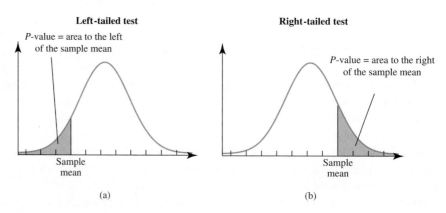

Figure 9.6 Graph of a normal distribution of sample means, showing the area that corresponds to a P-value for (a) a left-tailed test and (b) a right-tailed test.

One-Tailed Hypothesis Tests for Means

1. State H_0 in the form $\mu =$ claimed value and H_a in the form $\mu <$ claimed value (left-tailed test) or $\mu >$ claimed value (right-tailed test).

2. Identify the relevant sample statistics: the sample size, n; the sample mean, $\bar{x}$; and the sample standard deviation, s.

3. Find the standard deviation of the distribution of sample means,

$$\frac{\sigma}{\sqrt{n}} \approx \frac{s}{\sqrt{n}}$$

and find the standard score for the observed mean,

$$z = \frac{\bar{x} - \mu}{\sigma/\sqrt{n}}$$

Recall that we can use the standard deviation, s, to estimate σ as long as the sample is large ($n > 30$).

4. Determine the outcome of the test by either comparing the standard score with the critical value given in Table 9.1 or computing a P-value. If the P-value is less than or equal to 0.05, then the test is significant at the 0.05 level; there is sufficient evidence to reject H_0. If the P-value is greater than 0.05, H_0 cannot be rejected.

EXAMPLE 4   Driving by Women in Different Age Groups

According to data from the Federal Highway Administration, women aged 25–33 drive an average of 7,125 miles per year. The 720 women in a sample of 16–24-year-old drivers have a mean annual driving mileage of 6,047 miles with a standard deviation of 2,944 miles. Based on this sample, test the claim that the mean annual mileage for the 16–24 age category is less than the population mean for 25–33-year-olds.

SOLUTION   We use the four-step procedure. First, notice that the population in question is women drivers 16–24 years old; the sample is taken from this population.

Step 1: The null hypothesis is the claim that the mean annual mileage, μ, for women aged 16–24 is 7,125 miles:

$$H_0\colon \mu = 7{,}125 \text{ miles}$$

The alternative hypothesis is that the mean annual driving mileage for the 16–24 age category is less than 7,125 miles:

$$H_a\colon \mu < 7{,}125 \text{ miles}$$

Step 2: The sample statistics are the sample size, $n = 720$; the sample mean, $\bar{x} = 6{,}047$ miles; and the sample standard deviation, $s = 2{,}944$ miles.

Step 3: With s used to approximate σ, the standard deviation of the distribution of sample means is

$$\frac{\sigma}{\sqrt{n}} \approx \frac{s}{\sqrt{n}} = \frac{2{,}944}{\sqrt{720}} = 110$$

The standard score of the sample mean is

$$z = \frac{\bar{x} - \mu}{\sigma/\sqrt{n}} = \frac{6{,}047 - 7{,}125}{110} = -9.8$$

Step 4: The mean of the sample is almost 10 standard deviations below the claimed population mean, which is far less than the critical value of $z = -1.645$ for significance at the 0.05 level (see Table 9.1). We have very strong evidence to reject the null hypothesis. It is reasonable to conclude that the mean annual driving distance is less for 16–24-year-olds than for 25–33-year-olds. ∎

Time out to think

Can you identify reasons why 16–24-year-old women would drive less than 25–33-year-old women?

EXAMPLE 5   Average Farm Size

A sample of $n = 88$ farms in an Iowa county has a mean area of 466 acres and a standard deviation of 248 acres. County officials want to make the case that farms in the county are significantly larger than the national average of 434 acres. Test this claim at the 0.05 significance level.

SOLUTION   We again follow the four-step procedure.

Step 1: The null hypothesis is the claim that the mean farm size in the county equals the national average, which is $\mu = 434$ acres:

$$H_0: \mu = 434 \text{ acres}$$

The alternative hypothesis is the claim that the mean farm size in the county is greater than the national average, or

$$H_a: \mu > 434 \text{ acres}$$

Note that this will be a right-tailed test.

Step 2: The sample statistics are the sample size, $n = 88$; the sample mean, $\bar{x} = 466$ acres; and the sample standard deviation, $s = 248$ acres.

Step 3: With s used to approximate σ, the standard deviation of the distribution of sample means is

$$\frac{\sigma}{\sqrt{n}} \approx \frac{s}{\sqrt{n}} = \frac{248}{\sqrt{88}} = 26.4$$

The standard score of the sample mean is

$$z = \frac{\bar{x} - \mu}{\sigma/\sqrt{n}} = \frac{466 - 434}{26.4} = 1.2$$

Step 4: The mean of the sample is 1.2 standard deviations above the claimed population mean. This standard score does not satisfy the critical condition of $z \geq 1.645$ required for significance at the 0.05 level. Thus, the test is not significant at the 0.05 level. We cannot claim that the average farm size in the county is above the national average. ∎

Time out to think

In the previous example, would the test be more or less likely to be significant if a larger sample were chosen, with the same values of $\bar{x}$ and s?

Statistical Significance and Practical Significance

While we have spent a lot of time discussing statistical significance, there are other kinds of significance that should be included in interpreting a hypothesis test.

Consider a weight loss program that guarantees weight loss after two days in the program. Suppose a random sample of thousands of people in the program has a mean weight loss of 0.37 pound after two days, and this mean weight loss proves to be significant at the 0.05 significance level (in other words, the null hypothesis of no weight loss is rejected). Despite this *statistical* significance, the mean loss of 0.37 pound is so small that it has virtually no *practical* significance. Who would ever bother to enroll in a program that results in an average weight loss of only 0.37 pound?

On the other hand, consider a large company in which the employees claim they are underpaid compared to the national average for workers in the same position. Suppose a sample of 50 employees has a mean monthly salary of $2,500, compared to a national mean salary of $2,950. A significance test gives a *P*-value of 0.072, which is *not* significant at the 0.05 level. Because of factors such as large variation and a small sample, the difference of $450 might not be statistically significant, but it could have *practical significance* for the employees.

There is another subtle aspect to interpreting significance. You may have noticed that the standard score of a sample mean actually *increases* in magnitude (becomes either more positive or more negative) as the sample size, n, increases, while everything else remains the same. Therefore, in the previous example of salary differences, a test that is statistically insignificant with a sample size of $n = 50$ could become significant with a larger sample size of, say, $n = 100$. A small difference can be statistically significant if the sample size is large enough.

Review Questions

1. Explain how the standard score of a sample mean indicates whether a test is significant.

2. In testing an alternative hypothesis having the form H_a: $\mu <$ claimed value, what condition must the standard score of the sample mean satisfy in order for a test to be significant? What condition must be satisfied for an alternative hypothesis of the form H_a: $\mu >$ claimed value?

3. Explain the meaning of a *P*-value. How do we interpret *P*-values in hypothesis tests? If you chose a 0.05 significance level, how would you interpret a *P*-value of 0.02?

4. Describe the four-step process for a one-tailed hypothesis test for a population mean. Give an example.

5. Explain the difference between statistical significance and practical significance.

Exercises

BASIC SKILLS AND CONCEPTS

SENSIBLE STATEMENTS? For Exercises 1–4, determine whether the given statement is sensible and explain why it is or is not.

1. In a hypothesis test, the test statistic is the same as the P-value.

2. When you are testing a claim about a population mean, a test statistic of $z = 8.26$ leads to rejection of the null hypothesis.

3. In a hypothesis test, a P-value of 0.00001 indicates that you should reject the null hypothesis.

4. When you are testing a claim at the 0.05 significance level, a P-value of 0.2 indicates that you should reject the null hypothesis and support the stated claim.

REJECT OR NOT REJECT? In Exercises 5–10, find the value of the standard score, z, and determine whether the alternative hypothesis is supported.

5. H_a: $\mu < 25$, $\bar{x} = 24$, $s = 9$, claimed population mean = 25, $n = 100$

6. H_a: $\mu < 2.040$, $\bar{x} = 2.01$, $s = 0.1$, claimed population mean = 2.040, $n = 400$

7. H_a: $\mu < 12.55$, $\bar{x} = 12.4$, $s = 2.3$, claimed population mean = 12.55, $n = 900$

8. H_a: $\mu > 25$, $\bar{x} = 26.2$, $s = 9$, claimed population mean = 25, $n = 100$

9. H_a: $\mu > 2.040$, $\bar{x} = 2.045$, $s = 0.1$, claimed population mean = 2.040, $n = 400$

10. H_a: $\mu > 12.55$, $\bar{x} = 12.75$, $s = 2.3$, claimed population mean = 12.55, $n = 900$

FINDING P-VALUES. In Exercises 11–18, use Table 5.1 to find the P-values that correspond to the standard scores. In each case, give an interpretation of the P-value you found.

11. $z = -0.9$ for H_a: $\mu <$ claimed value

12. $z = -1.9$ for H_a: $\mu <$ claimed value

13. $z = 2.1$ for H_a: $\mu >$ claimed value

14. $z = 1.8$ for H_a: $\mu >$ claimed value

15. $z = 2.3$ for H_a: $\mu >$ claimed value

16. $z = 3.1$ for H_a: $\mu >$ claimed value

17. $z = -2.5$ for H_a: $\mu <$ claimed value

18. $z = -1.6$ for H_a: $\mu <$ claimed value

FURTHER APPLICATIONS

19. **INTERPRETING P-VALUES.** Suppose you were testing an alternative hypothesis of the form H_a: $\mu >$ claimed value and the sample mean had a standard score of $z = -4.0$. Without using tables, decide whether this test is significant. Explain.

20. **INTERPRETING P-VALUES.** Suppose you were testing an alternative hypothesis of the form H_a: $\mu <$ claimed value and the sample mean had a standard score of $z = 4.0$. Without using tables, decide whether this test is significant. Explain.

HYPOTHESIS TESTS FOR MEANS. For Exercises 21–28, carry out a full hypothesis test using the four-step process described in the text. State your conclusion for the 0.05 significance level.

21. A Roper poll used a sample of 100 randomly selected car owners. Within the sample, the mean time for ownership of a single car was 7.01 years with a standard deviation of 3.74 years. Test the claim of the owner of a large dealership that the mean time of ownership for all cars is less than 7.5 years.

22. According to a study by the Centers for Disease Control, the national mean hospital stay after childbirth is 2.0 days. Reviewing records at her own hospital, a hospital administrator calculates that the mean hospital stay for a sample of 81 women after childbirth is 2.2 days with a standard deviation of 1.2 days. Assuming that the patients represent a random sample of the population, test the claim that this hospital keeps new mothers longer than the national average. How does the result change if the standard deviation of the sample is 0.8 day?

23. According to the Energy Information Administration (Federal Highway Administration data), the average gas mileage of all automobiles is 21.4 miles per gallon. Suppose that within a random sample of 40 sport utility vehicles (SUVs), the mean gas mileage is 19.8 miles per gallon with a standard deviation of 3.5 miles per gallon. Test the claim that the mean mileage of all SUVs is less than 21.4 miles per gallon.

24. For a random sample of 1,700 households with VCRs, the mean household income is $41,182 with a standard deviation of $19,900 (based on Nielsen Media Research data). Test the claim that the mean household income of VCR owners across the population is greater than $40,000.

25. According to the Bureau of Economic Analysis, the (mean) per capita income in the United States is $25,598. The following data are gathered from four states (standard deviations are estimates). For each state, test the claim that the per capita income is greater than (for Illinois and Washington) or less than (for Alabama and Georgia) the national mean. Assume that a random sample of 900 residents is used in each state.

State	Per capita income	Standard deviation
Alabama	$20,842	$12,123
Illinois	$28,202	$18,302
Georgia	$24,061	$15,309
Washington	$26,718	$14,823

26. According to the Energy Information Administration (Federal Highway Administration data), the average annual mileage for automobiles in the United States is 11,725 miles. The owner of a rental car company selects a random sample of 225 cars from his entire fleet. Within the sample, the mean annual mileage is 12,145 miles with a standard deviation of 3,000 miles. Test the claim that the mean mileage of the fleet is greater than the national average.

27. The Department of Energy and many utility companies use heating degree-days to measure the outdoor air temperature and the demand for heating. The following table shows the mean heating degree-days for each month over all years since 1950, averaged over the entire United States.

Month	Degree-days	Month	Degree-days	Month	Degree-days
Jan	948	May	150	Sep	69
Feb	768	Jun	36	Oct	271
Mar	611	Jul	7	Nov	528
Apr	339	Aug	13	Dec	836

SOURCE: U.S. Department of Commerce, National Oceanic and Atmospheric Administration.

Suppose that for the year 2000, $n = 50$ weather stations, one in each state, record the heating degree-days for April and September. The sample statistics for these two months are given in the table below. Do the data suggest that the mean heating degree-days were greater than normal in April of 2000 and less than normal in September of 2000?

Month	Mean	Standard deviation
Apr	355	70
Sep	60	35

28. According to the Current Population Survey by the Bureau of the Census, men with a bachelor's degree earn an average of $46,700. Suppose a random sample of 400 women with a bachelor's degree has mean earnings of $28,700 (the national mean for women) with a standard deviation of $16,500. Test the claim that the mean for women's earnings is less than the mean for men's earnings.

PROJECTS FOR THE WEB AND BEYOND

For useful links, select "Links for Web Projects" for Chapter 9 at www.aw.com/bbt.

29. COMPARISONS WITH NATIONAL AVERAGES. Choose several variables that are relatively easy to measure in a class or sample of students. The variables should involve a quantity that can be averaged (for example, height, weight, family size, blood pressure, heart rate, reaction time). Use the Web or other references to determine national averages for these variables (by age categories, if appropriate). Collect data on the variables, using a random sample of at least 50 individuals. Carry out the relevant hypothesis test to determine whether the sample mean differs significantly from the population mean.

30. COUNTY DATA. The *Statistical Abstract of the United States* and the Current Population Survey provide an inexhaustible supply of social, economic, and vital statistics at the county, state, and local levels. Use their Web sites to compare state data to national data in the following way.
    a. Choose a variable of interest that involves a mean for a particular state (for example, the mean household size in Illinois).
    b. Find the current national value for that variable (for example, the national mean household size).
    c. Choose a particular county within the state and obtain the corresponding data for the county, as well as the sample size (for example, the mean household size in Cook County, Illinois).
    d. Assuming that the county is a random sample of the state, test the claim that the state is above or below the national level in terms of that variable.
    e. Discuss and interpret your results. The hypothesis test depends on the sample (the county) being a random sample of the population (the state). Be sure to discuss this factor in your conclusions.

9.4 Hypothesis Testing: Further Considerations

So far, we've discussed how to carry out a basic, one-tailed hypothesis test and how to state a conclusion in terms of a 0.05 significance level and a *P*-value. In this section, we will explore two-tailed tests, along with two common types of error that occur in hypothesis testing.

Two-Tailed Tests

One-tailed tests involve alternative hypotheses with one of the following two forms:

$$H_a: \mu < \text{claimed value (left-tailed)} \quad \text{or} \quad H_a: \mu > \text{claimed value (right-tailed)}$$

We now extend our discussion to two-tailed tests, in which the alternative hypothesis takes the form

$$H_a: \mu \neq \text{claimed value}$$

Consider a drug company that seeks to be sure that its 500-milligram aspirin tablets really contain 500 milligrams of aspirin. If the tablets contain less than 500 milligrams, consumers are not getting the advertised dose. If the tablets contain more than 500 milligrams, consumers are getting too much of the drug. The null hypothesis says that the population mean of the aspirin content is 500 milligrams:

$$H_0: \mu = 500 \text{ milligrams}$$

The drug company is interested in the possibility that the mean weight is *either* less than *or* greater than 500 milligrams. Thus, the alternative hypothesis becomes

$$H_a: \mu \neq 500 \text{ milligrams}$$

This form of H_a leads to a *two-tailed test,* because we consider means on both sides of the claimed mean of 500 milligrams. We can extend the results found for one-tailed tests: If the standard score of the observed mean is less than or equal to -1.96 or greater than or equal to 1.96, the test is significant at the 0.05 level (Figure 9.7). Note that this critical value for two-tailed tests is different from the critical values (± 1.645) for one-tailed tests, shown in Table 9.1.

P-values for two-tailed tests also are found in a way that is different from the methods for one-tailed tests. For a two-tailed test, the *P*-value is the probability of selecting a sample with a mean at least as extreme as the observed mean in *either* direction. For this reason, the *P*-value is *twice* the value for a one-tailed test.

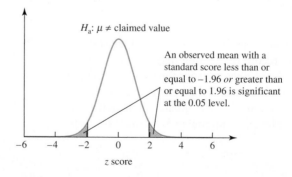

Figure 9.7 For a two-tailed test to be significant at the 0.05 level, the standard score of the observed mean must be either less than or equal to -1.96 or greater than or equal to 1.96.

> ### Two-Tailed Tests (H_a: $\mu \neq$ Claimed Value)
>
> The test is significant at the 0.05 level if the standard score of the observed mean is less than or equal to -1.96 or greater than or equal to 1.96.
>
> The P-value is *twice* the area of the extreme region bounded by the standard score of the observed mean.

Aside from these differences in determining significance and P-values, the procedure for two-tailed hypothesis tests is the same as the four-step procedure introduced in Section 9.3 for one-tailed tests.

EXAMPLE 1   Two-Tailed Tests

The standard score of a sample mean in a two-tailed test is $z = 2.1$. Is the test significant at the 0.05 level? What is the P-value in this case?

SOLUTION   The standard score is greater than the critical value of $z = 1.96$, so the test is significant at the 0.05 level and we can reject the null hypothesis. By Table 5.1, a standard score of $z = 2.1$ is the 98.2 percentile, which corresponds to a probability of 0.982. Thus, the probability of selecting a sample mean with a standard score *greater than* 2.1 is $1 - 0.982 = 0.018$ (Figure 9.8). Because this is a two-tailed test, the P-value is twice this probability, or 0.036.

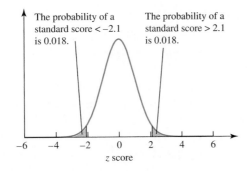

Figure 9.8 In a two-tailed test, a sample mean with a standard score of $z = 2.1$ has a P-value of 0.036 and is significant at the 0.05 level. ∎

EXAMPLE 2   What Is Human Mean Body Temperature?

Consider again the study in which University of Maryland researchers measured body temperatures in a sample of $n = 106$ healthy adults (see Example 4 in Section 8.2). They found a sample mean body temperature of $\bar{x} = 98.20°F$ with a standard deviation of $s = 0.62°F$. Use the four-step procedure for hypothesis testing to determine whether this sample provides evidence for rejecting the common belief that mean human body temperature is $\mu = 98.6°F$.

SOLUTION

Step 1:   The null hypothesis is the claim that mean human body temperature is 98.6°F, or H_0: $\mu = 98.6°F$. We want to test the alternative hypothesis that mean body temperature is *not* 98.6°F, or H_a: $\mu \neq 98.6°F$. This will be a two-tailed test.

Step 2:   The sample statistics are given in the problem: $n = 106$; $\bar{x} = 98.20°F$; and $s = 0.62°F$.

Step 3: With s used to approximate σ, the standard deviation of the distribution of sample means is

$$\frac{\sigma}{\sqrt{n}} \approx \frac{s}{\sqrt{n}} = \frac{0.62}{\sqrt{106}} = 0.06$$

The standard score of the sample mean is

$$z = \frac{\bar{x} - \mu}{\sigma/\sqrt{n}} = \frac{98.20 - 98.60}{0.06} = -6.67$$

Figure 9.9 shows the graph of the sampling distribution.

Step 4: The standard score for the sample mean is much less than the critical value of -1.96 for significance at the 0.05 level. Thus, there is very strong evidence for rejecting the null hypothesis; according to this test, human mean body temperature is not equal to 98.6°F.

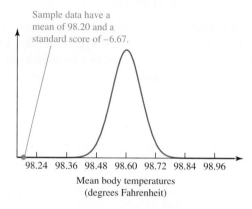

Sample data have a mean of 98.20 and a standard score of −6.67.

Mean body temperatures (degrees Fahrenheit)

Figure 9.9 Assuming the null hypothesis is true, the distribution of sample means has a mean of 98.6°F and a standard deviation of 0.06°F. A mean temperature of 98.20°F is more than 6 standard deviations below the mean.

Errors in Hypothesis Testing

Even if we apply all of our methods correctly, two common types of error occur in decision making with hypothesis tests. They are easy to understand with the legal analogy described in Section 9.1. Recall that the null hypothesis in the legal analogy is H_0: The defendant is innocent. One type of error occurs if we reject the null hypothesis and conclude that the defendant is guilty when, in reality, he or she is innocent. In this case, the null hypothesis has been wrongly rejected. The other type of error occurs if we find the defendant *not guilty*—that is, if we do *not* reject the null hypothesis—when he or she actually is guilty. In this case, we have wrongly failed to reject the null hypothesis.

For a statistical example, consider again the pharmaceutical company testing the claim that the mean amount of aspirin in its tablets is 500 milligrams (H_0: $\mu = 500$ milligrams). In drawing a conclusion from the test, the company could make the following two types of error:

- The company might reject the null hypothesis and conclude that the mean amount is not 500 milligrams, when it really is. This error can lead to wasted time and money trying to fix a process that isn't broken. An error of this type, in which H_0 is *wrongly rejected*, is called a **type I error**.

- The company might fail to reject the null hypothesis when, in fact, the mean amount of aspirin is not 500 milligrams. In this case, the company will distribute tablets that have

If a man will begin with certainties, he shall end in doubts; but if he will be content to begin with doubts, he shall end in certainties.

—Francis Bacon

too much or too little aspirin; consumers could suffer or sue. An error of this type, in which we *wrongly fail to reject H_0*, is called a **type II error**.

Table 9.3 summarizes the four possible cases.

Table 9.3 Decision Table for H_0 and H_a		Reality	
		H_0 true	H_a true
Decision	Reject H_0	Type I error	Correct decision
	Do not reject H_0	Correct decision	Type II error

Time out to think

Explain the effects of type I and type II errors in the medical analogy given in Section 9.1. Which error type would you consider more serious in this case?

EXAMPLE 3   Errors in Body Temperature Test

Consider the null hypothesis from Example 2—that mean body temperature equals 98.6°F (H_0: $\mu = 98.6$°F).

a. What correct decisions are possible with this null hypothesis?

b. Explain the meaning of type I and type II errors in this case.

SOLUTION

a. Any hypothesis test has two possible correct decisions. In this case, one correct decision occurs if the mean body temperature really is 98.6°F and we do *not* reject H_0. The other correct decision occurs if the mean body temperature is *not* 98.6°F and we reject H_0.

b. A type I error occurs if we reject H_0 when it is actually true. In this case, a type I error occurs if mean body temperature really is 98.6°F, but we conclude that it is not. A type II error occurs if we do not reject H_0 when it is actually false. In this case, a type II error occurs if the mean body temperature is not 98.6°F, but we fail to reach this conclusion. ∎

Choosing Hypotheses and Expressing the Conclusion

A hypothesis test is used because there is an original claim that needs to be investigated. Depending on the particular situation and *who is conducting the test*, the original claim may appear either as the null hypothesis or as the alternative hypothesis. For example, consider a situation in which a factory is investigated for releasing pollutants into a nearby stream. Suppose there is a maximum allowed level of pollutants specified by the government. In order to determine whether the factory is violating the regulations, samples of water are collected from the stream and analyzed. It is hoped to determine, based on the samples, whether the mean level of pollutants released from the factory is less than the maximum allowed level.

If *factory officials* conduct the test, they would be inclined to make the claim that the mean level of pollutants released by the factory is *less than* the maximum allowed level. The hypotheses should be chosen as follows:

$$H_0: \text{mean level of pollutants} = \text{maximum level allowed}$$
$$H_a: \text{mean level of pollutants} < \text{maximum level allowed}$$

TECHNICAL NOTE

Some statisticians take the position that the claim to be proved must always appear as the alternative hypothesis. However, many claims appear directly or indirectly as the null hypothesis.

If H_0 is rejected, then H_a is supported, and based on the samples taken, the mean level of pollutants is less than the allowed maximum. This outcome is a conclusive decision in *favor* of the factory officials. If the test indicates that H_0 should not be rejected, then the test is inconclusive. The company has not yet demonstrated that it complies with regulations.

On the other hand, suppose *an environmental group* conducts the test using the same sample statistics. If the environmental group hopes to prove that the factory is in violation, it would claim that the mean level of pollutants released by the factory is *greater than* the maximum allowed level. This claim becomes the alternative hypothesis, and we have the following hypotheses:

$$H_0: \text{mean level of pollutants} = \text{maximum level allowed}$$
$$H_a: \text{mean level of pollutants} > \text{maximum level allowed}$$

If H_0 is rejected, then H_a is supported, and based on the samples taken, the mean level of pollutants is greater than the allowed maximum. Now there is a conclusive decision against the factory. If we do not reject H_0, this is a favorable decision for the factory, but again, it's not proof that the factory is actually in compliance.

Time out to think

What is the effect of a type I error (wrongly rejecting the null hypothesis) in the test that factory officials might conduct? What is the effect of a type I error in the test that the environmental group might conduct? Contrast the implications of the errors in these two situations.

These two examples suggest a general rule: If you want conclusive support for a claim, then state the claim in such a way that it becomes the alternative hypothesis. Rejecting the null hypothesis gives conclusive support for the claim. Conversely, if you want to reject or disprove a claim, then state the claim in such a way that it becomes the null hypothesis. If the null hypothesis is rejected, then the claim is refuted conclusively.

Unfortunately, there are some limitations in choosing hypotheses because of the requirement that the null hypothesis must involve equality. For example, suppose that, by law, a baker must package loaves of bread that have a mean weight of 1 pound. To test the claim that the mean weight of the loaves is 1 pound, the hypotheses would be

$$H_0: \text{mean weight} = 1 \text{ pound}$$
$$H_a: \text{mean weight} \neq 1 \text{ pound}$$

These hypotheses lead to a two-tailed test. If H_0 is rejected, there is sufficient evidence to conclude that the baker is not complying with the regulations. If we fail to reject H_0, there is some evidence (but not proof) that the baker is meeting the requirements.

EXAMPLE 4   Mining Gold

The success of precious metal mines depends on the purity (or grade) of ore removed and the market price for the metal. Suppose the purity of gold ore must be at least 0.5 ounce of gold per ton of ore in order to keep a particular mine open. Samples of gold ore are used to estimate the purity of the ore for the entire mine. Discuss the impacts of type I and type II errors on two of the possible alternative hypotheses:

$$H_a: \text{purity} < 0.5 \text{ ounce per ton}$$
$$H_a: \text{purity} > 0.5 \text{ ounce per ton}$$

SOLUTION   For the left-tailed case, the null and alternative hypotheses are

H_0: purity = 0.5 ounce per ton

H_a: purity < 0.5 ounce per ton

The two possible decisions in this case are

- Rejecting H_0, which means concluding that the purity of the ore is less than that needed to keep the mine open—so the mine is closed.

- Not rejecting H_0, which means we have insufficient evidence to conclude that the purity of the ore is less than that needed—so the mine stays open.

A type I error (wrongly rejecting a true H_0) means the mine is closed when, in fact, the purity of the gold ore is sufficient to operate the mine. For the mine's operators, this means the loss of potential profits from the mine; for the employees, it means unnecessary loss of jobs. A type II error (failing to reject a false H_0) means that the mine continues to operate, but is actually unprofitable.

Now suppose that the mine will be kept open only if the purity is greater than 0.5 ounce per ton. For this right-tailed case, the null and alternative hypotheses are

H_0: purity = 0.5 ounce per ton

H_a: purity > 0.5 ounce per ton

The two possible decisions are

- Rejecting H_0, which means concluding that the purity of the ore is *more* than that needed to keep the mine open—so the mine stays open.

- Not rejecting H_0, which means there is insufficient evidence to conclude that the purity of the ore is high enough to keep the mine open—so the mine is closed.

A type I error (wrongly rejecting a true H_0) means the mine is left open when it is actually unprofitable. A type II error (failing to reject a false H_0) means closing the mine when it is actually profitable—putting the employees out of work for no reason. ■

Significance and Type I Errors

There is an important connection between the significance level of a hypothesis test and type I errors (wrongly rejecting H_0). Recall that a hypothesis test is always done under the assumption that the null hypothesis is true. The significance level (say, 0.05) determines the region within which the standard score for the sample mean must lie in order to reject the null hypothesis (see Figure 9.4). The probability that a random sample mean lies in this region is 0.05. Thus, the significance level is also the probability that the null hypothesis is rejected when, in fact, it is true. That is,

The significance level is the probability of making a type I error.

If there are serious consequences of making a type I error (wrongly rejecting the null hypothesis), then it is advisable to choose a small significance level. Similarly, if H_0 is a claim that is generally believed to be true, it will take extremely convincing evidence to reject it; thus, a small value for the significance level is required. Conversely, if H_0 is a controversial claim, less evidence may be needed to reject it and a larger significance level may suffice. The probability of making a type II error is more complicated and is beyond the scope of this book.

Review Questions

1. What form should the null and alternative hypotheses have in a two-tailed test?

2. Explain how to use the standard score of a sample mean to determine significance in a two-tailed test. How does the critical value for significance at the 0.05 level in two-tailed tests compare to the critical value for one-tailed tests?

3. Explain how to find a P-value in a two-tailed test. How does the computation of this P-value differ from the computation in one-tailed tests?

4. Explain the meaning of type I and type II errors. Give an example of each.

5. Describe how the selection of hypotheses can depend on whose claim is being tested.

6. Explain how the significance level of a hypothesis test is related to the probability of making a type I error.

Exercises

BASIC SKILLS AND CONCEPTS

SENSIBLE STATEMENTS? For Exercises 1–4, determine whether the given statement is sensible and explain why it is or is not.

1. The significance level in a hypothesis test is the probability of making a type I error.

2. If it is extremely important not to reject a true null hypothesis wrongly, the test should be conducted with a very small significance level.

3. With a very small significance level, a possible conclusion of a hypothesis test is that the null hypothesis is true.

4. With a very small significance level, a possible conclusion of a hypothesis test is that the alternative hypothesis is supported.

FORMULATING HYPOTHESES. In Exercises 5–10, formulate the null and alternative hypotheses for a hypothesis test. State in clear terms the conclusion that should be given if the null hypothesis is rejected.

5. A consumer group claims that the mean price of a new model automobile sold in Cedar County is not the same as the manufacturer's recommended price.

6. The Food and Drug Administration claims that a pharmaceutical company fails to put the recommended amount of ephedrine in its decongestants.

7. The local public utility company claims that the mean annual energy consumption in Glendale is not equal to 11,000 kilowatt-hours, which is the average for the state.

8. A breakfast cereal manufacturer wishes to test the claim that the mean weight of Oats Deluxe packaged in each box does not equal 32 ounces as advertised

9. Researchers following a group of patients immediately after heart surgery want to determine whether the mean blood pressure for the group is either higher or lower than that of a peer group with no heart problems.

10. Manufacturers of standard weights and measures wish to determine whether their standard 1-kilogram weights really have a mean weight of 1 kilogram.

TWO-TAILED SIGNIFICANCE. Exercises 11 and 12 give standard scores for two-tailed tests (H_a: $\mu \neq$ claimed value). Determine whether each test is significant at the 0.05 level, and use Table 5.1 to find its P-value.

11. a. $z = 2.3$
    b. $z = -1.8$
    c. $z = 1.2$
    d. $z = -2.05$

12. a. $z = 0.9$
    b. $z = -1.4$
    c. $z = -1.3$
    d. $z = 1.90$

REJECT OR NOT REJECT? In Exercises 13–15, find the value of the standard score, z, and, without computing a P-value, determine whether the null hypothesis should be rejected in a two-tailed test.

13. $\bar{x} = 24.1$, $s = 9$, claimed population mean $= 25$, $n = 100$

14. $\bar{x} = 2.05$, $s = 0.1$, claimed population mean = 2.04, $n = 400$

15. $\bar{x} = 12.44$, $s = 2.3$, claimed population mean = 12.55, $n = 900$

FURTHER APPLICATIONS

16. **GROUP LEARNING.** Researchers for the Central Valley Education Center want to compare new group learning methods to the traditional lecture format that was used for many years. A longstanding course is taught to 100 students with no changes in its traditional structure (for example, no changes to course content, tests, or grading) *except* for the introduction of group learning methods. At the end of the semester, the 100 students have a mean score of $\bar{x} = 78$ with a standard deviation of $s = 18$. We know that when the course was taught using traditional methods, the mean score was 76. The goal is to determine whether group learning methods have any effect on student performance.

    a. State the null and alternative hypotheses for the test.
    b. Is there a significant difference in mean score for the group learning class at the 0.05 level?
    c. What is the P-value for the test?
    d. What would the conclusion of the test be if $\bar{x} = 79$?
    e. What would the conclusion of the test be if $\bar{x} = 78$ and $s = 9$?
    f. What would the conclusion of the test be if $\bar{x} = 79$ and $s = 25$?
    g. With $s = 18$, what is the minimum sample mean above 76 that produces a significant test at the 0.05 level?
    h. With $s = 18$, what is the minimum sample mean above 76 that produces a significant test at the 0.01 level?

TWO-TAILED HYPOTHESIS TESTS. For Exercises 17–22, carry out a full two-tailed hypothesis test using the four-step procedure described in the text. State your conclusion for the 0.05 significance level.

17. The makers of a leading brand of pasta need to be sure that the amount of pasta in every "16-ounce" package is neither too large (which leads to wasted product) nor too small (which leads to customer dissatisfaction). The packages in a sample of $n = 144$ packages of pasta have a mean weight of 15.8 ounces with a standard deviation of 1.6 ounces. Test the claim that the 16-ounce packages actually have a mean weight of 16 ounces.

18. The manufacturers of axles for motorcycles must produce axles that meet the specified dimensions. In particular, the diameters of the axles must be 8.50 centimeters. The axles in a sample of $n = 64$ axles have a mean diameter of 8.56 centimeters with a standard deviation of 0.24 centimeter. Test the claim that the axles actually have the specified mean diameter.

19. The cold medicine Dozenol lists 600 milligrams of acetominophen per fluid ounce as an active ingredient. The Food and Drug Administration tests 65 one-ounce samples of the medicine and finds that the mean amount of acetominophen for the sample is 589 milligrams with a standard deviation of 21 milligrams. Test the claim of the FDA that the medicine does not contain the required amount of acetominophen.

20. The mean household income in Wasatch County, Utah, recently was $41,045 with a standard deviation of $1,605. Assume that the 13,267 residents of the county provide a random sample of all Utah residents. Based on this sample, test the claim that the mean household income in Utah differs from the national mean household income of $35,100.

21. According to the U.S. Department of the Treasury, the mean weight of a quarter is 5.670 grams. A random sample of 50 quarters has a mean weight of 5.622 grams with a standard deviation of 0.068 gram. Test the claim that the mean weight of quarters in circulation is 5.670 grams.

22. The mean birth weight of male babies born to 121 mothers on a vitamin supplement is 3.67 kilograms with a standard deviation of 0.66 kilogram (based on data from the New York State Department of Health). Test the claim that the mean birth weight of all babies born with the vitamin supplement is equal to 3.39 kilograms, which is the mean for the population of all male babies.

TYPE I AND TYPE II ERRORS. In Exercises 23–30, null and alternative hypothesis are given. For each problem,
a. Describe the consequence of making a type I error.
b. Describe the consequence of making a type II error.
c. State a correct conclusion of the test if H_0 is rejected.
d. State a correct conclusion of the test if H_0 is not rejected.

23. H_0: The patient is free of a particular disease.

    H_a: The patient has the disease.

24. H_0: The defendant is not guilty.

    H_a: The defendant is guilty.

25. H_0: The lottery is fair.

    H_a: The lottery is biased.

26. H_0: The mean length of a bolt in the suspension system of new Audi cars is 3.456 centimeters.

    H_a: The mean length of a bolt in the suspension system of new Audi cars is not equal to 3.456 centimeters.

27. H_0: The mean time to failure of a microchip in the regulating system of a nuclear power plant is 4.5 years.

    H_a: The mean time to failure of a microchip in the regulating system of a nuclear power plant is less than 4.5 years.

28. H_0: The mean verbal achievement test score of all eighth-graders in Wisconsin is equal to the national average.

    H_a: The mean verbal achievement test score of all eighth-graders in Wisconsin is less than the national average.

29. H_0: The mean weight of the active ingredient in every tablet of an antidepressant is 500 milligrams.

    H_a: The mean weight of the active ingredient in every tablet of an antidepressant is not equal to 500 milligrams.

30. H_0: Customers get an average of 16 ounces of peanuts in every jar of peanut butter.

    H_a: Customers get less than an average of 16 ounces of peanuts in every jar of peanut butter.

PROJECTS FOR THE WEB AND BEYOND

For useful links, select "Links for Web Projects" for Chapter 9 at www.aw.com/bbt.

31. STUDENT *t*-DISTRIBUTION. The methods used in this chapter assume that sample sizes are large enough to justify replacing the population standard deviation, σ, by the sample standard deviation, s. A method for small sample sizes was published in 1908 in a paper by William Gosset, an employee of the Guinness Brewery in Dublin, Ireland (W. S. Gosset, "The Probable Error of a Mean," *Biometrika,* Vol. 6, 1908, pp. 1–25). Investigate the history of this method. Why is it called the student *t*-distribution? How is it used for hypothesis tests of means? Find Gosset's original (small) data set on corn yield from two different kinds of seed. How are other scientists such as Charles Darwin, Francis Galton, and R. A. Fisher involved in the story?

9.5 Hypothesis Testing: Population Proportions

We now turn to the problem of hypothesis testing with *proportions* (rather than means). All of the ideas from previous sections apply, except we will need a different method for calculating the standard deviation of the sampling distribution. Let's begin with an example.

A political candidate commissions a poll in advance of a close election. Using a random sample of $n = 400$ likely voters, the poll finds that 204 people support the candidate. Should the candidate be confident of winning? In Chapter 8, we discussed how to determine the margin of error and confidence interval for this poll. We now cast the question as a hypothesis test.

For a hypothesis test, we ask whether the poll results (which are the sample statistics) support the hypothesis that the candidate has more than 50% of the vote. As usual, we let p represent the proportion of people in the voting *population* who favor the candidate, and we let $\hat{p}$ denote the proportion of people in the *sample* who favor the candidate. Because 204 of the 400 people in the sample support the candidate, the sample proportion is

$$\hat{p} = \frac{204}{400} = 0.510$$

We can now formulate the null and alternative hypotheses. Because we always set up our null hypothesis with a single claimed value, we write

$$H_0: p = 0.5 \text{ (50\% of voters favor the candidate)}$$

The alternative hypothesis is that the candidate has *more than* 50% of the vote:

$$H_a: p > 0.5 \text{ (more than 50\% of voters favor the candidate)}$$

When I was a boy, I was told that anybody could become President; I'm beginning to believe it.

—Clarence Darrow

How do we determine whether there is enough evidence in the sample to reject the null hypothesis? Thinking as we did with sample means, we imagine selecting many samples of size $n = 400$. For each sample, we compute the proportion of people who favor the candidate. We know (from Section 8.3) that this distribution of sample proportions should be very close to a normal distribution. Assuming that the null hypothesis is true (that the proportion of people in the population who favor the candidate is 0.5), this distribution has a mean of 0.5. We also need the standard deviation of the sampling distribution, which is found from the following formula:

$$\text{standard deviation of distribution of sample proportions} = \sqrt{\frac{p(1 - p)}{n}}$$

In this case, the standard deviation is

$$\sqrt{\frac{0.5(1 - 0.5)}{400}} = 0.025$$

Figure 9.10 shows this distribution of sample proportions. Note that the particular sample selected by the candidate is just one of these many samples; it is shown with its value of $\hat{p} = 0.510$.

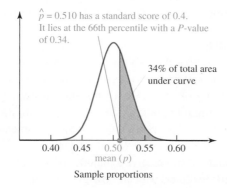

Figure 9.10 The distribution of sample proportions for an election poll with a mean of $p = 0.5$ and a standard deviation of 0.025. The sample proportion $\hat{p} = 0.510$ has a standard score of 0.4 and a P-value of 0.34.

Proceeding much as we did for hypothesis tests with means, we now ask, What is the probability of randomly selecting a sample from this distribution with a proportion of at least $\hat{p} = 0.510$? To answer this question, we need to compute the standard score for the sample proportion, which is

$$z = \frac{\hat{p} - p}{\sqrt{p(1 - p)/n}}$$
$$= \frac{0.510 - 0.5}{0.025} = 0.4$$

In other words, the sample used in the poll has a proportion that is 0.4 standard deviation above the mean.

Using Table 5.1, we see that this standard score lies at the 66th percentile. Assuming the null hypothesis that 50% of all voters favor the candidate is true, the probability of selecting a sample in which 51% or *fewer* favor the candidate is about 0.66. This probability corresponds to the area to the left of 0.51 in Figure 9.10. Conversely, the probability of selecting a sample in which *at least* 51% favor the candidate is about $1 - 0.66 = 0.34$. This probability is the P-value; it corresponds to the shaded area to the right of 0.51 in Figure 9.10. Because the P-value of 0.34 is greater than 0.05, the test is not significant at the 0.05 level and we should not reject the null hypothesis. Thus, the candidate cannot be confident of victory.

By the Way...

In the 1948 presidential election, most pollsters and newspapers predicted a large victory for the Republican Dewey over the Democrat Truman. In part because of an intense whistle-stop campaign, Truman won by a popular vote margin of 49.5% to 45.1%.

Now, suppose that in a different election poll 222 people in a sample of $n = 400$ respondents favor the candidate. In this case, the sample proportion is

$$\hat{p} = \frac{222}{400} = 0.555$$

The null hypothesis is still the claim that the population proportion is 0.5. The standard deviation for the distribution of sample proportions is still 0.025. Therefore, the standard score for $\hat{p} = 0.555$ is

$$z = \frac{\hat{p} - p}{\sqrt{p(1 - p)/n}}$$

$$= \frac{0.555 - 0.5}{0.025} = 2.20$$

Note that the proportion for this sample is more than 2 standard deviations above the population proportion, suggesting a significant result. More precisely, because this standard score is greater than the critical value of 1.645 for significance at the 0.05 level (see Table 9.1), this test is significant at the 0.05 level and there is sufficient evidence to reject the null hypothesis.

We can find the P-value by noting that, from Table 5.1, a standard score of 2.20 is the 98.6 percentile. Thus, the P-value is about $1 - 0.986 = 0.014$. This low P-value also gives good reason to reject the null hypothesis.

Let's summarize the results of the two polls. In the first poll, the proportion of the sample favoring the candidate ($\hat{p} = 0.510$) was very close to the proportion claimed by the null hypothesis ($p = 0.5$). The result of this poll was not statistically significant and did not provide evidence for rejecting the null hypothesis. In the second poll, the sample proportion ($\hat{p} = 0.555$) was farther from the claimed proportion ($p = 0.5$) and turned out to be statistically significant. For this reason, the second poll provided evidence for rejecting the null hypothesis and believing that the candidate would receive a majority of the vote.

Putting all of our work together, we can now give a four-step procedure for hypothesis testing with proportions. Note that the process is the same as for a population mean, except in the computation of the standard deviation for the sampling distribution.

Hypothesis Testing for Proportions

1. Formulate H_0 in the form p = claimed value. Formulate H_a in the form $p <$ claimed value (left-tailed), $p >$ claimed value (right-tailed), or $p \neq$ claimed value (two-tailed).

2. Identify the relevant sample statistics: the sample size, n, and the sample proportion, $\hat{p}$.

3. Using the claimed value of the population proportion, p, find the standard deviation of the distribution of sample proportions,

$$\sqrt{\frac{p(1 - p)}{n}}$$

and the standard score for the observed proportion,

$$z = \frac{\hat{p} - p}{\sqrt{p(1 - p)/n}}$$

4. Determine the outcome of the test either by comparing the standard score with the critical value given in Table 9.1 or by computing a P-value. If the P-value is less than or equal to 0.05, then the test is significant at the 0.05 level; there is sufficient evidence to reject H_0. If the P-value is greater than 0.05, H_0 cannot be rejected.

TECHNICAL NOTE

The procedure given here requires that $np \geq 5$ and $n(1-p) \geq 5$. Some more advanced texts include procedures for dealing with situations in which these two conditions are not both met.

EXAMPLE 1   Local Unemployment Rates

The October national unemployment rate is 3.5%. In a survey of $n = 450$ people in a rural Wisconsin county, 22 people are found to be unemployed. County officials apply for state aid based on the claim that the local unemployment rate is significantly higher than the national average. Test this claim at a 0.05 significance level.

SOLUTION   We use the four-step procedure. Note that the unemployment rate is a *proportion*; it is the fraction of the labor force not at work.

Step 1: The null hypothesis states that the unemployment rate in the county is the national rate:

$$H_0: p = 0.035$$

The alternative hypothesis states that the unemployment rate is higher than the national average of 0.035:

$$H_a: p > 0.035$$

Step 2: The sample statistics are the sample size, $n = 450$, and the proportion of unemployed people in the sample,

$$\hat{p} = \frac{22}{450} = 0.0489$$

Step 3: The distribution of sample proportions (Figure 9.11) has a mean of 0.035 (the claimed value in H_0) and a standard deviation of

$$\sqrt{\frac{p(1-p)}{n}} = \sqrt{\frac{0.035(1-0.035)}{450}} = 0.0087$$

The standard score for the sample proportion ($\hat{p} = 0.0489$) is

$$z = \frac{\hat{p} - p}{\sqrt{p(1-p)/n}}$$

$$= \frac{0.0489 - 0.035}{0.0087} = 1.598$$

Step 4: The standard score is about 1.60, which does not satisfy the condition $z \geq 1.645$. Thus, the test is not significant at the 0.05 level, so we do not reject the null hypothesis. The standard score corresponds to the 94.5 percentile (refer to Table 5.1), so the P-value is $1 - 0.945 = 0.055$. This P-value is not less than or equal to 0.05 and therefore does not warrant rejection of the null hypothesis. Thus, this sample does not quite provide

> **TECHNICAL NOTE**
>
> The standard deviation of sample proportions shown here uses the value of the claimed proportion, p. When confidence intervals are being constructed for a population proportion, there is no claimed value that can be used for p, so the calculation of the standard deviation is different from $\sqrt{p(1-p)/n}$.

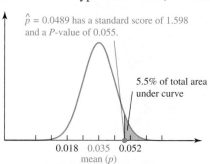

$\hat{p} = 0.0489$ has a standard score of 1.598 and a P-value of 0.055.

5.5% of total area under curve

0.018   0.035   0.052
mean (p)

Sample unemployment rates

Figure 9.11   A sample proportion of $\hat{p} = 0.0489$ has a standard score of about 1.60, which does not meet the critical condition of $z \geq 1.645$ needed for significance at the 0.05 level. The P-value of 0.055 is the area of the shaded region to the right of $\hat{p} = 0.0489$.

sufficient evidence to support the claim that the county unemployment rate is above the national average. ∎

EXAMPLE 2　Left-Handed Population

A random sample of $n = 750$ people is selected, of whom 92 are left-handed. Use these sample data to test the claim that 10% of the population is left-handed.

SOLUTION　We again follow the four-step procedure.

Step 1: The null hypothesis is the claim that 10% of the population is left-handed:

$$H_0: p = 0.1$$

To test this claim, we need to account for the possibility that the actual population proportion is either less than *or* greater than 10%. Therefore, the alternative hypothesis is

$$H_a: p \neq 0.1$$

This is a two-tailed test.

Step 2: The sample statistics are the sample size, $n = 750$, and the proportion of left-handed people in the sample,

$$\hat{p} = \frac{92}{750} = 0.123$$

Step 3: The distribution of sample proportions (Figure 9.12) has a mean of 0.1 (the claimed value in H_0) and a standard deviation of

$$\sqrt{\frac{p(1-p)}{n}} = \sqrt{\frac{0.1(1-0.1)}{750}} = 0.011$$

The standard score for the sample proportion ($\hat{p} = 0.123$) is

$$z = \frac{\hat{p} - p}{\sqrt{p(1-p)/n}} = \frac{0.123 - 0.1}{0.011} = 2.09$$

Step 4: The standard score is greater than the critical value for a two-tailed test, which is $z = 1.96$ (see box on p. 385), so the test is significant at the 0.05 level and we should reject the null hypothesis. We find the *P*-value by noting that a standard score of 2.09

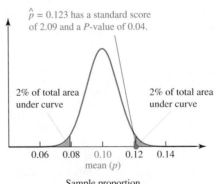

$\hat{p} = 0.123$ has a standard score of 2.09 and a *P*-value of 0.04.

2% of total area under curve

2% of total area under curve

0.06　0.08　0.10　0.12　0.14
mean (*p*)

Sample proportion of left-handed people

Figure 9.12 Distribution of sample proportions for estimating the proportion of left-handed people. A sample proportion of $\hat{p} = 0.123$ has a standard score of 2.09, which is greater than the critical value of 1.96 for a two-tailed test. The *P*-value is 0.04 and the test is significant at the 0.05 level.

corresponds to the 98th percentile (see Table 5.1); thus, the probability of a standard score of 2.09 or more is $1 - 0.98 = 0.02$. Because this is a two-tailed test, the P-value is *twice* this probability, or 0.04. Again, the P-value says that we have sufficient evidence to reject the null hypothesis. We conclude that the proportion of the population that is left-handed is *not* equal to 10%.   ∎

Review Questions

1. Give an example of a situation that requires testing a population *proportion* for significance.

2. In testing a population proportion, what does the null hypothesis claim? What are the three possible forms of the alternative hypothesis?

3. Describe the four-step procedure for testing a population proportion. How does this procedure differ from the procedure for a population mean?

Exercises

BASIC SKILLS AND CONCEPTS

SENSIBLE STATEMENTS? For Exercises 1–4, determine whether the given statement is sensible and explain why it is or is not.

1. In a test of the claim that a majority of Americans favor registration of all handguns, the null hypothesis is $p = 0.5$.

2. In a test of the claim that a majority of Americans favor registration of all handguns, the alternative hypothesis is $p > 0.5$.

3. In a two-tailed hypothesis test of a claim about a proportion, the P-value is the area to the right of the standard score, z.

4. The claim that $p > 0.5$ can never be supported if the sample proportion, $\hat{p}$, is less than 0.5.

FURTHER APPLICATIONS

5. **BREAD RECIPE.** A baker wants to compare a new bread recipe to the traditional recipe that he has used for many years. He sells the new bread for a week and collects customer approval ratings. Of the first 100 ratings, $\hat{p} = 92\%$ are favorable, compared to $p = 91.5\%$ for the traditional recipe (established over many years). The goal is to determine whether customer approval is significantly higher for the new recipe than for the traditional recipe.

a. State the null and alternative hypotheses for the test.
b. Is there a significant increase at the 0.05 level in the proportion of customers who approve of the new recipe?
c. What is the P-value for the test?
d. What would the conclusion of the test be if $\hat{p} = 93\%$?
e. What would the conclusion of the test be if $\hat{p} = 97\%$?
f. What is the minimum sample proportion, $\hat{p}$, that produces a significant test at the 0.05 level?
g. What is the minimum sample proportion, $\hat{p}$, that produces a significant test at the 0.01 level?

HYPOTHESIS TESTS. For Exercises 6–16, carry out a hypothesis test using the four-step procedure described in the text. State your conclusion for the 0.05 significance level.

6. In a pre-election poll, a candidate for district attorney receives 205 of 400 votes. Assuming that those people polled represent a random sample of the voting population, test the claim that a majority of the voters support the candidate.

7. According to the U.S. Census Bureau, 58% of college students 25 years of age or older are women. Suppose that at Clarion College, within a random sample of 2,100 students who are 25 years of age or older, 1,234 are women. Test the claim that the percentage of older women at the college is above the national average.

8. A study commissioned by the U.S. Department of Education, based on responses from 1,015 randomly selected teenagers, concluded that 44% of teenagers cite grades as the greatest source of pressure. Test the claim that fewer than half of all teenagers in the population feel that grades are the greatest source of pressure.

9. In 2000, 115.7 million adults, or 57.3% of the adult American population, were married and living with a spouse. Suppose that in a small New England town with 3,200 adults, 55.9% of all adults are married. Test the claim that this sample comes from a population with a married percentage less than 57.3%.

10. A Department of Health and Human Services study of illegal drug use among 12–17-year-olds reported a decrease in use from 11.4% in 1997 to 9.9% now. Suppose a survey in a large high school revealed that, within a random sample of 1,050 students, 98 reported using illegal drugs. Test the principal's claim that illegal drug use in her school is less than the current national average.

11. According to recent estimates, 12.1% of the 4,342 people in Custer County, Idaho, lived in poverty. Assume this county represents a random sample of all people in Idaho. Based on this sample, test the claim that the poverty rate in Idaho is less than the national rate of 13.3%.

12. The annual survey of first-year college students, conducted by the Higher Education Research Institute at UCLA, asks approximately 276,000 students about their attitudes on a variety of subjects. According to a recent survey, 51% of first-year students believe that abortion should be legal (down from 65% in 1990). Test the claim that over half of all first-year students believe that abortion should be legal.

13. In an ABC/*Washington Post* survey, 1,526 randomly selected Americans were asked to list the most important issues for the 2000 elections. Fifty-six percent of those surveyed (44% of men and 67% of women) claimed that gun control was a very important issue. Test the claim that a majority of all Americans felt that gun control was an important issue.

14. A study by Alfred University researchers estimated that 80% of student-athletes in America are subjected to some form of hazing. Assume that this study correctly estimates the population proportion. Suppose that at a single college, 125 out of 160 athletes surveyed report having experienced hazing. Assuming that this sample is a random sample from the entire population of all athletes at the college, test the claim that the rate of hazing at this campus is less than the national average.

15. The smoking rate for the entire U.S. population of adults (over 21 years of age) is 32% (U.S. National Institute on Drug Abuse). Suppose that for a sample of 75 fine arts students over 21 years old, the smoking rate is 35%. Test the claim that the smoking rate for all fine arts students is significantly (at the 0.05 level) higher than the national average.

16. According to the Energy Information Administration, 53.0% of households nationwide used natural gas for heating in 1997. Recently, a survey of 3,600 randomly selected households showed that 54.0% used natural gas. Use a 0.05 significance level to test the claim that the 53.0% national rate has changed.

 # PROJECTS FOR THE WEB AND BEYOND

For useful links, select "Links for Web Projects" for Chapter 9 at www.aw.com/bbt.

17. Given the claim that 10% of Americans are left-handed, randomly select at least 50 students at your college and determine whether they are left-handed. Test the claim with a formal hypothesis test.

18. Use the Web or library references to determine the proportion of Americans who smoke. Test the claim that the proportion of students at your college who smoke is different from the proportion for all Americans. Collect sample data from at least 50 randomly selected students.

19. Use the Web or library references to determine the proportion of college students in the United States who are women. Test the claim that the proportion of women students at your college is different from the proportion for U.S. college students. Collect sample data from at least 100 randomly selected students.

20. COUNTY DATA. The *Statistical Abstract of the United States* and the Current Population Survey provide an extensive supply of social, economic, and vital statistics at the county, state, and local levels. Use their Web sites to compare state data to national data in the following way.
   a. Choose a variable of interest that involves a proportion for a particular state (for example, the percentage of people living in poverty in Arizona).
   b. Find the current national value for that variable (for example, the national poverty rate).
   c. Choose a particular county within the state and obtain the corresponding data for the county (for example, the poverty rate in Pima County, Arizona).

d. Assuming that the county is a random sample of the state, test the claim that the state is above or below the national level in terms of that variable.

e. Discuss and interpret your results. The hypothesis test depends on the sample (the county) being a random sample of the population (the state). Be sure to discuss this factor in your conclusions.

IN THE NEWS

1. **HYPOTHESIS TESTING IN THE NEWS.** Find a news article or research report that describes (perhaps not explicitly) a hypothesis test for a population proportion. Attach the article and summarize the method used.

Chapter Review Exercises

1. Randomly selected cans of Coke are measured for the amount of cola in ounces. The sample values listed below have a mean of 12.19 ounces and a standard deviation of 0.11 ounce. Assume that we want to use a 0.05 significance level to test the claim that cans of Coke have a mean amount of cola greater than 12 ounces.

12.3	12.1	12.2	12.3	12.2	12.3
12.0	12.1	12.2	12.1	12.3	12.3
11.8	12.3	12.1	12.1	12.0	12.2
12.2	12.2	12.2	12.2	12.2	12.4
12.2	12.2	12.3	12.2	12.2	12.3
12.2	12.2	12.1	12.4	12.2	12.2

   a. What is the null hypothesis?
   b. What is the alternative hypothesis?
   c. What is the value of the standard score for the sample mean of 12.19 ounces?
   d. What is the critical value?
   e. What is the *P*-value?
   f. What do you conclude? (Be sure to address the original claim that the mean is greater than 12 ounces.)
   g. Describe a type I error for this test.
   h. Describe a type II error for this test.
   i. Find the *P*-value if the test is modified to test the claim that the mean is *different from* 12 ounces (instead of being greater than 12 ounces).

2. In a survey of 1,002 people, 701 (or 70%) said that they voted in the 2000 presidential election (based on data from ICR Research Group). We want to use a 0.05 significance level to test the claim that, when surveyed, the proportion of people who say that they voted is different from 0.61, which is the proportion of people who actually did vote.

   a. What is the null hypothesis?
   b. What is the alternative hypothesis?
   c. What is the value of the standard score for the sample proportion of 701/1,002?
   d. What is the critical value?
   e. What is the *P*-value?
   f. What do you conclude? (Be sure to address the original claim that the proportion is different from 0.61.)
   g. Describe a type I error for this test.
   h. Describe a type II error for this test.
   i. Find the *P*-value if the test is modified to test the claim that the mean is *greater than* 0.61 (instead of being different from 0.61).

3. We want to test the claim that the Clarke method of gender selection is effective in increasing the likelihood that a baby will be born a girl. In a random sample of 80 couples who use the Clarke method, it is found that among the 80 newborn babies there are 35 girls. What should we conclude about the claim? Why is it *not* necessary to go through all of the steps of a formal hypothesis test?

4. Mike Shanley wants to test the hypothesis that adult men have a mean weight that is greater than 150 pounds. He surveys the adult male members of his family and obtains 40 sample values, which lead him to support the alternative hypothesis that the population mean is greater than 150 pounds. What is the fundamental flaw in his procedure?

FOCUS ON HISTORY

Where Did Statistics Begin?

The origins of many disciplines are lost in antiquity, but the roots of statistics can be identified with some certainty. Systematic record keeping began in London in 1532 with weekly data collection on deaths. Later in the same decade, official data collection on baptisms, deaths, and marriages began in France. In 1608, the collection of similar vital statistics began in Sweden. Canada conducted the first official census in 1666.

Of course, statistics is more than the collection of data. If there is a founder of statistics, that person must be someone who worked with the data in clever and systematic ways and who used the data to reach conclusions that were not previously evident. Many experts believe that an Englishman named John Graunt deserves the title of the founder of statistics.

John Graunt was born in London in 1620. As the eldest child in a large family, he took up his father's business as a draper (a dealer in clothing and dry goods). He spent most of his life as a prominent London citizen, until he lost his house and possessions in the Fire of London in 1666. Eight years later, he died in poverty.

It's not clear how John Graunt became interested in the weekly records of baptisms and burials—known as bills of mortality—that had been kept in London since 1563. In the preface of his book *Natural and Political Observations on the Bills of Mortality,* he noted that others "made little other use of them" and wondered "what benefit the knowledge of the same would bring to the World." He must have worked on his statistical projects for many years before his book was first published in 1662.

Graunt worked primarily with the annual bills, which were year-end summaries of the weekly bills of mortality. Figure 9.13 shows the annual bill for 1665 (the year of the Great Plague). The top third of the bill shows the numbers of burials and baptisms (christenings) in each parish. Total burials and baptisms are noted in the middle of the bill, with deaths due to the plague recorded separately. The lower third of the bill shows deaths due to a variety of other causes, with totals given for males and females.

Graunt was aware of rough estimates of the population of London that were made periodically for taxation purposes, but he must have been skeptical of one estimate that put the population of London at 6 or 7 million in 1661. Using the annual bills, comparing burials and baptisms, and estimating the density of families in London (with an average family size of eight), he arrived at a population estimate of 460,000 by three different methods—quite a change from 6 or 7 million! He also found that the population of London was increasing while the populations of towns in the countryside were decreasing, showing an early trend toward urbanization. He raised awareness of the high rates of infant mortality. He also refuted a popular theory that plagues arrive with new kings.

Graunt's most significant contribution may have been his construction of the first life table. Although detailed data on age at death were not available, Graunt knew that of 100 new

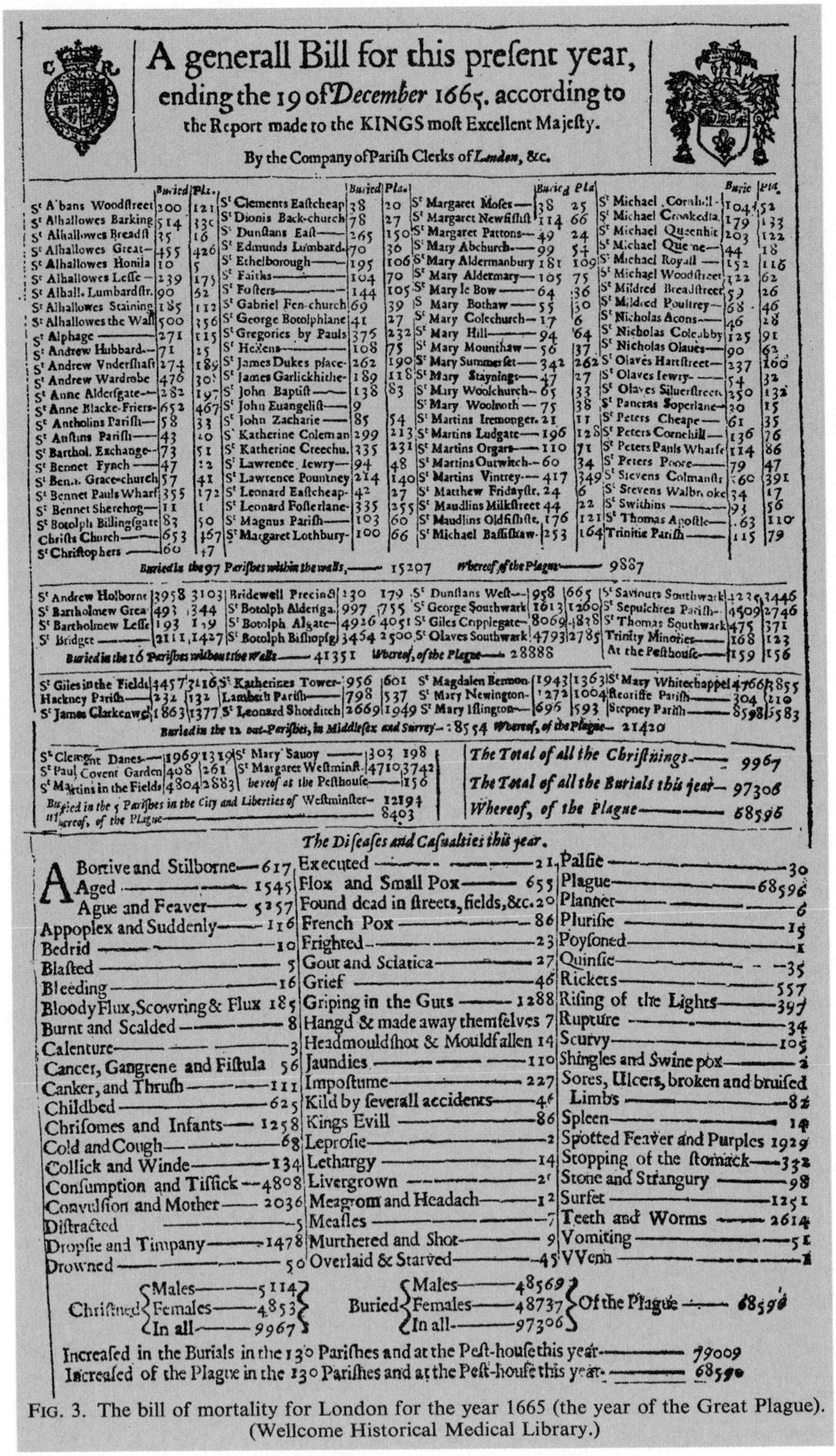

A generall Bill for this present year, ending the 19 of *December* 1665. according to the Report made to the KINGS most Excellent Majesty.

By the Company of Parish Clerks of *London*, &c.

	Buried	Pla.		Buried	Pla.		Buried	Pla.		Buried	Pla.
St Albans Woodstreet	200	121	St Clements Eastcheap	38	20	St Margaret Moses	38	25	St Michael Cornhill	104	52
St Alhallowes Barking	514	330	St Dionis Back-church	78	27	St Margaret Newfishst	114	66	St Michael Crookedla.	179	133
St Alhallowes Breadst	35	16	St Dunstans East	265	150	St Margaret Pattons	49	24	St Michael Queenhit	203	122
St Alhallowes Great	455	426	St Edmunds Lumbard.	70	36	St Mary Abchurch	99	54	St Michael Quene	44	18
St Alhallowes Honila	10	5	St Ethelborough	195	106	St Mary Aldermanbury	181	109	St Michael Royall	152	116
St Alhallowes Lesse	239	175	St Faiths	104	70	St Mary Aldermary	105	75	St Michael Woodstreet	122	62
St Alhall. Lumbardstr.	90	62	St Fosters	144	105	St Mary le Bow	64	36	St Mildred Breadstreet	59	26
St Alhallowes Staining	185	112	St Gabriel Fen-church	69	39	St Mary Bothaw	55	30	St Mildred Poultrey	68	46
St Alhallowes the Wall	500	356	St George Botolphlane	41	27	St Mary Colechurch	17	6	St Nicholas Acons	46	28
St Alphage	271	115	St Gregories by Pauls	375	232	St Mary Hill	94	64	St Nicholas Coleabby	125	91
St Andrew Hubbard	71	25	St Hellens	108	75	St Mary Mounthaw	56	37	St Nicholas Olaues	90	62
St Andrew Vndershaft	274	189	St James Dukes place	262	190	St Mary Summerset	342	262	St Olaves Hartstreet	237	160
St Andrew Wardrobe	476	308	St James Garlickhithe	189	118	St Mary Staynings	47	27	St Olaves Iewry	54	32
St Anne Aldersgate	282	197	St John Baptist	138	83	St Mary Woolchurch	65	33	St Olaves Siluerstreet	250	132
St Anne Blacke-Friers	652	467	St John Euangelist	9		St Mary Woolnoth	75	38	St Pancras Soperlane	30	15
St Antholins Parish	58	33	St John Zacharie	85	54	St Martins Iremonger.	21	11	St Peters Cheape	61	35
St Austins Parish	43	20	St Katherine Coleman	299	213	St Martins Ludgate	196	128	St Peters Cornhill	136	76
St Barthol. Exchange	73	51	St Katherine Creechu.	335	231	St Martins Orgars	110	71	St Peters Pauls Wharfe	114	86
St Bennet Fynch	47	22	St Lawrence Iewry	94	48	St Martins Outwitch	60	34	St Peters Poore	79	47
St Bennet Grace-church	57	41	St Lawrence Pountney	214	140	St Martins Vintrey	417	349	St Stevens Colmanstr	560	391
St Bennet Pauls Wharf	355	172	St Leonard Eastcheap	42	27	St Matthew Fridaystr.	24	6	St Stevens Walbrooke	34	17
St Bennet Sherehog	11	1	St Leonard Fosterlane	335	255	St Maudlins Milkstreet	44	22	St Swithins	93	56
St Botolph Billingsgate	83	50	St Magnus Parish	103	60	St Maudlins Oldfishstr.	176	121	St Thomas Apostle	163	110
Christs Church	653	467	St Margaret Lothbury	100	66	St Michael Bassishaw	253	164	Trinitie Parish	115	79
St Christophers	60	47									

Buried in the 97 Parishes within the walls, — 15207 *Whereof, of the Plague* — 9887

St Andrew Holborne	3958	3103	Bridewell Precinct	230	179	St Dunstans West	958	665	St Saviours Southwark	4235	3446
St Bartholmew Great	493	344	St Botolph Aldersga.	997	755	St George Southwark	1613	1260	St Sepulchres Parish	4509	2746
St Bartholmew Lesse	193	139	St Botolph Algate	4926	4051	St Giles Cripplegate	8069	4838	St Thomas Southwark	475	371
St Bridget	2111	1427	St Botolph Bishopsg.	3454	2500	St Olaves Southwark	4793	2785	Trinity Minories	168	123
									At the Pesthouse	159	156

Buried in the 16 Parishes without the Walls — 41351 *Whereof, of the Plague* — 28888

St Giles in the Fields	4457	3216	St Katherines Tower	956	601	St Magdalen Bermon.	1943	1363	St Mary Whitechappel	4766	3855
Hackney Parish	232	132	Lambeth Parish	798	537	St Mary Newington	1272	1004	Rederiffe Parish	304	210
St James Clarkenwel	1863	1377	St Leonard Shorditch	2669	1949	St Mary Islington	696	593	Stepney Parish	8598	6583

Buried in the 12 out-Parishes, in Middlesex and Surrey — 18554 *Whereof, of the Plague* — 21420

St Clement Danes	1969	1319	St Mary Sauoy	303	198
St Paul Covent Garden	408	261	St Margaret Westminst.	4710	3742
St Martins in the Fields	4804	2883	whereof at the Pesthouse		156

Buried in the 5 Parishes in the City and Liberties of Westminster — 12194
Whereof, of the Plague — 8403

The Total of all the Christnings — 9967
The Total of all the Burials this year — 97306
Whereof, of the Plague — 68596

The Diseases and Casualties this year.

Abortive and Stilborne	617	Executed	21	Palsie	30
Aged	1545	Flox and Small Pox	655	Plague	68596
Ague and Feaver	5257	Found dead in streets, fields, &c.	20	Planner	6
Appoplex and Suddenly	116	French Pox	86	Plurisie	15
Bedrid	10	Frighted	23	Poysoned	1
Blasted	5	Gout and Sciatica	27	Quinsie	35
Bleeding	16	Grief	46	Rickets	557
Bloody Flux, Scowring & Flux	185	Griping in the Guts	1288	Rising of the Lights	397
Burnt and Scalded	8	Hangd & made away themselves	7	Rupture	34
Calenture	3	Headmouldshot & Mouldfallen	14	Scurvy	105
Cancer, Gangrene and Fistula	56	Jaundies	110	Shingles and Swine pox	2
Canker, and Thrush	111	Impostume	227	Sores, Ulcers, broken and bruised Limbs	82
Childbed	625	Kild by severall accidents	46	Spleen	14
Chrisomes and Infants	1258	Kings Evill	86	Spotted Feaver and Purples	1929
Cold and Cough	68	Leprosie	2	Stopping of the stomack	332
Collick and Winde	134	Lethargy	14	Stone and Strangury	98
Consumption and Tissick	4808	Livergrown	20	Surfet	1251
Convulsion and Mother	2036	Meagrom and Headach	12	Teeth and Worms	2614
Distracted	5	Measles	7	Vomiting	51
Dropsie and Timpany	1478	Murthered and Shot	9	Wenn	1
Drowned	50	Overlaid & Starved	45		

Christned: Males — 5114; Females — 4853; In all — 9967
Buried: Males — 48569; Females — 48737; In all — 97306 — Of the Plague — 68596

Increased in the Burials in the 130 Parishes and at the Pest-house this year — 79009
Increased of the Plague in the 130 Parishes and at the Pest-house this year — 68590

FIG. 3. The bill of mortality for London for the year 1665 (the year of the Great Plague). (Wellcome Historical Medical Library.)

Figure 9.13 Reproduction of 1665 annual bill of mortality. SOURCE: *Journal of the Royal Statistical Society,* Vol. 126, part 4, 1963, pp. 537–557.

babies, "36 of them die before they be six years old, and that perhaps but one surviveth 76." With these two data points, he filled in the intervening years as shown in Table 9.4, using methods that he did not fully explain.

Table 9.4	Graunt's Life Table: Number of Deaths by Age for 100 People
Age	Deaths
Within the first six years	36
The next ten years	24
The second decade	15
The third decade	9
The fourth	6
The next	4
The next	3
The next	2
The next	1

Table 9.5	Graunt's Table of Survivors at Various Ages
Age	Survivors
At sixteen years end	40
At twenty six	25
At thirty six	16
At forty six	10
At fifty six	6
At sixty six	3
At seventy six	1
At eighty	0

With estimates of deaths for various ages, he was able to make the companion table of survivors shown in Table 9.5. While some modern statisticians have doubted the methods used to construct these tables, Graunt appears to have appreciated the value of these tables and anticipated actuarial tables used by life insurance companies of the future. It wasn't until 1693 that Edmund Halley, of comet fame, constructed life tables using age-based mortality rates.

QUESTIONS FOR DISCUSSION

1. Is it surprising that accurate estimates of the population of London were not available in 1660? How would you have suggested making such estimates at that time?

2. Do you think that records of burials and baptisms would have given accurate counts of actual births and deaths? Why or why not?

3. Assuming that the estimates of deaths in Table 9.4 are accurate, are the estimates of survivors in Table 9.5 consistent? Note that the numbers in Table 9.5 do not total to 100; should they? Explain.

4. Based on Graunt's tables, estimate the average life expectancy in 1660. Explain your reasoning.

SUGGESTED READING

Johnson, N., and Kotz, S. (Eds.). *Leading Personalities in the Statistical Sciences*. John Wiley and Sons, 1997.

Sutherland, I. "John Graunt: A Tercentenary Tribute." *Journal of the Royal Statistical Society,* Vol. 126, part 4, 1963, pp. 537–557.

FOCUS ON AGRICULTURE

Are Genetically Modified Foods Safe?

British newspapers call them "Frankenfoods" (a word play on *Frankenstein*). Europeans, by and large, don't eat them at all. But by 2000, many foods sold in the United States already fell in this category, including 36% of American corn and 55% of American soybeans. We are speaking of "genetically modified foods," or GM foods. GM foods are a fairly recent agricultural invention, dating only from about the mid-1990s. Nevertheless, they are at the forefront of one of the biggest debates in agricultural history.

A genetically modified organism is an organism into which scientists have inserted a gene that does not naturally exist in the organism or its close relatives. Genetically modified organisms are being tested and used for many purposes. For example, scientists have developed bacteria containing genes that produce drugs like insulin.

In agriculture, genetic modification allows scientists to create traits in crops that would be difficult or impossible to achieve by traditional breeding techniques. One of the first widely used genetically modified crops, often called "Bt corn," illustrates the idea behind genetic modification. Corn is usually susceptible to destruction by a variety of insect pests. As a result, farmers usually spray crops with pesticides. Unfortunately, many pesticides are toxic to animals besides the insect pests and hence can cause environmental damage (and may not be ideal for humans to consume).

Here's where Bt, a bacterium known as *bacillus thuringiensis,* comes in. Bt lives naturally in soil. As early as 1911, scientists discovered that Bt produces a toxin that kills certain types of insects. Different strains of Bt kill different insects, but are generally harmless to other animals and humans. By the 1960s, these traits had led to the use of Bt as an "environmentally friendly" pesticide that could be sprayed on crops (in the form of killed bacteria). Unfortunately, Bt pesticide proved to be expensive and often ineffective, mainly because it works only if insects eat it (some other pesticides kill on contact). And, because it breaks down rapidly in sunlight and washes off plants in rain, it kills pests only if its application is timed just right.

Genetic modification solves the problems inherent in Bt pesticide. The pesticide action of Bt bacteria arises from particular proteins that the bacteria produce. Scientists identified the genes responsible for these proteins, then learned to transfer these genes into plants such as corn. Once the corn contains the necessary genes, it produces the same pest-killing proteins as the Bt bacteria. The application of a sprayed pesticide is no longer necessary because the corn itself is now toxic to the pests. Moreover, because the corn continually produces the pest-killing proteins, there are no more concerns about pesticide breaking down or washing away.

The advantages of the Bt corn over traditional strains of corn are clear, but are there any disadvantages? This is where the great debate over GM foods begins. On one side, many scientists argue that GM foods are completely safe and that their benefits will help improve

By the Way...

More than 50 new GM crops have been approved for sale in the United States, including corn and soybeans that produce their own pesticide and tomatoes engineered for longer shelf lives. GM crops under development include potatoes that resist bruising, better-tasting soybeans, and grains containing vitamins and other nutrients that could improve nutrition in impoverished countries.

nutrition for people throughout the world. On the other side stand the people, including some scientists, who label GM foods as "Frankenfoods" and argue that they are one of the most dangerous technologies ever invented.

Broadly speaking, the issue of GM food safety can be broken down into three major questions:

1. Do GM foods have any toxic effects in humans?

2. Do the new proteins contained in GM foods cause allergic reactions in some people?

3. Can the GM crops cause any unforeseen environmental damage, such as transferring their genes into weeds (thereby making "superweeds") or killing animals besides the insect pests?

These questions can be addressed through hypothesis testing. In each case, we begin with a null hypothesis that states that there is no safety difference between traditional foods and GM foods. For example, the null hypothesis for Question 1 says that GM foods are no more toxic than traditional foods (which often contain low levels of toxic chemicals). The alternative hypothesis states that there is a difference in toxicity. Scientists then develop experiments to test the hypotheses. If the evidence provided by the experiments reveals a significant difference in toxicity between the food groups (beyond what would be expected by chance alone), there is reason to reject the null hypothesis.

Unfortunately, experiments to date have been unable to resolve the controversy. For example, some genetically modified crops *do* contain products toxic to humans—and therefore never receive approval for sale. To proponents of GM foods, this fact provides an argument in their favor, because it appears to show that current regulatory procedures (for example, requiring approval of GM foods by the U.S. Food and Drug Administration) adequately ensure that only safe foods enter the marketplace. Similarly, while some of the proteins in GM foods undoubtedly can cause allergic reactions, many scientists think they understand these reactions well enough to ensure that only safe foods are approved.

Opponents of GM foods use the very same experimental results to support their case. They argue that even if scientists have identified toxicity that is obvious in experiments, they may still be missing long-term effects that might not show up in the population for many years. Similarly, they claim that we cannot be sure we understand *all* allergic reactions and therefore might inadvertently approve GM foods that could have severe allergic consequences in at least some people.

The environmental issues are even more difficult to study. For example, one study has shown that Bt corn is toxic to the beautiful Monarch butterfly, a species that is not a pest and that no one wants to kill. But the study was conducted in a laboratory and may not accurately represent what occurs in the real environment. Similarly, the possibility of "gene jumping," in which the Bt genes might spread from corn to other plants, is not well understood and therefore is very difficult to study.

The debate over GM foods is likely to continue, and even to intensify, for many years to come. After all, when it comes to food, everyone has an interest.

By the Way...

Opponents of GM foods often point to the pesticide DDT as an example of how hard it may be to discover long-term toxic or environmental effects. DDT was used for some 60 years before its detrimental effects were discovered.

QUESTIONS FOR DISCUSSION

1. Propose an experiment that could be used to test the safety of a GM food product, such as corn. Describe your experiment in detail, and discuss any practical difficulties that might be involved in carrying it out or interpreting its results.

2. Ignoring any safety issues of GM foods, make a list of as many ways as you can think of in which GM foods might be beneficial to humanity. Then, ignoring any benefits of GM foods, make a list of as many ways as you can think of in which GM foods might be

dangerous. Overall, do you think that further use of GM foods should be encouraged or discouraged? Defend your opinion.

3. Some people advocate giving the choice about GM foods to consumers by requiring labeling on all products that contain genetically modified ingredients. Do you think this is a good idea? Why or why not?

4. Investigate recent developments in the debate over GM foods. Does the new information alter any of the major arguments in the debate? Explain.

SUGGESTED READING

The debate over genetically modified foods generates headlines almost daily. Besides the articles listed below, find recent news articles by using a Web news service such as Yahoo! Full Science Coverage (go to Yahoo! and search on "GM foods").

Belsie, Laurent. "Superior Crops or 'Frankenfood'?" *Christian Science Monitor,* March 1, 2000.

Butler, D., and Reichhardt, T. "Briefing: Long-Term Effect of GM Crops Serves Up Food for Thought." *Nature,* Vol. 398, April 22, 1999, pp. 651–656.

MacKenzie, Debora. "Unpalatable Truths." *New Scientist,* April 17, 1999.

Pollack, Andrew. "We Can Engineer Nature. But Should We?" *New York Times,* February 6, 2000.

Simon, Stephanie. "Bioengineered Crops on Shaky Ground." *Los Angeles Times,* March 5, 2000.

Yoon, Carol K. "Pollen from Genetically Altered Corn Threatens Monarch Butterfly, Study Finds." *New York Times,* May 20, 1999.

CHAPTER 10

The web of this world is woven of necessity and chance. Woe to him who has accustomed himself to find something capricious in what is necessary, and who would ascribe something like reason to chance.

—Johann Goethe

Further Applications of Statistics

We have explored many core ideas and applications of statistics in Chapters 1–9, but you will undoubtedly run across many more applications in the future. In this final chapter, we will explore three particularly common applications of statistics, chosen to give you a taste of what you may encounter in future course work and your daily life. First, we will examine the ideas of risk and life expectancy, which are crucial to decisions we make almost every day. Next, we will discuss some common paradoxes that can arise in statistics and which can be confusing unless you analyze them with great care. Finally, we will investigate an application of hypothesis testing to two-way tables, which will allow us to look for a relationship between two categorical variables, such as gender and college major.

LEARNING GOALS

1. Compute and interpret various measures of risk as they apply to travel, disease, and life expectancy.

2. Investigate and understand some common paradoxes, including Simpson's paradox, that arise in many practical situations.

3. Interpret and carry out hypothesis tests for independence of variables in two-way tables.

10.1 Ideas of Risk and Life Expectancy

Our lives are filled with risk, but we have some control over how much risk. For example, we can decide whether to drive or fly, whether to wear a seat belt, and how much health insurance to carry. To make such decisions, we need ways of quantifying and comparing risks. In this section, we will explore some common measures of risk used in evaluating travel options and mortality. We will also investigate the important idea of life expectancy in some detail.

Risk and Travel

Are you safer in a small car or in a sport utility vehicle? Are cars today safer than those 30 years ago? If you need to travel across country, are you safer flying or driving? To answer these and many similar questions, we must quantify the risk involved in travel. We can then make decisions appropriate for our own personal circumstances.

Travel risk is often expressed in terms of an **accident rate** or **death rate.** For example, suppose an annual accident rate is 750 accidents per 100,000 people. This means that, within a group of 100,000 people, on average 750 will have an accident over the period of a year. The statement also gives a relative frequency probability; it says that the probability of a person being involved in an accident is 750 in 100,000, or 0.0075.

However, caution is needed in working with accident and death rates. For example, travel risks are sometimes stated *per 100,000 people,* as above, but other times they are stated *per trip* or *per mile.* If we use death rates *per trip* to compare the risks of flying and driving, we neglect the fact that airplane trips are typically much longer than automobile trips. Similarly, if we use accident rates *per person,* we neglect the fact that most automobile accidents involve only minor injuries.

EXAMPLE 1   Is Driving Getting Safer?

Figure 10.1 shows the number of automobile fatalities and the total number of miles driven (among all Americans) for each year over a period of more than three decades. In terms of death rate per mile driven, how has the risk of driving changed?

SOLUTION   Figure 10.1a shows that the annual number of fatalities decreased from about 51,000 in 1966 to about 42,000 in 2000. Meanwhile, Figure 10.1b shows that the number of miles driven increased from about 900 billion (9×10^{11}) to about 2,750 billion (2.75×10^{12}). Thus, the death rates per mile for the beginning and end of the period were

$$1966: \quad \frac{51,000 \text{ deaths}}{9 \times 10^{11} \text{miles}} \approx 5.7 \times 10^{-8} \text{ death per mile}$$

$$2000: \quad \frac{42,000 \text{ deaths}}{2.75 \times 10^{12} \text{ miles}} \approx 1.5 \times 10^{-8} \text{ death per mile}$$

Note that, because $10^8 = 100$ million, 5.7×10^{-8} death per mile is equivalent to 5.7 deaths per 100 million miles. Thus, over 34 years, the death rate per 100 million miles dropped from 5.7 to 1.5. By this measure, driving became much safer over the period. Most researchers believe the improvements resulted from better automobile design and from safety features, such as shoulder belts and air bags, that are much more common today.

By the Way...

Roughly 115 people die in automobile accidents per day, or 1 every 13 minutes. Automobile accidents are the leading cause of death in all age groups between 6 and 27. Vehicle occupants account for 85% of automobile fatalities, while pedestrians and cyclists account for 15%.

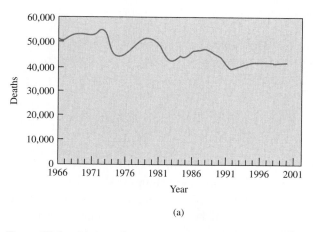
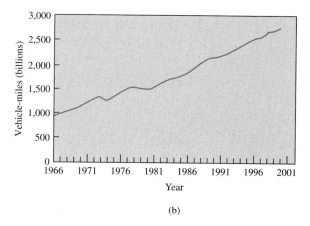

Figure 10.1   (a) Annual automobile fatalities. (b) Total miles driven annually. Both sets of data are for the United States only. SOURCE: National Transportation Safety Board.

EXAMPLE 2   Which Is Safer: Flying or Driving?

Figure 10.2 shows the numbers of deaths in commercial airline accidents in recent years. (This figure is for accidents and does not include the deaths in terrorist attacks.) In 2000, airplane passengers in the United States traveled a total of about 7 billion miles. Use the 2000 numbers of about 100 deaths and 7 billion miles to calculate the death rate per mile of air travel. Compare the risk of flying to the risk of driving.

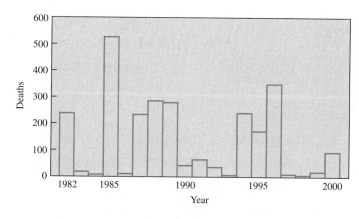

Figure 10.2   Fatalities in commercial airline accidents, U.S. carriers. SOURCE: National Transportation Safety Board.

The cost of living is going up and the chance of living is going down.
—Flip Wilson, comedian

SOLUTION   Assuming 100 deaths and 7 billion miles in a year, the risk of air travel is

$$\frac{100 \text{ deaths}}{7 \times 10^9 \text{ miles}} \approx 1.4 \times 10^{-8} \text{ death per mile}$$

This risk is equivalent to 1.4 deaths per 100 million miles, which is roughly the same as the risk of 1.5 deaths per 100 million miles for driving (see Example 1). Note that, because the average air trip covers a considerably longer distance than the average driving trip, the risk *per trip* is much higher for air travel, although the risk *per mile* is about the same.

Time out to think

Based on your driving and flying patterns, what is your risk of dying in a car? In an airplane? Is there any way that you could change the likelihood of dying in a car? In an airplane? How do the airplane fatalities of September 11, 2001, affect these results?

Vital Statistics

Only those who risk going too far can possibly find out how far one can go.
— T. S. Eliot

Data concerning births and deaths of citizens, often called *vital statistics*, are very important to understanding risk-benefit tradeoffs. For example, insurance companies use vital statistics to assess risks and set rates. Health professionals study vital statistics to assess medical progress and decide where research resources should be concentrated. Demographers use birth and death rates to predict future population trends.

One important set of vital statistics, shown in Table 10.1, concerns causes of death. These data are extremely general; a more complete table would categorize data by age, sex, and race. Vital statistics are often expressed in terms of deaths per person or per 100,000 people, which makes it easier to compare the rates for different years and for different states or countries.

Table 10.1 Leading Causes of Death in the United States (in a single recent year)

Cause	Deaths	Cause	Deaths
Heart Disease	725,192	Diabetes	68,399
Cancer	549,838	Pneumonia/Influenza	63,730
Stroke	167,366	Alzheimer's Disease	44,536
Pulmonary Disease*	124,181	Kidney Disease	35,525
Accidents	97,860	Septicemia (blood poisoning)	30,680

*Includes diseases linked to smoking, such as chronic bronchitis and emphysema.
SOURCE: Centers for Disease Control.

By the Way...

For college-age students, alcohol consumption presents one of the most serious health risks. According to a report from the National Institutes of Health, each year drinking contributes to 1,400 deaths among college students (many of these through car crashes in which alcohol was a factor), as well as 500,000 injuries and 70,000 cases of sexual assault.

EXAMPLE 3 Interpreting Vital Statistics

Assuming a U.S. population of 281 million, find and compare risks per person and per 100,000 people for pneumonia and cancer.

SOLUTION We find the risk per person by dividing the number of deaths by the total population of 281 million:

$$\text{Pneumonia:} \quad \frac{63,730 \text{ deaths}}{281,000,000 \text{ people}} \approx 0.00023 \text{ death per person}$$

$$\text{Cancer:} \quad \frac{549,838 \text{ deaths}}{281,000,000 \text{ people}} \approx 0.0020 \text{ death per person}$$

We can interpret these numbers as empirical probabilities: The probability of death by pneumonia is about 2.3 in 10,000, while the probability of death by cancer is about 20 in 10,000. To put them in terms of deaths per 100,000 people, we simply multiply the per person rates by 100,000. We get a pneumonia death rate of 23 deaths per 100,000 people and a cancer death rate of 200 deaths per 100,000 people. Thus, the probability of death by cancer is almost nine times greater than the probability of death by pneumonia. ∎

Time out to think
Table 10.1 suggests that the risk of death by stroke is about 70% higher than the risk of death by accident, but these data include all age groups. How do you think the risks of stroke and accident would differ between young people and old people? Explain.

Life Expectancy

One of the most commonly cited vital statistics is *life expectancy*, which is often used to compare overall health at different times or in different countries. However, the idea of life expectancy is surprisingly subtle, so let's investigate it. Figure 10.3a shows the overall U.S. death rate (or mortality rate), in deaths per 1,000 people, for different age groups. Note that there is an elevated risk of death near birth, after which the death rate drops to very low levels. At about 15 years of age, the death rate begins a gradual rise.

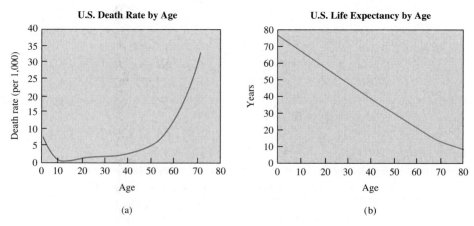

(a)   (b)

Figure 10.3   (a) The overall U.S. death rate (deaths per 1,000 people) for different ages. (b) Life expectancy for different ages. SOURCE: U.S. National Center for Health Statistics.

Figure 10.3b shows the **life expectancy** of Americans of different ages, which is defined as the number of years a person of a given age can expect to live on average. As we would expect, life expectancy is higher for younger people, because, on average, they have longer left to live. At birth, the life expectancy of Americans today is about 77 years (74.1 years for men and 79.5 years for women).

Definition

Life expectancy is the number of years a person with a given age today can expect to live on average.

The subtlety in interpreting life expectancy comes from changes in medical science and public health. Life expectancies are calculated by studying *current* death rates. For example, when we say that the life expectancy of infants born today is 77 years, we mean that the average baby will live to age 77 *if there are no future changes* in medical science or public health. Thus, while life expectancy provides a useful measure of current overall health, it should not be considered a *prediction* of future life spans.

In fact, because of advances in both medical science and public health, life expectancies increased significantly during the past century. As shown in Figure 10.4, U.S. life expectan-

By the Way...

Life expectancies vary widely around the world, as well as by sex and race. Canada and Australia have the highest at 78 years (both sexes combined). Among the lowest are Sierra Leone (34 years) and Uganda (43 years).

cies rose to their current values from less than 50 years in 1900. If this trend continues, today's infants are likely to live much longer than 77 years on average. On the other hand, declining public health can lead to falling life expectancies. For example, war-torn nations often experience precipitous drops in life expectancy as public health deteriorates. Note that the change in life expectancy shown in Figure 10.4 is truly remarkable: In less than 100 years, life expectancy increased by about 60%. There may be no better single measure of the changes that science and technology have brought to society.

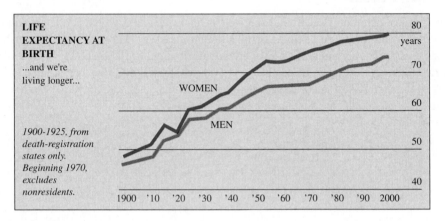

Figure 10.4   Changes in U.S. life expectancy during the 20th century. SOURCE: *New York Times* and National Center for Health Science Statistics.

Time out to think

Using Figure 10.4, compare the life expectancies of men and women. Briefly discuss these differences. Do they have any implications for social policy? For insurance rates? Explain.

EXAMPLE 4   Life Expectancies

Using Figure 10.3b, find the life expectancies of a 20-year-old person and a 60-year-old person. Are the numbers consistent? Explain.

SOLUTION   The graph shows that the life expectancy at age 20 is about 57 years and at age 60 is about 21 years. This means that an average 20-year-old can expect to live about 57 more years, to age 77. An average 60-year-old can expect to live about 21 more years, to age 81.

It might at first seem strange that 60-year-olds have a longer average life span than 20-year-olds (81 years versus 77 years). But remember that life expectancies are based on *current* data. If there were no changes in medicine or public health, a 60-year-old would have a greater probability of reaching age 81 than a 20-year-old simply because he or she has already made it to age 60. However, if medicine and public health continue to improve, today's 20-year-olds may live to older ages than today's 60-year-olds. ∎

CASE STUDY   Life Expectancy and Social Security

Because of the changing age makeup of the U.S. population, the number of retirees collecting Social Security benefits is expected to become much larger in the future, while the number of wage earners paying Social Security taxes is expected to remain relatively constant. Thus, one of the biggest challenges to the future of Social Security is finding a way to make sure there is enough money to pay benefits for future retirees.

Current projections show that, without significant changes, the Social Security program will be bankrupt and unable to pay benefits after about 2040. Social Security officials have proposed several different ways to solve this problem, including changes in the amounts of benefits paid, increases in the Social Security tax rate, changes in the retirement age, and partially or fully privatizing the Social Security program. Each of these proposals faces political obstacles. But, in addition, all of these proposals are based on assumptions about future life expectancy. Without corresponding changes in retirement age, longer lives mean more years of Social Security benefits.

More specifically, recent proposals by Social Security officials have assumed that life expectancy at birth will rise to 79.3 by 2030 and to 81.5 by 2070 (for both sexes). But these numbers may be far too pessimistic. For example, the Social Security projections assume that American women will not reach a life expectancy of 82 years until 2033—but women in France have *already* achieved this life expectancy. Will it really take the United States more than 30 years to catch up to life expectancies in France? One expert panel suggested that more realistic estimates of future life expectancy would be 81 in 2030 and 85.2 in 2070, meaning nearly four more years of Social Security benefits for the average person in 2070. And, if medical science achieves any major breakthroughs in combatting the effects of aging, the Social Security crisis could become far worse. ∎

Time out to think
According to some biologists, there is a good chance that 21st-century advances in medical science will allow most people to live to ages of 100 and more. How would that affect programs like Social Security? What other effects would you expect it to have on society? Overall, do you think large increases in life expectancy will be good or bad for society? Defend your opinion.

Review Questions

1. Give an example of rates that are used to measure risk in travel. Why are rates more useful than total numbers of deaths or accidents in measuring risk?

2. What are vital statistics? How are they usually described? Give a few examples.

3. Explain the meaning of the term *life expectancy*. How does life expectancy change with age? How is it affected by changes in the overall health of a population?

Exercises

BASIC SKILLS AND CONCEPTS

SENSIBLE STATEMENTS? For Exercises 1–4, determine whether the given statement is sensible and explain why it is or is not.

1. Every time someone dies in a shark attack, I hear about it in the news. Therefore, the risk of death by shark attack must be greater than the risk of death in an automobile.

2. The risk of having an accident riding a bicycle is less when the rider is wearing a helmet than when the rider is not wearing a helmet.

3. Your expected age of death decreases as you get older.

4. In 2000, the total numbers of deaths in the United States due to accidents and pneumonia were approximately equal. Therefore, the risks in deaths per 100,000 people are approximately equal.

CHANGING AUTOMOBILE SAFETY. Use the table on page 414 to answer the questions in Exercises 5–8.

Year	U.S population (millions)	Traffic fatalities	Licensed drivers (millions)	Vehicle miles (trillions)
1977	220	47,878	138	1.5
1987	242	46,390	162	1.9
2000	281	41,821	185	2.7

5. Express the 2000 fatality rate in deaths per 100 million vehicle-miles traveled.

6. Express the 2000 fatality rate in deaths per 100,000 population.

7. Express the 2000 fatality rate in deaths per 100,000 licensed drivers.

8. Use three different fatality rates (deaths per person, deaths per driver, and deaths per vehicle-mile) to compare the empirical risk of automobile travel in 1977 and 2000. Explain your conclusions in detail.

9. COMMERCIAL AIRLINE SAFETY. In 1994, there were 239 fatalities in commercial airline accidents. In that year, there were 13.1 million flight hours, 5.4 billion miles flown, and 8.2 million departures. In 2000, there were 92 fatalities in commercial airline accidents, 18.0 million flight hours, 7.1 billion miles flown, and 11.6 million departures. Compute the fatality rate in both years in terms of deaths per 100,000 flight hours, deaths per 1,000,000 miles flown, and deaths per 100,000 departures. Compare the risks of flying in these two years.

10. GENERAL AVIATION SAFETY. In 1982, there were 1,187 deaths and 3,233 accidents in general aviation (commercial, private, military) over 29,640,000 flight hours. In 2000, there were 592 deaths and 1,835 accidents in general aviation over 30,800,000 flight hours. Compare the safety of general aviation in these two years in terms of deaths and accidents per 100,000 flight hours.

11. DISEASE FATALITY RATES. Consider Table 10.1 and assume a U.S. population of 281 million.
    a. What is the empirical probability of death by diabetes during a single year? How much greater is the risk of death by diabetes than the risk of death by septicemia?
    b. How much greater is the risk of death by accident than the risk of death by kidney disease?
    c. What is the death rate due to kidney disease in deaths per 100,000 of the population?
    d. What is the death rate due to heart disease in deaths per 100,000 of the population?
    e. If you lived in a typical city of 500,000, how many people would you expect to die of a stroke in a year?

FURTHER APPLICATIONS

12. DEATH RATES. Use the graphs in Figure 10.3 to answer the following questions.
    a. Estimate the death rate for 50- to 55-year-olds.
    b. Assuming that there were about 15.8 million 50- to 55-year-olds, how many people in this bracket could be expected to die in a year?
    c. To what age could the average 50-year-old expect to live?
    d. To what age could the average 70-year-old expect to live?
    e. Suppose that a life insurance company insures 1 million 50-year-old people. The cost of the premium is $200 per year, and the death benefit is $50,000. What is the expected profit or loss for the insurance company?

13. DEATH RATES. Use the graphs in Figure 10.3 to answer the following questions.
    a. Estimate the death rate for 60- to 65-year-olds.
    b. Assuming that there were about 10.2 million 60- to 65-year-olds, how many people in this bracket could be expected to die in a year?
    c. To what age could the average 40-year-old expect to live?
    d. To what age could the average 60-year-old expect to live?
    e. Suppose that a life insurance company insures 1 million 60-year-old people. The cost of the premium is $200 per year, and the death benefit is $50,000. What is the expected profit or loss for the insurance company?

14. READING MORTALITY GRAPHS. Use the graphs in Figure 10.3 to answer the following questions.
    a. Approximately what is the death rate for 25- to 35-year-olds? If the population in this age category is 44 million, how many 25- to 35-year-olds can be expected to die in a year?
    b. At what average age will today's 60-year-olds die?
    c. Suppose that a life insurance company insures 5,000 40-year-old people. The cost of the premium is $200 per year, and the death benefit is $50,000. How much can the company expect to gain (or lose) in a year?

15. HIGH/LOW U.S. BIRTH RATES. The highest and lowest birth *rates* in the United States in 2000 were in Utah and Maine, respectively. Utah reported 47,368 births with a population of about 1.8 million people. Maine reported 13,603 births with a population of about 1.3 million people. Use these data to answer the following questions.
    a. How many people were born per day in Maine? In Utah?
    b. What was the birth rate in Utah in births per 1,000 people? In Maine?

16. **High/Low U.S. Death Numbers.** In 2000, there were 227,381 deaths in California, the highest number in the United States. The lowest number of deaths in any state was for Alaska, with 2,911 deaths. The populations of California and Alaska were approximately 30 million and 550,000, respectively.

    a. Compute the 2000 death *rates* for California and Alaska in deaths per 1,000.

    b. Based on the fact that California and Alaska had the highest and lowest death numbers, respectively, does it follow that California and Alaska had the highest and lowest death *rates*? Why or why not?

17. **U.S. Birth and Death Rates.** In 2000, the U.S. population was about 281 million. The overall birth rate was 14.8 births per 1,000, and the overall death rate was 8.6 deaths per 1,000.

    a. Approximately how many births were there in the United States?

    b. About how many deaths were there in the United States?

    c. Based on births and deaths alone (i.e., not counting immigration), about how much did the U.S. population rise during 2000?

    d. According to the U.S. Census Bureau, the U.S. population actually increased by about 3.0 million. Based on this fact and your results from part c, estimate how many people immigrated to the United States. What proportion of the overall population growth was due to immigration?

18. **Sex Ratios.** Figure 10.5 shows the ratio of men to women in various age categories since 1920. Discuss the patterns and trends that you see in this figure and offer explanations in terms of mortality and life expectancy.

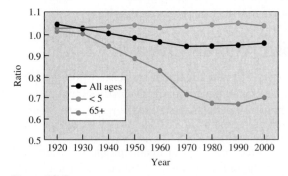

Figure 10.5

19. **Birth Statistics.** Figure 10.6 shows the number of births (in millions) and the fertility rate (births per 100 women ages 15 to 44). Use these data to answer the following questions.

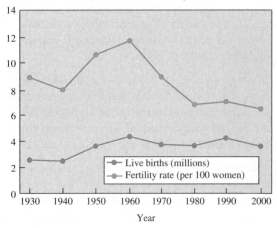

Figure 10.6

a. What was the fertility rate in 1997?

b. How many live births were there in 1998?

c. Estimate the number of women between the ages of 15 and 44 in 1990.

d. Explain the peaks in both curves in 1960 and 1990.

e. Explain the fact that the peaks in the total births curve in 1960 and 1990 are both at about the same level, while the corresponding peaks in the fertility rate curve are at significantly different levels.

 PROJECTS FOR THE WEB AND BEYOND

For useful links, select "Links for Web Projects" for Chapter 10 at www.aw.com/bbt.

20. **Crime Reduction.** According to a 1998 FBI report, the 1990s showed an overall decrease in crime in this country. Criminologists cannot agree on which of many factors are most important in accounting for this trend. Find annual data on both total crimes and crime rates in several categories of crime at the Census Bureau Web site. Discuss the trends you observe in these categories. Propose possible causes for these trends.

21. **Richter Scale for Risk.** The Royal Statistical Society has proposed a system of risk magnitudes and risk factors analogous to the Richter scale for measuring earthquakes. Go to the Web to learn how these measures of risk are defined and computed. Using these measures, discuss the risks of various activities and events.

22. **Life Expectancy Calculations.** There are many interesting and subtle aspects of computing life expectancies. Go to the Web and find as much information as possible

about how life expectancies are determined. Use several of the life expectancy calculators that are available on the Web. Do they seem to give accurate or realistic results? Explore the actual statistical techniques that are used to make life expectancy tables.

23. UPDATED CAUSES OF DEATH. Go the Web site for the Centers for Disease Control and find updated data on the 10 leading causes of death in the United States. Discuss how the risks have changed in comparison to the risks given in Table 10.1. Where does homicide rank on the list?

24. CANCER VS. HEART DISEASE. Heart disease and cancer are the two leading causes of death in this country. Based on mortality data for the last 10 years available from the Centers for Disease Control, make a case either for or against the claim that cancer will soon overtake heart disease as the leading cause of death.

25. UNDERSTANDING RISK. The book *Against the Gods: The Remarkable Story of Risk* by Peter Bernstein (John Wiley, 1996) is an award-winning account of the history of probability and risk assessment. Find the book in a library or bookstore (it's a worthwhile purchase) and identify a particular event that changed our understanding of risk. Write a two-page essay on both the history and the consequences of this particular event.

IN THE NEWS

1. **TRAVEL SAFETY.** Find a recent news article discussing some aspect of travel safety (such as risk of accidents in automobiles or airplanes, the efficacy of child car seats, or the effects of driving while talking on a cellular phone). Summarize any given statistics about risks, and give your overall opinion regarding the safety issue under discussion.

2. **VITAL STATISTICS.** Find a recent news report that gives new data about vital statistics or life expectancy. Summarize the report and the statistics, and discuss any personal or social implications of the new data.

10.2 Statistical Paradoxes

The government administers polygraph tests ("lie detectors") to new applicants for sensitive security jobs. The polygraph tests are reputed to be 90% accurate; that is, they catch 90% of the people who are lying and validate 90% of the people who are truthful. But how many people who fail a polygraph test are actually telling the truth? Most people guess that only about 10% of those who fail a test are falsely identified. In fact, the actual percentage of false accusations can be much higher. How can this be?

We'll discuss the answer soon, but first let's draw the important moral of this story: Statistics may not lie, but they can be deceiving if we do not interpret them carefully. In this section, we'll discuss several common ways in which statistics can deceive.

Better in Each Case, But Worse Overall

Suppose a pharmaceutical company creates a new treatment for acne. To decide whether the new treatment is better than an older treatment, the company gives the old treatment to 90 patients while the new treatment is given to 110 patients. Here are the results after four weeks of treatment (also summarized in Table 10.2):

- Among patients with mild acne, 10 received the old treatment and 90 received the new treatment. Among the 10 receiving the old treatment, 2 were cured, for a 20% cure rate. Among the 90 receiving the new treatment, 30 were cured, for a 33% cure rate. Thus, the new treatment had a higher cure rate among patients with mild acne.

- Among patients with severe acne, 80 received the old treatment and 20 received the new treatment. Among the 80 receiving the old treatment, 40 were cured, for a 50% cure rate. Among the 20 receiving the new treatment, 12 were cured, for a 60% cure rate. Thus, the new treatment had a higher cure rate among patients with severe acne.

Table 10.2 Results of Acne Treatments	Mild acne		Severe acne	
	Cured	Not cured	Cured	Not cured
Old treatment	2	8	40	40
New treatment	30	60	12	8

Notice that the new treatment had a higher cure rate both for patients with mild acne and for patients with severe acne. Is it fair for the company to claim that their new treatment is better than the old treatment?

At first, this might seem to make sense. But let's look at the overall results:

- Among the total of 90 patients receiving the old treatment, 42 were cured: 2 out of 10 with mild acne and 40 out of 80 with severe acne. Thus, the overall cure rate for the old treatment was 42/90 = 46.7%.

- Among the total of 110 patients receiving the new treatment, 42 were cured: 30 out of 90 with mild acne and 12 out of 20 with severe acne. Thus, the overall cure rate for the new treatment was 42/110 = 38.2%.

Overall, the old treatment had the higher cure rate, despite the fact that the new treatment had a higher rate for both mild and severe acne cases.

This example illustrates that it is possible for something to appear better in each of two or more group comparisons, but actually be worse overall. This situation is an example of **Simpson's paradox**. If you look carefully, you'll see that it occurs because of the way in which the overall results were divided into unequally sized groups (in this case, the groups were mild acne patients and severe acne patients). Simpson's paradox arises surprisingly often and is one way in which numbers can deceive unless they are examined with great care.

By the Way...

Simpson's paradox gets its name from Edward Simpson, who described it in 1951. However, the same idea was actually described much earlier, by Scottish statistician George Yule around 1900.

EXAMPLE 1   Who Played Better?

Table 10.3 gives the shooting performance of two players in each half of a basketball game. Shaq had a higher shooting percentage in both the first half (40% to 25%) and the second half (75% to 70%). Can Shaq claim that he had the better game?

Table 10.3 Basketball Shots	First Half			Second Half		
Player	Baskets	Attempts	Percent	Baskets	Attempts	Percent
Shaq	4	10	40%	3	4	75%
Kobe	1	4	25%	7	10	70%

SOLUTION   No, and we can see why by looking at the overall game statistics. Shaq made a total of 7 baskets (4 in the first half and 3 in the second half) on 14 shots (10 in the first half and 4 in the second half), for an overall shooting percentage of 7/14 = 50%. Kobe made a total of 8 baskets on 14 shots, for an overall shooting percentage of 8/14 = 57.1%. Thus, even though Shaq had a higher shooting percentage in both halves, Kobe had a better overall shooting percentage for the game.

Other Cases of Simpson's Paradox

Variations of Simpson's paradox arise in many situations. One famous case involves an investigation of possible gender discrimination in graduate programs at the University of California, Berkeley. Examining graduate admissions for 1973, university officials found that male applicants were admitted at a significantly higher rate than female applicants. Assuming that male and female applicants were equally qualified, the university appeared to be discriminating against women.

The university began investigating individual departments to see which ones were responsible for the discrimination. Table 10.4 shows the results for the six departments with the largest numbers of graduate applicants. (University policy requires that departments not be named, so we refer to them as Departments A through F.)

Table 10.4 Applicants and Admissions for Six Graduate Programs at UC Berkeley, 1973

Department	Men			Women		
	Applied	Admitted	Percent	Applied	Admitted	Percent
A	825	512	62%	108	89	82%
B	560	353	63%	25	17	68%
C	325	120	37%	593	202	34%
D	417	138	33%	375	131	35%
E	191	53	28%	393	94	24%
F	373	22	6%	341	24	7%
Total	**2,691**	**1,198**	**44.5%**	**1,835**	**557**	**30.4%**

SOURCE: P. Bickel and J. W. O'Connell, *Science*, Vol. 187, p. 398.

Note that the admission rates (percentages) for women were actually higher than those for men in all but Departments C and E, and the rates were quite close in those cases. Nevertheless, the totals confirm a much higher overall admission rate for men than for women. Thus, we see Simpson's paradox in action again: While women were admitted at a significantly lower rate overall, no individual department was guilty of this practice.

Time out to think

Look closely at the data in Table 10.4. Would it be reasonable to conclude that Department A discriminates against men? Why or why not?

By the Way...

A National Science Foundation study once found that, on average, women in science and engineering earned only 77% as much as men. But when the data were examined within each field, in every case the average salary for women was at least 92% of the average salary for men. The explanation for this case of Simpson's paradox: Women were more highly represented in life sciences and social sciences, which tend to be lower paying than physical sciences and engineering.

EXAMPLE 2 Does Smoking Make You Live Longer?

In the early 1970s, a medical study in England involved many adult residents from a district called Wickham. Twenty years later, a follow-up study looked at the survival rates of the people from the original study. The follow-up study found the following surprising results (D. R. Appleton, J. M. French, and M. P. Vanderpump, *American Statistician*, Vol. 50, 1996, pp. 340–341):

- Among the adults who smoked, 24% died during the 20 years since the original study.

- Among the adults who did not smoke, 31% died during the 20 years since the original study.

Do these results suggest that smoking can make you live longer?

SOLUTION   Not necessarily, because the given results don't tell us the ages of the smokers and nonsmokers. It turned out that, in the original study, the nonsmokers were older on average than the smokers. Thus, the higher death rate among nonsmokers simply reflected the fact that death rates tend to increase with age. When the results were analyzed by age groups, they clearly showed that for any given age group, nonsmokers had a higher 20-year survival rate than smokers. That is, 55-year-old smokers were less likely to reach age 75 than 55-year-old nonsmokers, and so on. Rather than suggesting that smoking prolongs life, a careful study of the data showed just the opposite.   ∎

Does a Positive Mammogram Mean Cancer?

Although we often associate tumors with cancers, many tumors are not cancers. Medically, any kind of abnormal swelling or tissue growth is considered a tumor. A tumor caused by cancer is said to be *malignant* (or *cancerous*); all others are said to be *benign*. When a medical practitioner spots a tumor, one of the first steps in treatment is determining whether the tumor is benign or malignant.

With that background, imagine you are a doctor or nurse treating a patient who has been referred to you because she has a breast tumor. The patient will be understandably nervous, but you can give her some comfort by telling her that only about 1 in 100 breast tumors turns out to be malignant. But, just to be safe, you order a mammogram to determine whether her tumor is one of the 1% that are malignant.

Now, suppose the mammogram comes back positive; that is, it indicates the tumor is likely to be malignant. Mammograms are not perfect, so even the positive result does not necessarily mean that your patient has breast cancer. More specifically, let's assume that the mammogram screening is 85% accurate: It will correctly identify 85% of malignant tumors as malignant and 85% of benign tumors as benign. When you tell your patient that her mammogram was positive, what should you tell her about the chance that she actually has cancer?

Because the mammogram screening is 85% accurate, most people guess that the positive result means that the patient probably has cancer. Studies have shown that in this situation most doctors also believe this to be the case, and would tell the patient to be prepared for cancer treatment. But a more careful analysis shows otherwise. In fact, the chance that the patient has cancer is still quite small—about 5%. We can see why by analyzing some numbers.

Consider a study in which mammograms are given to 10,000 women with breast tumors. Assuming that 1% of tumors are malignant, $1\% \times 10,000 = 100$ of the women actually have cancer; the remaining 9,900 women have benign tumors. Table 10.5 summarizes the mammogram results.

> **TECHNICAL NOTE**
>
> This mammogram example and the polygraph example that follows illustrate the use of conditional probabilities. The proper way of handling conditional probabilities was discovered by the Reverend Thomas Bayes (1702–1761) and is often called *Bayes rule*.

By the Way...

Most cancer tests have different accuracy rates for malignant and benign tumors. For example, the accuracy rate for malignant tumors (called the *sensitivity*) might be 80% and the accuracy rate for benign tumors (called the *specificity*) might be 90%.

Table 10.5 Summary of Results for 10,000 Mammograms (when in fact 100 tumors are malignant and 9,900 are benign)			
	Tumor is malignant	Tumor is benign	Total
Positive mammogram	85 true positives	1,485 false positives	**1,570**
Negative mammogram	15 false negatives	8,415 true negatives	**8,430**
Total	**100**	**9,900**	**10,000**

Note the following:

- The mammogram screening correctly identifies 85% of the 100 malignant tumors as malignant. Thus, it gives positive (malignant) results for 85 of the malignant tumors; these cases are called **true positives.** In the other 15 malignant cases, the result is negative, even though the women actually have cancer; these cases are **false negatives.**

- The mammogram screening correctly identifies 85% of the 9,900 benign tumors as benign. Thus, it gives negative (benign) results for $85\% \times 9,900 = 8,415$ of the benign tumors; these cases are **true negatives.** The remaining $9,900 - 8,415 = 1,485$ women get positive results in which the mammogram incorrectly identifies their tumors as malignant; these cases are **false positives**.

Overall, the mammogram screening gives positive results to 85 women who actually have cancer and to 1,485 women who do *not* have cancer. The total number of positive results is $85 + 1,485 = 1,570$. Because only 85 of these are true positives (the rest are false positives), the chance that a positive result really means cancer is only $85/1,570 = 0.054$, or 5.4%. Therefore, when your patient's mammogram comes back positive, you should reassure her that there's still only a small chance that she has cancer.

EXAMPLE 3   False Negatives

Suppose you are a doctor seeing a patient with a breast tumor. Her mammogram comes back negative. Based on the numbers in Table 10.5, what is the chance that she has cancer?

SOLUTION   For the 10,000 cases summarized in Table 10.5, the mammograms are negative for 15 women with cancer and for 8,415 women with benign tumors. The total number of negative results is $15 + 8,415 = 8,430$. Thus, the fraction of women with cancer who have false negatives is $15/8,430 = 0.0018$, or slightly less than 2 in 1,000. In other words, the chance that a woman with a negative mammogram has cancer is only about 2 in 1,000.   ∎

By the Way...

The accuracy of mammograms is improving with technology. In 2000, studies showed the overall accuracy rate to be over 91%, and some experimental technologies appeared to achieve accuracy rates of over 98%. The accuracy is lower, however, for women receiving hormone replacement therapy, which is common among women over 40.

Time out to think

While the chance of cancer with a negative mammogram is small, it is not zero. Therefore, it might seem like a good idea to biopsy all tumors, just to be sure. However, biopsies involve surgery, which means they can be painful and expensive, among other things. Given these facts, do you think that biopsies should be routine for all tumors? Should they be routine for cases of positive mammograms? Defend your opinion.

Polygraphs and Drug Tests

We're now ready to return to the question asked at the beginning of this section, about how a 90% accurate polygraph test can lead to a surprising number of false accusations. The explanation is very similar to that found in the case of the mammograms.

Suppose the government gives the polygraph test to 1,000 applicants for sensitive security jobs. Further suppose that 990 of these 1,000 people tell the truth on their polygraph test, while only 10 people lie. For a test that is 90% accurate, we find the following results:

- Of the 10 people who lie, the polygraph correctly identifies 90%, meaning that 9 fail the test (they are identified as liars) and 1 passes.

- Of the 990 people who tell the truth, the polygraph correctly identifies 90%, meaning that $90\% \times 990 = 891$ truthful people pass the test and the other $10\% \times 990 = 99$ truthful people fail the test.

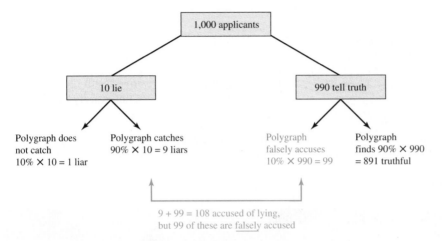

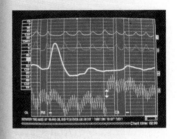

Figure 10.7   A tree diagram summarizes results of a 90% accurate polygraph test for 1,000 people, of whom only 10 are lying.

Figure 10.7 summarizes these results. The total number of people who fail the test is 9 + 99 = 108. Of these, only 9 were actually liars; the other 99 were falsely accused of lying. That is, 99 out of 108, or 99/108 = 91.7%, of the people who fail the test were actually telling the truth.

The percentage of people who are falsely accused in any real situation depends on both the accuracy of the test and the proportion of people who are lying. Nevertheless, for the numbers given here, we have an astounding result: Assuming the government rejects applicants who fail the polygraph test, then almost 92% of the rejected applicants were actually being truthful and may have been highly qualified for the jobs.

Time out to think
Imagine that you are falsely accused of a crime. The police suggest that, if you are truly innocent, you should agree to take a polygraph test. Would you do it? Why or why not?

EXAMPLE 4   High School Drug Testing

All athletes participating in a regional high school track and field championship must provide a urine sample for a drug test. Those who fail are eliminated from the meet and suspended from competition for the following year. Studies show that, at the laboratory selected, the drug tests are 95% accurate. Assume that 4% of the athletes actually use drugs. What fraction of the athletes who fail the test are falsely accused and therefore suspended without cause?

SOLUTION   The easiest way to answer this question is by using some sample numbers. Suppose there are 1,000 athletes in the meet. Then 4%, or 40 athletes, actually use drugs; the remaining 960 athletes do not use drugs. In that case, the 95% accurate drug test should return the following results:

- 95% of the 40 athletes who use drugs, or 0.95 × 40 = 38 athletes, fail the test. The other 2 athletes who use drugs pass the test.

- 95% of the 960 athletes who do not use drugs pass the test, but 5% of these 960, or 0.05 × 960 = 48 athletes, fail.

The total number of athletes who fail the test is 38 + 48 = 86. But 48 of these athletes who fail the test, or 48/86 = 56%, are actually nonusers. Despite the 95% accuracy of the drug test, more than half of the suspended students are innocent of drug use. ∎

Review Questions

1. What is Simpson's paradox? Give an example to show how, in individual cases, it can make something that is actually worse overall look better.

2. Describe an example that illustrates how Simpson's paradox can suggest discrimination in overall numbers even when no discrimination occurs.

3. Briefly explain why a positive result on a cancer test such as a mammogram does not necessarily mean that a patient has cancer.

4. Explain how it is possible for a very accurate polygraph or drug test to result in a large proportion of false accusations.

Exercises

BASIC SKILLS AND CONCEPTS

SENSIBLE STATEMENTS? For Exercises 1–4, determine whether the given statement is sensible and explain why it is or is not.

1. Ann and Bret are taking the same statistics course, in which the final grade is determined by assignments and exams. Ann's average on the assignments is higher than Bret's, and Ann's average on the exams is higher than Bret's. It follows that Ann's overall average in the course is higher than Bret's.

2. Ann's batting average for the first half of the season is higher than Bret's, and Ann's batting average for the second half of the season is higher than Bret's. It follows that Ann's batting average for the entire season is higher than Bret's.

3. The probability of having strep if your test is positive is the same as the probability of having a positive test if you have strep.

4. If a drug test is 90% accurate, 90% of those who test positive are actual drug users.

5. BATTING PERCENTAGES. The table below shows the batting records of two baseball players in the first half (first 81 games) and last half of a season.

| | First Half | | |
Player	Hits	At-bats	Batting average
Josh	50	150	.333
Jude	10	50	.200
	Second Half		
Player	Hits	At-bats	Batting average
Josh	35	70	.500
Jude	70	150	.467

Who had the higher batting average in the first half of the season? Who had the higher batting average in the second half? Who had the higher overall batting average? Explain how these results illustrate Simpson's paradox.

6. PASSING PERCENTAGES. The table below shows the passing records of two rival quarterbacks in the first half and second half of a football game.

| | First Half | | |
Player	Completions	Attempts	Percent
Allan	8	20	40%
Abner	2	6	33%
	Second Half		
Player	Completions	Attempts	Percent
Allan	3	6	50%
Abner	12	25	48%

Who had the higher completion percentage in the first half? Who had the higher completion percentage in the second half? Who had the higher overall completion percentage? Explain how these results illustrate Simpson's paradox.

7. TEST SCORES. The table below shows eighth-grade mathematics test scores in Nebraska and New Jersey. The scores are separated according to the race of the students. Also shown are the state averages for all races.

	White	Nonwhite	Average for all races
Nebraska	281	250	277
New Jersey	283	252	272

National Assessment of Educational Progress scores for 1992, from *Chance*, Spring 1999.

a. Which state had the higher scores in both racial categories? Which state had the higher overall average across both racial categories?

b. Explain how a state could score lower in both categories and still have a higher overall average.

c. Now consider the table below that gives the percentages of whites and nonwhites in each state. Use these percentages to verify that the overall average test score in Nebraska is 277, as claimed in the first table.

	White	Nonwhite
Nebraska	87%	13%
New Jersey	66%	34%

d. Use the racial percentages to verify that the overall average test score in New Jersey is 272, as claimed in the first table.

e. Explain briefly, in your own words, how Simpson's paradox appeared in this case.

8. **TEST SCORES.** Consider the following table comparing the grade point average (GPA) and mathematics SAT scores of high school students in 1988 and 1998.

GPA	% students 1988	% students 1998	SAT score 1988	SAT score 1998	Change
A+	4	7	632	629	−3
A	11	15	586	582	−4
A−	13	16	556	554	−2
B	53	48	490	487	−3
C	19	14	431	428	−3
Overall average			504	514	+10

Cited in *Chance*, Vol. 12, No. 2, 1999, from data in *New York Times*, September 2, 1999.

a. In general terms, how did the SAT scores of the students in the five grade categories change between 1988 and 1998?

b. How did the overall average SAT score change between 1988 and 1998?

c. How is this an example of Simpson's paradox?

9. **TUBERCULOSIS DEATHS.** The following table shows deaths due to tuberculosis (TB) in New York City and Richmond, Virginia, in 1910.

a. Compute the death rates for whites, nonwhites, and all residents in New York City.

b. Compute the death rates for whites, nonwhites, and all residents in Richmond.

c. Explain why this is an example of Simpson's paradox and explain how the paradox arises.

New York		
Race	Population	TB deaths
White	4,675,000	8400
Nonwhite	92,000	500
Total	74,767,000	8900

Richmond		
Race	Population	TB deaths
White	81,000	130
Nonwhite	47,000	160
Total	128,000	290

SOURCE: Cohen and Nagel, *An Introduction to Logic and Scientific Method*, Harcourt, Brace and World, 1934.

10. **WEIGHT TRAINING.** Two cross-country running teams participated in a (hypothetical) study in which a fraction of each team used weight training to supplement a running workout. The remaining runners did not use weight training. At the end of the season, the mean improvement in race times (in seconds) was recorded in the table below.

	Mean improvement (seconds)		
	Weight training	No weight training	Team average
Gazelles	10	2	6.0
Cheetahs	9	1	6.2

Describe how Simpson's paradox arises in this table. Resolve the paradox by finding the percentage of each team that used weight training.

11. **BASKETBALL RECORDS.** Consider the following hypothetical basketball records for Spelman and Morehouse Colleges.

	Spelman College	Morehouse College
Home games	10 wins, 19 losses	9 wins, 19 losses
Away games	12 wins, 4 losses	56 wins, 20 losses

a. Give numerical evidence to support the claim that Spelman College has a better team than Morehouse College.

b. Give numerical evidence to support the claim that Morehouse College has a better team than Spelman College.

c. Which claim do you think makes more sense? Why?

12. **BETTER DRUG.** Two drugs, A and B, were tested on a total of 2,000 patients, half of whom were women and half of whom were men. Drug A was given to 900 patients, and Drug B to 1,100 patients. The results appear in the table below.

	Women	Men
Drug A	5 of 100 cured	400 of 800 cured
Drug B	101 of 900 cured	196 of 200 cured

a. Give numerical evidence to support the claim that Drug B is more effective than Drug A.

b. Give numerical evidence to support the claim that Drug A is more effective than Drug B.

c. Which claim do you think makes more sense? Why?

13. **POLYGRAPH TEST.** Suppose that a polygraph is 90% accurate (it will correctly detect 90% of people who are lying and it will correctly detect 90% of people who are telling the truth). The 2,000 employees of a company are given a polygraph test during which they are asked whether they use drugs. All of them deny drug use, when, in fact, 1% of the employees actually use drugs. Assume that anyone whom the polygraph operator finds untruthful is accused of lying.

a. Verify that the entries in the table below follow from the given information. Explain each entry.

	Users	Nonusers	Total
Test finds employee lying	18	198	216
Test finds employee truthful	2	1,782	1,784
Total	20	1,980	2,000

b. How many employees are accused of lying? Of these, how many were actually lying and how many were telling the truth? What percentage of those accused of lying were falsely accused?

c. How many employees are found truthful? Of these, how many were actually truthful? What percentage of those found truthful really were truthful?

14. **DISEASE TEST.** Suppose a test for a disease is 80% accurate for those who have the disease (true positives) and 80% accurate for those who do not have the disease (true negatives). Within a sample of 4,000 patients, the incidence rate of the disease is the national average, which is 1.5%.

a. Verify that the entries in the table below follow from the information given and that the overall incidence rate is 1.5%. Explain.

	Disease	No disease	Total
Test positive	48	788	836
Test negative	12	3,152	3,164
Total	60	3,940	4,000

b. Of those with the disease, what percentage test positive?

c. Of those who test positive, what percentage have the disease? Compare this result to the one in part b and explain why they are different.

d. Suppose a patient tests positive for the disease. As a doctor using this table, how would you describe the patient's chance of actually having the disease? Compare this figure to the overall incidence rate of the disease.

FURTHER APPLICATIONS

15. **HIRING STATISTICS.** (This problem is based on an example in "Ask Marilyn," *Parade Magazine*, April 28, 1996.) A company decided to expand, so it opened a factory, generating 455 jobs. For the 70 white-collar positions, 200 males and 200 females applied. Of the females who applied, 20% were hired, while only 15% of the males were hired. Of the 400 males applying for the blue-collar positions, 75% were hired, while 85% of the 100 females who applied were hired. How does looking at the white-collar and blue-collar positions separately suggest a hiring preference for women? Do the overall data support the idea that the company hires women preferentially? Explain why this is an example of Simpson's paradox and how the paradox can be resolved.

16. **DRUG TRIALS.** (This problem is based on an example in "Ask Marilyn," *Parade Magazine*, April 28, 1996.) A company runs two trials of two treatments for an illness. In the first trial, Treatment A cures 20% of the cases (40 out of 200) and Treatment B cures 15% of the cases (30 out of 200). In the second trial, Treatment A cures 85% of the cases (85 out of 100) and Treatment B cures 75% of the cases (300 out of 400). Which treatment had the better cure rate in the two trials individually? Which treatment had the better overall cure rate? Explain why this is an example of Simpson's paradox and how the paradox can be resolved.

17. **HIV RISKS.** The New York State Department of Health estimates a 10% rate of HIV for the at-risk population and a 0.3% rate for the general population. Tests for HIV are 95% accurate in detecting both true negatives and true positives. Random selection of 5,000 at-risk people and 20,000 people from the general population results in the following table.

	At-risk population	
	Test positive	Test negative
Infected	475	25
Not infected	225	4,275
	General population	
	Test positive	Test negative
Infected	57	3
Not infected	997	18,943

a. Verify that incidence rates for the general and at-risk populations are 0.3% and 10%, respectively. Also verify that detection rates for the general and at-risk populations are 95%.

b. Consider the at-risk population. Of those with HIV, what percentage test positive? Of those who test positive, what percentage have HIV? Explain why these two percentages are different.

c. Suppose a patient in the at-risk category tests positive for the disease. As a doctor using this table, how would you describe the patient's chance of actually having the disease? Compare this figure to the overall incidence rate of the disease.

d. Consider the general population. Of those with HIV, what percentage test positive? Of those who test positive, what percentage have HIV? Explain why these two percentages are different.

e. Suppose a patient in the general population tests positive for the disease. As a doctor using this table, how would you describe the patient's chance of actually having the disease? Compare this figure to the overall incidence rate of the disease.

PROJECTS FOR THE WEB AND BEYOND

For useful links, select "Links for Web Projects" for Chapter 10 at www.aw.com/bbt.

18. **POLYGRAPH ARGUMENTS.** Visit Web sites devoted to either opposing or supporting the use of polygraph tests. Summarize the arguments on both sides, specifically noting the role that false negative rates play in the discussion.

19. **DRUG TESTING.** Explore the issue of drug testing either in the workplace or for athletic competitions. Discuss the legality of drug testing in these settings and the accuracy of the tests that are commonly conducted.

20. **CANCER SCREENING.** Investigate recommendations concerning routine screening for some type of cancer (for example, breast cancer, prostate cancer, colon cancer). Explain how the accuracy of the screening test is measured. How is the test useful? How can its results be misleading?

IN THE NEWS

1. **POLYGRAPHS.** Find a recent article in which someone or some group proposes a polygraph test to determine whether a person is being truthful. In light of what you know about polygraph tests, do you think the results will be meaningful? Why or why not?

2. **DRUG TESTING AND ATHLETES.** Find a new report concerning drug testing of athletes. Summarize how the testing is being used, and discuss whether the testing is reliable.

10.3 Hypothesis Testing with Two-Way Tables

Up until about 20 years ago, most college students were men. Today, however, the statistics are very different. To quote a headline in *U.S. News and World Report,*

Nearly 60 percent of college students are women: have men lost their minds?

The proportion of women's and men's bachelor's degrees varies from one major to another. However, in terms of the total degrees awarded, the crossover point (when women and men were awarded the same number of degrees) occurred in 1982 (Figure 10.8). Since then, the gap between the numbers of degrees awarded to women and men has widened steadily. Projections indicate that by 2010 women will receive almost 40% more bachelor's degrees than men.

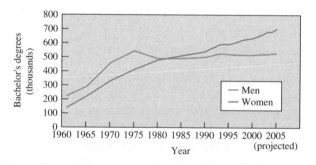

Figure 10.8   Numbers of bachelor's degrees awarded to men and women at all U.S. colleges and universities since 1961. SOURCE: *U.S. News and World Report.*

Table 10.6 shows the numbers of bachelor's degrees awarded to a random sample of men and women in two disciplines, biology and business. This display is an example of a **two-way table** (also called a **contingency table**) because it displays two variables—*major* and *gender.* In this case, each variable takes on just two values (biology and business for *major,* men and women for *gender*). There are four primary *cells* in the table, one for each combination of the two variables. All cells contain *frequencies* (or *counts*). For example, we see that 32 bachelor's degrees were awarded to women in biology and 87 bachelor's degrees were awarded to men in business. Notice that, without giving them a name, we encountered two-way tables during our discussion of Simpson's paradox in Section 10.2.

> I cannot do it without counters.
> —William Shakespeare,
> The Winter's Tale

Two-Way Tables

A **two-way table** shows the relationship between two variables by listing one variable in the rows and the other variable in the columns. The entries in the table's cells are called *frequencies* (or *counts*).

Table 10.6 also includes totals for the row and column entries. For example, we can see that a total of 142 women received degrees (last entry in first row) and a total of 53 biology degrees were awarded within this sample (last entry in first column). The lower right cell gives the grand total of all degrees in the two disciplines awarded to both men and women. You should confirm that the entries in both the row labeled "Total" and the column labeled "Total" sum to 250.

Table 10.6 Two-Way Table for Biology and Business Degrees

	Biology	Business	Total
Women	32	110	**142**
Men	21	87	**108**
Total	**53**	**197**	250

By the Way...

Across all American colleges and universities, men outnumber women in declared majors of engineering, computer science, and architecture. Women outnumber men in psychology, fine art, accounting, biology, and elementary education.

Time out to think

Answer the following questions to make sure you understand how to read Table 10.6. How many business degrees were awarded to men? How many business degrees were awarded to men and women? Does the total of all degrees awarded to men and women equal the total of all degrees awarded in business and biology? Explain why these totals should (or should not) agree.

EXAMPLE 1   A Two-Way Table for a Survey

Table 10.7 shows the results of a pre-election survey on gun control.

Table 10.7 Two-Way Table for Gun Control Survey (with totals)

	Favor stricter laws	Oppose stricter laws	Undecided	Total
Democrat	456	123	43	**622**
Republican	332	446	21	**799**
Total	**788**	**569**	**64**	1,421

a. What percentage of Democrats favored stricter laws?

b. What percentage of all voters favored stricter laws?

c. What percentage of those who opposed stricter laws were Republicans?

SOLUTION   Note that the total of the row totals and the total of the column totals are equal.

a. Of the 622 Democrats polled, 456 favored stricter laws. The percentage of Democrats favoring stricter laws is $456/622 = 0.733$, or 73.3%.

b. Of the 1,421 people polled, 788 favored stricter laws. The percentage of all respondents favoring stricter laws is 788/1,421 = 0.555, or 55.5%.

c. Of the 569 people polled who opposed stricter laws, 446 were Republicans, and 446/569 = 0.783, or 78.3%. ∎

Finding the Frequencies Expected by Chance

Our goal is to determine whether there is a relationship between the variables in a two-way table. In order to do this, we need to find the frequency we would expect for each cell *if there were no relationship* between the variables, which is equivalent to the frequency expected by chance alone.

As an example, let's find the frequency we would expect by chance for male business majors in Table 10.6. We start by finding the fraction of all students in the sample who received business degrees:

$$\frac{\text{total business degrees}}{\text{total degrees}} = \frac{197}{250}$$

As discussed in Chapter 6, we can interpret this result as a *relative frequency probability*. That is, if we select a student *at random* from the sample, the probability that he or she earned a business degree is 197/250. Using the notation for probability, we write

$$P(\text{business}) = \frac{197}{250}$$

Similarly, we can compute the fraction of total degrees awarded to men and interpret this fraction as a probability. That is, if we select a student at random from the sample, the probability that this student is a man is

$$P(\text{man}) = \frac{\text{men's degrees}}{\text{total degrees}} = \frac{108}{250}$$

We now have all the information needed to find the frequency we would expect by chance for male business majors. Recall from Chapter 6 that if two events A and B are independent (so that the outcome of one does not affect the probability of the other), then

$$P(A \text{ and } B) = P(A) \times P(B)$$

We can apply this rule to determine the probability that a student is *both* a man and a business major, *provided a student's gender is independent of major*:

$$P(\text{man and business}) = P(\text{man}) \times P(\text{business}) = \frac{108}{250} \times \frac{197}{250} \approx 0.3404$$

We now take one final step. The probability we have found is equivalent to the fraction of the total students whom we expect to be male business majors *if there is no relationship* between gender and major. Thus, by multiplying this probability by the total number of students in the sample (250), we can find the number (or frequency) of male business majors that we expect by chance:

$$\frac{108}{250} \times \frac{197}{250} \times 250 \approx 85.104$$

Notice that while the observed frequencies in Table 10.6 must be whole numbers because they come from counting students, the expected frequencies needn't be whole numbers.

Let's recap. We used three "total" entries from Table 10.6 to calculate that we would expect 85.104 male business majors if there were no relationship between gender and major.

> Humanity did not take control of society out of the realm of Divine Providence . . . to put it at the mercy of the laws of chance.
>
> —Maurice Kendall

This **expected frequency** agrees fairly well with the observed frequency of 87 shown in Table 10.6 for male business majors, so in this case the assumption of no relationship between gender and major seems reasonable. If we had instead found a significant difference between the expected and observed frequencies, we might reject the assumption of no relationship. (We'll discuss what constitutes a "significant difference" shortly.)

Frequencies Expected by Chance

In a two-way table, the **expected frequencies** are the frequencies we would expect by chance *if there were no relationship* between the row and column variables. The expected frequency in any cell is calculated in two steps:

1. Calculate the probability for the cell under the assumption of independence (no relationship) between the variables. That is, multiply

$$\frac{\text{row total}}{\text{grand total}} \times \frac{\text{column total}}{\text{grand total}}$$

2. Multiply the probability from step 1 by the total number of subjects (the grand total).

If the expected frequencies differ significantly from the observed frequencies, then we have evidence to reject the assumption of no relationship.

TECHNICAL NOTE

A general formula for finding the expected frequency for a particular cell is

$$\frac{\text{row total} \times \text{column total}}{\text{grand total}}$$

EXAMPLE 2   Frequencies Expected by Chance

Find the frequencies expected by chance for the other three cells in Table 10.6. Then construct a table showing both observed frequencies and frequencies expected by chance.

SOLUTION   We follow the same procedure used to find the expected number of male business majors, for which we found $P(\text{man})$ and $P(\text{business})$. We'll also need $P(\text{woman})$ and $P(\text{biology})$:

$$P(\text{woman}) = \frac{142}{250} = 0.5680$$

$$P(\text{biology}) = \frac{53}{250} = 0.2120$$

We now find the probability for each of the remaining cells:

$$P(\text{woman and business}) = P(\text{woman}) \times P(\text{business}) = \frac{142}{250} \times \frac{197}{250} \approx 0.4476$$

$$P(\text{man and biology}) = P(\text{man}) \times P(\text{biology}) = \frac{108}{250} \times \frac{53}{250} \approx 0.0916$$

$$P(\text{woman and biology}) = P(\text{woman}) \times P(\text{biology}) = \frac{142}{250} \times \frac{53}{250} \approx 0.1204$$

We have accounted for all possible events. Therefore, we should confirm that the sum of the probabilities in all cells is 1:

$$0.3404 + 0.4476 + 0.09158 + 0.1204 = 1.0000$$

To compute the frequency expected by chance for the three remaining cells in Table 10.6, we simply multiply the probabilities for each cell (in the form of fractions) by the total number of students (250):

I'm not smart, but I like to observe. Millions saw the apple fall, but Newton was the one who asked why.

—Bernard Baruch

$$\text{Frequency of women business majors} = 250 \times \frac{142}{250} \times \frac{197}{250} \approx 111.896$$

$$\text{Frequency of men biology majors} = 250 \times \frac{108}{250} \times \frac{53}{250} \approx 22.896$$

$$\text{Frequency of women biology majors} = 250 \times \frac{142}{250} \times \frac{53}{250} \approx 30.104$$

Again, it's useful to check that the sum of the four frequencies equals the total number of students in the sample, or 250:

$$85.104 + 111.896 + 22.896 + 30.104 = 250.000$$

Table 10.8 shows both the observed frequencies and the expected frequencies.

Table 10.8 Observed Frequencies and Frequencies Expected by Chance (in parentheses) for Table 10.6

	Biology	Business	Total
Women	32 (30.104)	110 (111.896)	**142 (142.000)**
Men	21 (22.896)	87 (85.104)	**108 (108.000)**
Total	**53 (53.000)**	**197 (197.000)**	**250 (250.000)**

Note that the values in the "Total" row and "Total" column are the same for both the observed frequencies and the frequencies expected by chance. This should always be the case, providing a good check on your work.

TECHNICAL NOTE

When all individuals in the sample are selected from a single population, the resulting test is a *test of independence*. When a two-way table is generated by selecting individuals from more than one population, the resulting test is called a *test of homogeneity*.

EXAMPLE 3   A Two-Way Table for Demographics

Hawaii and Idaho have nearly equal populations of about 1.3 million people. However, the ethnic compositions of the two states are quite different. A random sample of people in the two states results in the data in Table 10.9.

For the data in Table 10.9, find the expected number of people in each category under the assumption that there is no relationship between state and ethnic composition.

Table 10.9 Two-Way Table for Ethnic Populations in Hawaii and Idaho

	Non-Hispanic white	African American	Native American	Asian/Pacific Islander	Hispanic	Total
Hawaii	396	35	7	749	95	**1,282**
Idaho	1,174	7	16	13	86	**1,296**
Total	**1,570**	**42**	**23**	**762**	**181**	**2,578**

SOLUTION We need to find the frequencies expected by chance for the ten cells in Table 10.9. Consider the cell for Asian/Pacific Islanders in Hawaii. Proceeding as in Example 2, we find that

$$P(\text{Hawaii}) = \frac{1,282}{2,578} = 0.4973 \quad \text{and} \quad P(\text{Asian}) = \frac{762}{2,578} = 0.2956$$

Assuming that there is no relationship between state and ethnicity, we would expect the number of Asian/Pacific Islanders in Hawaii to be

$$\frac{1,282}{2,578} \times \frac{762}{2,578} \times 2,578 = 378.931$$

Continuing in this manner, we can fill the table with expected frequencies. Table 10.10 shows the result.

Table 10.10 Racial and Ethnic Populations in Hawaii and Idaho						
	Non-Hispanic white	African American	Native American	Asian/Pacific Islander	Hispanic	Total
Hawaii	396 (780.737)	35 (20.886)	7 (11.438)	749 (378.931)	95 (90.009)	**1,282** **(1.282.001)**
Idaho	1,174 (789.263)	7 (21.114)	16 (11.562)	13 (383.069)	86 (90.991)	**1,296** **(1,295.999)**
Total	**1,570** **(1,570.000)**	**42** **(42.000)**	**23** **(23.000)**	**762** **(762.000)**	**181** **(181.000)**	**2,578** **(2,578.000)**

NOTE: Expected frequencies are in parentheses.

Notice that the row and column sums agree for both the observed frequencies and the expected frequencies, as they should. Making a visual comparison of the two sets of frequencies, we see that some cells involve sizable differences, suggesting that the assumption of no relationship is false and that there is a relationship between ethnic composition and state. ∎

Time out to think

Explain how Table 10.10 shows that the number of Asian/Pacific Islanders in Hawaii is almost double what we would expect if there were no relationship between ethnic composition and state. How does the number of Asian/Pacific Islanders in Idaho compare to what we would expect if there were no relationship? Are you surprised that the assumption of no relationship between ethnic composition and state has proven false in this case? Why or why not?

Identifying the Hypotheses

We are now ready to be more precise about whether observed and expected frequencies are significantly different. Look again at the data in Table 10.6. Based on these sample data, is there a relationship between the variables *gender* and *major* within the population, or is the major *independent* of gender? In other words, do the same proportions of men and women choose each major, or do the numbers in the table suggest that men and women each favor particular majors? This sort of question can arise with almost any two-way table. In all cases, the question takes this general form:

Do the data in the table suggest a relationship between the two variables, or could these data simply have arisen by chance in the absence of any relationship?

We can visualize this question with the bar graph in Figure 10.9, showing the percentages of biology and business degrees for men, women, and all students. If there were no relationship between gender and major, then the segments in the three bars would have approximately the same height, with differences resulting from random fluctuations. When the heights of the segments are not the same, we need to determine whether the differences are due to random fluctuations or whether the differences are significant, suggesting some relationship between the two variables.

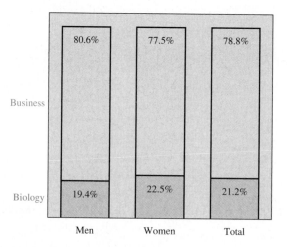

Figure 10.9   Bar graph showing the percentages of biology and business degrees for men, women, and all students in the sample.

We can approach this question of independence by using a slight variation on the hypothesis testing procedure introduced in Chapter 9. We first pose two hypotheses. The **null hypothesis,** H_0, always states that the variables in the table are *independent* (there is *no relationship* between them). For the data in Table 10.6, the null hypothesis states that the gender of a student is independent of whether she/he earns a business or biology degree. The **alternative hypothesis,** H_a, states the opposite: There is a relationship between gender and major.

Null and Alternative Hypotheses for Two-Way Tables

The **null hypothesis,** H_0, states that the variables in the table are independent (there is *no relationship* between them).

The **alternative hypothesis,** H_a, states that there is a relationship between the two variables in the table.

It is important to remember that even if we find a relationship between the variables, we have not proved that one variable *causes* the other variable. The alternative hypothesis suggests *some* relationship between the variables, but not necessarily a causal relationship.

EXAMPLE 4 Null and Alternative Hypotheses

Look again at Table 10.9, which displays the ethnic composition of two random samples taken in Hawaii and Idaho. State the null and alternative hypotheses for testing the independence of the variables *state* and *ethnicity* in this table.

SOLUTION The question arising from the table is whether there is a relationship between the variables *state* and *ethnicity*. The null hypothesis states that there is no relationship between state and ethnicity; that is, people of different ethnic groups live in Hawaii or Idaho at random. The alternative hypothesis states that there is a relationship between the variables, meaning that the ethnic proportions are not chance occurrences; instead, they arise because people of different ethnic groups tend to favor one state over the other. ∎

The Idea of the Hypothesis Test

As with all hypothesis tests, we begin by assuming that the null hypothesis is true; that is, the variables are independent and there is no relationship between them. Using this assumption, we compute the expected frequency for each cell, as in Example 2. If the null hypothesis is true, the observed and expected frequencies should be roughly the same. If the observed and expected frequencies differ significantly, we reject the null hypothesis and support the alternative hypothesis of a relationship between the variables.

Hypothesis Testing with Two-Way Tables

The basic idea of a hypothesis test with a two-way table is as follows:

1. Assume the null hypothesis is true; that is, there is *no relationship* between the two variables in the table. Determine the frequencies that we would expect by chance under this assumption.

2. Compare the observed frequencies to the expected frequencies.

3. If the observed frequencies are significantly different from the expected frequencies, reject the null hypothesis and conclude that there is a relationship between the two variables. Otherwise, continue to assume that the null hypothesis is true.

The methods we use to determine whether the differences are significant are similar to those we used in past chapters. In particular, our tests of significance depend on three factors:

• The size of the differences between the observed frequencies and the frequencies expected by chance. The larger the differences, the less likely that they could have occurred by chance.

• The size of the sample. The larger the sample, the less likely that a given difference could have occurred by chance.

• The value we select for the level of significance.

Let's continue with the example about gender and major (Table 10.6). We begin by stating formally the null and alternative hypotheses.

• **Null hypothesis, H_0:** Within the population, there is no relationship between gender and major; the proportions of degrees earned by men and women in business and biol-

ogy are what we would expect by chance. Gender and major are independent of each other.

- **Alternative hypothesis, H_a:** Within the population, there is some relationship between gender and major; the proportions of degrees earned by men and women in business and biology are not what we would expect by chance.

Our goal is to test the null hypothesis and determine whether there is sufficient evidence in the sample data to warrant its rejection. We have already determined the expected frequency (assuming the null hypothesis is true) for each cell in the table (see Table 10.8). We must now somehow measure the differences between the observed frequencies and those expected by chance.

The Chi-Square Statistic

Let's denote the observed frequencies by O and the expected frequencies by E. With this notation, $O - E$ ("*O minus E*") tells us the difference between the observed frequency and the expected frequency for each cell. We are looking for a measure of the *total* difference for the whole table. We cannot get such a measure by simply adding the individual differences, $O - E$, because they always sum to zero. Instead, we consider the *square* of the difference in each cell, $(O - E)^2$. We then make each value of $(O - E)^2$ a relative difference by dividing it by the corresponding expected number; this gives us the quantity $(O - E)^2/E$ for each cell. Summing the individual values of $(O - E)^2/E$ gives us the **chi-square statistic,** denoted χ^2 and pronounced "ki-square." To do this calculation in an organized way, it's best to make a table such as Table 10.11, with a row for each of the four cells in the original two-way table.

Finding the Chi-Square Statistic

1. For each cell in the two-way table, identify O as the observed frequency and find the value of E, the expected frequency.
2. Compute the value $(O - E)^2/E$ for each cell.
3. Sum the values from step 2 to get the chi-square statistic; that is,

$$\chi^2 = \text{sum of all values of } \frac{(O - E)^2}{E}$$

Time out to think
Why must the numbers in the $O - E$ column always sum to zero?

The larger the value of χ^2, the greater the average difference between the observed and expected frequencies in the cells. Thus, the value of χ^2 gives us a direct way of testing the null hypothesis of no relationship between the variables. If the value of χ^2 is small, then the differences between the observed frequencies and the frequencies expected by chance are *not* great enough to warrant rejecting the null hypothesis; in this case, there is no evidence of a relationship between the variables. By contrast, if the value of χ^2 is sufficiently large, we can

Table 10.11 Calculation of χ^2 Statistic for Data in Table 10.8

Outcome	O	E	$O - E$	$(O - E)^2$	$(O - E)^2/E$
Women/business	110	111.896	−1.896	3.595	0.032
Women/biology	32	30.104	1.896	3.595	0.119
Men/business	87	85.104	1.896	3.595	0.042
Men/biology	21	22.896	−1.896	3.595	0.157
Totals	250	250.000	0.000	14.380	$\chi^2 = 0.350$

reject the null hypothesis of independence. The only question is, How large is sufficiently large?

Table 10.12 answers this question by giving the critical values of χ^2 for two significance levels, 0.05 and 0.01, and several different table sizes. If the calculated value of χ^2 is *less than* the critical value, the differences between the observed and expected values are small and there is not enough evidence to reject the null hypothesis. If the calculated value of χ^2 is *greater than or equal to* the critical value, then there is enough evidence in the sample to reject the null hypothesis.

We can now apply the significance test to the data about gender and major. According to Table 10.12, for a 2 × 2 table we can reject the null hypothesis at a 0.05 significance level only if the chi-square statistic is greater than or equal to 3.841. However, for the gender/major data, we have $\chi^2 = 0.350$. Thus, we cannot reject the null hypothesis. Of course, failing to reject the null hypothesis does not *prove* that major and gender are independent. It simply means that we do not have enough evidence to justify rejecting the null hypothesis of independence.

TECHNICAL NOTE

The χ^2 test statistic is technically a discrete variable, whereas the actual χ^2 distribution is continuous. That discrepancy does not cause any substantial problems as long as the expected frequency for every cell is at least 5. We will assume that this condition is met for all the examples in this book.

Table 10.12 Critical Values of χ^2: Reject H_0 Only If $\chi^2 >$ Critical Value

Table size (rows × columns)	Significance level 0.05	0.01
2 × 2	3.841	6.635
2 × 3 or 3 × 2	5.991	9.210
3 × 3	9.488	13.277
2 × 4 or 4 × 2	7.815	11.345
2 × 5 or 5 × 2	9.488	13.277

EXAMPLE 5   Vitamin C Test

A study seeking to determine whether vitamin C has an effect in preventing colds involved a sample of 220 people. Within the sample, 105 randomly selected people took a vitamin C pill daily for a period of 10 weeks and the remaining 115 people took a placebo daily for 10 weeks. At the end of 10 weeks, the number of people who got colds was recorded. The results of the study are summarized in Table 10.13. Determine whether there is a relationship between taking vitamin C and getting colds.

	Cold	No cold	Total
Vitamin C	45	60	**105**
Placebo	75	40	**115**
Total	**120**	**100**	**220**

Table 10.13 Two-Way Table for Observed Number in Each Category

SOLUTION   We begin by stating the null and alternative hypotheses:

H_0 (null hypothesis): There is no relationship between taking vitamin C and getting colds; that is, vitamin C has no more effect on colds than the placebo.

H_a (alternative hypothesis): There is a relationship between taking vitamin C and getting colds; that is, the numbers of colds in the two groups are not what we would expect if vitamin C and the placebo were equally effective.

As always, we assume that the null hypothesis is true and calculate the expected frequency for each cell in the table. Noting that the sample size is 220 and proceeding as in Example 2, we find the following expected frequencies:

$$\text{Vitamin C and cold: } 220 \times \underbrace{\frac{105}{220}}_{P(\text{vit. C})} \times \underbrace{\frac{120}{220}}_{P(\text{cold})} = 57.273$$

$$\text{Vitamin C and no cold: } 220 \times \underbrace{\frac{105}{220}}_{P(\text{vit. C})} \times \underbrace{\frac{100}{220}}_{P(\text{no cold})} = 47.727$$

$$\text{Placebo and cold: } 220 \times \underbrace{\frac{115}{220}}_{P(\text{placebo})} \times \underbrace{\frac{120}{220}}_{P(\text{cold})} = 62.727$$

$$\text{Placebo and no cold: } 220 \times \underbrace{\frac{115}{220}}_{P(\text{placebo})} \times \underbrace{\frac{100}{220}}_{P(\text{no cold})} = 52.273$$

Table 10.14 shows the two-way table with the frequencies expected by chance in parentheses.

Table 10.14 Observed and Expected Frequencies for Vitamin C Study

	Cold	No cold	Total
Vitamin C	45 (57.273)	60 (47.727)	**105 (105.000)**
Placebo	75 (62.727)	40 (52.273)	**115 (115.000)**
Total	**120 (120.000)**	**100 (100.000)**	**220 (220.000)**

Now, we must compute chi-square and determine whether it gives us reason to reject the null hypothesis. To compute chi-square for these data, we organize our work as shown in Table 10.15.

Table 10.15 Table for Computing χ^2 Statistic for Vitamin C Study

Outcome	O	E	$O - E$	$(O - E)^2$	$(O - E)^2/E$
Vitamin C/cold	45	57.273	−12.273	150.627	2.630
Vitamin C/no cold	60	47.727	12.273	150.627	3.156
Placebo/cold	75	62.727	12.273	150.627	2.401
Placebo/no cold	40	52.273	−12.273	150.627	2.882
Totals	220	220.000	0		$\chi^2 = 11.069$

According to Table 10.12 (for two rows and two columns), we reject the null hypothesis at a 0.01 significance level if chi-square is greater than 6.635. With $\chi^2 = 11.069$, we reject the null hypothesis and conclude that there is a relationship between vitamin C and colds. Based on the data from this sample, there is reason to believe that whether a person gets a cold is dependent on whether the person takes vitamin C or a placebo. Again, this does not imply that vitamin C causes a reduction in colds; it implies only that there is some relationship which may or may not be causal.

By the Way...

Dozens of careful studies have been conducted on the question of vitamin C and colds. Some have found high levels of confidence in the effects of vitamin C, but others have not. Because of these often conflicting results, the issue of whether vitamin C prevents colds remains controversial.

EXAMPLE 6   To Plead or Not to Plead

The two-way table in Table 10.16 shows how a plea of guilty or not guilty affected the sentence in 1,028 selected burglary cases in the San Francisco area. Test the claim, at the 0.05 significance level, that the sentence (prison or no prison) is independent of the plea.

Table 10.16 Observed Frequencies for Plea and Sentence

	Prison	No prison	Total
Guilty plea	392	564	**956**
Not-guilty plea	58	14	**72**
Total	**450**	**578**	**1,028**

SOURCE: *Law and Society Review*, Vol. 16, No. 1.

SOLUTION   The null and alternative hypotheses for the problem are

H_0 (null hypothesis): The sentence in burglary cases is independent of the plea.

H_a (alternative hypothesis): The sentence in burglary cases depends on the plea.

We find the following *expected* number of people in each category, assuming that the row variables are independent of the column variables:

$$\text{Guilty and prison: } 1{,}028 \times \frac{956}{1{,}028} \times \frac{450}{1{,}028} = 418.482$$

$$\text{Guilty and no prison: } 1{,}028 \times \frac{956}{1{,}028} \times \frac{578}{1{,}028} = 537.518$$

$$\text{Not guilty and prison: } 1{,}028 \times \frac{72}{1{,}028} \times \frac{450}{1{,}028} = 31.518$$

$$\text{Not guilty and no prison: } 1{,}028 \times \frac{72}{1{,}028} \times \frac{578}{1{,}028} = 40.482$$

Table 10.17 summarizes the observed frequencies and expected frequencies.

Table 10.17 Observed Frequencies and Frequencies Expected by Chance (in parentheses) for Plea and Sentence

	Prison	No prison	Total
Guilty	392 (418.482)	564 (537.518)	**956**
Not guilty	58 (31.518)	14 (40.482)	**72**
Total	**450**	**578**	**1,028**

As usual, we calculate χ^2 by organizing our work as shown in Table 10.18, where O denotes observed frequency and E denotes expected frequency. We are asked to use the 0.05 significance level, for which the critical value (for a 2×2 table) is $\chi^2 = 3.841$. The computed value of χ^2 is greater than the critical value, meaning that we have evidence to reject the null hypothesis at the 0.05 significance level. Based on these data, there is reason to believe that the sentence given in a burglary case is associated with the plea. Specifically, of the people who pled guilty, fewer actually went to prison than expected (by chance) and more avoided prison than expected. Of the people who pled not guilty, more actually went to prison than expected and fewer avoided prison than expected. Remember that the test does not prove a causal relationship between the plea and the sentence.

Table 10.18 Calculation of χ^2

Outcome	O	E	$O - E$	$(O - E)^2$	$(O - E)^2/E$
Guilty/prison	392	418.482	−26.482	701.296	1.676
Guilty/no prison	564	537.518	26.482	701.296	1.305
Not guilty/prison	58	31.518	26.482	701.296	22.251
Not guilty/no prison	14	40.482	−26.482	701.296	17.324
Totals	1,028	1,028.000	0		$\chi^2 = 42.556$

Time out to think

If you were the lawyer for a burglary suspect, how might the results of the previous example affect your strategy in defending your client? Explain.

Review Questions

1. What is a *two-way table*? Give an example of a two-way table. Be sure to identify the two variables and their possible values.

2. What do we mean by the *frequencies* in a two-way table? Explain how to find the totals in the rows and columns.

3. Explain how to find the *frequency expected by chance* in a cell in a two-way table.

4. Explain what it means for the variables in a two-way table to be *independent*. Why do we assume that the variables are independent when calculating frequencies expected by chance?

5. Describe the overall goal of a hypothesis test with a two-way table.

6. What do the null and alternative hypotheses claim in a test of a two-way table?

7. Explain the basic idea behind using a hypothesis test to investigate a two-way table.

8. Describe the procedure for calculating a chi-square statistic. What do the symbols O and E represent in the formula for calculating chi-square? Does a larger value of χ^2 mean larger or smaller differences?

9. For a two-way table with two rows and two columns, how do we use chi-square to determine whether to reject the null hypothesis? Give your answer for both the 0.05 and the 0.01 level of significance.

Exercises

BASIC SKILLS AND CONCEPTS

SENSIBLE STATEMENTS? For Exercises 1–6, determine whether the given statement is sensible and explain why it is or is not.

1. I'll count the number of people in my statistics class who have been to Paris and the number who have not been to Paris and summarize the data in a two-way table.

2. Suppose the 20 people in a room can be categorized as 10 men, 10 women, 10 citizens, and 10 noncitizens. If there are 10 male citizens, then there is a significant relationship between gender and citizenship.

3. Suppose the 20 people in a room can be categorized as 10 men, 10 women, 10 citizens, and 10 noncitizens. If there are no female citizens, then the variables *gender* and *citizenship* are independent.

4. If the null hypothesis of no relationship between the variables is true, then there should be large differences between the observed and expected frequencies.

5. If you compute $\chi^2 = 589$ with a 2 × 2 table, you should conclude that the variables are most likely independent.

6. If you compute $\chi^2 = 0.023$ with a 3 × 3 table, you should conclude that there is a relationship between the variables.

7. **MISSING DATA.** Fill in the missing counts in the following two-way table.

	A	B	Total
C	47		**56**
D		23	
Total	**65**		

8. **MISSING DATA.** Fill in the missing counts in the following two-way table.

	Full-time	Part-time	Total
Women		37	
Men	38		**72**
Total	**82**		

9. **COLLEGE DEMOGRAPHICS.** The following table (consistent with national data) shows the distribution of part-time and full-time college students for a sample of 148 men and women.

	Full-time	Part-time	Total
Women	47	37	
Men	38	26	
Total			148

a. Find the totals for the rows and columns. Check that the totals agree with the grand total.
b. What is the percentage of women in the sample?
c. What is the percentage of part-time students in the sample?
d. What is the probability that a randomly selected student is a man?
e. What is the percentage of women who are part-time?
f. What is the probability that a randomly selected man is full-time?
g. What percentage of part-time students are women?
h. What is the probability that a randomly selected full-time student is a man?
i. Does it appear to you that the full- or part-time status of a student is related to his/her gender?

10. **VOTER TURNOUT.** The following table shows the number of citizens in a sample who voted in the last presidential election according to gender (consistent with national population data).

	Vote	No vote	Total
Women	139	111	
Men	132	118	
Total			500

a. Find the totals for the rows and columns. Check that the totals agree with the grand total.
b. What is the percentage of women in the sample?
c. What is the percentage of voters in the sample?
d. What is the probability that a randomly selected woman did not vote?
e. What is the percentage of men who voted?
f. What is the probability that a randomly selected voter is a woman?
g. What percentage of nonvoters are men?
h. What is the probability that a randomly selected man is a nonvoter?
i. Does it appear to you that whether a person chooses to vote is related to his/her gender?

11. **AUTOMOBILE AND MOTORCYCLE DRIVERS.** The following table gives the numbers of automobiles and motorcycles registered by random samples of Oregon and Connecticut citizens (two states of nearly equal population). The data are consistent with statewide data. Answer the following questions, assuming that all selections are made from this sample and that no one in the sample has registered both an automobile and a motorcycle.

	Automobiles	Motorcycles	Total
OR	150	6	
CT	193	5	
Total			354

a. Find the totals for the rows and columns. Check that the totals agree with the grand total.
b. What is the percentage of Oregonians in the sample?
c. What is the percentage of registered motorcycles in the sample?
d. Based on the sample, what is the probability that a randomly selected Connecticut resident has registered an automobile?
e. What is the percentage of Oregonians in the sample who have registered a motorcycle?
f. What is the probability that a randomly selected motorcycle in the sample is registered in Connecticut?
g. What percentage of registered automobiles in the sample are from Oregon?
h. Why is the assumption that no one owns both an automobile and a motorcycle needed?

12. **HOUSEHOLDS AND VEHICLES.** The following table gives data from a sample of households in Arkansas and Kansas, two states of nearly equal population (the data are representative of population data for the entire state). It shows the numbers of households that have either zero, one, or more than one vehicle.

	Zero vehicles	One vehicle	More than one vehicle	Total
Arkansas	88	303	501	892
Kansas	60	302	583	945
Total	148	605	1,084	1,837

SOURCE: *Statistical Abstract of the United States.*

a. Check the totals for the rows and columns. Check that the totals agree with the grand total.
b. What is the percentage of households in the sample in Arkansas?

c. What percentage of households in the sample have zero vehicles?

d. Based on the sample, what is the probability that a randomly selected household from either state owns zero vehicles?

e. Based on the sample, what is the percentage of households in Arkansas with one vehicle?

f. What is the probability that a randomly selected household in the Kansas sample has more than one vehicle?

g. Based on the sample, what percentage of households in each state own at least one vehicle?

h. Based on the sample, what is the probability that a randomly selected household from either state has more than one vehicle?

13. **FINDING EXPECTED FREQUENCIES.** The following two-way table shows whether men and women from a sample chose to attend a school play or a soccer game. The events occurred at the same time, so only one choice could be made. The table gives the observed frequencies (counts).

	Play	Soccer	Total
Men	6	14	
Women	9	10	
Total			

a. Find the row and column totals and check that they agree with the grand total.

b. Find the probability that a randomly selected person from the sample is a man; is a woman.

c. Find the probability that a randomly selected person from the sample attended the soccer game; attended the play.

d. Assuming that gender and school event are independent, use the multiplication rule for independent events to find P(man and play).

e. Find the frequencies expected by chance for each category in the table, assuming independence of the variables.

f. Comment on the apparent difference between the observed frequencies and the frequencies expected by chance.

14. **FINDING EXPECTED FREQUENCIES.** The following two-way table shows the age category of men and women who attended an environmental conference. The table gives the observed frequencies (counts).

	Under 25	Over 25	Total
Men	16	24	
Women	25	12	
Total			

a. Find the row and column totals and check that they agree with the grand total.

b. Find the probability that a randomly selected person from the sample is a man.

c. Find the probability that a randomly selected person from the sample is under 25.

d. Assuming independence between gender and age category, use the multiplication rule for independent events to find P(man and under 25).

e. Find the frequencies expected by chance for each category in the table, assuming independence of the variables.

f. Comment on the apparent difference between the observed frequencies and the frequencies expected by chance.

15. **ARTHRITIS TREATMENT.** Of the 98 participants in a drug trial who were given a new experimental treatment for arthritis, 56 showed improvement. Of the 92 participants given a placebo, 49 showed improvement. Make a two-way table for these data and determine the frequency expected by chance for each category, assuming independence of the variables. Comment on the apparent difference between the observed frequencies and the frequencies expected by chance.

16. **EXPECTED FREQUENCIES FOR COLLEGE DEMOGRAPHICS.** Assuming independence of the variables, find the frequencies expected by chance for the college demographics table in Exercise 9. Comment on the apparent difference between the observed frequencies and the frequencies expected by chance.

17. **EXPECTED FREQUENCIES FOR VOTER TURNOUT.** Assuming independence of the variables, find the frequencies expected by chance for the voter turnout table in Exercise 10. Comment on the apparent difference between the observed frequencies and the frequencies expected by chance.

18. **EXPECTED FREQUENCIES FOR AUTOMOBILE AND MOTORCYCLE REGISTRATION.** Assuming independence of the variables, find the frequencies expected by chance for the table in Exercise 11. Comment on the apparent difference between the observed frequencies and the frequencies expected by chance.

19. **ETHNIC GROUPS IN NEVADA AND NEBRASKA.** Random samples of people in Nebraska and Nevada have the following ethnic composition (consistent with statewide demographics). Assuming independence of the variables, find the frequency expected by chance for each category. Comment on the apparent difference between the observed frequencies and the frequencies expected by chance.

	Non-Hispanic white	African American	Native American	Asian/Pacific Islander	Hispanic
Nebraska	1,555	66	15	21	68
Nevada	1,448	125	30	74	253

20. **AGE GROUPS IN IOWA AND KANSAS.** Random samples of people in Iowa and Kansas have the following compositions of age groups (consistent with statewide demographics). Assuming independence of the variables, find the frequency expected by chance for each category. Comment on the apparent difference between the observed frequencies and the frequencies expected by chance.

	<5	5–17	18–24	25–44	45–64	>65
Iowa	182	540	277	808	624	431
Kansas	182	515	262	773	542	354

21. **AGE AND SMOKING.** The data in the table below show the numbers of smokers and nonsmokers in various age groups in a randomly selected sample of people (based on data from the National Center for Health Statistics). Assuming independence of the variables, find the frequency expected by chance for each category. Comment on the apparent difference between the observed frequencies and the frequencies expected by chance.

	20–24	25–34	35–44	45–64
Smokers	18	15	17	15
Nonsmokers	32	35	33	35

FURTHER APPLICATIONS

For Exercises 22–30, carry out the following steps.

a. State a null and alternative hypothesis for the study.
b. Determine the frequency expected by chance for each cell of the table.
c. Determine the χ^2 statistic for the problem.
d. Using the 0.05 or 0.01 significance level in Table 10.12, interpret the result carefully in words and discuss your conclusions.

22. **TWO-WAY TABLE FOR GRADES AND STUDY.** A study of 400 students in first-year biology was designed to determine whether grades could be raised from the first semester to the second semester by attending a special study session during the second semester. The results are summarized in the following two-way table.

	Improvement	No improvement	Total
Study session	145	55	**200**
No study session	75	125	**200**
Total	**220**	**180**	**400**

23. **TWO-WAY TABLE FOR FLU TREATMENTS.** A study of 500 people was designed to determine whether flu shots were effective in preventing flu during the winter months. The results are summarized in the following two-way table.

	Flu	No flu	Total
Flu shots	100	200	**300**
No flu shots	100	100	**200**
Total	**200**	**300**	**500**

24. **GINKGO AND MEMORY IMPROVEMENT.** A (hypothetical) study of the effects of the herb ginkgo on memory improvement was conducted with 360 participants. All subjects were given a pre-test of memory skills. Then half of the subjects were put on a daily regimen of ginkgo, while the other half of the subjects were given a placebo.

After six months, all subjects were given a memory post-test. The results of the study are shown in the table below.

	Memory improvement	No improvement	Total
Ginkgo	95	85	**180**
Placebo	94	86	**180**
Total	**189**	**171**	**360**

25. **DRINKING AND PREGNANCY.** The table below shows the drinking status of a sample of pregnant women in two age categories.

	Drinking	Nondrinking
Age < 25	13	1,239
Age ≥ 25	37	1,992

SOURCE: U.S. National Center for Health Statistics, 1995 data, in *Statistical Abstract of the United States*, 1998.

26. **POVERTY AND MAMMOGRAMS.** Women over 40 years of age are encouraged to have regular mammograms. About 43% of women living below the poverty level received a mammogram in the last two years, compared to 64% for all other women. Approximately 11% of adults in this country live below the poverty level. The data in the following table, taken from a random sample of 500 women, are consistent with these figures.

	Mammogram	No mammogram
Poverty	21	29
No poverty	288	162

27. **POVERTY AND HIGH SCHOOL GRADUATION.** For a random sample of 345 adults, the following table shows the numbers of high school graduates among those who live above and below the poverty line (based on U.S. Census Bureau data).

	No HS diploma	HS diploma
Poverty	23	15
No poverty	46	261

28. **COFFEE AND SMOKING.** A byproduct of a study on the effect of coffee on gall stones was data for the subjects on smoking and coffee consumption. The following table represents these data. The category "Coffee" means more than four cups of caffeinated coffee per day. The category "No coffee" means no caffeinated coffee.

	Current smoker	Nonsmoker
No coffee	3	57
Coffee	48	192

SOURCE: *Journal of the American Medical Association*.

29. **HOME TEAM ADVANTAGE.** Data collected for four sports show the following record of home team wins and visiting team wins.

	Home team wins	Visiting team wins
Basketball	127	71
Baseball	53	47
Hockey	50	43
Football	57	42

SOURCE: *Chance*, Vol. 5, No. 3–4.

30. **DRINKING AND CRIME.** The sample data given below were used by the English statistician Karl Pearson in 1909 to analyze the dependence of various crimes on the drinking habits of the criminal.

	Drinker	Abstainer
Arson	50	43
Rape	88	62
Violence	155	110
Stealing	379	300
Coining	18	14
Fraud	63	144

PROJECTS FOR THE WEB AND BEYOND

For useful links, select "Links for Web Projects" for Chapter 10 at www.aw.com/bbt.

31. **CONSTRUCTING TWO-WAY TABLES.** Choose two variables that appear to have a relationship that is worth investigating. One variable should have at least two categories of individuals—for example, two or more age categories, racial categories, or geographical locations. The other variable should have at least two categories for some social, economic, or health factor—for example, two or more income categories, drinking categories, or educational attainment categories. Find the required population or sample data needed to fill in a two-way table for the two variables. Discuss whether there appears to be a relationship between the variables. A good place to start is the Web site for the *Statistical Abstract of the United States* of the U.S. Bureau of the Census.

32. **ANALYZING TWO-WAY TABLES.** Choose two variables that appear to have a relationship that is worth investigating. One variable should have at least two categories of individuals—for example, two or more age categories, racial categories, or geographical locations. The other variable should have at least two categories for some social, economic, or health factor—for example, two or more income categories, drinking categories, or educational attainment categories. Find the frequency data needed to fill in a two-way table for the two variables. Carry out a hypothesis test to determine whether there is a relationship between the variables. A good data source is the Web site for the *Statistical Abstract of the United States* of the U.S. Bureau of the Census.

IN THE NEWS

1. **TWO-WAY TABLES IN THE NEWS.** It's unusual (but not impossible) to see a two-way table in a news article. But often a news story provides information that could be expressed in a two-way table. Find an article that discusses a relationship between two variables that could be expressed in a two-way table. Create the table.

2. **HYPOTHESIS TESTING IN THE NEWS.** News reports often describe results of statistical studies in which the conclusions came from a hypothesis test involving two-way tables. However, the reports rarely give the actual table or describe the details of the hypothesis test. Find a recent news report in which you think the conclusions *probably* were based on a hypothesis test with a two-way table. Assuming you are correct, describe in words how the hypothesis test probably worked. That is, describe the null and alternative hypotheses and the procedure by which the researchers probably carried out their test.

3. **YOUR OWN HYPOTHESIS TEST.** Think of an example of something you'd like to know that could be tested with a hypothesis test on a two-way table. Without actually collecting data or doing any calculations, describe how you would go about conducting your study. That is, describe how you would collect the data, explain how you would organize them into a two-way table, state the null and alternative hypotheses that would apply, and describe how you would conduct the hypothesis test and reach a conclusion.

Chapter Review Exercises

1. In a recent year, there were approximately 97,000 deaths in the United States (total population 281 million) due to accidents. Approximately 25,000 of these accidental deaths involved people age 75 or older. Roughly 42,000 of these deaths were due to automobiles and 17,000 involved falls.
   a. Find the overall death rate for accidents in units of deaths per 100,000 people.
   b. What is the probability that a person is less than 75 years of age and dies from an accident during the course of the year?
   c. Approximately how much more likely is it that you will die in an automobile accident than from a fall?
   d. What is the death rate for falls in units of deaths per 1 million people?

2. Suppose an airport scanning device has 98% accuracy in detecting certain kinds of metal objects (98% of people carrying these objects are correctly identified as carriers and 98% of people not carrying them are correctly identified as non-carriers). Suppose also that 1% of all travelers actually carry such metal objects. For the 100,000 people who pass through the scanning device each week, do the following:
   a. Find the number of people not carrying metal objects who are identified as carriers (false positives).
   b. Find the number of non-carriers who are identified as non-carriers (true negatives).
   c. Find the number of carriers who are identified as non-carriers (false negatives).
   d. Find the number of carriers who are identified as carriers (true positives).
   e. What percentage of the people identified as carriers are falsely identified?

3. In a study of store checkout scanning systems, samples of purchases were used to compare the scanned prices to the posted prices. The accompanying table summarizes results for a sample of 819 items.

	Regular-priced items	Advertised-special items
Undercharge	20	7
Overcharge	15	29
Correct price	384	364

   SOURCE: Ronald Goodstein, "UPC Scanner Pricing Systems: Are They Accurate?" *Journal of Marketing*, Vol. 58.

   a. Among regular-priced items, what percentage are overcharged? Among advertised-special items, what percentage are overcharged? Compare these two results.
   b. Identify the null and alternative hypotheses for a test of the claim that the two variables (*charge* and *pricing*) are independent.
   c. Find the expected value for each cell by assuming that the charge (under, over, correct) is independent of the pricing (regular, advertised special).
   d. Find the value of the χ^2 statistic for a hypothesis test of the claim that the two variables (*charge* and *pricing*) are independent.
   e. Based on the result from part d and the size of the table, refer to Table 10.12 and determine what is known about the *P*-value.
   f. Based on the preceding results, what can you conclude from the hypothesis test about whether the two variables (*charge* and *pricing*) are independent?

FOCUS ON CRIMINOLOGY

Can You Tell a Fraud When You See One?

Suppose your professor gives you a homework assignment in which you are supposed to toss a coin 200 times and record the results in order. The two data sets below represent the results turned in by two students. Now suppose you learn that one of the students really did the assignment, while the other faked the data. Can you tell which one is fake?

Data Set 1 (H = heads; T = tails)

```
H  T  H  T  H  H  T  T  T  T  T  T  H  T  H  T  T
T  T  H  H  H  T  T  T  T  H  T  T  H  T  T  H  H
H  T  T  H  H  T  H  T  H  T  H  H  H  H  T  T  H
T  H  T  H  H  H  H  T  T  H  H  T  T  H  H
H  T  T  T  T  T  H  H  T  H  T  T  H  T  H  T
T  H  H  T  T  H  T  H  T  H  H  T  T  H  T  T  T
H  T  H  T  H  H  T  T  T  H  H  T  T  H  H  H  H
T  H  T  H  H  T  H  T  T  T  T  H  T  T  T  H  T
H  T  H  H  T  H  T  H  H  H  H  T  H  T  H  H  H
T  T  T  H  T  T  H  T  H  T  T  T  H  T  H  H  T
H  H  H  H  T  T  T  H  H  T  T  H  T  H  H  T  T
H  T  H  H  H  T  H  H  T  H  T  T  T  H
```

Data Set 2 (H = heads; T = tails)

```
T  H  H  T  T  H  H  T  T  H  T  H  H  T  T  H  H
T  H  T  H  T  T  H  T  H  H  T  T  H  T  T  T
H  H  T  H  T  H  H  H  T  T  H  T  H  H  T  H  T
T  H  T  T  H  H  T  T  H  H  T  H  H  T  T  H  T
H  T  H  H  T  H  T  H  T  H  H  T  H  T  H  T  T
H  T  T  H  H  T  T  H  H  T  H  T  H  H  T  T  H
T  H  H  T  H  T  H  T  H  T  H  T  H  T  H  H
T  H  T  H  T  T  H  T  H  H  T  H  T  H  T  H  T
H  H  T  H  T  T  H  T  H  T  H  H  T  T  H  T  T
H  T  H  H  T  H  H  H  T  H  H  T  T  H  T  H  T
H  T  H  T  H  H  T  H  T  T  T  H  H  T  H  T  H
H  T  H  H  T  T  H  H  T  H  T  H  T
```

To make the job a little easier, the following table summarizes a few characteristics of the two data sets that might help you decide which one is fake.

Characteristics of Data Set 1	Characteristics of Data Set 2
Total of 97 H, 103 T Two cases of 6 T in a row Five cases of 4 H in a row Three cases of 4 T in a row	Total of 101 H, 99 T No case of more than 3 H or 3 T in a row

If you are like most people, you will probably guess that Data Set 1 is the fake one. After all, its total numbers of heads and tails are farther from the 100 of each that many people expect, and it has two cases in which there were 6 tails in a row, plus several more cases in which there were 4 heads or tails in a row.

But consider this: The probability of getting 6 heads in a row is $(1/2)^6$, or 1 in 64. The chance of 6 consecutive tails is also 1 in 64. Thus, with 200 tosses, the chance of getting at least one case of 6 heads or tails in a row is quite good, so the strings of consecutive heads and tails in Data Set 1 really are not surprising. In contrast, Data Set 2 has no string of even 4 heads or tails in a row, making it almost certain that it is a fake.

This simple example reveals an important application of statistics in criminology: It is often possible to catch people who have faked data of any kind. For example, statistics helps bank regulators and financial auditors to catch fraudulent financial statements, the Internal Revenue Service to identify people with fraudulent tax returns, and scientists to catch other scientists who have falsified data.

One of the most powerful tools for detecting fraud was identified by physicist Frank Benford. In the 1930s, Benford noticed that tables of logarithms (which scientists and engineers used regularly in the days before calculators) tended to be more worn on the early pages, where the numbers started with the digit 1, than on later pages. Following up on this observation, he soon discovered that many sets of numbers from everyday life, such as stock market values, baseball statistics, and the areas of lakes, include more numbers starting with the digit 1 than with the digit 2, and more starting with 2 than with 3, and so on. He eventually published a formula describing how often numbers begin with different digits, and this formula is now called *Benford's law.* Figure 10.10 shows what his law predicts for the first digits

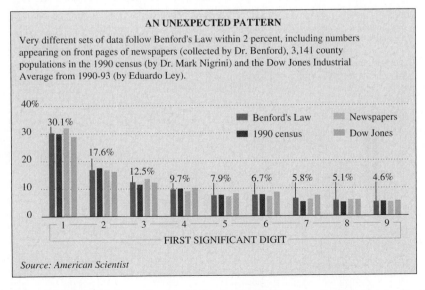

AN UNEXPECTED PATTERN

Very different sets of data follow Benford's Law within 2 percent, including numbers appearing on front pages of newspapers (collected by Dr. Benford), 3,141 county populations in the 1990 census (by Dr. Mark Nigrini) and the Dow Jones Industrial Average from 1990-93 (by Eduardo Ley).

Benford's Law   Newspapers
1990 census   Dow Jones

FIRST SIGNIFICANT DIGIT

Source: American Scientist

Figure 10.10   SOURCE: Malcolm W. Browne, "Following Benford's Law, or Looking Out for No. 1," *New York Times,* August 4, 1998, p. B10.

of numbers, along with actual results from several real sets of numbers. Notice how well Benford's law describes the results. (Interestingly, Benford's law was first discovered more than 50 years earlier, and published by astronomer and mathematician Simon Newcomb in 1881. However, Newcomb's article had been forgotten by the time Benford did his work.)

Benford's law is surprising because most people guess that every digit (1 through 9) would be equally likely as a starting digit. Indeed, this is the case for sets of random numbers such as lottery numbers, which are no more likely to start with 1 than with any other digit. Thus, Benford's law should not be used for picking lottery numbers. However, because Benford's law does apply to many *real data* sets, it can be used to detect fraud. As shown in Figure 10.11, the data from real tax forms follow Benford's law closely. In contrast, the data from financial statements in a 1995 study do not follow Benford's law. Based on this fact, the District Attorney suspected fraud, which he eventually was able to prove. The "random guess data" came from students of Professor Theodore P. Hill at Georgia Institute of Technology (the source of much of the information for this Focus). Note that the guesses do not follow Benford's law at all, which is why people who try to fake data can be caught easily.

Benford's law mystified scientists and mathematicians for decades. Today it seems to be fairly well understood, though still difficult to explain. Here is one explanation of why Benford's law applies to the Dow Jones Industrial Average, thanks to Dr. Mark J. Nigrini of Southern Methodist University (as reported in the *New York Times*): Imagine that the Dow is at 1,000, so the first digit is a 1, and rises at a rate of about 20% per year. The doubling time at this rate of increase is a little less than four years, so the Dow would remain in the 1,000s, still with a first digit of 1, for almost four years until it hit 2000. It would then have a first digit of 2 until it hit 3,000. However, moving from 2,000 to 3,000 requires only a 50% increase, which takes only a little over two years. Thus, the first digit of 2 would occur for only a little more than half the number of days that a first digit of 1 occurred. Subsequent changes of first digit take even shorter times. By the time the Dow hits 9,000, it takes only an 11% increase and just seven months to reach the 10,000 mark, so a first digit of 9 occurs for only seven months. At

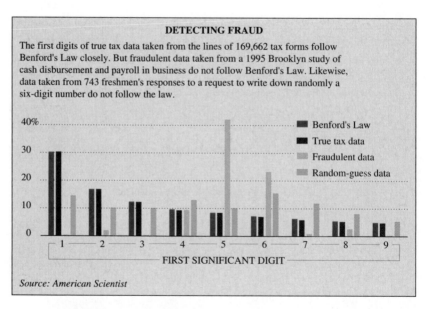

Figure 10.11   SOURCE: Malcom W. Browne, "Following Benford's Law, or Looking Out for No. 1," *New York Times*, August 4, 1998, p. B10.

the 10,000 mark, however, the Dow is back to a first digit of 1 again, and this would not change until the Dow doubled again to 20,000, which means another almost four years at the rate of increase of 20% per year. Thus, if you graphed the number of days the Dow had each starting digit 1 through 9, you'd find that the number 1 would be the starting digit for longer periods of time than the number 2, and so on down the line.

In summary, Benford's law shows that numbers don't always arise with the frequencies that most people would guess. As a result, it not only helps explain a lot of mysteries about numbers (such as those in the Dow), but also has become a valuable tool for the detection of criminal fraud.

QUESTIONS FOR DISCUSSION

1. Try this experiment with friends: Ask one friend to record 200 actual coin tosses and another friend to try to fake data for 200 coin tosses. Have them give you their results anonymously, so that you don't know which sheet came from which friend. Using the ideas in this Focus, try to determine which set is real and which one is fake. After you make your guess, check with your friends to see whether you were right. Discuss your ability to detect the fakes.

2. Briefly discuss how Benford's law might be used to detect fraudulent tax returns.

3. Can Benford's law alone prove that data are fraudulent? Or can it only point to data that should be investigated further? Explain.

4. Find a set of data similar to the sets shown in Figure 10.10 and make a bar chart showing the frequencies of the first digits 1 through 9. Do the data follow Benford's law? Explain.

SUGGESTED READING

Browne, Malcolm W. "Following Benford's Law, or Looking Out for No. 1." *New York Times,* August 4, 1998.

Hill, Theodore P. "The First Digit Phenomenon." *American Scientist,* Vol. 86, No. 358, July–August 1998.

Nigrini, Mark J. *Digital Analysis Using Benford's Law.* Global Audit Publications, Vancouver, 2000.

FOCUS ON EDUCATION

What Can a Fourth-Grader Do with Statistics?

Nine-year-old Emily Rosa was in fourth grade, trying to decide what to do for her school science fair project. She was thinking of doing a project on the colors of M&M candies, when she noticed her mother, a nurse, watching a videotape about a practice called "therapeutic touch" or "TT." TT is a leading alternative medical treatment, practiced throughout the world and taught at many schools of nursing. But no statistically valid test had ever clearly demonstrated whether it actually works. Emily told her mother that she had an idea for testing TT and wanted to make it her science fair project.

Despite the name, TT therapists do *not* actually touch their patients. Instead, they move their hands a few inches above a patient's body. Therapeutic touch supporters claim that these hand movements allow trained therapists to feel and manipulate what they call a "human energy field." By doing these manipulations properly, the therapists can supposedly cure many different ailments and diseases. Emily Rosa's science fair project sought to find out whether trained TT therapists could really feel a human energy field.

To do her project, Emily recruited 21 TT therapists to participate in a simple experiment. Each therapist sat across a table from Emily, laying his or her arms out flat, palms up. Emily then put up a cardboard partition with cutouts for the therapist's arms. This prevented Emily and the therapist from seeing each other's face, but allowed Emily to see the therapist's hands.

Emily then placed one of her hands a few inches above *one* of the therapist's two hands. If the therapist could truly feel Emily's "human energy field," then the therapist should have been able to tell whether his or her right or left hand was closest to Emily's hand. Thus, each trial of the experiment ended with Emily's recording whether the therapist was right or wrong in identifying the hand.

Emily took several precautions to make sure her experiment would be statistically valid. For example, to ensure that her choices between the two hands were random, Emily used the outcome of a coin toss to determine whether she placed her hand over the therapist's left or right hand in each case. And to make sure she had enough data to evaluate statistical significance, 14 of the 21 therapists got 10 tries each, while 7 got 20 tries each.

The results were a miserable failure for the TT therapists. Because there were only two possible answers in each trial—left hand or right hand—by pure chance the therapists ought to have been able to guess the correct hand about 50% of the time. But the overall results showed that they got the correct answer only 44% of the time. Moreover, none of the therapists performed better than expected by chance in a statistically significant way. Emily also checked to see whether therapists with more experience did better than those with less experience. They did not. Emily's conclusion: If there is such a thing as a "human energy field" (which she doubts), the TT therapists can't feel it. In that case, anyone who is "healed" by TT must be responding to a placebo effect, rather than a real treatment.

One of the most interesting aspects of this study was that Emily was able to do it at all. Other skeptics of TT had hoped to conduct similar studies in the past, but TT therapists had refused to participate. One famous skeptic, magician James Randi, had even offered a $1.1 million prize to any TT therapist who could pass a test similar to Emily's. Only one person accepted Randi's challenge, and she succeeded in only 11 of 20 trials, about the same as would be expected by chance. So why was Emily able to succeed where more experienced researchers had failed? Apparently, the therapists agreed to participate in Emily's experiment because they did not feel threatened by a fourth-grader.

The novelty of Emily's science fair project drew media attention, and it was not long before word reached retired Pennsylvania psychiatrist Stephen Barrett. Dr. Barrett specialized in debunking "quack" therapies, and he convinced Emily and her mother to report her results in a medical research paper. The paper was published in the *Journal of the American Medical Association* (April 1, 1998) when Emily was 11, making her the youngest-ever author of a paper in that prestigious journal.

QUESTIONS FOR DISCUSSION

1. After Emily's results were published, many TT supporters claimed that her experiment was invalid because she and her mother were biased against TT. Based on the way her experiment was designed, do you think that her personal bias could have affected her results? Why or why not?

2. Another objection to Emily's experiment was that it was only single-blind rather than double-blind. That is, the therapist could not see what Emily was doing, but Emily could see what the therapist was doing. Do you think this objection is valid in this case? Can you think of a way that Emily's experiment might be repeated but be made double-blind?

3. Emily's experiment was not a direct test of whether TT treatment works, because it did not check to see whether patients actually improved when treated by TT. Suggest a statistically valid way to test whether TT is more effective than a placebo.

4. Based on the results of Emily's study, skeptics now say that TT is so clearly invalid that it should no longer be used or funded. Do you agree? Why or why not?

SUGGESTED READING

Ball, T. S., and Alexander, D. D. "Catching Up with Eighteenth Century Science in the Evaluation of Therapeutic Touch." *Skeptical Inquirer,* July/August 1998.

Kolata, Gina. "4th Grader Challenges Alternative Therapy." *New York Times,* April 2, 1998.

Rosa, E. "TT and Me." *Skeptic Magazine,* September 1998.

Rosa, L., Rosa, E., Sarner, L., and Barrett, S. "A Close Look at Therapeutic Touch." *Journal of the American Medical Association,* Vol. 279, No. 1005, April 1, 1998.

Epilogue: A Perspective on Statistics

A single introductory statistics course cannot transform you into an expert statistician. After studying statistics in this book, you may feel that you have not yet mastered the material to the extent necessary to use statistics confidently in real applications. Nevertheless, by now you should understand enough about statistics to interpret critically the reports of statistical research that you see in the news and to converse with experts in statistics when you need more information. And, if you go on to take further course work in statistics, you should be well prepared to understand important topics that are beyond the scope of this introductory book—topics such as t tests, tests comparing two populations, analysis of variance (ANOVA), and nonparametric methods.

Most importantly, while this book is not designed to make you an expert statistician, it is designed to make you a better-educated person with improved job marketability. You should know and understand the basic concepts of probability and chance. You should know that in attempting to gain insight into a data set, it's important to investigate measures of center (such as mean and median), measures of variation (such as range and standard deviation), the nature of the distribution (via a frequency table or graph), and the presence of outliers. You should know and understand the importance of estimating population parameters (such as a population mean or proportion), as well as testing hypotheses about population parameters. You should understand that a correlation between two variables does not necessarily imply that there is also some cause-and-effect relationship. You should know the importance of good sampling. You should recognize that many surveys and polls obtain very good results, even though the sample sizes might seem to be relatively small. Although many people refuse to believe it, a nationwide survey of only 1,700 voters can provide good results if the sampling is carefully planned and executed.

There once was a time when a person was considered educated if he or she could read, but we are in a new millennium that is much more demanding. Today, an educated person must be able to read, write, understand the significance of the Renaissance, operate a computer, and apply statistical reasoning. The study of statistics helps us see truths that are sometimes distorted by a failure to approach a problem carefully or concealed by data that are disorganized. Understanding statistics is now essential for both employers and employees—for all citizens. H.G. Wells once said, "Statistical thinking will one day be as necessary for efficient citizenship as the ability to read and write." That day is now.

Appendix A   State Data

State	Population density[1] (people/sq. mi.)	% urban population[2]	% urban land[3]	Median age[4]	% foreign born[5]	Incarceration rate[6] (prisoners/ 100,000 residents)
AL	85	60.4	5.3	35.5	1.1	519
AK	1	67.5	0.1	31.5	4.5	413
AZ	40	87.5	1.8	34.6	7.6	507
AR	48	53.5	2.3	35.8	1.1	415
CA	207	92.6	5.2	33.3	21.7	483
CO	38	82.4	1.3	35.5	4.3	357
CT	675	79.1	25.9	37.0	8.5	372
DE	374	73.0	10.7	35.7	3.3	429
FL	272	84.8	9.5	38.3	12.9	447
GA	129	63.2	4.9	33.8	2.7	502
HI	185	89.0	10.0	36.2	14.7	307
ID	15	57.4	0.4	33.3	2.9	330
IL	214	84.6	5.5	34.9	8.3	357
IN	164	64.9	5.0	35.2	1.7	321
IA	51	60.6	2.0	36.6	1.6	258
KS	32	69.1	1.1	35.2	2.5	310
KY	98	51.8	2.7	35.6	0.9	379
LA	100	68.1	3.7	33.9	2.1	736
ME	40	44.6	2.3	37.4	3.0	125
MD	521	81.3	16.1	35.6	6.6	418
MA	781	84.3	27.4	36.2	9.5	275
MI	172	70.5	4.7	35.3	3.8	466
MN	59	69.9	2.3	35.2	2.6	117
MS	58	47.1	2.4	33.4	0.8	574
MO	78	68.7	2.7	35.8	1.6	457
MT	6	52.5	0.2	37.5	1.7	310
NE	22	66.1	0.5	35.3	1.8	215

Per capita personal income[7] ($)	% below poverty[8]	Average teacher salary[9] ($)	Per capita state tax[10] ($)	Expenditure per pupil[11] ($)	Automobiles per capita[12]
21,442	14.9	32,549	1,318	5,255	0.44
25,676	8.5	50,647	1,932	8,900	0.37
23,060	18.9	33,350	1,488	4,048	0.40
20,346	18.5	30,319	1,598	4,172	0.34
27,503	16.8	43,474	2,073	5,284	0.48
28,657	9.4	36,271	1,485	5,147	0.49
37,598	10.2	50,426	2,869	8,376	0.60
29,814	9.1	41,436	2,665	7,086	0.54
25,852	14.3	33,889	1,509	5,427	0.50
25,020	14.7	35,596	1,517	5,585	0.49
26,137	13.0	35,842	2,662	5,720	0.37
21,081	13.3	31,818	1,674	4,500	0.40
28,873	11.7	42,125	1,641	5,423	0.49
24,219	8.2	38,876	1,652	5,886	0.55
23,925	9.6	33,272	1,678	5,720	0.57
24,981	10.5	35,802	1,768	5,493	0.44
21,506	16.5	33,797	1,807	5,346	0.41
21,346	18.4	28,347	1,392	4,527	0.44
22,952	10.7	33,676	1,905	6,385	0.51
29,943	9.4	41,148	1,790	6,547	0.51
32,797	11.2	43,806	2,357	7,069	0.62
25,857	10.8	48,238	2,210	6,654	0.52
27,510	9.7	38,281	2,434	6,041	0.49
18,958	18.7	27,720	1,578	4,269	0.46
24,427	10.7	33,143	1,512	4,949	0.47
20,172	16.3	29,958	1,513	5,380	0.51
24,754	10.0	31,768	1,584	5,250	0.48

(cont.)

State	Population density[1] (people/sq. mi.)	% urban population[2]	% urban land[3]	Median age[4]	% foreign born[5]	Incarceration rate[6] (prisoners/ 100,000 residents)
NV	15	88.3	0.9	35.2	8.7	542
NH	131	51.0	5.7	35.7	3.7	182
NJ	1,085	89.4	32.7	36.7	12.5	382
NM	14	73.0	0.7	34.0	5.3	271
NY	384	84.3	7.2	35.9	15.9	397
NC	152	50.4	4.6	35.2	1.7	358
ND	9	53.3	0.2	35.8	1.5	128
OH	273	74.1	8.8	35.8	2.4	432
OK	48	67.7	2.7	35.6	2.1	622
OR	34	70.5	0.9	36.7	4.9	260
PA	268	68.9	6.7	37.6	3.1	303
RI	945	86.0	28.5	36.4	9.5	220
SC	125	54.6	4.7	35.2	1.4	550
SD	10	50.0	0.3	35.2	1.1	329
TN	130	60.9	5.7	35.9	1.2	325
TX	74	80.3	2.9	33.0	9.0	724
UT	25	87.0	0.9	26.7	3.4	205
VT	64	32.2	1.5	36.7	3.1	188
VA	170	69.4	5.5	35.2	5.0	399
WA	84	76.4	2.7	35.3	6.6	247
WV	75	36.1	1.6	38.6	0.9	192
WI	95	65.7	2.9	35.7	2.5	334
WY	5	65.0	0.2	35.7	1.7	327
Nat'l	75.7	75.2	2.5	35.2	7.9	461

[1]Population density: U.S. Bureau of the Census, 1997.
[2]Percent urban population: U.S. Bureau of the Census, 1991.
[3]Percent urban land: U.S. Bureau of the Census, 1991.
[4]Median age: U.S. Bureau of the Census, 1998.
[5]Percent of population foreign-born: U.S. Bureau of the Census, 1999.
[6]Incarceration rate (prisoners sentenced to more than 1 year): Bureau of Justice Statistics, 1998.
[7]Per capita personal income: Bureau of Economic Analysis, 1998.
[8]Percent of population below poverty level: U.S. Bureau of the Census, 1999.
[9]Average teacher salary: National Education Association, 1997.
[10]Per capita state tax (total: property, sales, licenses, income): U.S. Bureau of the Census, 1998.
[11]Expenditures per pupil in public education: National Education Association, 1997.
[12]Automobiles per capita: U.S. Department of Transportation, 1997.

Per capita personal income[7] ($)	% below poverty[8]	Average teacher salary[9] ($)	Per capita state tax[10] ($)	Expenditure per pupil[11] ($)	Automobiles per capita[12]
27,200	9.6	37,340	1,848	4,998	0.39
29,022	7.8	36,029	851	6,014	0.63
33,937	9.3	49,349	1,923	9,455	0.52
19,936	23.4	29,685	2,058	4,927	0.44
31,734	16.6	48,000	1,989	8,658	0.44
24,036	11.8	31,286	1,838	5,028	0.46
21,675	12.3	27,711	1,690	4,667	0.52
25,134	11.9	38,831	1,574	5,527	0.60
21,072	15.2	30,369	1,584	4,187	0.46
24,766	11.7	40,960	1,523	5,988	0.48
26,792	11.4	47,147	1,719	6,967	0.50
26,797	11.9	43,019	1,805	7,665	0.52
21,309	13.1	32,830	1,482	5,105	0.47
22,114	14.2	26,764	1,129	4,680	0.51
23,559	15.1	34,222	1,288	4,898	0.51
24,957	16.7	33,038	1,246	5,551	0.35
21,019	8.3	31,750	1,647	3,837	0.41
24,175	11.0	37,200	1,621	6,503	0.49
27,385	12.5	35,837	1,552	5,920	0.53
27,961	10.6	37,860	2,075	5,805	0.48
19,362	17.5	33,257	1,663	6,406	0.41
25,079	8.5	39,057	2,135	6,521	0.49
23,167	12.7	31,721	1,779	6,036	0.45
26,412	11.0	38,611	1,761	5,889	0.48

Appendix B   Table of Random Numbers

For many statistical applications it is useful to generate a set of randomly chosen numbers. You can generate such numbers with most calculators and computers, but sometimes it is easier to use a table such as the one given below. This table was generated by randomly selecting one of the digits 0, 1, 2, 3, 4, 5, 6, 7, 8, or 9 for each position in the table; that is, each of these digits is equally likely to appear in any position. Thus, you can generate a sequence of random digits simply by starting at any point in the table and taking the digits in the order in which they appear. A larger set of random numbers is available on the text Web site (www.aw.com/bbt).

Example 1: Generate a random list of yes/no responses.

Solution: Start at an arbitrary point in the table. If the digit is 0, 1, 2, 3, or 4, call it a *yes* response. If the digit is 5, 6, 7, 8, or 9, call it a *no* response. Continue through the table from your starting point, using each digit shown to determine either a *yes* or a *no* response for your list.

Example 2: Generate a random list of letter grades A, B, C, D, or F.

Solution: Let 0 or 1 be an A, 2 or 3 be a B, 4 or 5 be a C, 6 or 7 be a D, and 8 or 9 be an F. Start at an arbitrary point in the table and use each digit shown to determine a grade for your list.

```
9 9 3 2 7   5 6 0 8 1   6 0 2 3 2   8 3 3 1 2   4 7 6 3 4
9 7 1 8 1   6 6 7 6 6   5 4 4 7 7   6 8 1 7 1   0 8 4 9 9
8 1 7 5 0   7 8 5 2 0   9 4 3 9 0   7 6 1 9 1   0 8 7 3 4
0 5 1 0 2   8 7 4 0 3   9 2 6 2 5   8 4 2 2 5   1 9 8 3 4
2 7 8 0 8   1 8 5 6 6   4 4 5 5 4   9 3 5 2 8   6 5 5 4 3

4 8 8 3 3   8 4 6 9 1   8 2 5 7 6   9 7 1 2 3   6 5 1 8 2
5 4 5 8 7   3 8 4 5 7   4 5 2 2 2   7 7 0 2 3   0 2 4 8 6
4 1 9 4 3   0 7 1 9 0   7 3 1 4 0   8 3 2 8 0   5 0 1 0 1
2 8 7 4 6   5 7 7 6 0   0 8 9 5 5   4 0 7 3 9   1 6 3 3 2
5 8 6 8 6   9 6 7 7 5   1 3 5 2 9   7 6 6 3 5   9 4 6 0 5

8 0 9 4 8   5 0 5 6 9   1 0 6 9 5   9 0 7 8 9   9 4 8 3 7
6 1 0 4 1   7 4 0 0 3   5 6 4 2 1   5 1 1 9 0   0 2 5 0 7
3 4 0 9 1   9 3 9 5 1   0 7 4 8 1   1 9 7 0 7   1 4 5 2 6
9 9 6 5 0   7 8 6 1 0   4 9 8 7 7   4 4 7 4 0   7 8 6 4 9
1 5 0 7 8   1 6 3 6 9   5 4 9 5 4   2 4 6 0 4   3 2 6 8 4

5 7 2 8 5   8 3 1 6 4   4 2 2 3 7   0 6 6 3 2   3 5 0 4 6
1 2 3 5 7   1 9 7 7 6   5 5 3 1 9   6 0 5 1 6   3 8 0 3 4
3 4 9 3 5   0 3 7 4 6   2 1 8 5 1   4 1 7 0 2   1 4 9 4 9
2 4 2 6 6   9 9 7 2 1   7 7 3 2 0   6 7 0 7 3   9 3 3 7 5
9 9 1 5 2   8 3 9 9 4   8 9 6 0 6   7 4 6 3 1   3 6 9 8 3

2 6 3 3 9   3 4 1 9 3   0 6 9 1 2   4 1 9 9 8   7 3 2 6 9
1 3 7 8 0   1 0 6 1 6   5 4 9 3 0   7 0 7 2 3   1 7 8 9 2
8 8 4 8 4   0 2 8 5 5   7 3 7 1 2   1 8 3 5 2   0 1 5 3 2
4 2 9 2 8   8 8 9 1 2   4 6 7 1 7   5 1 5 1 9   3 2 2 8 0
1 3 7 2 1   0 6 4 7 6   1 0 8 4 8   9 7 6 3 5   5 1 2 2 9
```

Suggested Readings

Against the Gods: The Remarkable Story of Risk, P. Bernstein, Wiley, 1996.

All the Math That's Fit to Print: Articles from the Manchester Guardian, K. Devlin, Mathematical Association of America, 1994.

The Arithmetic of Life, G. Shaffner, Ballantine Books, 1999.

The Arithmetic of Life and Death, G. Shaffner, Ballantine Books, 2001.

The Bell Curve Debate, R. Jacoby and N. Glauberman (eds.), Times Books, 1995.

Beyond Numeracy: Ruminations of a Number Man, J. A. Paulos, Vintage, 1992.

Beyond the Limits: Confronting Global Collapse, Envisioning a Sustainable Future, D. H. Meadows, D. L. Meadows, and J. Randers, Chelsea Green Publishing Company, 1992.

Billions and Billions, C. Sagan, Random House, 1997.

The Broken Dice and Other Mathematical Tales of Chance, I. Ekeland, University of Chicago Press, 1993.

Can You Win?, M. Orkin, Freeman, 1991.

The Cartoon Guide to Statistics, L. Gonick and W. Smith, HarperCollins, 1993.

The Complete How to Figure It, D. Huff, Norton, 1996.

Damned Lies and Statistics, J. Best, University of California Press, 2001.

Ecological Numeracy: Quantitative Analysis of Environmental Issues, R. Herendeen, Wiley, 1998.

Emblems of Mind: The Inner Life of Music and Mathematics, E. Rothstein, Random House, 1995.

Envisioning Information, E. Tufte, Connecticut Graphics Press, Cheshire, 1983.

Flaws and Fallacies in Statistical Thinking, S. Campbell, Prentice-Hall, 1974.

Games, Gods, and Gambling, F. N. David, Dover, 1998.

Go Figure! The Numbers You Need for Everyday Life, N. Hopkins and J. Mayne, Visible Ink Press, 1992.

The Golden Mean, C. F. Linn, Doubleday, 1974.

The Honest Truth About Lying with Statistics, C. Holmes, Charles C Thomas, 1990.

How Many People Can the Earth Support, J. E. Cohen, Norton, 1995.

How to Lie with Statistics, D. Huff, Norton, 1993.

How to Tell the Liars from the Statisticians, R. Hooke, Dekker, 1983.

How to Use (and Misuse) Statistics, G. Kimble, Prentice-Hall, 1978.

Innumeracy, J. A. Paulos, Hill and Wang, 1988.

The Jungles of Randomness, I. Peterson, Wiley, 1998.

Lady Luck: The Theory of Probability, W. Weaver, Anchor Books, 1963.

The Lady Tasting Tea: How Statistics Revolutionized Science in the 20th Century, D. Salsburg, Freeman, 2001.

Life by the Numbers, K. Devlin, Wiley, 1998.

The Mathematical Tourist, I. Peterson, Freeman, 1988.

A Mathematician Reads the Newspaper, J. A. Paulos, Basic Books, 1995.

Mathematics, the Science of Patterns: The Search for Order in Life, Mind, and the Universe, K. Devlin, Scientific American Library, 1994.

The Mismeasure of Man, S. J. Gould, Norton, 1981.

The Mismeasure of Woman, C. Tavris, Touchstone Books, 1993.

Misused Statistics: Straight Talk for Twisted Numbers, A. Jaffe and H. Spirer, Dekker, 1987.

Nature's Numbers, I. Stewart, Basic Books, 1995.

Number: The Language of Science, T. Dantzig, Macmillan, 1930.

Once Upon a Number, J. A. Paulos, Basic Books, 1998.

Overcoming Math Anxiety, S. Tobias, Houghton Mifflin, 1978. Revised edition, Norton, 1993.

Pi in the Sky: Counting, Thinking, and Being, J. D. Barrow, Little, Brown, 1992.

The Population Explosion, P. R. Ehrlich and A. H. Ehrlich, Simon and Schuster, 1990.

Probabilities in Everyday Life, J. McGervey, Ivy Books, 1989.

The Psychology of Judgment and Decision Making, S. Plous, McGraw-Hill, 1993.

Randomness, D. Bennett, Harvard University Press, 1998.

The Statistical Exorcist: Dispelling Statistics Anxiety, M. Hollander and F. Proschan, Dekker, 1984.

Statistics with a Sense of Humor, F. Pyrczak, Fred Pyrczak Publisher, 1989.

Tainted Truth: The Manipulation of Fact in America, C. Crossen, Simon and Schuster, 1994.

The Tipping Point: How Little Things Can Make a Big Difference, M. Gladwell, Little, Brown, 2000.

200% of Nothing, A. K. Dewdney, Wiley, 1993.

The Universe and the Teacup: The Mathematics of Truth and Beauty, K. C. Cole, Harcourt Brace, 1998.

The Visual Display of Quantitative Information, E. Tufte, Connecticut Graphics Press, Cheshire, 1983.

Vital Signs, compiled by the Worldwatch Institute, Norton, 1998 (updated annually).

What Are the Odds? Chance in Everyday Life, L. Krantz, Harper Perennial, 1992.

Credits

Figure and Text Credits

Chapter 1 Figure 1.1: *New York Times*, November 1, 1999. Copyright © 1999 by the New York Times Co. Reprinted by permission.

Chapter 2 Figure 2.3: *New York Times*, April 7, 1995. Copyright © 1995 by the New York Times Co. Reprinted by permission. Figure 2.4: From Bennett/Briggs, *Using and Understanding Mathematics*, p. 225 (with data from the *Washington Post*). © 1999 Addison Wesley Longman Inc. Reprinted by permission of Pearson Education.

Chapter 3 Figure 3.15: From *The Wall Street Journal Almanac* by Editors of *The Wall Street Journal*, 1999, p. 540. Copyright © 1999 by Dow Jones & Company, Inc. Reprinted by permission of Ballantine Books, a Division of Random House, Inc. Figure 3.16: From *The Wall Street Journal Alamanc* by Editors of *The Wall Street Journal*, 1999, p. 694. Copyright © 1999 by Dow Jones & Company, Inc. Reprinted by permission of Ballantine Books, a Division of Random House, Inc. Figure 3.17: Portions from *The Wall Street Journal Almanac* by Editors of *The Wall Street Journal*, 1999, p. 662. Copyright © 1999 by Dow Jones & Company, Inc. Reprinted by permission of Ballantine Books, a Division of Random House, Inc. Figure 3.18: From *The Wall Street Journal Almanac* by Editors of *The Wall Street Journal*, 1999, p. 238. Copyright © 1999 by Dow Jones & Company, Inc. Reprinted by permission of Ballantine Books, a Division of Random House, Inc. Figure 3.19: *New York Times*, September 30, 1995. Copyright © 1995 by the New York Times Co. Reprinted by permission. Figure 3.20: From *The Wall Street Journal Almanac* by Editors of *The Wall Street Journal*, 1999, p. 351. Copyright © 1999 by Dow Jones & Company, Inc. Reprinted by permission of Ballantine Books, a Division of Random House, Inc. Figure 3.22: Portions from *The Wall Street Journal Almanac* by Editors of *The Wall Street Journal*, 1999, p. 114. Copyright © 1999 by Dow Jones & Company, Inc. Reprinted by permission of Ballantine Books, a Division of Random House, Inc. Figure 3.23: From Bennett/Briggs, *Using and Understanding Mathematics*, 2nd ed., p. 328. Copyright © 2002 Pearson Education, Inc. Reprinted by permission of Pearson Education. Figure 3.25: From Bennett/Briggs, *Using and Understanding Mathematics*, 2nd ed., p. 330. © 2002 Pearson Education, Inc. Reprinted by permission of Pearson Education. Figure 3.26: From Bennett/Briggs, *Using and Understanding Mathematics,* 2nd ed., p. 339. © 2002 Pearson Education, Inc. Reprinted by permission of Pearson Education. Figure 3.28: *New York Times*, October 3, 1995. Copyright © 1995 by the New York Times Co. Reprinted by permission. Figure 3.29: *New York Times,* August 20, 2000. Copyright © 2000 by the New York Times Co. Reprinted by permission. Figure 3.30: From *The Wall Street Journal Almanac* by Editors of *The Wall Street Journal*, 1999, p. 591. Copyright © 1999 by Dow Jones & Company, Inc. Reprinted by permission of Ballantine Books, a Division of Random House, Inc. Figure 3.31: From *The Wall Street Journal Almanac* by Editors of *The Wall Street Journal*, 1999, p. 328. Copyright © 1999 by Dow Jones & Company, Inc. Reprinted by permission of Ballantine Books, a Division of Random House, Inc. Figure 3.32: Reprinted with permission of Time-Life Books. Figure 3.33: From Bennett/Briggs, *Using and Understanding Mathematics*, 2nd ed., p. 338. © 2002 Pearson Education, Inc. Reprinted by permission of Pearson Education. Figure 3.34: From Bennett/Briggs, *Using and Understanding Mathematics*, 2nd ed., p. 338. © 2002 Pearson Education, Inc. Reprinted by permission of Pearson Education. Figure 3.36: From Bennett/Briggs, *Using and Understanding Mathematics*, 2nd ed., p. 330. © 2002 Pearson Education, Inc. Reprinted by permission of Pearson

Education. Figure 3.37: *New York Times*, April 2, 2000. Copyright © 2000 by the New York Times Co. Reprinted by permission. Figure 3.38: © 1996, *USA Today*. Reprinted with permission. Figure 3.41: From *The Wall Street Journal Almanac* by Editors of *The Wall Street Journal*, 1999, p. 323. Copyright © 1999 by Dow Jones & Company, Inc. Reprinted by permission of Ballantine Books, a Division of Random House, Inc. Figure 3.42: From Bennett/Briggs, *Using and Understanding Mathematics*, 2nd ed., p. 334. © 2002 Pearson Education, Inc. Reprinted by permission of Pearson Education. Figure 3.43: From Bennett/Briggs, *Using and Understanding Mathematics*, 2nd ed., p. 334. © 2002 Pearson Education, Inc. Reprinted by permission of Pearson Education. Figure 3.44: From *The Cosmic Perspective*, 2nd ed., by Jeffrey Bennett, Megan Donahue, Nicholas Schneider, and Mark Voit. Copyright © 2002 Pearson Education, Inc. Reprinted by permission. Figure 3.46: *New York Times*, October 6, 1999. Copyright © 1999 by the New York Times Co. Reprinted by permission. Figure 3.47: *USA Today*, October 18, 1999. Copyright © 1999, *USA Today*. Reprinted with permission. Figure 3.52: From Mortensen, Pedersen, Westergaard, Wohlfahrt, Ewald, Mors, Andersen, Melbye, "Effects of Family History and Place and Season of Birth on the Risk of Schizophrenia," *The New England Journal of Medicine*, 2/25/99, p. 606. Copyright © 1999 Massachusetts Medical Society. All rights reserved. Figure 3.51: From *The Wall Street Journal Almanac* by Editors of *The Wall Street Journal*, 1999, p. 577. Copyright © 1999 by Dow Jones & Company, Inc. Reprinted by permission of Ballantine Books, a Division of Random House, Inc. Figure 3.53: From *The Wall Street Journal Almanac* by Editors of *The Wall Street Journal*, 1999, p. 193. Copyright © 1999 by Dow Jones & Company, Inc. Reprinted by permission of Ballantine Books, a Division of Random House, Inc. Figure 3.55: From *The Wall Street Journal Almanac* by Editors of *The Wall Street Journal*, 1999, p. 235. Copyright © 1999 by Dow Jones & Company, Inc. Reprinted by permission of Ballantine Books, a Division of Random House, Inc. Figure 3.56: From *The Wall Street Journal Almanac* by Editors of *The Wall Street Journal*, 1999, p. 545. Copyright © 1999 by Dow Jones & Company, Inc. Reprinted by permission of Ballantine Books, a Division of Random House, Inc. Figure 3.57: Edward R. Tufte, *The Visual Display of Quantitative Information*, Cheshire, CT: Graphics Press, 1983. Reprinted with permission. Figure 3.58: From *The Cosmic Perspective*, 2nd ed., by Jeffrey Bennett, Megan Donahue, Nicholas Schneider, and Mark Voit, p. 371. Copyright © 2002 Pearson Education, Inc. Reprinted by permission. Figure 3.59: *New York Times*, December 1, 1997, p. D1. Copyright © 1997 by the New York Times Co. Reprinted by permission.

Chapter 4 Figure 4.1: From Mario F. Triola, *Elementary Statistics*, 7th ed., p. 61. © 1998 Addison Wesley Longman Inc. Reprinted by permission of Pearson Education. Figure 4.9: From Bennett/Briggs, *Using and Understanding Mathematics*, 2nd ed., p. 369. © 2002 Pearson Education, Inc. Reprinted by permission of Pearson Education. Figure 4.10: From Bennett/Briggs, *Using and Understanding Mathematics*, 2nd ed., p. 369. © 2002 Pearson Education, Inc. Reprinted by permission of Pearson Education. Figure 4.11: From Bennett/Briggs, *Using and Understanding Mathematics*, 2nd ed., p. 370. © 2002 Pearson Education, Inc. Reprinted by permission of Pearson Education. Figure 4.15: From Bennett/Briggs, *Using and Understanding Mathematics*, 2nd ed., p. 232. © 2002 Pearson Education, Inc. Reprinted by permission of Pearson Education.

Chapter 5 Figure 5.27: From Mario F. Triola, *Elementary Statistics*, 7th ed., p. 256. © 1998 Addison Wesley Longman, Inc. Reprinted by permission of Pearson Education. Figure 5.28: From Bennett/Briggs, *Using and*

Understanding Mathematics, p. 524. © 1999 Addison Wesley Longman. Reprinted by permission of Pearson Education. Figure 5.29: *New York Times*, February 24, 1998. Copyright © 1998 by the New York Times Co. Reprinted by permission.

Chapter 6 Epigraph: *Ogden Nash Poems & Stories* by Linell Nash Smith & Isabel Nash Eberstadt. Copyright © 1935 by Ogden Nash, renewed. Reprinted by permission of Curtis Brown, Ltd. and Andre Deutsch, Ltd. Figure 6.3: From *The Virginia Pilot*, 9/14/99. Reprinted with permission. Figure 6.15: From *Statistical Science*, May 15, 1994. Reprinted by permission of the Copyright Clearance Center.

Chapter 7 Figure 7.25: From Bennett/Briggs, *Using and Understanding Mathematics*, p. 501. © 1999 Addison Wesley Longman Inc. Reprinted by permission of Pearson Education.

Chapter 8 Figure 8.14: *New York Times*, December 22, 1998. Copyright © 1998 by the New York Times Co. Reprinted by permission.

Chapter 10 Figure 10.4: Portions from *New York Times*, December 11, 1999. Copyright © 1999 by the New York Times Co. Reprinted by permission. Figure 10.10: *New York Times*, August 4, 1998, p. B10. Copyright © 1998 by the New York Times Co. Reprinted by permission. Figure 10.11: *New York Times*, August 4, 1998, p. B10. Copyright © 1998 by the New York Times Co. Reprinted by permission.

Photo Credits

Chapter 1 p. 1, DigitalVision/PictureQuest; p. 3, AFP/Corbis; p. 6, PhotoDisc; p. 8, AFP/Corbis; p. 12, Beth Anderson; p. 14, Smithsonian Institution/NASM; p. 16, PhotoDisc; p. 23, Corbis RF; p. 34, PhotoDisc; p. 35, PhotoDisc; p. 38 (top) PhotoDisc, (bottom) From *The Cosmic Perspective*, 2nd ed., by Jeffrey Bennett, Megan Donahue, Nicholas Schneider, and Mark Voit, p. 11. Copyright © 2002 Pearson Education, Inc. Reprinted by permission; p. 42, Michael Okoniewski; p. 46, PhotoDisc; p. 49, PhotoDisc

Chapter 2 p. 51, AFP/Corbis; p. 58, PhotoDisc; p. 61, PhotoDisc; p. 62, NASA; p. 68, Peter Turnley/Corbis; p. 69, PhotoDisc; p. 76, Bettmann/Corbis; p. 81, PhotoDisc; p. 83, PhotoDisc

Chapter 3 p. 87, PhotoDisc; p. 90, Reuters Newsmedia Inc./Corbis; p. 102, Roger Ressmeyer/Corbis; p. 131, Charles & Josette Lenars/Corbis; p. 140, Archivo Iconographico, S.A./Corbis; p. 143, PhotoDisc

Chapter 4 p. 147, PhotoDisc; p. 151, Fernando Medina/AllSport; p. 163, AP/WW; p. 168, United Features Syndicate; p. 170, Beth Anderson; p. 175, Honda web; p. 181, John-Marshall Mantel/Corbis; p. 184, James L. Amos/Corbis

Chapter 5 p. 187 (left) Dimitri Lundt; TempSport/Corbis, (right) Bettmann/Corbis; p. 191, Beth Anderson; p. 200, PhotoDisc; p. 205, Leif Skoogfors/Corbis; p. 218, Beth Anderson; p. 221; PhotoDisc

Chapter 6 p. 225, PhotoDisc; p. 233, PhotoDisc; p. 237, Gary Holscher/Stone; p. 246, NOAA; p. 255, Bettmann/Corbis; p. 266, Beth Anderson; p. 269, Corbis RF

Chapter 7 p. 273, Bill Aron/Photo Edit; p. 275, PhotoDisc; p. 292, PhotoDisc; p. 296, PhotoDisc; p. 300, Vittoriano Rastelli/Corbis; p. 306, Bill Losh/FPG; p. 311, PhotoDisc; p. 313, PhotoDisc; p. 314, Associated Press; p. 319, PhotoDisc; p. 322, Stone

Chapter 8 p. 325, Corbis RF; p. 327, AFP/Corbis; p. 332, James A. Sugar/Corbis; p. 338, PhotoDisc; p. 350, PhotoDisc; p. 355, Steve Chenn/Corbis; p. 357, Bob Giuliani

Chapter 9 p. 359, PhotoDisc; p. 360, PhotoDisc; p. 365, PhotoDisc; p. 369, PhotoDisc; p. 389, Layne Kennedy/Corbis; p. 401, PhotoDisc; p. 402, Journal of the Royal Statistical Society; p. 404, PhotoDisc

Chapter 10 p. 407, PhotoDisc; p. 408, PhotoDisc; p. 421, Richard T. Nowitz/Corbis; p. 431, Joseph Sohm, Chromosohm, Inc./Corbis; p. 433, PhotoDisc; p. 446, Beth Anderson; p. 450, Yellow Dog Productions/Image Bank

Glossary

absolute change The actual increase or decrease from a reference value to a new value:

absolute change = new value − reference value

absolute difference The actual difference between the compared value and the reference value:

absolute difference = compared value − reference value

absolute error The actual amount by which a measured value differs from the true value:

absolute error = measured value − true value

accident rate The number of accidents due to some particular cause, expressed as a fraction of all people at risk for the same cause. For example, an accident rate of "5 per 1,000 people" means that an average of 5 in 1,000 people suffer an accident from this particular cause.

accuracy How closely a measurement approximates a true value. An accurate measurement is very close to the true value.

alternative hypothesis (H_a) A statement that can be accepted only if the null hypothesis is rejected.

***and* probability** The probability that event A *and* event B will both occur. How it is calculated depends on whether the events are independent or dependent. Also called *joint probability*.

***a priori* method** See *theoretical method*.

bar graph A diagram consisting of bars representing the frequencies (or relative frequencies) for particular categories. The bar lengths are proportional to the frequencies.

best-fit line The line on a scatter diagram that lies closer to the data points than all other possible lines (according to a standard statistical measure of closeness). Also called *regression line*.

bias In a statistical study, any problem in the design or conduct of the study that tends to favor certain results. See also *participation bias*; *selection bias*.

bimodal distribution A distribution with two peaks, or modes.

binning Grouping data into categories (bins), each of which covers a range of possible data values.

blinding The practice of keeping experimental subjects and/or experimenters in the dark about who is in the treatment group and who is in the control group. See also *double-blind experiment*; *single-blind experiment*.

boxplot A graphical display of a five-number summary. A number line is used for reference, the values from the lower to the upper quartiles are enclosed in a box, a line is drawn through the box for the median, and two "whiskers" are extended to the low and high data values. Also called *box-and-whisker plot*.

case-control study An observational study that resembles an experiment because the sample naturally divides into two (or more) groups. The participants who engage in the behavior under study form the cases, like the treatment group in an experiment. The participants who do not engage in the behavior are the *controls*, like the control group in an experiment.

causality A relationship present when one variable is a cause of another.

census The collection of data from every member of a population.

Central Limit Theorem Theorem stating that, for random samples (all of the same size) of a variable with any distribution (not necessarily a normal distribution), the distribution of the means of the samples will, as the sample size increases, tend to be approximately a normal distribution.

chi-square statistic (χ^2) A number used to determine the statistical significance of a hypothesis test in a contingency table (or two-way table). If it is less than a critical value (which depends on the table size and the desired significance level), the differences between the observed frequencies and the expected frequencies are not significant.

cluster sampling Dividing the population into groups, or clusters; selecting some of these clusters at random; and then obtaining the sample by choosing all the members within each cluster.

coefficient of determination (R^2) A number that describes how well data fit a best-fit equation found through multiple regression.

compared value A number that is compared to a reference value in computing a relative difference.

complement For an event A, all outcomes in which A does *not* occur, expressed as $\overline{A}$. Its probability is $P(\overline{A}) = 1 - P(A)$.

conditional probability The probability of one event given the occurrence of another event, written $P(B$ given $A)$ or $P(B|A)$.

confidence interval A range of values associated with a confidence level, such as 95%, that is likely to contain the true value of a population parameter.

confounding Confusion in the interpretation of statistical results that occurs when the effects of different factors are mixed such that the effects of the individual factors being studied cannot be determined.

confounding factors Any factors or variables in a statistical study that can lead to confounding. Also called *confounding variables*.

Consumer Price Index (CPI) An index number designed to measure the rate of inflation. It is computed and reported monthly, based on a sample of more than 60,000 goods, services, and housing costs.

contingency table See *two-way table*.

continuous data Quantitative data that can take on any value in a given interval.

contour map A map that uses curves (contours) to connect geographical regions with the same data values.

control group The group of subjects in an experiment who do not receive the treatment being tested.

convenience sampling Selecting a sample that is readily available.

correlation A statistical relationship between two variables. See also *negative correlation; no correlation; positive correlation.*

correlation coefficient (*r*) A measure of the strength of the relationship between two variables. Its value is always between -1 and 1 (that is, $-1 \leq r \leq 1$).

cumulative frequency For any data category, the number of data values in that category and all preceding categories.

death rate The number of deaths due to some particular cause, expressed as a fraction of all people at risk for the same cause. For example, a death rate of "5 per 1,000 people" means that an average of 5 in 1,000 people die from this particular cause.

dependent events Two events for which the outcome of one affects the probability of the other.

deviation How far a particular data value lies from the mean of a data set, used to compute standard deviation.

discrete data Quantitative data that can take on only particular values and not other values in between (for example, the whole numbers 0, 1, 2, 3, 4, 5).

distribution The way the values of a variable are spread over all possible values. It can be displayed with a table or with a graph.

distribution of sample means The distribution that results when the means ($\bar{x}$) of all possible samples of a given size are found.

distribution of sample proportions The distribution that results when the proportions ($\hat{p}$) in all possible samples of a given size are found.

dotplot A diagram similar to a bar graph except that each individual data value is represented with a dot.

double-blind experiment An experiment in which neither the participants nor the experimenters know who belongs to the treatment group and who belongs to the control group.

either/or probability The probability that *either* event A *or* event B will occur. How it is calculated depends on whether the events are overlapping or non-overlapping.

empirical method See *relative frequency method.*

event In probability, a collection of one or more outcomes that share a property of interest. See also *outcome.*

expected frequency In a two-way table, the frequency one would expect in a given cell of the table if the row and column variables were independent of each other.

expected value The mean value of the outcomes for some random variable.

experiment A study in which researchers apply a treatment and then observe its effects on the subjects.

experimenter effect An effect that occurs when a researcher or experimenter somehow influences subjects through such factors as facial expression, tone of voice, or attitude.

five-number summary A description of the variation of a data distribution in terms of the minimum value, lower quartile, median, upper quartile, and maximum value.

frequency For a data category, the number of times data values fall within that category.

frequency table A table that lists all the categories of data in one column and the frequency for each category in another column.

gambler's fallacy The mistaken belief that a streak of bad luck makes a person "due" for a streak of good luck.

geographical data Data that can be assigned to different geographical locations.

histogram A bar graph showing a distribution for quantitative data (at the interval or ratio level of measurement). The bars have a natural order, and the bar widths have specific meaning.

hypothesis In statistics, a claim about a population parameter, such as a population proportion, p, or population mean, μ. See also *alternative hypothesis*; *null hypothesis.*

hypothesis test A standard procedure for testing a claim about the value of a population parameter.

independent events Two events for which the outcome of one does not affect the probability of the other.

index number A number that provides a simple way to compare measurements made at different times or in different places. The value at one particular time (or place) must be chosen to be the reference value (or base value). The index number for any other time (or place) is

$$\text{index number} = \frac{\text{value}}{\text{reference value}} \times 100$$

inflation The increase over time in prices and wages. Its overall rate is measured by the CPI.

interval level of measurement A level of measurement for quantitative data in which differences, or intervals, are meaningful but ratios are not. Data at this level have an arbitrary starting point.

joint probability See *and probability.*

law of large numbers An important result in probability that applies to a process for which the probability of an event A is $P(A)$ and the results of repeated trials are independent. It states: If the process is repeated through many trials, the larger the number of trials, the closer the proportion should be to $P(A)$. Also called *law of averages.*

left-skewed distribution A distribution in which the values are more spread out on the left side.

left-tailed test A hypothesis test that involves testing whether a population parameter lies to the left (lower values) of a claimed value.

level of measurement See *nominal level of measurement*; *ordinal level of measurement*; *interval level of measurement*; *ratio level of measurement*.

life expectancy The number of years a person of a given age today can be expected to live, on average. It is based on current health and medical statistics and does not take into account future changes in medical science or public health.

line chart A graph showing the distribution of quantitative data as a series of dots connected by lines. The horizontal position of each dot corresponds to the center of the bin it represents and the vertical position corresponds to the frequency value for the bin.

lower quartile See *quartile, lower*.

margin of error The maximum likely difference between an observed sample statistic and the true value of a population parameter. Its size depends on the desired level of confidence.

mean The sum of all values divided by the total number of values. It's what is most commonly called the average value.

median The middle value in a sorted data set (or halfway between the two middle values if the number of values is even).

median class For binned data, the bin into which the median data value falls.

meta-analysis A study in which researchers analyze many individual studies (on a particular topic) as a combined group, with the aim of finding trends that were not evident in the individual studies.

middle quartile See *quartile, middle*.

mode The most common value (or group of values) in a distribution.

multiple bar graph A simple extension of a regular bar graph, in which two or more sets of bars allow comparison of two or more data sets.

multiple line chart A simple extension of a regular line chart, in which two or more lines allow comparison of two or more data sets.

multiple regression A technique that allows the calculation of a best-fit equation that represents the best fit between one variable (such as price) and a *combination* of two or more other variables (such as weight and color).

negative correlation A correlation in which the two variables tend to change in opposite directions, with one increasing while the other decreases.

no correlation Absence of any apparent relationship between two variables.

nominal level of measurement A level of measurement for qualitative data that consist of names, labels, or categories only and cannot be ranked or ordered.

nonlinear relationship A relationship between two variables that cannot be expressed with a linear (straight-line) equation.

non-overlapping events Two events for which the occurrence of one precludes the occurrence of the other.

normal distribution A special type of symmetric, bell-shaped distribution with a single peak that corresponds to the mean, median, and mode of the distribution. Its variation can be characterized by the standard deviation. See also *68-95-99.7 rule*.

null hypothesis (H_0) A specific claim (such as a specific value for a population parameter) against which an alternative hypothesis is tested.

observational study A study in which researchers observe or measure characteristics of the sample members, but do not attempt to influence or modify these characteristics.

***of* versus *more than (less than)* rule** A rule for comparisons. It states: If the compared value is *P% more than* the reference value, then it is $(100 + P)\%$ *of* the reference value. If the compared value is *P% less than* the reference value, then it is $(100 - P)\%$ *of* the reference value.

one-tailed test See *left-tailed test*; *right-tailed test*.

ordinal level of measurement A level of measurement for qualitative data that can be arranged in some order. It generally does not make sense to do computations with the data.

outcome In probability, the most basic possible result of an observation or experiment. See also *event*.

outlier A value in a data set that is much higher or much lower than almost all other values.

overlapping events Two events that could possibly both occur.

Pareto chart A bar graph of data at the nominal level of measurement, with the bars arranged in frequency order.

participants People (as opposed to objects) who are the subjects of a study.

participation bias Bias that occurs any time participation in a study is voluntary.

peer review A process by which several experts in a field evaluate a research report before it is published.

percentiles Values that divide a data distribution into 100 segments, each representing about 1% of the data values.

pictograph A graph embellished with artwork.

pie chart A circle divided so that each wedge represents the relative frequency of a particular category. The wedge size is proportional to the relative frequency, and the entire pie represents the total relative frequency of 100%.

placebo Something that lacks the active ingredients of a treatment but is identical in appearance to the treatment. Thus, participants in a study cannot distinguish the placebo from the real treatment.

placebo effect An effect in which patients improve simply because they believe they are receiving a useful treatment, when in fact they may be receiving only a placebo.

population The complete set of people or things being studied.

population mean The true mean of a population, denoted by the Greek letter μ (pronounced "mew").

population parameters Specific characteristics of the population that a statistical study is designed to estimate.

population proportion The true proportion of some characteristic in a population, denoted by p.

positive correlation A type of correlation in which two variables tend to increase (or decrease) together.

practical significance In a statistical study, significance in the sense that the result is associated with some meaningful course of action.

precision The amount of detail in a measurement.

probability For an event, the likelihood that the event will occur. The probability of an event, written as P(event), is always between 0 and 1 inclusive. A probability of 0 means the event is impossible and a probability of 1 means the event is certain. See also *relative frequency method; subjective method; theoretical method.*

probability distribution The complete distribution of the probabilities of all possible events associated with a particular variable. It may be shown as a table or as a graph.

***P*-value** In a hypothesis test, the probability of selecting a sample at least as extreme as the observed sample, assuming that the null hypothesis is true.

qualitative data Data consisting of values that describe qualities or nonnumerical categories.

quantitative data Data consisting of values representing counts or measurements. Quantitative data may be either discrete or continuous.

quartile, lower The median of the data values in the lower half of a data set. Also called *first quartile.*

quartile, middle The overall median of a data set. Also called *second quartile.*

quartile, upper The median of the data values in the upper half of a data set. Also called *third quartile.*

quartiles Values that divide a data distribution into four equal parts.

random errors Errors that occur because of random and inherently unpredictable events in the measurement process.

randomization The process of ensuring that the subjects of an experiment are assigned to the treatment or control group at random and in such a way that each subject has an equal chance of being assigned to either group.

range For a distribution, the difference between the lowest and highest data values.

range rule of thumb A guideline stipulating that, for a data set with no outliers, the standard deviation is approximately equal to range/4.

rare event rule Rule stating that it is appropriate to conclude that a given assumption (such as the null hypothesis) is probably not correct if the probability of a particular event at least as extreme as the observed event is very small.

ratio level of measurement A level of measurement for quantitative data in which both intervals and ratios are meaningful. Data at this level have a true zero point.

raw data The actual measurements or observations collected from a sample.

reference value The number that is used as the basis for a comparison.

regression line See *best-fit line.*

relative change The size of an absolute change in comparison to the reference value, expressed as a percentage:

$$\text{relative change} = \frac{\text{new value} - \text{reference value}}{\text{reference value}} \times 100\%$$

relative difference The size of an absolute difference in comparison to the reference value, expressed as a percentage:

$$\text{relative difference} = \frac{\text{compared value} - \text{reference value}}{\text{reference value}} \times 100\%$$

relative error The relative amount by which a measured value differs from the true value, expressed as a percentage:

$$\text{relative error} = \frac{\text{measured value} - \text{true value}}{\text{true value}} \times 100\%$$

relative frequency For any data category, the fraction or percentage of the total frequency that falls in that category:

$$\text{relative frequency} = \frac{\text{frequency in category}}{\text{total frequency}}$$

relative frequency method A method of estimating a probability based on observations or experiments by using the observed or measured relative frequency of the event of interest. Also called *empirical method.*

representative sample A sample in which the relevant characteristics of the members are generally the same as the characteristics of the population.

right-skewed distribution A distribution in which the values are more spread out on the right side.

right-tailed test A hypothesis test that involves testing whether a population parameter lies to the right (higher values) of a claimed value.

rounding rule For statistical calculations, the practice of stating answers with one more decimal place of precision than is found in the raw data. For example, the mean of 2, 3, and 5 is 3.3333 . . . , which would be rounded to 3.3.

sample A subset of the population from which data are actually obtained.

sample mean The mean of a sample, denoted $\bar{x}$ ("x-bar").

sample proportion The proportion of some characteristic in a sample, denoted $\hat{p}$ ("p-hat").

sample statistics Characteristics of the sample that are found by consolidating or summarizing the raw data.

sampling The process of choosing a sample from a population.

sampling distribution The distribution of a sample statistic, such as a mean or proportion, taken from all possible samples of a particular size.

sampling error Error introduced when a random sample is used to estimate a population parameter; the difference between a sample result and a population parameter.

sampling methods See *cluster sampling*; *convenience sampling*; *simple random sampling*; *stratified sampling*; *systematic sampling*.

scatter diagram A graph, often used to investigate correlations, in which each point corresponds to the values of two variables. Also called *scatterplot*.

selection bias Bias that occurs whenever researchers select their sample in a biased way. Also called *selection effect*.

self-selected survey A survey in which people decide for themselves whether to be included. Also called *voluntary response survey*.

simple random sampling A sample of items chosen in such a way that every possible sample of the same size has an equal chance of being selected.

Simpson's paradox A statistical paradox that arises when the results for a whole group seem inconsistent with those for its subgroups; it can occur whenever the subgroups are unequal in size.

single-blind experiment An experiment in which the participants do not know whether they are members of the treatment group or the control group but the experimenters do know—or, conversely, the participants do know but the experimenters do not.

single-peaked distribution A distribution with a single mode. Also called *unimodal distribution*.

68-95-99.7 rule Guideline stating that, for a normal distribution, about 68% (actually, 68.3%) of the data values fall within 1 standard deviation of the mean, about 95% (actually, 95.4%) of the data values fall within 2 standard deviations of the mean, and about 99.7% of the data values fall within 3 standard deviations of the mean.

skewed See *left-skewed distribution*; *right-skewed distribution*.

stack plot A type of bar graph or line chart in which two or more different data sets are stacked vertically.

standard deviation A single number commonly used to describe the variation in a data distribution, calculated as

$$\text{standard deviation} = \sqrt{\frac{\text{sum of all (deviations from the mean)}^2}{\text{total number of data values} - 1}}$$

standard score For a particular data value, the number of standard deviations (usually denoted by z) between it and the mean of the distribution:

$$z = \text{standard score} = \frac{\text{data value} - \text{mean}}{\text{standard deviation}}$$

Also called *z-score*.

statistical significance A measure of the likelihood that a result is meaningful.

statistically significant result A result in a statistical study that is unlikely to have occurred by chance. The most commonly quoted levels of statistical significance are the 0.05 level (the probability of the result's having occurred by chance is 5% or less, or less than 1 in 20) and the 0.01 level (the probability of the result's having occurred by chance is 1% or less, or less than 1 in 100).

statistics (plural) The data that describe or summarize something.

statistics (singular) The science of collecting, organizing, and interpreting data.

stem-and-leaf plot A graph that looks much like a histogram turned sideways, with lists of the individual data values in place of bars. Also called *stemplot*.

stratified sampling A sampling method that addresses differences among subgroups, or strata, within a population. First the strata are identified, and then a random sample is drawn within each stratum. The total sample consists of all the samples from the individual strata.

subjective method A method of estimating a probability based on experience or intuition.

subjects In a statistical study, the people or objects chosen for the sample. See also *participants*.

symmetric distribution A distribution in which the left half is a mirror image of the right half.

systematic errors Errors that occur when there is a problem in the measurement system that affects all measurements in the same way.

systematic sampling Using a simple system to choose the sample, such as selecting every 10th or every 50th member of the population.

theoretical method A method of estimating a probability based on a theory, or set of assumptions, about the process in question. Assuming that all outcomes are equally likely, the theoretical probability of a particular event is found by dividing the number of ways the event can occur by the total number of possible outcomes. Also called *a priori method*.

time-series diagram A histogram or line chart in which the horizontal axis represents time.

treatment Something given or applied to the members of the treatment group in an experiment.

treatment group The group of subjects in an experiment that receive the treatment being tested.

two-tailed test A hypothesis test that involves testing whether a population parameter lies to either side of a claimed value.

two-way table A table showing the relationship between two variables by listing the values of one variable in its rows and the values of the other variable in its columns. Also called *contingency table*.

type I error In a hypothesis test, the mistake of rejecting the null hypothesis, H_0, when it is true.

type II error In a hypothesis test, the mistake of failing to reject the null hypothesis, H_0, when it is false.

uniform distribution A distribution in which all data values have the same frequency.

unimodal distribution See *single-peaked distribution*.

unusual values In a data distribution, values that are not likely to occur by chance, such as those values that are more than 2 standard deviations away from the mean.

upper quartile See *quartile, upper.*

variable Any item or quantity that can vary or take on different values.

variables of interest In a statistical study, the items or quantities that the study seeks to measure.

variation How widely data are spread out about the center of a distribution. See also *five-number summary*; *range*; *standard deviation.*

vital statistics Data concerning births and deaths of people.

voluntary response survey See *self-selected survey.*

weighted mean A mean that accounts for differences in the relative importance of data values. Each data value is assigned a weight, and then

$$\text{weighted mean} = \frac{\text{sum of (each data value} \times \text{its weight)}}{\text{sum of all weights}}$$

z-score See *standard score.*

Answers

Section 1.1

1. Not sensible   3. Sensible   5. Not sensible

7. Population: registered voters in California. Sample: the 1,026 people interviewed. Raw data: responses from the 1,026 interviews. Sample statistics: consolidated measures of opinions, such as the percentage of people in the sample who support a specific candidate. Population parameters: measures of opinions about candidates in the population of all registered voters.

9. Population: all new computers of this particular model from Dell. Sample: the single computer tested. Raw data: results of all tests. Sample statistics: measures that describe test results for the sample computer in terms of speed and other benchmarks. Population parameters: measures that describe speed and other benchmarks for all computers in the population.

11. **a.** 4.1 to 4.9 months   **b.** 94.9% to 95.1%
    c. 455.8 to 458.2 miles

13. If the poll is accurate, it is very likely that the Republican will win because her likely percentage of the vote is between 50.5% and 55.5%.

15. Based on this poll, we would expect that between 67% and 73% of the population voted, which is significantly higher than the actual 61% found from the voting records. This suggests that at least some people in the sample claimed to have voted when actually they had not. However, because we get only a likely range from the sample, it is still possible that this particular sample was unusual and everyone told the truth.

17. **a.** Goal: determine whether people think they must rely on themselves above others. Population: all people (in the United States). Population parameter: percentage of people in the population who agree with the given statement.
    b. Sample: 4,000 people surveyed. Raw data: individual responses to the question. Sample statistic: percentage of people in the sample who agree with the given statement.
    c. 68.4% to 71.6%

19. **a.** Goal: determine the mean height of male Marine recruits. Population: all male Marine recruits between ages 18 and 24. Population parameter: mean height of the population of recruits.
    b. Sample: 772 people measured. Raw data: 772 individual height measurements. Sample statistic: mean height of the 772 people in the sample.
    c. 69.5 to 69.9 inches

21. **a.** Goal: determine how many people keep money in regular savings accounts. Population: all adults (in the United States). Population parameter: percentage of adults who keep money in regular savings accounts.

b. Sample: 2,000 people surveyed. Raw data: individual responses to the survey question. Sample statistic: percentage of people in the sample who keep money in regular savings accounts.
c. 62% to 66%

23. *Step 1.* Goal: predict winner of the election. Population: all students who will vote in the election. *Step 2.* Choose a sample of students who are likely to vote. *Step 3.* Interview the students in the sample to find out whom they plan to vote for. Determine what the election outcome would be if it were based solely on your sample. *Step 4.* Use techniques of statistical science to infer the likely results for the entire population. *Step 5.* Based on the likely population results, draw a conclusion about who will win the election.

25. *Step 1.* Goal: learn about tipping habits. Population: all tips left, perhaps in a particular city or restaurant (depending on your interests). *Step 2.* Choose a sample of restaurant bills from your population. *Step 3.* Find the percentage tip on each bill in your sample. Then calculate a mean tip for all the bills. *Step 4.* Use techniques of statistical science to infer the likely range of tips for the population. *Step 5.* Draw conclusions about tipping habits.

27. *Step 1.* Goal: determine percentage of high school students who are vegetarians. Population: all high school students, perhaps in a particular city or school (depending on your interests). *Step 2.* Choose a sample of students from your population. *Step 3.* Interview each student in your sample to learn whether she or he is a vegetarian. Then calculate the percentage of students in your sample who are vegetarians. *Step 4.* Use techniques of statistical science to infer the likely range of percentages of all students in your population who are vegetarians. *Step 5.* Draw conclusions about how many high school students are vegetarians.

Section 1.2

1. Not sensible   3. Sensible   5. Not sensible

7. A census should be practical, although you will need a reliable way of getting the data, such as viewing the transcripts of all 50 students.

9. A census would require determining the mean energy cost of every home in Missouri, which would require gathering data from every home's utility bills. These data should exist in principle, but would probably be very difficult to collect in practice.

11. The entire team from your school is probably the best sample. The other groups are specialized subsets of the track team, and their eating habits may be different from those of typical team members.

13. The critic may be under pressure to give the film a favorable review, since she works for the company that produced the film.

To decide whether this bias is a problem, you might look to see whether there appears to be a pattern of Disney films' receiving better-than-average reviews from this reviewer.

15. The Book Review section generates revenue from advertisements, and therefore has a financial interest in making advertisers happy. This could translate into giving better reviews than a book would otherwise get. To decide whether this bias is a problem, you might look at past reviews of books that were advertised in the section and see whether they seem to be better than the reviews given to books that were not advertised.

17. The university scientists are receiving funding from Monsanto, which might make them anxious to please Monsanto in hopes of getting similar funding opportunities in the future. Thus, they may have bias toward giving Monsanto the results it wants, even though they do not work directly for the company. To decide whether this bias is a problem, you will need to explore their scientific methods and conclusions carefully.

19. There is no reason to think that every 100th chip is different from the others, so the sample is likely to be representative. However, if there is a systematic defect in the manufacturing process, then this sampling method could be biased.

21. The poll involves only very early morning voters, who may not be representative of all voters because they may be voting early for some reason (such as having strong opinions or having children in school or being employed in certain industries).

23. Because students choose whether to return surveys, those with strong opinions are more apt to respond. This survey is very unlikely to be representative.

25. The employees of the company have a vested interest in the product's success, and this bias makes them a poor sample.

27. Customers who shop between 10:00 A.M. and noon tend to be people who do not work, so this sample is unlikely to be representative of all customers.

29. This is a convenience sample, and one that is unlikely to be representative because people with strong feelings are more apt to return the survey. The magazine probably chose this sampling method because it was easy; the magazine might even be interested in the opinions of those with the strongest feelings.

31. This is stratified sampling in which the strata are different age groups. This method probably was chosen to see whether sleep habits vary with age.

33. This is cluster sampling, with the clusters on the three streets chosen. Perhaps these three streets were chosen because they were deemed to be "typical" and likely to produce a representative sample.

35. This is stratified sampling in which the strata are different states. Presumably, the study is looking for income differences between states.

37. Simple random sampling should be adequate. However, you might want to use stratified sampling in which the strata consist of different ethnic groups. This will enable you to discover any differences in blood type percentages among ethnic groups.

39. Simple random sampling of lung cancer victims should be adequate. However, you might want to do cluster sampling, with clusters in different regions, in case there are regional differences in causes of lung cancer.

41. You will need stratified sampling in which you study both people who drink three cups of herbal tea daily and people who do not. This study would be best done as a controlled experiment.

Section 1.3

1. Sensible   3. Not sensible   5. Not sensible

7. This is an observational, case-control study. The cases were the identical twins and the controls were the fraternal twins.

9. This is an observational, case-control study. The cases were the melanoma patients and the controls were the cancer-free patients.

11. This is an observational study (a poll).

13. This is an observational, case-control study. The cases are the women who exercised regularly and the controls are those who did not.

15. This is a meta-analysis.   17. This is a meta-analysis.

19. This experiment should be effective. It might be advisable for the experiment to be double blind so that the researchers evaluating the growth do not know which group received which food.

21. This experiment is poor because the two experimental groups live in different states with different driving conditions. The two groups should have been randomized so that people from both states could end up in each group.

23. The idea behind this experiment is reasonable, but measurement difficulties introduce many opportunities for confounding. For example, it's hard to be sure that people set their alarms for 10% less sleep than usual, and it's difficult to evaluate whether people are "more or less likely" to have temper tantrums.

25. The control group should be students who listen to no music. Treatment groups could include students who listen to jazz or other music while studying. The study cannot be single- or double-blind.

27. The control group should be swimmers who do not play soccer. The treatment group should be swimmers who do play soccer. The study cannot be single- or double-blind.

29. In an experiment repeated with many people, the mind reader should be asked to describe something the person is thinking. The experiment should be double-blind. The mind reader should not know the identity of the person he is "reading," and the person, when asked what he or she is thinking, should not know what the mind reader has said.

Section 1.4

1. Guideline 2, regarding bias of researchers, is at issue.

3. Guideline 4 should at least be checked; the variable of interest is not clear and may be difficult to measure.

5. Guideline 6 is at issue because the question is leading.

7. The word *wrong* in the first question could be misleading. For example, some people might think that abortion is wrong but still favor choice. The second question could also be confusing, as some people might think "advice of her doctor" means that the woman's life is in danger, which could alter their opinion of whether abortion is justified in the situation. Groups opposed to abortion would be likely to cite the results of the first question, while groups favoring choice would be likely to cite the results of the second question.

9. The headline says "drugs," whereas the story says "drug use, drinking, or smoking." Because "drugs" is usually taken to mean drugs *other than* cigarettes or alcohol, the headline is very misleading.

11. The first question requires a study of blind dates generally, while the second examines only people who are married to see whether their first date was blind. The second group is much more limited than the first.

13. The first question asks what percentage of the time teenagers run red lights. The second might be taken to be asking what percentage of red light runners are teenagers. Interpreted this way, the two questions are very different and require different samples to be studied.

15. The report seems to be making an implication about restaurant quality in New York (the "Big Apple"), but much information is missing. What about restaurants receiving a score of 30 or 28? What criteria were used for the ratings? Without much more information, it would be difficult to act on these data.

Chapter 1 Review Exercises

1. **a.** 92.6% to 95.2%   **b.** All United States households
   c. A survey is an observational study.
   d. The reported value is a sample statistic because it was measured only for the sample.
   e. No; this would make the sample self-selected and therefore unlikely to be representative.
   f. Selecting a sample of households in each state
   g. Selecting all the households in a few randomly selected election precincts
   h. Selecting every 10th household, by address, on each street in a city

2. **a.** A sample chosen in such a way that every sample of the same size has an equal chance of being selected
   b. No; this is a cluster sample.
   c. Write the name of each student on a slip of paper, put the slips in a hat, and then draw five names from the hat.

3. **a.** No; there is no control group.
   b. The placebo might make subjects feel better just because they're taking something, not because of a specific effect of the drug.
   c. Split the 50 people into two groups of 25, with one group receiving the real drug and the other receiving a placebo.
   d. Blinding for the participants means that they do not know who gets the real drug and who gets the placebo.
   e. An experimenter effect occurs if the experimenter somehow alerts the participants to whether they are receiving the real

drug or the placebo. Experimenter effect can be minimized by making the study double-blind.

4. **a.** It is a convenience sample, because interviewing people in Marion is more convenient than trying to reach people nationwide.
   b. Many answers are possible. One example: Rate the weather where you live on a scale of 1 = poor to 5 = great.

Section 2.1

1. Sensible   3. Not sensible   5. Colors are qualitative.

7. Waiting times are quantitative.

9. Days of the week are qualitative.

11. Scores are quantitative.

13. *Yes* or *no* responses are qualitative.

15. Amounts of money are quantitative.

17. Names of TV shows are qualitative.

19. The number of defective computer components is discrete.

21. The number of taxicabs is discrete.

23. The number of exits is discrete.

25. The number of miles is continuous because it can include fractions.

27. Speeds are continuous.   29. Numbers of stars are discrete.

31. Average prices are continuous because an average or mean can have any value.

33. Names of parties are at the nominal level of measurement.

35. Celsius temperatures are at the interval level of measurement.

37. Car types are at the nominal level of measurement.

39. Safety ratings are at the ordinal level of measurement.

41. Course grades are at the ordinal level of measurement.

43. City temperatures, reported in Fahrenheit or Celsius, are at the interval level of measurement.

45. Breeds of horses are at the nominal level of measurement.

47. Not meaningful   49. Meaningful   51. Meaningful

53. Not meaningful

55. Quantitative; ratio level of measurement; discrete

57. Quantitative; ratio level of measurement; discrete

59. Quantitative; interval level of measurement; discrete

61. Qualitative; ordinal level of measurement

Section 2.2

1. Not sensible   3. Not sensible   5. Not sensible

7. Mistakes should lead to random errors. Dishonesty should lead to systematic errors that systematically benefit the taxpayer.

9. This is a systematic error that makes all altimeter readings too low.

11. This is a systematic error that makes all weights too high by 1.2 pounds.

13. Example of a random error: mistakes in counting. Example of a systematic error: misclassifying pickup trucks as SUVs.

15. Example of a random error: random mistakes in the data on the returns. Example of a systematic error: systematic dishonesty on the returns.

17. Example of a random error: random mistakes in reading the scale. Example of a systematic error: including the weight of a paper plate.

19. Example of a random error: random mistakes in counting the kernels. Example of a systematic error: counting unpopped kernels as popped.

21. Absolute error: too high by 50¢. Relative error: too high by 2.7%.

23. Absolute error: too low by $17. Relative error: too low by 26%.

25. a. Random
   b. The average is the better choice for minimizing random errors.
   c. Systematic errors might include, for example, a problem with the measuring device or a problem in defining the "length" of the room.
   d. Averaging measurements will *not* help any systematic errors.

27. The laser device is more precise, but the tape measure is more accurate (assuming that your actual height is what you thought).

29. The digital scale at the gym is more precise and more accurate (assuming that your actual weight is what you thought).

31. No one could possibly know the exact population of the United States today, let alone in 1860, so the claim is not believable.

33. This projection for the future sounds reasonable, so the claim is believable. Of course, only time will tell if it is correct.

35. The average temperature is presumably based on daily temperature measurements, which should be reliable, so the claim is believable.

37. Since no one keeps lists of cell phone users, no one could know this number so precisely; thus, the claim is not believable.

Section 2.3

1. Not sensible   **3.** Not sensible   **5.** Sensible

7. a. 1/4, 0.25, 25%   **b.** 9/20, 0.45, 45%
   c. 1/3, 0.33333..., 33.33%   **d.** 23/100, 0.23, 23%

9. a. 55.56%   **b.** 55.56%   **c.** 44.44%   **d.** 28.89%
   e. 17.78%

11. −33.3%   **13.** 50.0%   **15.** −17.1%

17. The circulation of the *Wall Street Journal* (compared value) is 60.6% more than the circulation of the *New York Times* (reference value).

19. Atlanta's airport (compared value) handles 25.8% more passengers than London's airport (reference value).

21. Saudi Arabia (compared value) produced 35.5% more oil than the United States (reference value).

23. Women (compared value) had 68.7% more knee replacement operations than men (reference value).

25. 124%   **27.** 55%

29. 34%, with a margin of error of 6 percentage points

31. −7.6 percentage points; −28.0%

33. 21 percentage points; 77.8%

Section 2.4

1. Not sensible   **3.** Sensible   **5.** Sensible   **7.** 255.7

9.

Year	Price	Price as a percentage of 1985 price	Price index (1985 = 100)
1955	29.1¢	24.3%	24.3
1965	31.2¢	26.1%	26.1
1975	56.7¢	47.4%	47.4
1985	119.6¢	100%	100
1995	120.5¢	100.8%	100.8
2000	155.0¢	129.6%	129.6

11. $3.92

13. The private college cost rose by 301%, while the CPI rose by 109% in the same period.

15. The typical price of a home rose by 26.9%, while the CPI rose by 22.8% in the same period.

17. Palo Alto: $629,300; Sioux City: $81,000; Boston: $313,800

19. Spokane: $260,000; Denver: $290,000; Juneau: $333,300

Chapter 2 Review Exercises

1. a. 706   **b.** Discrete; a person cannot fractionally survive.
   c. 23.89%   **d.** 64   **e.** Ratio   **f.** Nominal

2. a. Very precise
   b. It is based on a small sample and can only approximately represent the population; thus, it should not be so precise.
   c. $45,534
   d. This is a systematic error because the lack of low-income respondents systematically makes the average too high.
   e. Absolute error: too high by $4,405.95. Relative error: too high by 11.2%.

3. a. 105.6   **b.** 79.2
   c. Absolute change: −20. Relative change: −6%.
   d. Interval   **e.** Ratio

Section 3.1

1. Not sensible   **3.** Not sensible   **5.** Sensible

7.

Grade	Frequency	Relative frequency	Cumulative frequency
A	4	0.167	4
B	7	0.292	11
C	8	0.333	19
D	3	0.125	22
F	2	0.083	24
Total	**24**	**1**	**24**

9. a.

Jump length (feet)	Frequency	Relative frequency	Cumulative frequency
21–21.99	6	0.500	6
22–22.99	5	0.417	11
23–23.99	1	0.083	12
Total	**12**	**1**	**12**

b.

Jump length (feet)	Frequency	Relative frequency	Cumulative frequency
20–21.99	6	0.50	6
22–23.99	6	0.50	12
Total	**12**	**1**	**12**

Total return in 2000	Frequency (number of companies)
−70 to −60.1%	1
−60 to −50.1%	0
−50 to −40.1%	0
−40 to −30.1%	3
−30 to −20.1%	7
−20 to −10.1%	4
−10 to −0.1%	3
0 to 9.9%	3
10 to 19.9%	2
20 to 29.9%	3
30 to 39.9%	0
40 to 49.9%	1
50 to 59.9%	0
60 to 69.9%	1
70 to 79.9%	0
80 to 89.9%	0
90 to 99.9%	0
100% or more	1
Total	**29**

11.

Weight (pounds)	Frequency	Relative frequency	Cumulative frequency
0.7900–0.7949	1	1/36	1
0.7950–0.7999	0	0	1
0.8000–0.8049	1	1/36	2
0.8050–0.8099	3	3/36	5
0.8100–0.8149	4	4/36	9
0.8150–0.8199	17	17/36	26
0.8200–0.8249	6	6/36	32
0.8250–0.8299	4	4/36	36
Total	**36**	**1**	**36**

13.

Category	Frequency	Relative frequency
A	12	0.24
B	9	0.18
C	12	0.24
D	11	0.22
F	6	0.12
Total	**50**	**1**

15. As you can see from the table, about one-third of the stocks outperformed a bank account with 3% annual return (10 out of 29).

17.

Age	Frequency	Age	Frequency
20–29	0	50–59	5
30–39	12	60–69	3
40–49	13	70–79	1

19.

Tax	Frequency	Relative frequency
$500–999	1	0.02
$1,000–1,499	8	0.16
$1,500–1,999	31	0.62
$2,000–2,499	7	0.14
$2,500–2,999	3	0.06
Total	**50**	**1**

21.

Teacher salary	Frequency	Relative frequency
$24,000–27,999	3	0.06
$28,000–31,999	10	0.20
$32,000–35,999	15	0.30
$36,000–39,999	9	0.18
$40,000–43,999	7	0.14
$44,000–47,999	1	0.02
$48,000–51,999	5	0.10
Total	**50**	**1**

23. a. Age is quantitative at the ratio level of measurement. Transportation is qualitative at the nominal level of measurement.

b.

		Transportation				
		1	**2**	**3**	**4**	**5**
	1	2	1	1	1	0
	2	1	2	0	0	0
Age	**3**	1	0	1	1	2
	4	0	0	1	1	0
	5	1	1	0	0	3

c. Because the bar graph in part a has the categories ordered by frequency, it is also a Pareto chart.

9. a.

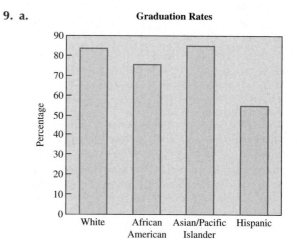

Graduation Rates

Section 3.2

1. Not sensible   **3.** Not sensible   **5.** Not sensible

7. a.

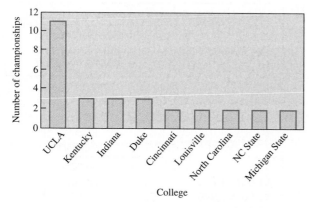

NCAA Basketball Champions

b.

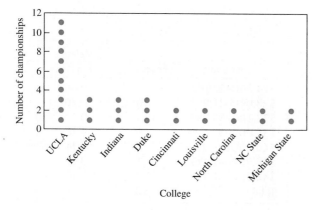

NCAA Basketball Champions

b.

Graduation Rates

11. a.

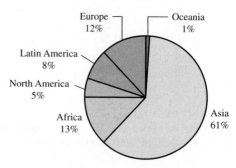

Population by Continent

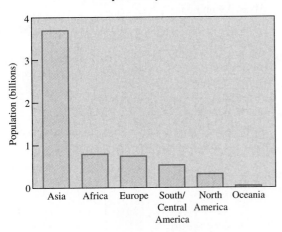

Population by Continent

b. A line chart would not be appropriate because the categories are at the nominal level of measurement.

13.

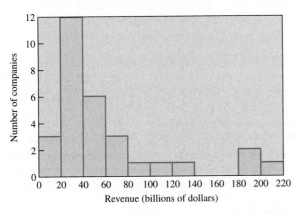

Revenue of DJIA Stocks

15. a.

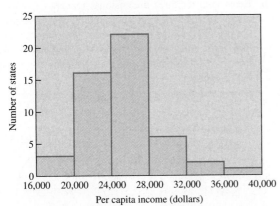

Per Capita Income by State

b.

Per capita income (dollars)	
16,000–19,999	MS WV NM
20,000–23,999	MT AR UT OK ID SC LA AL KY ND SD ME AZ WY TN IA
24,000–27,999	NC VT IN MO NE OR TX KS GA WI OH AK FL MI HI PA RI NV VA CA MN WA
28,000–31,999	CO IL NH DE MD NY
32,000–35,999	MA NJ
36,000–39,999	CT

17. a.

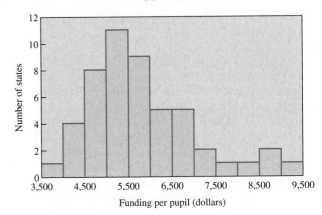

Funding per Pupil by State

b.

Funding per pupil (dollars)	
3,500–3,999	UT
4,000–4,499	AZ AR OK MS
4,500–4,999	ID LA ND SD TN NM MO NV
5,000–5,499	NC SC CO NE AL CA KY MT IL FL KS
5,500–5,999	OH TX GA HI IA WA IN VA OR
6,000–6,499	NH WY MN ME WV
6,500–6,999	VT WI MD MI PA
7,000–7,499	MA DE
7,500–7,999	RI
8,000–8,499	CT
8,500–8,999	NY AK
9,000–9,499	NJ

19.

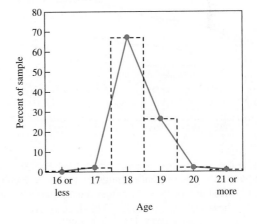

Age of First-Year College Students

21.

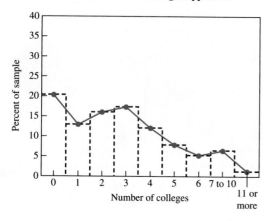

Number of Other Colleges Applied To

23. a. 1982: 25,000; 2000: 16,652   **b.** Decline by 33%
   c. 1982: 57%; 2000: 40%

25. Pareto chart or pie chart, to make it easy to see which colors are the most popular

27. Time-series line chart, because there is only one mean price for each year, but many years of data to show

Section 3.3

1. Not sensible   **3.** Not sensible

5. a. Teen smoking rates have generally risen for all ages, with a slight decline from 1996 to 1997 among 8th and 10th graders.

   b.

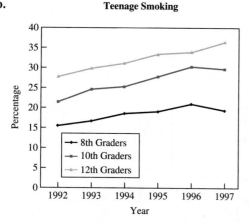

Teenage Smoking

7. a. The purpose is to show that earnings for people with college and advanced degrees have increased more rapidly than earnings for those with less education. The three sets of bars represent the three years 1975, 1985, and 1995.

   b. The three-dimensional appearance is cosmetic and makes the values somewhat difficult to read.

   c. For people with bachelor's degrees, earnings rose from about $10,000 to more than $30,000. For people who did not graduate from high school, earnings rose from about $6,000 to about $12,000.

9. a. About 1.8 million   **b.** About 0.9 million
   c. About 3.9 million

11. a. 1930: men, 75,000; women, 50,000. 2000: men, 500,000; women, 650,000
   b. 1980: slightly more men than women received degrees; 2000: 30% more women than men received degrees.
   c. During the 1960s
   d. 1950: 450,000; 2000: 1,150,000
   e. The stack plot makes it easy to see the changing total, but a double bar chart (one bar for men and one for women) would make the male/female comparison easier.

13.

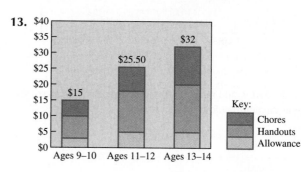

15. There are significant regional differences. For example, the probability that a black student has white classmates is generally much higher in the north than in the south and in rural areas than in urban areas.

17. A Pareto chart for the men's data and another one for the women's data will best illustrate the leading causes of cancer deaths.

19. A multiple line chart with two vertical scales will work here. One vertical scale measures number of daily newspapers and the other measures circulation.

Section 3.4

1. a, b. Because of the three-dimensional appearance of the pie charts, the sizes of the wedges on the page do not match the percentages. Instead they show how the wedges would look if the entire pie were tilted at an angle. This distortion makes it difficult to see the true relationships among the categories.

   c.

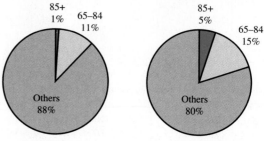

1990 Age Distribution   **2050 Age Distribution**

   d. In 2050, there will be relatively more older people and fewer younger people in the U.S. population than in 1990. Note, however, that because of population growth, all the age groups are expected to be larger in number in 2050 than in 1990.

3. The vertical scale does not begin from zero, which in this case exaggerates the difference in pay between men and women. Here is the same graph drawn more fairly.

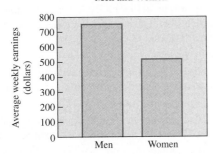

Average Weekly Pay for Men and Women

5. With the graph recast, we see the dramatic rise in world population that has occurred in recent years.

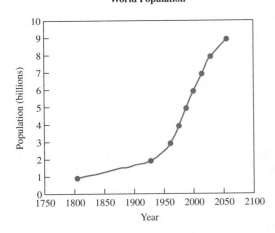

World Population

7. **a.** If we start the graph from zero, the changes look much smaller.
   b. A horizontal (flat) line is consistent with the error bars, so the claim of a seasonal variation is not justified by this graph.

9. Greatest change: 1990; least change: 1986. Prices increased every year, but the increases in the latter 1990s were generally smaller than those over the previous several years.

11. The first baby boom peaked in the late 1950s, and the second peaked around 1990.

Chapter 3 Review Exercises

1. a.

Sunday rainfall amount	Frequency
0.00–0.09	44
0.10–0.19	0
0.20–0.29	5
0.30–0.39	1
0.40–0.49	1
0.50–0.59	0
0.60–0.69	0
0.70–0.79	0
0.80–0.89	0
0.90–0.99	0
1.00–1.09	0
1.10–1.19	0
1.20–1.29	1

b.

Wednesday rainfall amount	Frequency
0.00–0.09	45
0.10–0.19	3
0.20–0.29	1
0.30–0.39	1
0.40–0.49	0
0.50–0.59	0
0.60–0.69	2

c. Both tables have the highest frequency for low rainfall values, but the Sunday data vary over a greater range than the Wednesday data.

2. a.

Rainfall amount	Relative frequency
0.00–0.09	44/52
0.10–0.19	0/52
0.20–0.29	5/52
0.30–0.39	1/52
0.40–0.49	1/52
0.50–0.59	0/52
0.60–0.69	0/52
0.70–0.79	0/52
0.80–0.89	0/52
0.90–0.99	0/52
1.00–1.09	0/52
1.10–1.19	0/52
1.20–1.29	1/52
Total	1

b.

Rainfall amount	Cumulative frequency
0.00–0.09	44
0.10–0.19	44
0.20–0.29	49
0.30–0.39	50
0.40–0.49	51
0.50–0.59	51
0.60–0.69	51
0.70–0.79	51
0.80–0.89	51
0.90–0.99	51
1.00–1.09	51
1.10–1.19	51
1.20–1.29	52

3. a.

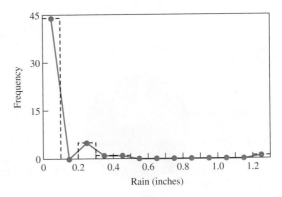

Sunday Rainfall Amount

b.

Wednesday Rainfall Amount

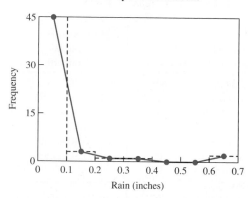

c. The histograms are qualitatively similar, suggesting that there is little difference between weekend and weekday rainfalls.

4. See the line charts overlaid on the histograms. Based on these charts, there does not appear to be an appreciable difference between weekend and weekday rainfall, with the exception of the one high rainfall day that occurred on a Sunday.

5. Networking appears to be the best approach to finding a job.

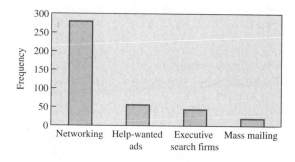

6. The pie chart is more useful for determining the *relative* importance of different job sources.

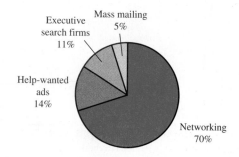

Section 4.1

1. Not sensible　　**3.** Sensible　　**5.** Not sensible

7. Mean: 671.37; median: 672.2; mode: 672.2

9. Mean: 0.188; median: 0.165; mode: 0.16

11. Mean: 9.5; median: 10; mode: 5, 10, 15

13. Mean: 0.919; median: 0.920; mode: none

15. a. Mean: 157,586; median: 104,100
　　b. Alaska is an outlier on the high end. Without Alaska, the mean is 81,317 and the median is 78,650.
　　c. Connecticut is an outlier on the low end. Without Connecticut (but with Alaska), the mean is 182,933 and the median is 109,050.

17. a. 73.75　　**b.** 80
　　c. No; a mean of 80 for five quizzes requires a total of 400 points, which you cannot reach even with 100 on the fifth quiz.

19. With a 90, your new mean is 81.4. The best you could do is score 100, which would make your new mean 82.9. The worst you could do would be to score 0, which would make your new mean 68.6.

21. 0.323

23. Mean height on your team is 6′8.6″. Median height on your team is 6′6″. So it depends on the meaning of "average." If it is the mean, then your team is above average height. If it is the median, then your team is below average height.

25. Either the mean or the median should work, as they would most likely be nearly equal.

27. The mean is best; on most flights, the number of lost pieces of luggage is very small.

29. Either the mean or the median should work, as they would most likely be nearly equal.

31. No; the classes have different numbers of students.

33. 81.25

35. No, unless the number of at-bats in all previous games is the same (only 4!) as in the most recent game.

37. a. 500　**b.** 929　**c.** Yes; it is possible (though not easy!) to average more than 1 point per at-bat.

39. 2.71

Section 4.2

1. Not sensible　　**3.** Sensible　　**5.** Not sensible

7. One mode; right-skewed; moderate variation

9. a. Right-skewed　　**b.** 150 (half of 300)
　　c. No; it depends on the precise distribution.

11. a. One mode　　**b.** Nearly symmetric

13. a. One mode　　**b.** Symmetric

15. a. Two modes　　**b.** Symmetric

17. a. One mode　　**b.** Symmetric

19. a. One mode　　**b.** Right-skewed

21. a. One mode   **b.** Right-skewed

23. a. One mode   **b.** Symmetric

25. a. One mode   **b.** Symmetric

27. a. One mode   **b.** Right-skewed

29. a. One mode   **b.** Right-skewed

Section 4.3

1. Not sensible   **3.** Not sensible   **5.** Sensible

7. Mean and median are 7.2 minutes.

9. a. 83   **b.** 92   **c.** 12

11. a. 61   **b.** 121   **c.** 167

13. Answers for the first data set only:
  a. Histogram has a frequency of 7 for a data value of 6; no other values.
  b. Low = 6, lower quartile = 6, median = 6, upper quartile = 6, high = 6
  c. 0

15. a. Faculty: mean = 2, median = 2, range = 4; Student: mean = 6.18, median = 6, range = 9
  b. Faculty: low = 0, lower quartile = 1, median = 2, upper quartile = 3, high = 4; Student: low = 1, lower quartile = 4, median = 6, upper quartile = 9, high = 10
  c. Faculty: 1.18; Student: 3.03   **d.** Faculty: 1.0; Student: 2.25

17. a. Stanley Cup: mean = 5.1, median = 4.5, range = 3; NBA: mean = 5.7, median = 6, range = 3
  b. Stanley Cup: low = 4, lower quartile = 4, median = 4.5, upper quartile = 6, high = 7; NBA: low = 4, lower quartile = 5, median = 6, upper quartile = 6, high = 7
  c. Stanley Cup: 1.29; NBA: 0.82
  d. Stanley Cup: 0.75; NBA: 0.75

19. a. Beethoven: mean = 38.8, median = 36, range = 42; Mahler: mean = 75, median = 80, range = 44
  b. Beethoven: low = 26, lower quartile = 29, median = 36, upper quartile = 45, high = 68; Mahler: low = 50, lower quartile = 62, median = 80, upper quartile = 87.5, high = 94
  c. Beethoven: 13.13; Mahler: 15.44
  d. Beethoven: 10.5; Mahler: 11

21. No; there are no negative percentiles.

23. Order from the first shop (standard deviation of 3 minutes).

25. Lower standard deviation means lower risk in this case.

27. Players' batting averages generally are closer to the mean of .260 today than they were in the past. High batting averages (such as .350) are therefore less common today.

Chapter 4 Review Exercises

1. a. Regular Coke: mean = 0.81518, median = 0.81675, range = 0.0363; Diet Coke: mean = 0.78459, median = 0.7852, range = 0.0165
  b. Regular Coke: low = 0.7901, lower quartile = 0.80945, median = 0.81675, upper quartile = 0.82275, high = 0.8264; Diet Coke: low = 0.7758, lower quartile = 0.78195, median = 0.7852, upper quartile = 0.78755, high = 0.7923

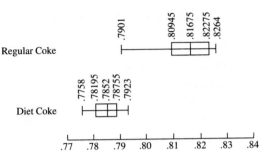

  c. Regular Coke: 0.0094; Diet Coke: 0.0044
  d. By the range rule of thumb, the standard deviation for Regular Coke should be about 0.0091, which is quite close to the actual standard deviation. For Diet Coke, this rule says that the standard deviation should be approximately 0.0041, also quite close to the actual value.
  e. Regular Coke has both a higher center and a higher variation than Diet Coke.
  f. Cans of Diet Coke weigh less than cans of Regular Coke. Assuming the cans are all the same size, it is likely that one of the ingredients in Regular Coke—perhaps the sugar—makes the liquid in Regular Coke slightly more dense than the liquid in Diet Coke.

2. a. 0
  b. The batteries with a standard deviation of 2 months are more likely to achieve their expected lifetimes.
  c. The outlier pulls the mean either up or down, depending on whether it is above or below the mean, respectively.
  d. The outlier has no effect on the median.
  e. The outlier increases the range.
  f. The outlier increases the standard deviation.

Section 5.1

1. Not sensible   **3.** Not sensible   **5.** Sensible

7. Distribution b is not normal. Distribution c has the larger standard deviation.

9. Normally distributed, because test scores are the result of many different factors

11. Normally distributed, because there will be small random variations in weight both above and below the advertised weight

13. Normally distributed, because there will be random variations in travel time both above and below the mean time

15. Normally distributed, because there will be small random variations in weight both above and below the mean weight

17. If prices at the dealers are independent, they will be nearly normally distributed (price wars or price fixing would lead to other distributions).

19. Not normal; this variable will not have a normal distribution because movie lengths have a minimum length (zero) and no maximum length.

21. Nearly normal; this variable should have a normal distribution because the deviations from the mean will be evenly distributed around the mean.

23. (b) and (c) are closest to normal.

25. a. The total area under the curve is 1. **b.** 0.50 **c.** 0.30 **d.** 0.70 **e.** 0.20

27. a. The mean is 155. **b.** 20% **c.** 20% **d.** 45%

29. a. 12% **b.** 0.03 **c.** 88% **d.** 0.44

Section 5.2

1. Not sensible **3.** Sensible **5.** Not sensible

7. a. 50% **b.** 0.84 **c.** 97.5% **d.** 16% **e.** 0.025
f. 16% **g.** 2.5% **h.** 0.84 **i.** 68% **j.** 81.5%

9. Cents: 2.44 g to 2.56 g; nickels: 4.88 g to 5.12 g; dimes: 2.208 g to 2.328 g; quarters: 5.530 g to 5.810 g; half dollars: 11.060 g to 11.620 g. In all cases, 5% of coins are rejected.

11.

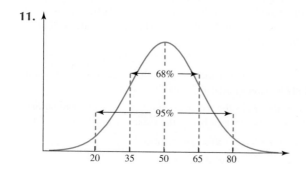

13. a. $z = 0.5$, 69.15th percentile
b. $z = -1.0$, 15.87th percentile
c. $z = 2$, 97.72th percentile
d. $z = 1.9$, 97.13th percentile

15. a. 1.0 standard deviation above the mean
b. 1.3 standard deviations below the mean
c. 0.1 standard deviation above the mean
d. 0.75 standard deviation below the mean

17. a. $z = -0.53$, approximately 30th percentile
b. $z = 0.13$, approximately 56th percentile
c. $z = -0.87$, approximately 19th percentile
d. $z = -0.20$, approximately 42nd percentile

19. a. Approximately 90.5th percentile
b. A score of approximately 687

21. 38%

23. a. 0.47% **b.** Approximately 4%
c. Below 30.055 in.; above 30.745 in.
d. The best estimate would be the mean of the readings, or 30.4 in.

25. Approximately 96.33%

27. a. 68% **b.** 66.8% **c.** 95% **d.** 95.3%

Section 5.3

1. With $n = 100$, the mean is approximately 100 and the standard deviation is approximately 1.6; with $n = 400$, the mean is approximately 100 and the standard deviation is approximately 0.8. With larger sample sizes, the distribution of sample means is narrower (smaller standard deviation).

3. With $n = 36$, the mean is approximately 6.5 and the standard deviation is approximately 0.58; with $n = 100$, the mean is approximately 6.5 and the standard deviation is approximately 0.35. The distribution of sample means should be near normal with $n = 36$ and closer to normal with $n = 100$.

5. a. 0.36 **b.** 0.044 **c.** 0.036 **d.** 0.0019

7. 40%; 4% **9.** 29%; 99.96% **11.** 33%; 1.4%

13. 12%; 77%

15. a. 4.5% **b.** 24% **c.** 91% **d.** 60%

Chapter 5 Review Exercises

1. a. Not normally distributed; the numbers occur with equal likelihood.
b. Incomes are not generally normally distributed; there is no upper limit.
c. Test scores are often normally distributed.

2. a. 90th percentile **b.** 1.29
c. No; the data value lies less than 2 standard deviations from the mean.
d. 0.0065
e. The temperature is unusual; it lies more than 2 standard deviations above the mean.
f. 99.22°F **g.** 97.18°F
h. Fewer than 0.01%
i. The sample mean is 6.6 standard deviations below the mean; the chance of selecting such a sample is extremely small. The assumed mean (98.60°F) may be incorrect.

Section 6.1

1. Not sensible **3.** Sensible

5. Not sensible

7. Significant **9.** Not significant **11.** Significant

13. Not significant **15.** Significant

17. Significant; one would not expect a difference this large among 30 cars driven under identical conditions.

19. a. If 100 samples were selected, the mean temperature would be expected to be 98.20°F or less in 5 or fewer of the samples.
   b. Selecting a sample with a mean this small is extremely unlikely and would not be expected by chance.

21. Not significant; the probability is greater than 0.05.

Section 6.2

1. Not sensible   **3.** Sensible

5. Not sensible

7. a. One way   **b.** More than one way   **c.** One way
   d. More than one way   **e.** More than one way
   f. More than one way

9. 0.5   **11.** 1/7 = 0.14   **13.** 0.10   **15.** 1/24 = 0.04

17. 0.375   **19.** 5/6 = 0.833   **21.** 3/4   **23.** 3/4

25. 2/3 = 0.667   **27.** 0.22; 0.33; 0.44; 0.55   **29.** 20%

31. a. 0.125   **b.** 0.375   **c.** 0.125   **d.** 0.875   **e.** 0.5

33. 0.40   **35.** 0.01   **37.** 0.0000011

39. In 2000, the probability was 0.127; in 2050, the probability will be 0.200.

41. 0.52; 0.51   **43. a.** 0.045   **b.** 0.066

45. a. 1/42   **b.** 0.027   **c.** 0.019
   d. The deviations are what might be expected by chance.

Section 6.3

1. Not sensible   **3.** Not sensible

5. You should not expect to get exactly 5,000 heads. The proportion of heads should approach 0.5 as the number of tosses increases.

7. Expected value = $0.50. The outcome of one game cannot be predicted. After 100 games, you could expect to win $50.

9. 0.94; 0.74   **11.** 15 minutes   **13.** −$0.78; lose $285.02

15. Approximately 35 years

17. a. $1,000,000; $1,140,000; $110,000; $250,000
   b. Responses are not consistent in decision 1.
   c. People chose the certain outcome ($1,000,000) in decision 1.

19. a. 0.5; 0.5   **b.** Lost $10   **c.** Lost $16   **d.** Lost $20
   e. 45%; 46%; 48%. The percentage of even numbers approaches 50%, but the difference between the numbers of even and odd numbers increases.
   f. 60 even numbers

Section 6.4

1. Not sensible   **3.** Not sensible

5. Independent; 0.0059   **7.** Independent; 0.00001

9. Independent; 1/32   **11.** Independent; 1/16

13. a. 1/256   **b.** 1/1024   **c.** 1/8   **d.** 1/16   **e.** 1/60

15. 0.986

17. a. 0.20   **b.** 0.038   **c.** 0.400   **d.** 0.0083   **e.** 0.316

19. a. 0.733   **b.** 1   **c.** 0.643   **d.** 0.22

21. a. 0.563   **b.** 0.375   **c.** 0.0625

d.

Event	Probability
AA	0.5625
Aa	0.1875
aA	0.1875
aa	0.0625

   e. 0.9375

23. a. 0.17   **b.** 0.31   **c.** 23

Chapter 6 Review Exercises

1. 0.50   **2.** 0.74   **3.** 0.61   **4.** 0.86   **5.** 0.065

6. 0.25

7. a. 0.73   **b.** 0.073   **c.** 1.35
   d. 0.0014; you should doubt the stated yield.

8. a. 0.10   **b.** No   **9. a.** 0.026   **b.** Yes

10. a. 0.14   **b.** No

Section 7.1

1. Not sensible   **3.** Sensible

5. Weak positive correlation   **7.** Strong negative correlation

9. Strong positive correlation

11. Either no correlation or a weak positive correlation

13. Significant at the 0.05 level, but not significant at the 0.01 level

15. Significant at the 0.05 level, but not significant at the 0.01 level

17. Significant at the 0.05 and 0.01 levels

19. Significant at the 0.05 and 0.01 levels

21. Strong positive correlation; correlation coefficient is approximately $r = 0.8$. Much of the correlation is due to the fact that a large fraction of the grain produced is used to feed livestock.

23. a.

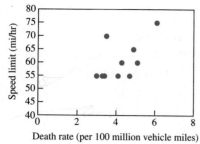

b. There is a moderate positive correlation ($r = 0.59$, exactly).

25. a.

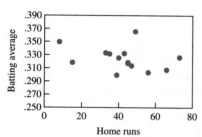

b. There is a negative correlation ($r = -0.29$).

27. a.

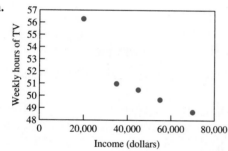

b. There is a strong negative correlation ($r = -0.86$).

29. a.

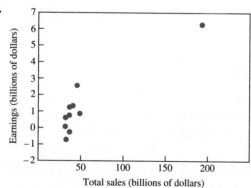

b. There is a strong positive correlation ($r = 0.92$).

31. There is a moderate negative correlation between depth and price, no correlation between table and price, and a moderate negative correlation between clarity and price. These correlations are either weaker than or opposite in direction to the correlation between weight and price.

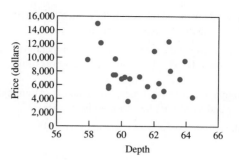

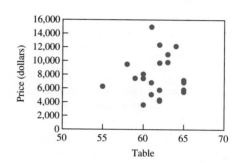

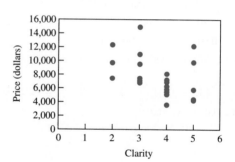

33.

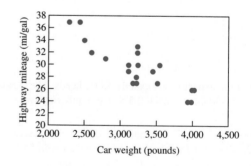

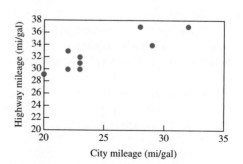

Highway mileage does appear to correlate reasonably well with weight. Highway and city mileage appear to correlate.

Section 7.2

1. Not sensible   3. Not sensible

5. There is a positive correlation between the crime rate and the number of people in prison. Either a direct cause or a common cause could explain the correlation.

7. There is a positive correlation between the miles of freeways and congestion. A direct cause could explain the correlation.

9. There is a positive correlation between grades and the age of the student's car. The correlation is most likely due to coincidence.

11. There is a positive correlation between gasoline prices and the number of airline passengers. A direct cause could explain the correlation.

13. There is a positive correlation between the number of ministers and priests and movie attendance. Both are likely to be the result of a common cause: increasing population.

15. a. The outlier is the upper left point (0.4, 1.0). Without the outlier, the correlation coefficient is 0.0.

   b. With the outlier, the correlation coefficient is −0.58.

17.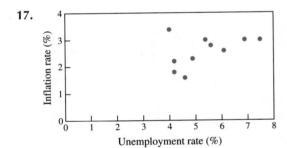

There is more correlation with the 1990 and 1991 points removed. The actual correlation coefficient with 1990 and 1991 is $r = 0.39$. The actual correlation coefficient without 1990 and 1991 is $r = 0.44$.

19. a.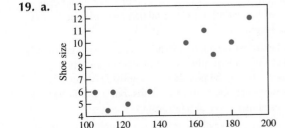

The actual correlation coefficient is $r = 0.92$. There appears to be a strong correlation.

   b. The data for the men and the women are not highly correlated when taken separately.

21. a. The actual correlation coefficient is $r = 0.77$, indicating strong correlation.

   b. The points on the left half correspond to relatively poor countries, such as Uganda. The points on the right half correspond to relatively affluent countries, such as Sweden.

   c. There appears to be a negative correlation between the variables for the poorer countries and a positive correlation between the variables for the wealthier countries.

Section 7.3

1. Not sensible   3. Not sensible

5. a.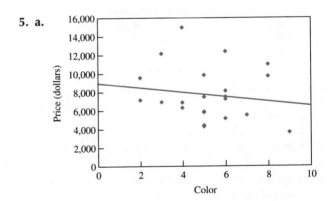

   b. Actual $r = -0.16$; $r^2 = 0.027$. About 3% of the variation can be explained by the best-fit line.

   c. The best-fit line should not be used to make predictions.

7. a.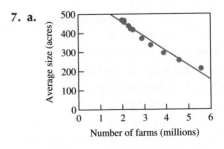

   b. Actual $r = -0.99$; $r^2 = 0.97$. About 97% of the variation can be explained by the best-fit line.

   c. The best-fit line could be used to make predictions.

9. a.

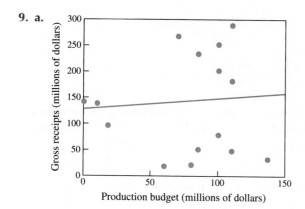

b. Actual $r = 0.05$, $r^2 = 0.003$; about 0.3% of the variation can be explained by the best-fit line.
c. The best-fit line should not be used to make predictions.

11. a. See the answer for Exercise 23 in Section 7.1.
b. Actual $r = 0.59$, $r^2 = 0.34$; 34% of the variation can be accounted for by the best-fit line.
c. (6.1, 75) and (3.5, 70) are both possible outliers.
d. Predictions based on the best-fit line are not reliable.

13. a. See the answer for Exercise 25 in Section 7.1.
b. Actual $r = -0.29$, $r^2 = 0.08$; 8% of the variation can be accounted for by the best-fit line.
c. (49, 0.366) is an outlier.
d. Predictions based on the best-fit line are not reliable.

15. a. See the answer for Exercise 27 in Section 7.1.
b. Actual $r = -0.86$, $r^2 = 0.74$; 74% of the variation can be accounted for by the best-fit line.
c. (20,000, 56.3) is an outlier.
d. Predictions based on the best-fit line could be reliable.

17. a. See the answer for Exercise 29 in Section 7.1.
b. Actual $r = 0.92$, $r^2 = 0.07$; 7% of the variation can be accounted for by the best-fit line.
c. (118, 3.53) is an outlier.
d. Predictions based on the best-fit line could be reliable.

Section 7.4

1. Not sensible   **3.** Not sensible

5. The physical model involves the physics of the internal combustion engine, which requires gas to run.

7. The physical model involves Newton's Law of Gravity.

9. Guideline 1; Guidelines 2 and 5; Guidelines 3 and 5; bad ventilation in the building is a plausible cause.

11. Smoking can only increase the risk already there.

13. This was an observational study. Later child bearing reflects an underlying cause. While it's possible that the conclusions are correct, there are other possible explanations for the findings. For example, it's also possible that the younger women lived during a time when having babies after the age of 40 was less likely (by choice). It is still possible for them to live to be 100.

15. Availability is not itself a cause. Social, economic, or personal conditions cause individuals to use the available weapons.

Chapter 7 Review Exercises

1. There appears to be a strong correlation between duration and interval.

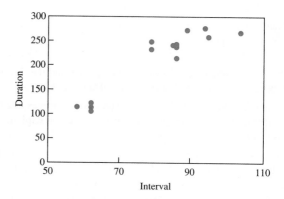

2. Height appears to be fairly independent of interval.

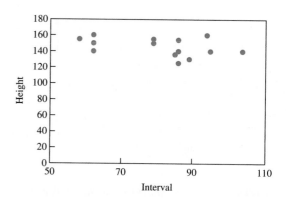

3. The correlation between interval and duration is significant at the 0.01 level. About 87% of the variation in interval can be explained by the variation in duration.

4. The correlation between interval and height is not significant. About 15% of the variation in interval can be explained by the variation in height.

5. Data would consist of the neighborhood's death rate and the distance between the neighborhood and power lines for many neighborhoods. It may be possible to establish a correlation between power lines and leukemia deaths, but it would be very difficult to establish a causal relationship.

6. The points on the scatter diagram lie on a straight line with negative slope (falling to the right).

7. Correlation alone never implies causation. Households with more disposable income can afford more trips to the dentist.

8. Location, age, size, condition, and amount of surrounding property certainly affect the value of a home. Location is often cited as the most important factor. The age of the previous homeowner is an example of a variable unrelated to the value of the home.

9. The data values that were collected were uncorrelated. It's still possible that the variables represented by the data values are related.

Section 8.1

1. Not sensible   3. Sensible

5. 13.54; sample mean is a little too high.

7. 17.8 hours; yes

9. a. 1.4   b. 0.0808

11. a. 0.528   b. 0.49   c. Sample proportion is a little too low.

13. 783; yes   15. a. -0.7 b. 0.2420

17. Each is 49.4, the same as the population mean.

Section 8.2

1. Not sensible   3. Not sensible

5. 1; $119 < \mu < 121$ 7. 0.8; $159.2 < \mu < 160.8$

9. 10,000   11. 2,500   13. $49.53 < \mu < 52.97

15. $5.10 < \mu < 5.20$

17. $34,700 < \mu < $35,300$; $26,240 < \mu < $26,760$

19. 172

21. a. 3.1   b. 1.4   c. 3.1   d. $2.6 < \mu < 3.6$

Section 8.3

1. Not sensible   3. Sensible

5. 0.05; $0.45 < p < 0.55$ 7. 0.012; $0.888 < p < 0.912$

9. 2,500   11. 625   13. $0.337 < p < 0.363$

15. 0.021; $0.779 < p < 0.821$

17. a. $0.153 < p < 0.191$ b. $0.118 < p < 0.146$
    c. $0.215 < p < 0.243$

19. a. 0.98   b. $0.966 < p < 0.994$
    c. No; the sample is biased.

21. Yes; the confidence interval is $0.538 < p < 0.587$.

23. $0.494 < p < 0.546$; $0.494 < p < 0.534$; $0.498 < p < 0.532$

    Because all of the confidence intervals include values less than 0.5, Martinez cannot yet be confident of winning a majority.

25. a. $50\% < p < 58\%$ b. 621

27. Factor of 4; if n increases by a factor of 4, $1/\sqrt{n}$ decreases by a factor of 2.

Chapter 8 Review Exercises

1. a. $133.3 < \mu < 135.7$
   b. We can be 95% confident that the limits of 133.3 and 135.7 actually contain the population mean.
   c. 1.2 millimeters   d. It is normal.
   e. They would get closer.
   f. The population mean is a fixed value, not a random variable; either it is contained within the limits or it is not, and there is no associated probability.

2. a. 256
   b. If it is larger than necessary, the estimate is better than it appears; if it is smaller than necessary, the estimate is worse than it appears.
   c. It is likely to be less than 16 because of the more homogeneous group. If we used the actual value, the answer would be smaller.

3. a. 0.0900   b. $0.0737 < p < 0.106$
   c. We can be 95% confident that the limits of 0.0737 and 0.106 actually contain the population proportion.

4. a. 625   b. No; the sample is likely to be biased.

Section 9.1

1. Sensible   3. Not sensible

5. a. Random fluctuations   b. Significant
   c. Answers will vary; perhaps 4 pounds

7. H_0: $p = 0.11$; H_a: $p < 0.11$

9. H_0: $\mu = 78$ years; H_a: $\mu > 78$ years   11. Yes; 0.007

13. No; 0.060   15. Yes; 0.006   17. No; 0.052

Section 9.2

1. Sensible   3. Not sensible

5. H_0: $\mu =$ listed amount; H_a: $\mu >$ listed amount. *Population*: All packages of potato chips of the brand in question. *Random sample*: Answers will vary. *Consumer Reports* researchers obtain samples at different times from different locations throughout the United States. *Conclusion*: There is sufficient evidence to

support the claim that the mean amount exceeds the amount listed on the packages.

7. H_0: $\mu = 1{,}675$ gal; H_a: $\mu > 1{,}675$ gal. *Population*: All households in the town. *Random sample*: Answers will vary. One approach is to construct a numbered list of all households and then use a computer to randomly generate numbers for households to be selected. *Conclusion*: There is sufficient evidence to support the claim that the mean usage exceeds 1,675 gal.

9. H_0: μ(Moon Valley) = mean for the nation; H_a: μ(Moon Valley) > mean for the nation. *Population*: All adult residents of Moon Valley. *Random sample*: Answers will vary. One approach is to use voter registrations, motor vehicle registrations, tax rolls, and any other lists necessary to develop a numbered list of all adult residents and then use a computer to randomly select individuals for a sample. *Conclusion*: There is sufficient evidence to support the claim that the mean per capita income for Moon Valley exceeds the national average.

11. H_0: $\mu = 275$ lb; H_a: $\mu > 275$ lb. *Population*: The linemen on the coach's team. *Random sample*: Answers will vary. One approach is to put the names of the linemen on individual index cards in a bowl, mix them up, and then select a sample. *Conclusion*: There is sufficient evidence to support the claim that the mean weight of his linemen is greater than 275 lb.

13. H_0: μ = critical EPA level; H_a: μ < critical EPA level. *Population*: All locations downstream from the plant. *Random sample*: Answers will vary. One approach is to partition the downstream locations into small sectors, number them, and then use a computer to randomly select a sample. *Conclusion*: There is sufficient evidence to support the claim that the mean amount of pollution is less than the critical amount specified by the EPA.

Section 9.3

1. Not sensible   3. Sensible

5. $z = -1.1$; H_a is not supported.

7. $z = -2.0$; H_a is supported.

9. $z = 1.0$; H_a is not supported.

11. 0.1841; do not reject H_0; do not support H_a.

13. 0.0179; reject H_0; support H_a.

15. 0.0107; reject H_0; support H_a.

17. 0.0062; reject H_0; support H_a.

19. The sample mean is significantly *below* the assumed mean, but the claim is that the population mean is *greater than* the assumed mean. Do not reject H_0, and do not support H_a.

21. H_0: $\mu = 7.5$ years; H_a: $\mu < 7.5$ years. Standard score for mean: $z = -1.3$. *P*-value = 0.0968. Critical value: $z = -1.645$. Do not reject H_0. There is not sufficient evidence to support the claim that the mean time of ownership is less than 7.5 years.

23. H_0: $\mu = 21.4$ mi/gal; H_a: $\mu < 21.4$ mi/gal. Standard score for mean: $z = -2.9$. *P*-value = 0.0019. Critical value: $z = -1.645$. Reject H_0. There is sufficient evidence to support the claim that the mean for SUVs is less than 21.4 miles per gallon.

25. *Alabama*: $z = -11.8$. Reject H_0. Support the claim that Alabama has a mean that is less than the national average. *Illinois*: $z = 4.3$. Reject H_0. Support the claim that Illinois has a mean that is greater than the national average. *Georgia*: $z = -3.0$. Reject H_0. Support the claim that Georgia has a mean that is less than the national average. *Washington*: $z = 2.3$. Reject H_0. Support the claim that Washington has a mean that is greater than the national average.

27. *April*: H_0: $\mu = 339$; H_a: $\mu > 339$. Standard score for mean: $z = 1.6$. *P*-value = 0.0548. Critical value: $z = 1.645$. Do not reject H_0. There is not sufficient evidence to support the claim that the mean is greater than 339. *September*: H_0: $\mu = 69$; H_a: $\mu < 69$. Standard score for mean: $z = -1.8$. *P*-value = 0.0359. Critical value: $z = -1.645$. Reject H_0. There is sufficient evidence to support the claim that the mean is less than 69.

Section 9.4

1. Sensible   3. Not sensible

5. H_0: μ = manufacturer's recommended price; H_a: $\mu \neq$ manufacturer's recommended price. *Conclusion*: Support the claim that the mean is not the same as the manufacturer's suggested retail price.

7. H_0: $\mu = 11{,}000$ kilowatt-hours; H_a: $\mu \neq 11{,}000$ kilowatt-hours. *Conclusion*: Support the claim that the mean is not equal to 11,000 kilowatt-hours.

9. H_0: μ = mean blood pressure for group with no heart problems; H_a: $\mu \neq$ mean blood pressure for group with no heart problems. *Conclusion*: Support the claim that the mean blood pressure for the heart surgery patients is not the same as the mean blood pressure for the group with no heart problems.

11. a. Significant; 0.0214   b. Not significant; 0.0718
    c. Not significant; 0.2302   d. Significant; less than 0.0456

13. Do not reject H_0. Standard score for mean: $z = -1$.

15. Do not reject H_0. Standard score for mean: $z = -1.4$.

17. H_0: $\mu = 16$ oz; H_a: $\mu \neq 16$ oz. Standard score for mean: $z = -1.5$. *P*-value = 0.1336. Critical values: $z = -1.96, 1.96$. Do not reject H_0. There is not sufficient evidence to reject the claim that the mean is 16 oz.

19. H_0: $\mu = 600$ mg; H_a: $\mu \neq 600$ mg. Standard score for mean: $z = -4.2$. *P*-value is less than 0.0004. Critical values: $z = -1.96, 1.96$. Reject H_0. There is sufficient evidence to support the claim that the medicine does not contain the required amount of acetaminophen.

21. H_0: $\mu = 5.670$ g; H_a: $\mu \neq 5.670$ g. Standard score for mean: $z = -5.0$. *P*-value is less than 0.006. Critical values: $z = -1.96, 1.96$. Reject H_0. There is sufficient evidence to reject the claim that the mean weight of quarters in circulation is 5.670 g.

23. a. Treat a patient who doesn't need treatment.
  b. Fail to treat a diseased patient.
  c. Treat a diseased patient.
  d. Don't treat a patient who is free of disease.

25. a. Don't bet on a lottery (thinking it is biased) when it is actually fair.
  b. Bet on a lottery (thinking it is fair) when it is biased.
  c. Don't bet on a lottery (thinking it is biased) when it is actually biased.
  d. Bet on a lottery (thinking it is fair) when it is actually fair.

27. a. Replace the microchip before it needs to be replaced.
  b. Leave the microchip in place even though it needs to be replaced.
  c. Replace the chip when it needs to be replaced.
  d. Don't replace the chip before it needs to be replaced.

29. a. Tamper with the production process when it is OK.
  b. Distribute tablets that do not have the correct amount of the active ingredient.
  c. Correct a production process that is not working as it should be.
  d. Save time and money by not tampering with a production process that is OK.

Section 9.5

1. Sensible   **3.** Not sensible

5. a. $H_0: p = 0.915$; $H_a: p > 0.915$ **b.** No   **c.** 0.4207
  d. Not significant   **e.** Significant   **f.** 0.97
  g. 0.98. Using Table 5.1, we get 0.99, but the answer is 0.98 if we use a more precise table.

7. $H_0: p = 0.58$; $H_a: p > 0.58$. Standard score: $z = 0.7$. P value: 0.2420. Critical value: $z = 1.645$. Don't reject H_0. There is not enough evidence to support the claim that the percentage of older women at the college is above the national average.

9. $H_0: p = 0.56$; $H_a: p < 0.56$. Standard score: $z = -1.6$. P-value: 0.0548. Critical value: $z = -1.645$. Do not reject H_0. There is not sufficient evidence to support the claim that the sample comes from a population with a married percentage less than 56.0%.

11. $H_0: p = 0.133$; $H_a: p < 0.133$. Standard score: $z = -2.3$. P-value: 0.011. Critical value: $z = -1.645$. Reject H_0. There is sufficient evidence to support the claim that the poverty rate in Custer County is less than the national rate of 13.3%.

13. $H_0: p = 0.5$; $H_a: p > 0.5$. Standard score: $z = 4.7$. P-value is less than 0.0002. Critical value: $z = 1.645$. Reject H_0. Support the claim that a majority of Americans feel that gun control is an important issue.

15. $H_0: p = 0.32$; $H_a: p > 0.32$. Standard score: $z = 0.6$. P-value: 0.3085. Critical value: $z = 1.645$. Do not reject H_0. There is not sufficient evidence to support the claim that the smoking rate for the fine arts students is higher than the national average.

Chapter 9 Review Exercises

1. a. $H_0: \mu = 12$ oz   **b.** $H_a: \mu > 12$ oz
  c. $z = 10.4$ **d.** $z = 1.645$ **e.** Less than 0.0002
  f. Support the claim that the mean is greater than 12 ounces.
  g. Concluding that the mean is greater than 12 ounces when it really is not greater than 12 ounces
  h. Not supporting the claim that the mean is greater than 12 ounces when it really is greater than 12 ounces
  i. Less than 0.0004

2. a. $H_0: p = 0.61$ **b.** $H_a: p \neq 0.61$ **c.** $z = 5.8$
  d. $z = -1.96, 1.96$ **e.** Less than 0.0004
  f. Support the claim that the proportion is different from 0.61.
  g. Concluding that there is a difference when there really is not a difference
  h. Failing to support the claim of a difference when there really is a difference
  i. Less than 0.0002

3. The claim of an increase in the likelihood of a girl is not supported. With a sample proportion *less than* 0.5, there is no way we could ever support the claim that $p > 0.5$.

4. The sample is not random. It is very possible that the family is not representative of the population, so the results are biased.

Section 10.1

1. Not sensible   **3.** Not sensible

5. 1.5 deaths per 100 million vehicle-miles

7. 22.6 deaths per 100,000 drivers

9. 1994: 1.82 deaths per 100,000 flight hours; 0.044 death per 1,000,000 miles flown; 2.91 deaths per 100,000 departures. 2000: 0.51 death per 100,000 flight hours; 0.013 death per 1,000,000 miles flown; 0.79 death per 100,000 departures.

11. a. 0.00023; 2.2 times greater   **b.** 2.75 times greater
  c. 12.6 deaths per 100,000   **d.** 258 deaths per 100,000
  e. Approximately 298

13. a. Approximately 20 deaths per 1,000
  b. Approximately 2.0 million people   **c.** 80 years old
  d. 82 years old   **e.** Loss of $400 million

15. a. 37 births per day in Maine; 130 births per day in Utah
  b. 26.3 births per 1,000 people in Utah; 10.4 births per 1,000 people in Maine

17. a. Approximately 4.2 million births
  b. Approximately 2.4 million deaths   **c.** 1.8 million people
  d. The net number of people who immigrated (in-migration minus out-migration) was about 1.2 million; about 40% due to immigration.

19. a. 0.7 birth per 1,000 women   **b.** 3.8 million births
  c. 56 million
  d. The baby boom and the children of baby boomers
  e. Fertility rate depends on the population (number of women ages 15–44).

Section 10.2

1. Sensible   **3.** Not sensible   **5.** Josh; Josh; Jude

7. a. New Jersey; Nebraska

   b. The percentage of nonwhites is significantly lower in Nebraska than in New Jersey.

9. a. Whites: 0.18%; nonwhites: 0.54%; total: 0.19%
   b. Whites: 0.16%; nonwhites: 0.34%; total: 0.23%
   c. The rate for both whites and nonwhites was higher in New York than in Richmond, yet the overall rate was higher in Richmond than in New York. The percentage of nonwhites was significantly lower in New York than in Richmond.

11. a. Spelman has a better record for home games (34.5% vs. 32.1%) and away games (75.0% vs. 73.7%), individually.
   b. Morehouse has a better overall average (62.5% vs. 48.9%).
   c. Morehouse has a better team; teams are generally rated on their overall record.

13. b. 216 people were accused of lying; of these, 18 were actually lying and 198 were telling the truth; 91.7% were falsely accused.
   c. 1,784 people were telling the truth, according to the polygraph; of these, 1,782 were actually telling the truth and 2 were lying; 99.9% of those found to be telling the truth were actually telling the truth.

15. Within each category, the percentage of women hired was greater than the percentage of men hired. Overall, the hiring rate was higher for men (55%) than for women (41.7%). Note that the number of blue-collar jobs was over five times greater than the number of white-collar jobs.

17. b. 95% of those at risk for HIV test positive; 67.9% of those at risk who test positive have HIV.
   c. The chance of the patient's having HIV is 67.9%, which is greater than the overall incidence rate of 10%.
   d. 95% of those in the general population with HIV test positive; 5.4% of those in the general population who test positive have HIV.
   e. The chance of the patient's having HIV is 5.4%, compared to the overall incidence rate of 0.3%.

Section 10.3

1. Not sensible   **3.** Not sensible   **5.** Not sensible

7.

	A	B	Total
C	47	9	56
D	18	23	41
Total	65	32	97

9. a. Row 1: 84; row 2: 64; column 1: 85; column 2: 63
   b. 56.8%   **c.** 42.6%   **d.** 0.432   **e.** 44.0%
   f. 0.594   **g.** 58.7%   **h.** 0.447   **i.** No

11. a. Row 1: 156; row 2: 198; column 1: 343; column 2: 11
   b. 44.1%   **c.** 3.1%   **d.** 0.975   **e.** 3.8%
   f. 0.455   **g.** 43.7%

h. The overlapping data would allow some people to be counted twice.

13. a. Row 1: 20; row 2: 19; column 1: 15; column 2: 24; grand total: 39
   b. 0.51, 0.49   **c.** 0.62, 0.38   **d.** 0.197
   e. Men: 7.692, 12.308; women: 7.308, 11.692
   f. Differences are not too large.

15. Observed: 56, 42, 49, 43; expected: 54.158; 43.842, 50.842, 41.158. Differences appear to be small.

17. 135.5, 114.5, 135.5, 114.5. Differences appear to be fairly small.

19. 1,417.285, 90.144, 21.238, 44.836, 151.498, 1,585.715, 100.856, 23.762, 50.164, 169.502. Differences appear to be quite large.

21. 16.250, 16.250, 16.250, 16.250, 33.750, 33.750, 33.750, 33.750. Differences appear to be small.

23. a. H_0: getting a flu shot is independent of getting flu; H_a: getting a flu shot and getting flu are dependent
   b. 120, 180, 80, 120   **c.** $\chi^2 = 13.889$
   d. We reject the null hypothesis at the 0.01 level and conclude that getting a flu shot and getting flu are dependent.

25. a. H_0: age and drinking status are independent; H_a: age and drinking status are dependent
   b. 19.080, 1232.920, 30,920, 1998.080   **c.** $\chi^2 = 3.1814$
   d. We do not reject the null hypothesis at the 0.05 level, and we cannot conclude that age and drinking status are dependent.

27. a. H_0: poverty status is independent of whether a person has a high school diploma; H_a: poverty status and whether a person has a high school diploma are dependent
   b. 7.6, 30.4, 61.4, 245.6   **c.** $\chi^2 = 43.835$
   d. We reject the null hypothesis at the 0.01 level and conclude that poverty status and whether a person has a high school diploma are dependent.

29. a. H_0: the sport is independent of whether the wins are at home or away; H_a: the sport and whether the wins are at home or away are dependent
   b. 115.971, 82.029, 58.571, 41.429, 54.471, 38.529, 57.986, 41.014
   c. $\chi^2 = 4.737$
   d. We do not reject the null hypothesis at the 0.05 level, and we cannot conclude that the sport and whether the wins are at home or away are dependent.

Chapter 10 Review Exercises

1. a. 34.5 deaths per 100,000 people   **b.** 0.00027
   c. About 2.5 times more likely
   d. 60.5 deaths per 1 million people

2. a. 1,980   **b.** 97,020   **c.** 20   **d.** 980   **e.** 66.9%

3. a. 3.58%; 7.25%; there is considerable difference.
   b. H_0: charge and pricing are independent; H_a: charge and pricing are dependent
   c. 13.813, 13.187, 22.510, 21.490, 382.676, 365.324
   d. 10.814   **e.** It is less than 0.01.
   f. Reject independence; it appears that charge and pricing are somehow related.

Index